# THEORY OF VIBRATORY
# TECHNOLOGY

# THEORY OF VIBRATORY TECHNOLOGY

*Revised and Augmented Edition*

## I. F. Goncharevich

*Blagonravov Institute of Machine Science*
*USSR Academy of Sciences*

## K. V. Frolov

*Director, Mechanical Engineering Research Institute*
*USSR Academy of Sciences*

English Edition Editor

## E. I. Rivin

*Department of Mechanical Engineering*
*Wayne State University, Detroit*

⬤HEMISPHERE PUBLISHING CORPORATION

A member of the Taylor & Francis Group

New York    Washington    Philadelphia    London

**THEORY OF VIBRATORY TECHNOLOGY. Revised and Augmented Edition**

Copyright © 1990 by Hemisphere Publishing Corporation. All rights reserved. Printed in the United States of America. Except as permitted under the United States Copyright Act of 1976, no part of this publication may be reproduced or distributed in any form or by any means, or stored in a data base or retrieval system, without the prior written permission of the publisher. Originally published as Teoriya vibratsionnoi tekhniki i tekhnologii by Nauka, Moscow, 1981.

Translated by Jamil I. Ghojel.

1 2 3 4 5 6 7 8 9 0    E B E B    9 8 7 6 5 4 3 2 1 0

This book was set by Desktop Publishing.
The editor was Janine Ludlam.
Cover design by Sharon Martin DePass.
Edwards Brothers, Inc. was the printer and binder.

**Library of Congress Cataloging-in-Publication Data**

Goncharevich, Igor´ Fomich.
    [Teoriia vibratsionnoĭ tekhniki i tekhnologii. English]
    Theory of vibratory technology / I. F. Goncharevich, K. V. Frolov ;
English-edition edited by E. I. Rivin. — Rev. and augm. ed.
        p.    cm.
    Translation of: Teoriia vibratsionnoĭ tekhniki i tekhnologii.
    Includes index.

    1. Vibration. 2. Vibrators. I. Frolov, K. V. II. Rivin, Eugene
I. III. Title.
TA355.G59813    1990
620.3—dc20                                        89-15421
                                                  CIP

ISBN 0-89116-700-5

# CONTENTS

FOREWORD                                                    vii

**CHAPTER ONE**   **PHYSICAL PRINCIPLES OF VIBRATORY**
                  **TECHNOLOGY**                             1

1.1   Fundamental types of processes realized with the use of
      vibration. Vibration regimes                            1
1.2   Vibratory conveyance                                    2
1.3   Vibratory bunkering (hoppers)                          47
1.4   Vibratory crushing                                     51
1.5   Vibratory boiling layer and vibratory compaction       56


**CHAPTER TWO**   **THEORETICAL PRINCIPLES OF VIBRATORY**
                  **TECHNOLOGY**                             69

2.1   Phenomenological description of the process load of
      vibratory machines                                     69
2.2   Vibratory conveyance of bulk loads                     89
2.3   Vibratory processing of dispersed media and bulk loads 120
2.4   Vibration crushing of rock masses                     145
2.5   Casting and working of metals in vibratory environment 163
2.6   Parameter identification of the rheological models    195

CHAPTER THREE       BASIC SCHEMATICS AND DESIGN OF
                    VIBRATORY MACHINES                                   213

        3.1         Principal types of vibratory machines                213
        3.2         Vibratory mills and crushers                         216
        3.3         Design schematics of vibratory conveying and
                    process-conveying machines                           235
        3.4         Vibratory elevators                                  246
        3.5         Vibratory machines for compaction of granular media  252
        3.6         Vibratory machines for part strengthening            255
        3.7         Vibratory separators and mixers                      259


CHAPTER FOUR        VIBRATION EXCITERS (VIBRATORS): PRINCIPLE
                    OF OPERATION, GENERATION OF EXCITING
                    FORCE                                                273

        4.1         Types of vibration exciters (vibrators)              273
        4.2         Basic design features; formation of the exciting force  274
        4.3         Interaction of vibration exciter with oscillatory system  320


CHAPTER FIVE        DESIGN OF VIBRATORY MACHINES
                    CONSIDERING LOADING AND DRIVE
                    CHARACTERISTICS                                      415

        5.1         Formulation of design problems for system: vibratory
                    machine-load-motor                                   415
        5.2         Heavy duty vibratory conveying machines              416
        5.3         Vibratory crushers                                   434
        5.4         Optimization of structural and process parameters of
                    vibratory equipment                                  466

                    REFERENCES                                           529

                    INDEX                                                533

# FOREWORD

Vibratory technology today is associated with new, progressive scientific and production development methods. The use of vibratory technology leads to radical improvements in traditional production processes and mechanisms. The successes in the development of vibratory technology are largely predetermined by the implementation of the applied theory of vibration and vibration rheology.

The need for a further increase in the effectiveness of vibratory technology and the expansion of its application creates, on the one hand, a great demand for further study of vibratory processes in the course of various production processes with the purpose of establishing new physical effects. On the other hand, increasing the power of vibratory machines and the growth of their specific loading, dictated by the need to intensify production, strengthens the links between the machine, load, and the drive, leading to the necessity of considering them as a single load-machine-drive unit with strong interaction between the constituent subsystems. Such systems exhibit special effects that are not characteristic of traditional lightly loaded vibratory machines.

The presented methods of systemic consideration of load-machine-drive vibratory systems enabled the formulation of useful approaches to the utilization of resonance phenomena in the processed medium, which opened the possibility of a radical increase of power intensity of vibratory production processes, reduction of nonproductive energy losses, and an increase of productivity of the equipment.

In light of current demands for further development of vibratory technology, the authors devote considerable attention to the development of the theory of vibratory processes on the basis of phenomenological vibration rheology and the computational techniques for analysis of vibratory machines under load, taking into account the characteristics of the drive. Special attention is given to the development of methods of optimum development and identification of phenomenological models of various production processes.

New applications of vibratory technology and new approaches in traditional spheres of its application are considered in this book. Significant attention is given to the systematization and classification of vibratory machines, which, in the opinion of the authors, identifies the fundamental features of different modifications of each type of machinery, illustrates trends for further development, and indicates directions of their improvement.

Development of reliable methods of analysis of the process parameters and the characteristics of vibratory machines enables the development of application principles for multicriterial optimal design methods. This is illustrated by the numerous examples of characteristic vibratory machines.

The authors express the hope that the reader will find the book interesting and that it will prove useful in practical activities.

# PHYSICAL PRINCIPLES OF VIBRATORY TECHNOLOGY

## 1.1 FUNDAMENTAL TYPES OF PROCESSES REALIZED WITH THE USE OF VIBRATION. VIBRATION REGIMES

At the present time a wide range of different processes are being realized with the application of vibration in which vibration plays various roles. Therefore, it is difficult to construct a rigorous functional classification. However, if the main operation executed in the given process is adopted as the classification index, then the following main categories of operations realized with the application of vibration can be identified: conveyance, processing of disperse systems with the purpose of maintaining or increasing efficiency of one or another production process, cutting, and crushing.

Vibratory conveyance serves not only conveying purposes, but also constitutes the basis for several technological processes. The main types of vibratory conveying operations are conveyance along horizontal or slightly elevated (tilted) surfaces, lifting along a spiral load-carrying element or in devices of special construction, vibratory loading and discharge from containers, bunkering. The second field of application of vibration - processing of disperse systems with the purpose of technological treatment - is the most versatile. This includes, first of all, formation of a vibratory boiling layer to accelerate mass-transfer processes in which various types of chemical and physico-chemical reactions occur. Such processes include catalytic and solid-phase reactions, combustion, extraction, dissolution and leaching, reclamation of metals from ores, and many others. Realized in vibration-processed disperse systems also are mixing, classification and separation, compaction of poured mixtures and concretes, crystalization, pressure processing, hardening, drying, dewatering, granulation, washing, centrifuging, and other operations. Vibration is fairly widely used in the processes of cutting and crushing: these are vibration turning and drilling, abrasive forming, breaking and grinding, crushing of soil and rocks.

Since the efficiency of the given processes is dependent on the regime of oscillations of the working element of the machine, various types of vibrations are used in practice: harmonic and semi-harmonic rectilinear, two-component, and three-dimensional. The trajectory of motion is generally shaped by translational and torsional vibrations of the working elements. Parameters of trajectories and modes of vibrations of the working elements of vibratory machines are constantly being developed and improved. In many processing operations superposition of vibrations to the working element can increase the specific energy intensity and efficiency of the process. The range of frequencies used in modern machines starts with low-frequency mechanical oscillations (infrasound) and reaches high-frequency ultrasound. Large amplitudes correspond to low frequencies and lesser ones to high frequencies.

## 1.2   VIBRATORY CONVEYANCE

Due to influences of many factors, the process of conveyance by vibration of massive loads is very complicated. Of main interest when determining the efficiency of the regimes of conveying processing machines is the study of physical characteristics of the process and establishment of the dependence of conveyance speed, process energy-intensity, degree of speed transmission to the transported medium, intensity of its mixing, creating the state of vibratory boiling, on parameters of the oscillation regime. These parameters are the shape of the trajectory, frequencies, amplitudes, phase shift angle between the harmonic components of two-component oscillations, angle of vibration and inclination of the load-carrying element. Of no lesser significance is the study of the effect of the properties of the transported medium, degree of filling of the working element and the operation conditions on the enumerated performance parameters of vibratory conveyance/processing.

Let us consider the main correlations between the speed of vibratory conveyance under rectilinear harmonic oscillations on parameters of the working regime of the vibratory machine, such as vibration amplitude and frequency and also the angles of vibration and inclination of the working element.

The dependence of the conveyance speed of a reference of piled up product (sand) with a coefficient of transportability $k = 1$, layer thickness of up to 50 mm for a horizontal arrangement of the vibratory machine and angle of vibration of $20^{\circ}$ on frequency at various vibration amplitudes is presented in Figure 1.1a. For each constant value of amplitude the dependence is, to a certain extent, of parabolic character. Then at higher frequencies the curves become gently sloping and with further increase of frequency, they can pass through an extremum. Moreover, with increasing vibration amplitude the curves become straight and acquire a larger inclination.

As the graphs show, the larger the oscillation amplitude, the lower the frequencies at which the extreme values of the speed of conveyance are reached. It can also be noted that with increasing amplitude the value of the extremal speed of conveying increases. By comparing the shape of the curves of the speed – frequency dependence with the character of the vibratory conveyance regime (continuous, intermittent), it can be noticed that with increasing oscillation frequency in the

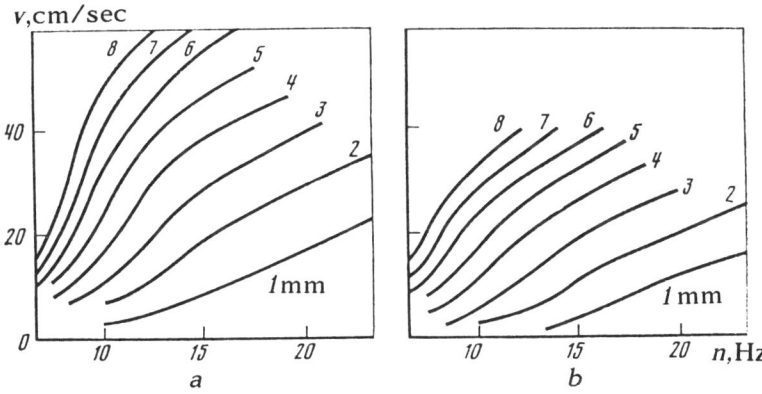

**Figure 1.1** Conveyance speed vs. the regime parameter under rectilinear harmonic oscillations a) sand; b) small-lump piled-up load.

region of continuous (no loss of contact) vibratory conveyance regimes, a slow but more or less uniform increase in speed takes place. Upon transition to a jolting regime, frequency increase entails more intensive growth of conveying speed. However, the latter takes place only within a limited range of frequencies which is narrower the greater the oscillation amplitude. Further increase of oscillation frequency at first causes only an insignificant increase of speed and then even a decrease in speed occurs. In this case, an unstable conveying process is observed which is caused by physical reasons associated with the violation of the energy transfer conditions from the working element to the transported medium.

For each amplitude value there exists a maximum speed obtained at a frequency which is lower the greater the oscillation amplitude. Thus, in order to attain maximum speeds, one must operate at a possibly larger amplitude adopting such frequencies and vibration angles which enable obtaining the required speed of vibration conveyance.

Analysis of experimental data indicates that to determine the speed of motion of a product of interest, the speed obtained from the graph for sand should be multiplied by the coefficient of transportability for the product.

Figure 1.1b shows the dependence of the conveying speed for a small-lump load.

For different loads at the same regimes magnitude of the conveying speed is different. When difficult to transport loads are moved, a sharper decrease in speed takes place beyond the extremal value.

Vibration angle has a considerable effect on the speed of vibratory conveyance. Figure 1.2a shows the dependence of the speed of vibration conveying of a 40 mm layer of sand on the vibration angle at acceleration amplitudes of vibrations of the load-carrying element from 2.4 to 6.35 $g$, and Fig. 1.2b shows the dependence for rocks up to 50 mm chunk size at $K = 7.8 - 14.5$. As can be seen, the correlation between the speed of motion on the vibration angle is very complex, its character is largely determined by the properties and thickness of the layer of the transported load, in particular by the vibration of the load-carrying element. Under limited oscillation intensity of the latter, the speed rises with increasing vibration angle. For easily transportable loads at small layer thickness this tendency is retained in a narrower range of accelerations than for loads that are difficult to transport and are moved in a thick layer. Under medium vibration intensity an opposite

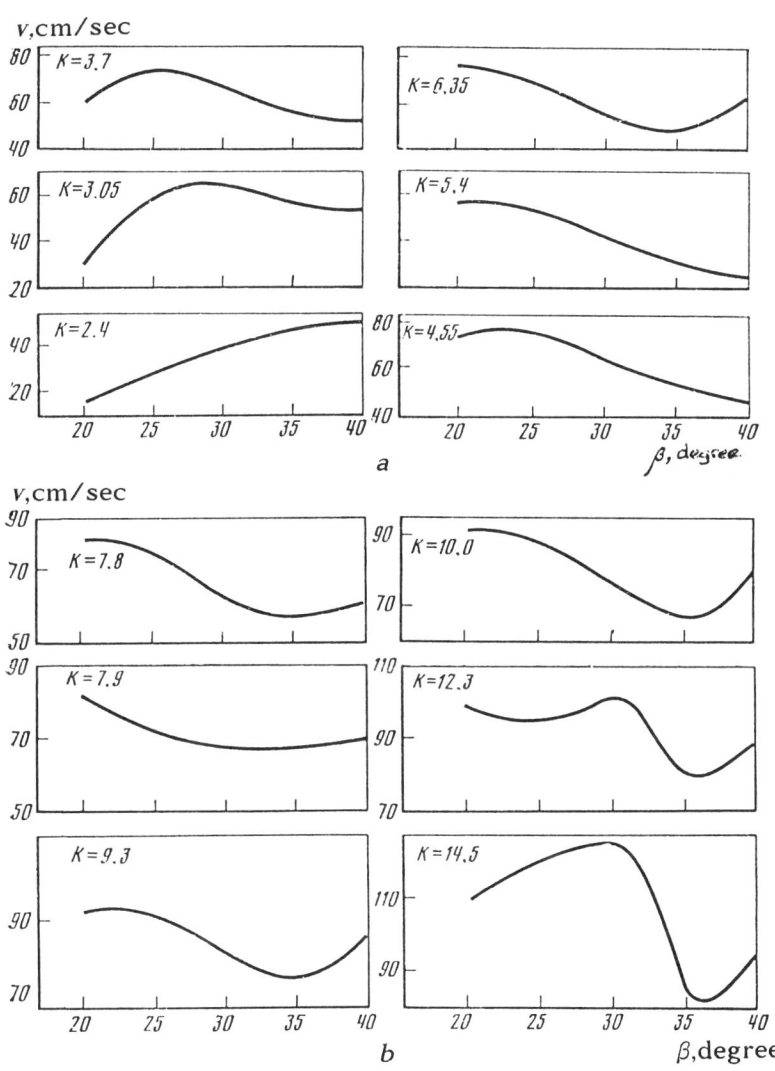

**Figure 1.2** Conveyance speed of piled-up loads vs. vibration angle under various accelerations of the load-carrying element. a) sand; b) rock of chunk size up to 50 mm.

tendency is observed - when the vibration angle is increased the speed of vibration motion decreases. In the region of high-intensity regimes increasing the vibration angle leads first to a decrease of the speed then to its increase again, although the absolute maximum speed is attained nonetheless at lesser vibration angles.

Hence, it follows that to increase the speed of vibratory conveyance at high-intensity regimes the vibration angle ought to be decreased and, conversely, in calm (low-intensity) regimes the angle ought to be increased. In order to attain higher speeds with deteriorating load transportability and increasing layer thickness the vibration angle must be increased.

Investigations indicate that with increasing angle of inclination the speed of conveyance is rising; furthermore, for slow regimes a more intensive increase of speed is observed. Decreasing the vibration angle also has favorable effect. However, increasing the angle of inclination does not ensure a sharp increase of speed. Upon conveying upwards along an elevation, the speed drops very significantly.

Thus, although vibration conveyance at an inclination increases the translational speed, it does not however significantly exceed, as a rule, the amplitude values of vibratory speed of the carrying element. To increase conveyance speeds under motion at an inclination, one must increase the vibration amplitude, and decrease the vibration angle and vibratory frequency.

The effect of layer thickness on the vibratory conveyance speed is very important. Experiments show that when investigating vibratory conveyance parameters, one must consider not the load itself as such but rather its thickness which is an intrinsic characteristic of transportability. In the majority of cases the layer thickness acts in the same way - it lowers the speed of vibratory conveyance. The only exception is transported media (loads) with round particles, for example potatos, when the lowest speed is developing while transporting a monolayer. This is explained by the fact that due to the round shape the individual particles in a monolayer are rolling; sliding friction is substituted by rolling friction which disturbs the transfer of energy from the carrying element to the transported load. When operating in high-intensity regimes, stability of conveyance is increasing with increasing layer thickness.

A somewhat different effect is exerted by the layer thickness on the output of the vibratory conveying machine.

Thus, despite the decrease of speed of motion with increasing layer thickness, the output of the plant is increased. This is explained by the fact that with increasing thickness, the cross-sectional area of the load layer is increasing, up to a certain limit, faster than the decrease of conveying speed. However, this tendency is gradually changing. At some point the increase of the cross-sectional area of the load layer begins to be fully compensated by the decrease of the conveyance speed, thus stabilizing the output. Further increase of the layer thickness can lead to a decrease of the output. It is established that for each type of load at a given configuration of the carrying element, there is an ultimately attainable output.

The presented experimental data are characterizing effects on the conveyance speed of amplitude and frequency of vibrations, vibration angle, and angle of inclination of the carrying element of the vibratory plant under rectilinear harmonic oscillations. Let us analyze a correlation between the speed of load conveyance, and vibratory velocity, and acceleration of the carrying element.

The dependence of the speed of vibratory conveyance of 50 mm thick sand layer on a horizontal vibratory conveyer and at vibration angle of $20^{\circ}$ on the amplitude of vibratory velocity of the carrying element is depicted in Fig. 1.3a. The experimental points plotted on the graph pertain to frequencies 4.2 – 23.4 Hz and amplitudes 0.55 – 25.0 mm. The region where the experimental points are located is shaded. The points corresponding to the same vibration frequency are joined by the lines. It is evident that various conveyance speeds correspond to a single vibratory velocity of the carrying element.

Analysis of experimental data indicates that the conveyance speed at invariable vibratory velocity of the carrying element is changing within a certain range with the change of frequency. The greater the frequency, the higher the speed of conveyance at a given constant vibratory velocity of the carrying element. This can be explained by the fact that the efficiency of the process of velocity transfer to the load in specific regimes is increasing with increasing vibratory acceleration of the carrying element.

The dependence of the speed of vibratory conveyance on the amplitude of vibratory acceleration of the carrying element, plotted from the same experimental data, is depicted in Fig. 1.3b, which shows that oscillation acceleration of the carrying

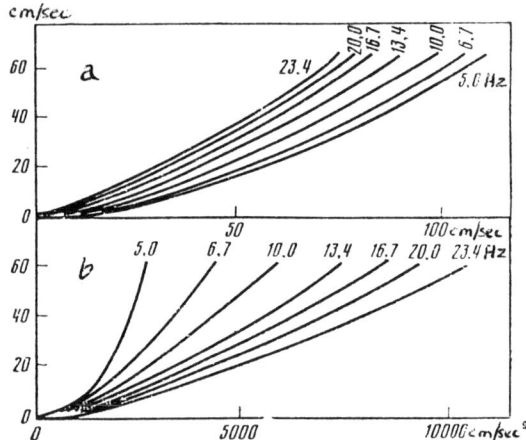

**Figure 1.3** Conveyance speed of sand vs. parameters of rectilinear harmonic vibrations a) velocity amplitude; b) acceleration amplitude.

element, just as vibratory velocity, does not uniquely determine the conveyance speed. The conveyance speed can vary within very wide limits at the same value of acceleration. A second parameter for the determination of the conveyance speed is vibration frequency of the carrying element. In the case under consideration, an effect opposite to the one which occurred for the dependence on the vibratory velocity of the carrying element is observed, namely, with increasing frequency the conveyance speed decreases. This is due to the fact that at a constant value of acceleration amplitude of the carrying element, velocity amplitude of its motion decreases with increasing frequency and, conversely, it rises with decreasing frequency. Hence, it follows that in order to increase the conveyance speed of the vibratory installation which is operated with a specified vibratory acceleration of the carrying element, it should operate at low frequencies and with increasing vibration amplitudes.

This allows one to conclude that neither vibratory velocity nor acceleration taken separately singularly determines the speed of vibratory conveyance. The vibration frequency of the load-carrying element affects magnitude of this speed. This is due to the fact that the conveyance speed is determined by two principal factors: vibratory velocity of the carrying element, and efficiency of the process of velocity transfer to the load. Since the vibratory velocity of the carrying element is proportional to the first power of the frequency, and the coefficient of velocity transfer is proportional to the second power of frequency, it can be assumed that their combined effect is proportional to the vibration frequency raised to an intermediate power. Indeed, the effect of frequency is diverse: if in the first case (see Fig. 1.3a) the speed of conveyance is increasing with increasing frequency, then in the second case (see Fig. 1.3b) this speed is decreasing. This allows one to presume that one can find some generalized parameter $A\omega^n$ which would uniquely determine the speed for the specified conditions of vibratory conveyance. The sought result for the presented experimental data was achieved at the exponent equal to 1.25. This parameter of the regime for a given load in the considered conditions of vibratory conveyance uniquely determines magnitude of the speed.

Experimental data show that the speed of the load motion in vibratory conveying machines is less than the velocity amplitude of the carrying element and is lower than the velocity component in the direction of motion.

The efficiency of the process of velocity transfer from the carrying element to the transported load is usually estimated by the value of the coefficient of velocity transfer which is defined as the ratio of the mean speed of load motion to the amplitude value of velocity of the carrying element. Figure 1.4 shows values of the coefficient of velocity transfer which were computed from experimental data obtained during the vibratory conveyance of 50 mm sand layer at a vibration angle of $20^\circ$ (curve 1) and 10 mm layer of crushed ore of 0 - 6 mm size and vibration angle of $22^\circ$ (curve 2). It is evident that the coefficient of velocity transfer rises with increasing vibratory acceleration of the carrying element, and approaching some constant value of 0.70 - 0.80 at large acceleration magnitudes. At more intensive oscillations, i.e., with increasing amplitude of acceleration, a decrease in the coefficient of velocity transfer takes place. Therefore, in order to increase the efficiency of the process of velocity transfer from the carrying element to the transported load, one must increase intensity of oscillations of the load-carrying element up to the limits ensuring stable regimes of vibratory conveyance.

The required conveyance speed can be obtained with various combinations of parameters of the operating regime of the plant, namely, amplitudes, frequencies, and the angle of vibration. The problem is to select such a combination which would ensure the optimum operating conditions. Depending on

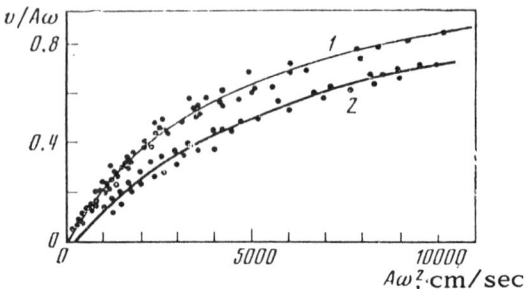

**Figure 1.4** Coefficient of velocity transfer vs. vibratory acceleration of the load-carrying element.

selection of the criteria determining the desired efficiency of operation of the vibratory conveyance installation, different combinations of parameters will be optimal.

The criteria of optimality of the operating regime can generally be very diverse for different plants, conditions of use, conveyance tasks, etc. As such criteria, one can adopt, for example, the provision for the minimum total vibratory acceleration (or minimum value of one of the component of the total acceleration) of the load carrying element, achieving minimum energy consumption for transporting the load, and so on. One can demand for several indicators to be minimized simultaneously, for example, minimum dynamic loading of the plant and the lowest power-consumption. For valuation of the process efficiency, attaining a minimum vibratory acceleration of the carrying element as the most general and essential condition of efficiency of the operating regime is often required. The fact of the matter is that the maximum acceleration of the carrying element determines magnitudes of dynamic loads on the drive components and on elastic links and, hence, the cost and reliability of the installation. For example, the conveying speed in the low-speed regimes of machine operation (low frequencies and relatively large vibration amplitudes) can be increased without changing vibratory acceleration by a correct selection of the vibration angle. Experiments show that the coefficient of velocity transfer at low-speed regimes is, practically, increasing in direct proportion to the increase of the vibration angle. Hence, it follows that at a constant machine operating regime its output can be increased by increasing only the vibration angle. In practice it is not always feasible to increase the speed of conveyance by increasing vibration frequency and amplitude of the carrying element. When finely dispersed loads with poor air permeability are transported, the increase of vibration intensity does not cause an increase of speed and it often results in disturbing the normal conveying regime. As a result of poor air permeability of the finely dispersed load, significant pressure fluctuations of the air bed take place in the process of vibrations of the carrying element. Furthermore, additional aerodynamic resistances to vibratory conveyance arise which can, in some regimes, considerably reduce the speed.

To increase efficiency of transporting finely dispersed loads, perforated load-carrying elements with film coating are being used. To determine the rational parameters of such carrying elements, the effect of the number and size of perforations and also the properties of the film coating on the efficiency of vibratory conveyance were experimentally established. Increasing size of the holes, on the one hand, causes a reduction in aerodynamic resistances to motion of the load and, on the other hand, seems to reduce the actual displacement of the conveying surface (as a consequence of the insufficiently rigid film coating above the hole). Investigations indicate that to increase the efficiency of vibratory conveyance of finely dispersed loads, it is expedient to increase the number of holes by decreasing their size. The use of a perforated carrying element with film coating substantially increases in all cases the speed of motion of the finely dispersed load. The use of perforated carrying elements with film coating is particularly effective in moving of thick layers when the conveying process in normal conditions comes totally to a halt.

It is of particular interest from the standpoint of increasing efficiency of vibratory conveying machines to study effect of special oscillation regimes of their carrying element on the performance characteristics. It was established that under elliptical oscillations the conveyance speed is mainly affected by the shape of the trajectory of the carrying element, which is determined by the ratio of amplitude components and the phase shift between them, the angle of inclination of the major axis of the ellipse to the horizontal, frequency and amplitude of vibrations, and also by the direction of traversing the trajectory.

The results of investigations of the conveyance speed as a function of phase shift between the components at various oscillation frequencies for quartz ballast and for sand are presented in Fig. 1.5. The maximum speed is reached at phase

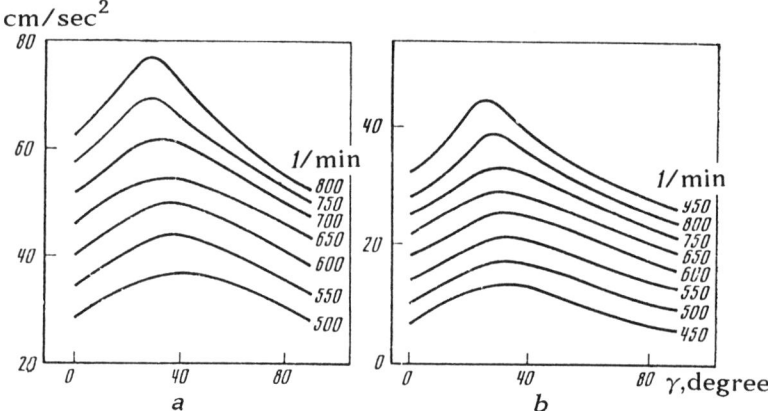

**Figure 1.5** Speed of conveying of quartz ballast (*a*) and sand (*b*) vs. phase shift between the components at various frequencies in the regime of elliptical oscillations. *a*) $A$ = 5 mm, $B$ = 10 mm; *b*) $A$ = 2.5 mm, $B$ = 5 mm.

shifts ranging from 20 to 30°. Analysis of the experimental data shows that the extremal speeds in the most high-frequency regimes occur at low values of the angle of phase shift. The character of the curves for sand and quartz ballast is similar, although the speed is higher in the latter case. The absolute maximum speed under elliptical oscillations of the carrying element is reached for an oblong ellipse with a 1:4 - 1:5 ratio of the minor and major axes.

Study of the conveyance speed of quartz ballast, river sand, fine coal, and manganese ore of 50 mm layer thickness on a horizontal carrying element as a function of its frequency indicated that increasing the frequency from 8.4 to 16.7 Hz at amplitudes of $A$ = 5 and $B$ = 10 mm leads to a sharp, almost linear, rise in the speed of vibratory conveyance. The greatest speed is attained when moving quartzite, the conveyance speeds for coal and sand being close. The high speed of motion of quartzite is explained by a lower vibration damping in the layer; the damping indicators are similar for sand and coal. The minimum speed for manganese ore, which is difficult to transport, is explained by the high adhesion forces with the carrying element. Adhesion of the particles to the element and coalescence of the particles increases both internal and external resistances, thus influencing speed of vibratory conveyance. The effect of the vibration frequency of the load-carrying element on the conveyance speed of all the enumerated loads is practically the same: with increasing frequency the speed is increasing, and the higher the vibration amplitude, the higher the rate of increase.

By comparing the speed of the same load in the regimes of rectilinear and elliptical vibrations at equal horizontal components of their amplitudes, it can be noticed that the conveyance speed under rectilinear oscillations is lower in magnitude and is sharply different from the speed under elliptical oscillations with respect to the character of its frequency dependence. With increasing frequency of the rectilinear vibrations the speed is increasing but then the dependence flattens and changes approximately in a parabolic law.

The extremal values of the conveyance speed under rectilinear harmonic oscillations are reached at lower frequencies and have lower magnitudes than the corresponding values of the speed under elliptical oscillations for equal horizontal components of their amplitudes. Consequently, under elliptical oscillations the speed can be increased compared with rectilinear harmonic oscillations of the same double amplitude, approximately 1.5 times. There is, however, a specific optimum beyond which an increase of oscillation amplitude of the vertical component is useless.

The speed of vibratory conveyance in the separation (jolting) regime at constant vibration frequency is mainly determined by the value of the horizontal component of the vibration amplitude of the carrying element. Investigation of this dependence enables one to conclude that with increasing horizontal component the speed first increases, then declines and the higher the oscillation frequency the greater the rate of decline. Hence, under elliptical oscillations, just as under rectilinear oscillations, it is expedient to operate at low frequencies when the amplitudes are large.

It follows from the presented data that it is useless to increase the horizontal component infinitely at a fixed value of the magnitude of the vertical component, since after a certain limit is reached (in our case at a ratio of amplitudes of roughly 1:4) the increase of the conveyance speed becomes insignificant. Limiting the amplitude ratio within the range 1:2 to 1:4 is advisable.

Under elliptical oscillations of the carrying element the speed of load is greatly affected by the direction of traversing the trajectory. The speed is greater when the direction of motion coincides with the traversing direction in the lower part of the trajectory. If the major axis of the ellipse is positioned with an inclination, the direction of motion of the load remains unaltered for any direction of the trajectory traversing.

At a horizontal position of the major axis of the ellipse the variation of the character of trajectory traversing causes a change in direction of load conveyance. It is evident that, all parameters being the same, the horizontal position of the axis of the ellipse causes a reduction of the conveyance speed.

Thus, to attain high speeds of vibratory conveyance in the regime of elliptical oscillations the traversing of the upper part of the trajectory must be counter to direction of the load motion. Elliptical oscillations enable the construction of conveying devices in which the change in the direction of motion is effected by means of reversing the drive motor.

The speed of vibratory conveyance is essentially dependent on the inclination angle of the major axis of the ellipse. An increase of the angle to approximately 20 – 25° increases the speed, and with further increase of the angle the speed is decreasing. The extremal values of the speed at various vibration frequencies are reached at different angles of inclination of the major axis of the ellipse. With increasing frequency the inclination angle of the ellipse corresponding to the extremal speed is somewhat decreasing. Increasing the angle of inclination in the ranges under consideration causes practically linearly correlated increase of speed. The speed of motion for quartzite is increasing more than for coal.

Increasing the angle of upward slope of the carrying element leads to a decline of speed. For quartzite, the speed declines more which can apparently be explained by a lower internal and external friction in the quartzite layer than in the coal layer. Vibratory conveyance downwards along a slope is associated with an increase of speed. In this case, it is advisable to increase the amplitude and reduce frequency. Starting with an angle of inclination of 15 – 16° load displacement due to gravity takes place, and then it becomes uncontrollable. When operating downhill the elliptical regimes of vibrations of the carrying element are more effective than other forms of vibrations and, conversely, are less effective when operating uphill.

When investigating the behavior of vibratory conveyance, one should take into account the thickness and condition (dry, wet and so on) of the layer of the transported load. The dependence of the speed of conveying quartzite on the layer thickness for the vibration components $A$ = 5 mm, $B$ = 10 mm, angle of inclination of the major axis of the ellipse 30°, and a horizontal position of the carrying element is depicted in Fig. 1.6. With increasing layer thickness the speed is decreasing. At high vibration frequencies the effect of thickness on the speed is less pronounced than at low frequencies.

The variation of transportability of the loads depending on the moisture content must be taken into account particularly when the installations are erected in the open, and the weather conditions can change humidity of the loads within wide ranges. It can significantly influence effectiveness of operation of the installation in such conditions.

Figure 1.6 also shows graphs characterizing effect of humidity on the conveyance speed of manganese ore with elliptical vibrations of the carrying element. Increasing humidity of manganese ore sharply reduces the conveyance speed. In the moist state the ore has high adhesive properties and sticks to the carrying element as a result of which the conditions of vibratory conveyance are distorted. As experimental investigations show, when manganese ore is conveyed, the regimes of rectilinear biharmonic and polyfrequency vibrations are more effective compared with the harmonic rectilinear and elliptical vibrations.

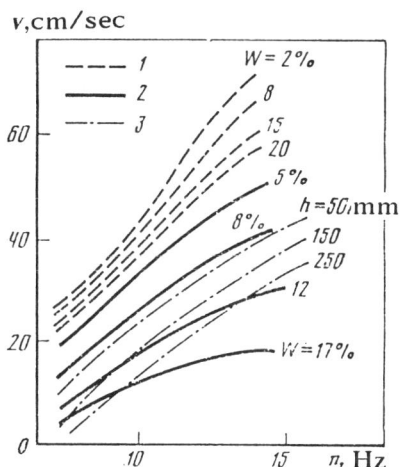

**Figure 1.6** Conveyance speed vs. regime of elliptical vibrations, thickness and humidity of the transported load. *1, 3)* quartz ballast; *2)* manganese ore.

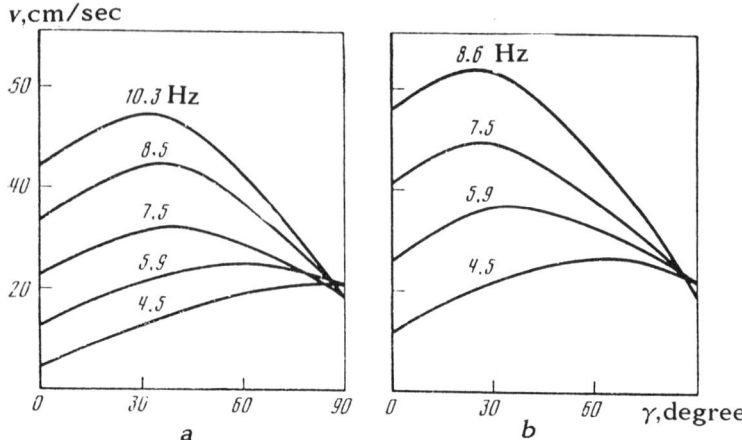

**Figure** 1.7 Conveyance speed of a large-chunk load vs. the regime parameters at half-wave elliptical vibrations. a) *A* = 10 mm, *B* = 2 mm; b) *A* = 15 mm, *B* = 3 mm

The above allows one to conclude that to increase effectiveness of the process of velocity transfer from the load-carrying element to the transported load, intensity of vibrations, at angles of inclination of the major axis of the ellipse of about 20°, should be increased up to the limits that ensure stable regimes of vibration conveying.

At the present time vibratory conveyers which use half-wave vibrations of the load-carrying element are being developed successfully. Effect of regime parameters on the conveyance speed in case of half-wave elliptical vibrations has been identified. The dependence of the speed of motion of a large-chunk load at different amplitudes and frequencies of the carrying element on the phase shift between vertical and horizontal components of the displacement is shown in Fig. 1.7. As can be seen, the speed is considerably dependent on the shape of trajectory of the carrying element which is determined by the amplitude ratio of the vertical and horizontal components of the vibrations and by the phase shift between them. At low frequencies the speed increases with increasing phase shift from 0 to 90°, at higher frequencies the extremum is moving towards lower values of the phase shifts. In no loss of contact regimes of vibratory conveyance the coefficient of velocity transfer increases with the decrease of the amplitude ratio of vibration components of the carrying element. In order to attain high conveyance speeds, one must operate at low-frequency regimes with large amplitudes of oscillations.

Experience shows that for moving finely dispersed loads and also loads that stick to the carrying element of the machine when wet, neither one-component nor two-component harmonic oscillations are sufficiently effective. In such cases use of biharmonic oscillations yields better results.

In order to find the best regimes of vibratory conveyance of moist, sticky loads under biharmonic oscillations, investigations were conducted to establish the effect on the speed of motion of amplitude and frequency of the first harmonic, ratio of amplitudes of the harmonics, their order (second or third), and phase shift between them.

Introduction of higher harmonics results in increase of the conveyance speed; it rises particularly considerably at an angle of phase shift of 90°. The third harmonic yields a more appreciable rise of speed in the region of low-speed regimes than the second harmonic. This rise of speed is attained at the expense of additional dynamic loading of the vibratory machine due to using higher harmonics, particularly, the third. In the case of transporting dry loads, at a certain level of dynamics load the same conveying speeds can be attained with rectilinear monoharmonic vibrations; therefore, biharmonic oscillations here do not have particular advantages as compared with harmonic vibrations. The advantages of biharmonic regimes are pronounced when operating with moist sticky loads.

For harmonic oscillations and ore humidity not exceeding 10%, no substantial difference in the conveyance speed is noticed; at higher humidity the speed starts to drop and at humidity of 20% vibratory conveyance is practically halted. It is pertinent to mention that at a humidity of 10 – 12% a tendency toward compaction of the layer of the transported load is observed during the process of conveying. At a humidity of 15% a wet sticky layer remains on the carrying element following the passage of manganese ore; this results in some increase of the resistance to motion. Under rectilinear harmonic oscillations it is not advisable to transport manganese ore with humidity exceeding 10%.

High-speed filming and analysis of oscillograms, obtained from strain gauges on the carrying element, reveal considerable differences in the affect of harmonic and biharmonic oscillations on wet manganese ore. While ore separation from the carrying element is absent in the former case, conveying regimes with separation (jolting) are established in the latter.

Under biharmonic oscillations at even 12% humidity the manganese oxide ore retains practically the same speed as at 2% humidity. At 15% humidity the speed changes slightly. Its reduction commences at a humidity of 18-20%; however, even at this humidity the conveying process does not come to a halt. As regards the adhesion of the ore to the surface of the carrying element, this phenomena is absent up to a humidity of 17%. Above 20%, adhesion of some chunks is observed, which hampers the motion of the whole mass.

The reason for low effectiveness of harmonic oscillations in preventing sticking of the conveyed load can be detected in the symmetry of the generated inertial forces. Therefore, if the conveyed load sticks strongly to the conveying surface and does not separate over one cycle of oscillations from the carrying element, then even multiple application of forces is found to be ineffective. As a result of the symmetry of dynamic forces under harmonic oscillations, the effect of the forces acting over one-half of a cycle towards separating the load, is compensated during the second half of the cycle by oppositely directed forces. If under asymmetric biharmonic oscillations larger effort is directed toward separation of the mass from the load-carrying element, then under multicyclic action summing of the desired effect takes place. Therefore, continuous action of asymmetric oscillations facilitates removal of the adhering load.

For the purpose of establishing the optimum parameters of biharmonic oscillations providing effective vibration conveying of finely dispersed loads, the effect of phase shift between the first and second harmonics on the speed of motion was investigated for various ratios of amplitudes and frequencies. It was established that the conveyance speed is essentially dependent on the phase shift between the harmonics and, in the investigated range of amplitudes and frequencies, the maximum values are reached at phase shift angle of $90^\circ$. For higher frequencies this optimum is more sharply pronounced. When the amplitude ratio is increased from 1.5 to 2.5, significant speed increment was observed. With a further ratio increase, speed increase is less substantial, which apparently associated with the decreasing effectiveness of the action of the second harmonic.

When finely dispersed loads are conveyed by vibration, the layer thickness is regarded as one of the main factors on which speed and output are dependent. Increasing layer thickness practically always leads to decreasing its displacement speed, and for considerable thicknesses the process can be disrupted and halted altogether. The effect of layer thickness on the output of vibratory conveyers is more complex: within some range, increasing the thickness causes an increase in output, and then leads to its decrease. Experiments show that with increasing layer thickness, the speed is decreasing at a lesser rate for biharmonic oscillations than for harmonic.

When working with finely dispersed loads, aerodynamic resistances of the medium have large influence. In order to study this effect investigations were carried out [1].

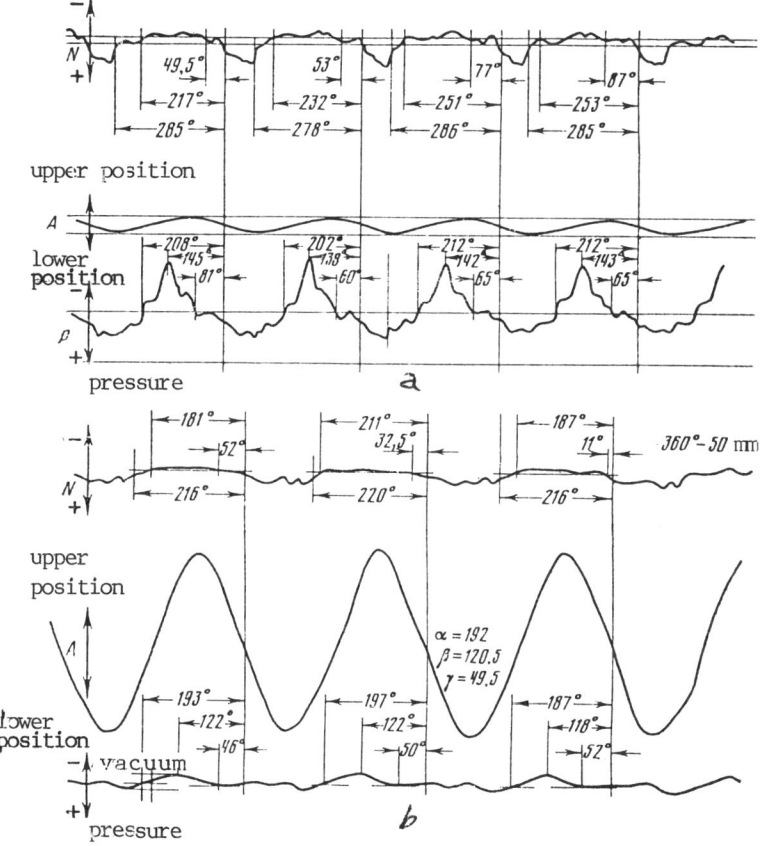

**Figure 1.8** Parameters of vibratory conveyance of finely dispersed loads. *a)* displacement of the load-carrying element, normal reactions of the load, air pressure ($A$ = 5 mm, $n$ = 15.8 Hz, $h$ = 90 mm); *b)* $A$ = 7 mm, $n$ = 10.8 Hz, $h$ = 80 mm

Figure 1.8 shows time histories of vibration amplitudes of the carrying element $A$, normal reaction of the material on the carrying surface $N$, and pressure fluctuations of the gas $P$ under the layer during conveying of a granular material with a large content of dust particles.

The amplitude value of gas rarefaction under the layer reaches a maximum during the flight stage, and the pressure – during the stage of joined motion with the carrying surface. At the moment of load separation, the gas pressure under the layer is a minimum. In the diagrams a wave-form variation of the mean effective pressure can be observed under the load layer which also moves in waves. The presented oscillograms enable one to determine sufficiently correctly the moments of separation and fall of the load.

The character of change of the gas pressure under the layer is the determining factor for parameters of vibratory conveyance process of a finely dispersed load in regimes with bouncing. The fluctuation of the gas pressure under the load layer is mostly affected by the height of the layer, vibration velocity and acceleration. The more finely the load is dispersed, the higher the amplitude values of the pressure jumps $p_-$ and $p_+$. The averaged pressure of the gas under the load layer for materials with poor air permeability assumes negative values.

Increase of the load layer thickness leads to increases both in rarefaction and in gas pressure under the layer. The averaged pressure for the loads having good air permeability shifts towards positive values and for loads with poor air permeability the pressure shifts towards negative values.

The aerodynamic loading of the load layer is characterized by the Schwab criterion Sb, which represents the ratio of the mean effective pressure of the gas under the load layer to the square of the vibratory velocity. Materials with large content of finely dispersed particles should be transported with a dynamic loading factor of the conveyor $\Gamma \le 1.5$. Under such conditions the Schwab criterion assumes positive values which favorably affects the process, since a lesser part of the vibration energy is expended on overcoming resistances at the flight stage.

To determine transportability of the conveyed loads, one can use the Strouhal criterion Sh which characterizes, in the general case, the homogeneous synchronism of motion of the disperse medium and, in our case, reflects the ratio of the vertical vibratory velocity of the vibrating layer to the conveyance speed.

When investigating the conveyance mechanisms for finely dispersed loads, the problem is reduced to obtaining regression equations for the dependences of random quantities ($\rho_-$, $\rho_+$, $\varphi_Q$, $\varphi_p$, and $v$) on nonrandom quantities-arguments (A, n, h) which are given. The values of the nonrandom quantities in their general set are, in their turn, selected in a random manner on the basis of the method of rational planning of experiments.

Parameters of the initial state of the loads and of the operating element of the vibratory conveyor are dispersion composition $d$, volumetric weight $\gamma$, conveyance output $Q$, and cross-sectional area of the load flow $S$.

The following parameters are adopted as controlling actions: vibration amplitude $A$, frequency $n$, height of load layer $h$, conveyer operation regime factor $\Gamma$. The parameters of the final state of the process of conveyance are: conveying speed, rarefaction and pressure of the gas under the load layer, phase angles of interaction of the load with the plane of the working element, the state criteria for vibratory conveyance of the layer of a granular load, namely, the Schwab and Strouhal criteria. Figure 1.8 depicts the dependences of the speed of vibratory conveyance of a material with a large content of finely dispersed particles on frequency at fixed amplitude values.

Analysis of the obtained plots indicates that the speed of vibratory conveyance rises dramatically up to some value of the operation regime factor. This is explained by resonance phenomena arising in the layer under forced vibrations. At the inflection point of the curve, layer compaction and speed reduction take place. A similar character of change is shown for correlation between speed and vibration amplitude at fixed arithmetic mean values of frequency and layer height. The speed of conveyance decreases with increasing layer height. This is explained by the fact that at the given parameters of vibration and layer height of 75 mm, the forces of resistance to motion in both materials are roughly equal. The equality of forces are, in their turn, explained by approaching of the ratio $h/d$ to such a value above which air permeability of the layer is the same for materials of the same physical nature.

Mechanisms of the process of vibratory conveyance ought to be considered in their correlation to several main factors: amplitude and frequency of vibration, height of layer and its granulometric composition, coefficient of friction, and amounts of dust-like particles. This correlation can be more fully assessed by using equations of multiple regression.

As a result of the processing of experimental data by methods of mathematical statistics with the aid of a computer, sufficiently accurate equations of multiple regression have been obtained describing changes of phase of motion and aerodynamic reaction of a layer of a granular load. These equations enable the forming of fully determined qualitative and quantitative representation of the character of behavior of a layer of finely dispersed load on a harmonically oscillating surface.

From the equations of multiple regression it follows that, on the average, for all loads the character of conveyance is affected mostly by the vibratory velocity. The experimental data presented in the coordinates $Sh = \varphi\ (\Gamma)$ vividly shows that the most transportable are the materials with high coefficient of friction and good air permeability. Finely dispersed materials have very low transportability: when being conveyed by vibration over a carrying element oscillating with an acceleration over 2.5 $g$, the power of vibrations is mainly expended to overcome the resistances.

From the analysis of experiments it follows that with increasing the operation regime factor $\Gamma$ above some value, the value of the Strouhal criterion Sh remains unchanged, and the energy additionally supplied to the layer is expended on nonproductive vertical oscillations without substantially increasing the conveyance speed. Therefore, increasing dynamic loading of the vibratory conveyer, which is characterized by factor $\Gamma$, does not make sense when moving materials with large content of finely dispersed particles: for river sand higher than $\Gamma = 2.1$ and for quartz sand $\Gamma = 1.9$, since criterion Sh here does not change substantially.

Materials having better air permeability are separated earlier from the vibrating surface. In particular, it was noticed that at a regime factor of $\Gamma = 1.6$, materials with large content of finely dispersed particles commence their flight stage earlier than quartz sand.

### 1.2.1 Investigation of Intralayer Processes during Vibratory Conveyance

The most complete representation of the physical picture of the process of vibratory conveyance can be derived from an analysis of the real laws of motion. To obtain these laws various methods can be used. Each of them has its merits and shortcomings and is characterized by a specific domain of application.

Use of special filming ensures high accuracy, which compensates to a large degree for the labor-intensive process of frame-by-frame processing of the film. However, reliability of the obtained diagrams can be assured only when investigating single pieces or lump loads. When working with a granular load, the task is complicated by the need to select the objects of the filming, namely the marked particles moving in the monolayer and whose size is within the bounds of resolution of the camera.

A less labor-intensive method is the study of the motion of the marked particles by means of a stroboscope. When the flash of a neon lamp is synchronized with the frequency of oscillations of the working element, the observer gets the impression that the latter is stationary. Furthermore, the relative displacements of the particles which are considerably slower than the oscillating motion become pronounced. On the basis of preliminary experiments, round pieces of thick paper 7 mm in diameter with the transported load (sand) pasted over them with the contrast point at the center were selected as marked particles for filming at the transparent wall and bottom. For the experiment with X-rays filming, 2 mm diameter balls made from barium powder with glue were used. The balls possess the same bulk weight and coefficient of friction as the investigated load, demonstrating at the same time the necessary quality for the experiment, namely, weak penetrability for the X-rays.

To conduct the filming, transparent sections were cut in the wall and bottom of the trough of the vibratory conveyer. The vibration parameters were recorded simultaneously with the filming. A contact breaker mounted on the drive shaft gave marks on the oscillogram corresponding to the initial position of the crank. The oscillograms which were viewed simultaneously with the film enabled one to identify the start of the steady motion of the device. The control panel was equipped with a time relay enabling the starting of the vibratory conveyor and the movie camera for a strictly fixed period of time with a specified speed. Filming was done with the aid of a high speed camera SKS-1M. In addition to the marked particles. a tachometer and time indicator were located in the frame.

Recording of the laws of motion of the vibrating plane and transported object can also be made on a continuously moving film. For this purpose, a special device was constructed comprising a lens, a housing with a drum, cassettes, and film. The rear wall of the housing has a slit covered by an enclosure with a helium flash lamp which is the time marker. The driving drum is driven by a synchronous electric motor with a reduction gear to ensure that the film moves at a constant speed. Rotation to the receiving cassette is transmitted from the driving drum by means of gearing with a friction clutch.

As a marked particle, a contrast point on the moving object or a point source of light (bulk SMN-6, 3x20-1) of 3 mm diameter fixed to the object can be used.

The device is mounted rigidly near the object of the filming and is focused. The moving film registers the absolute displacements of the characteristic points, namely the particles and the vibrating working element.

The greatest difficulty is presented by studies of the true law of motion of the particles deep in the nontransparent layer of the load, which are not subjected to the influence of friction near the wall and the bottom. In order to conduct the investigations, the X-ray device KhIRODUR 125 V (Czechoslovakia) was used. This diagnostic X-ray device has a sufficiently powerful beam of X-rays and is equipped with the high-speed camera Admira 16RT2.

Glass windows transparent to X-rays, a fixed lead ruler (at the vibration angle), and a travelling lead (triangular) particle which is linked with the trough are mounted in the zone of the frame. Additionally, for convenience of interpretation of the pictures, a lead grid is mounted in the path of the X-rays. In the filming zone, which has a size of 180x240 mm, marked particles are placed on the left border at different distances from the bottom with pitch of 10 - 20 mm. In order to reveal the effect of near-wall friction the distance of the marking particles from the walls of the trough was varied in different tests.

To contruct displacement diagrams, the film is projected frame by frame on a plotting screen with moving millimeter paper and fixed tracing paper. Positions of centers of the marked particles and the points on the trough are plotted. Before projecting the next frame, the paper is shifted by 3 - 7 mm to allow for time base.

To investigate the intralayer processes one can use the method of paraffin impregnation enabling one to register the "instantaneously frozen" picture of intralayer displacement processes. For this purpose, prior to the formation of the layer in the removable section of the vibratory trough, a batch of initially dyed material is added through an opening (5 - 8 mm wide) between two vertical partitions. The partitions are then withdrawn and the device is switched on after setting the time relay in the drive circuit in such a way that the marked particles would be displaced by 250 - 300 mm. Then part of the granular body in the removable section of the trough is cut out by partitions and held in melted paraffin till saturation. Following solidification, a slice is cut out along the new boundary of the marked particles. The obtained slice is the spatial profile of the displacement of the granular body and in the same time represents in some scale the velocity profile of the load. The shape of the working element and the regime of oscillations can be different and, hence, the form of the profiles will change.

For the disclosure of the physical nature of intralayer phenomena it is of interest to record the process of displacement of the granular body "in ideal conditions", i.e., under the section of an inertial force which is constant in magnitude and direction. This is done in a device comprising a trolley with a trough. The trolley is propelled at low acceleration, then its motion is slowed down providing substantial constant-magnitude acceleration. As a result, a series of slices-profiles is obtained for different layer sizes at different accelerations giving an idea about the character of distribution of the dynamic linking conditions between the particles inside the layer.

The investigation of the motion of the layer can also be conducted with the aid of embedded sensors. Only acceleration transducers not requiring stationary bases are applicable. When working with chunk loads, the transducers can be mounted directly inside some chunks. When investigating the behavior of granular loads the transducers must be mounted so that their specific weight is equal to the bulk weight of the load. For unchanged orientation and "blending" of the transducers in the surrounding monolayer, they must be equipped with motion stabilizers and be covered with the tested material. Naturally, the smaller the size of the transducer the more accurately it reproduces the law of displacement of a given specific region of the layer. Acceleration senors DU-5 and D-14 are adequately small. The piezoelectric transducer MS-579A-1 has even smaller dimension ($\varphi$ 7 mm and weighs 3 g).

Embedded sensors can be used to obtain the mean integral law of motion for the layer or part of the layer.

The described experiments enable one to obtain the initial information for the subsequent determination of the parameters of computational rheological models of various transportable loads. The most accurate results can be obtained by achieving correlation between theoretical and experimental laws of motion of the layer which is the optimum criterion of validation of the mathematical model. To this end, adequate techniques for obtaining the experimental mean law of conveying a piled up load are of particular significance.

It is feasible to obtain the indicated characteristics of the medium using a mechanical phenomenological layer model. This method is successfully used to obtain phenomenological parameters of a number of piled-up loads [2]. The model is comprised of elastic elements, i.e., variable-stiffness springs equivalently modeling the elastic characteristics of the layer. Viscous properties are reproduced by piston dampers with adjustable cross section of by-pass orifice. Plastic deformations in the vertical direction are simulated by a spring-loaded wedge.

The model and the container with the investigated load are placed on the table of a vertical shaker. The signals from the sensor "floating" in the load and on the lumped mass of the model are recorded. The coefficients of elasticity and viscous resistances corresponding to a close coincidence of the recordings are adopted as the parameters of the phenomenological model of the load.

A quick determination of parameters of the phenomenological models can be realized by using a mechanical identifiable model of the medium which is equipped with adjustment means and instruments recording characteristics of its elements. Signals from accelerometers can be recorded by a two-beam oscillograph, and the parameters of the model are being adjusted until a coincidence of its laws of motion with the law of motion of the medium is achieved. As an analog of the medium, one can also use an electronic model.

Using the described methods, many diagrams have been obtained of the true laws of motion of the marked particles that are located at various levels and distances from the wall of the trough. Experiments cover the operating ranges of vibratory machines corresponding to both separation and nonseparation regimes which are most frequently encountered in practice.

As an example, Fig. 1.9 presents motion diagrams of a dry sand layer 100 mm high along a trough oscillating at frequency $\omega$ = 7.8 Hz, amplitude $A$ = 8.5 mm, vibration angle $\beta$ = 30° (obtained with the aid of an X-ray device). The operation regime factor $\Gamma$ = $(A\omega^2\sin\beta)/g$ = 1.05 corresponds to the boundary between separation and nonseparation regimes. Curve *1* characterizes the law of motion of the marked particles inside the layer in the middle section at the bottom of the trough, curve *2* at a depth of 40 mm, curve *3* at the surface of the layer at a depth of 10 mm; curves *2'* and *3'* represent, respectively, the laws of motion of the marked particles at the same levels, but at the transparent wall of the load–carrying element, curve *4* shows time history of vibrations of the working element of the conveyor. The averaged law of motion of the layer in this case practically coincides with curve *2*.

Figure 1.9*b* shows the diagram of motion of particles inside the layer at levels *1*–20, *2*–50, *3*–75 mm when $A$ = 8.5 mm and $\omega$ = 105 Hz, which corresponds to $\Gamma$ = 1.9.

Figure 1.9*c* depicts the diagrams of vertical components of the displacement of the marked particles located at levels *1*–10, *2*–50, *3*–100 mm. The experiment was conducted for $h$ = 150 mm, $A$ = 7 mm, $\omega$ = 9.4 Hz, $\beta$ = 30°, $\Gamma$ = 1.25.

Analysis of the displacement diagram of the load in the regime with separation shows that the intermediate monolayers can receive a force impulse not directly from the working element, but from collision with underlying monolayers. A portion of the kinetic energy is expended on this, which is manifested in local distortion of the curves at the moments corresponding to the collisions. This enables one to record the sequence and duration of such contacts.

Horizontal components of the displacements of the marked particles are also identified and presented in the form of diagrams. With the aid of these curves one can identify the different phases of the translational motion and periods of relative rest. In the latter phase the curves of displacements of the particles and trough are equidistant, which is easily revealed by means of a template, i.e., the trajectory of motion of the trough plotted on a tracing paper which is displaced along the axis. Thus one can obtain the moments of start and end of the phases of motion in straight and reverse directions and corresponding to their phase angles and identify the sequence of mutual displacements of monolayers.

The intensity of decay of the elastic pulse can be judged from the analog of force action which is represented by the acceleration of the monolayers positioned at different levels. For this purpose, at the known moments of commencement of specific phases of motion, double differentiation of the diagrams of displacements was conducted. Comparison of the diagrams of vertical and horizontal components of accelerations of the monolayers enabled revealing that the horizontal components of force pulses are damped more intensely than the vertical components. This is also apparent from the results of study of monolayer acceleration using a compact accelerometer embedded in the load.

The profiles of layer displacement obtained by means of paraffin impregnation when recording the kinematic vibration parameters on an oscillograph during the test enble to determine velocity of any point of the granular body $V_i$. In other words, for a known time $t_s$, of the steady-state motion of the layer, the profiles represent spatial profiles of its speed with the scale factor $\mu v_i = 1/t_s$. Figure 1.10 presents a number of sections of the slices of a dry sand layer of height $h = 65$ mm by horizontal planes passing at various distances $z_i$ from the free surface at $A = 7$ mm, $\beta = 30°$, $n = 6.1$ Hz. The character of load displacement is affected not only by vibration intensity, but also by the height of the layer. This is evident in particular from Fig. 1.10 where a number of velocity profiles are depicted for the middle section of sand layers of different height at the same vibration parameters. A set of profiles thus obtained indicates that motion of a granular load is accompanied, even in the nonseparation regime, by intensive intralayer shear phenomena. In a number of cases, velocity change is observed on some levels of the layer. As follows from the analysis of the high speed films, a relatively sharp change of the law of the load motion takes place at these boundaries, so that two closely positioned particles at the same height can move in different regimes. Apparently, particles with close dynamic coupling conditions are combined in separate layers - families and travel in one of the possible regimes. The boundaries of these layers, which are the surfaces composed from particles with identical dynamic coupling conditions, are close at the walls of the trough to the slip surfaces. Families of such surfaces, which can be the boundaries for individual layers, were identified in a device which imparts constant horizontal acceleration to a loaded trough during a certain period of time. In this test under smooth acceleration of the trough, energy dissipation in the layer thickness is practically excluded, and the effect of friction at the walls and bottom of the working element on the behavior of individual parts of the layer is revealed in a pure form.

Spatial displacement profiles for marked particles under the action of a constant inertial force were obtained by the described method. The front of these profiles is a convex in the direction of slip surface generated by straight lines which belongs to the category of cylindroids [3]. Displacement magnitude of each individual particle is apparently inversely proportional to the intensity of its dynamic coupling conditions.

Curves that are the generatrixes of the sought surfaces with equal coupling conditions can be found by sectioning the front of the spatial profile by a number of vertical planes.

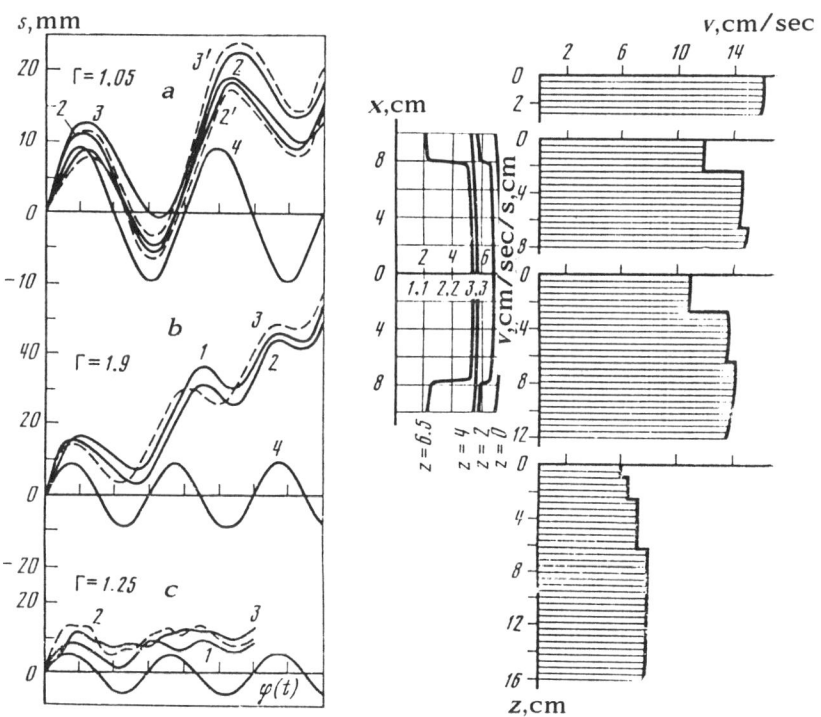

**Figure 1.9**  Diagrams of absolute motion and its vertical components of the particles at the wall of the trough and inside the layer.

**Figure 1.10**  Cross sections of mould-slices of displacement by horizontal planes, and velocity profiles of sand layers of various heights in the mid-section.

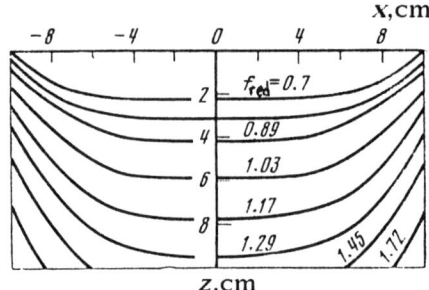

**Figure 1.11** Curves of equal resistances and of monolayer boundaries.

This pattern for sand in the transversal and longitudinal cross sections of the trough is presented in Fig. 1.11. Each curve corresponds to a constant value of the (reduced) coefficient of shear resistance $f_{red}$.

It is pertinent to note that for these profiles, just as in the tests on a vibratory conveyor, there is a distinctive behavior of the upper part of the layer bounded by the surface of equal coupling conditions (the upper curve in Fig. 1.11) which is passing through the extremal points of the air boundary of the granular body. This part of the layer is not subjected to the action of wall friction, cannot be practically compacted, and behaves as a unit.

The above is pertinent mainly to nonseparation regimes. However, as experiments show, stratified shear strains with stepwise speed change take place at some layer levels even under more intensive oscillations. Apparently, this phenomenon can be explained as follows. Firstly, by the attenuation of the elastic pulse in the load layer. For parts of the load conditions of nonseparation regime can be created in which a number of subregimes are possible: motion with two or one instantaneous stops, with two prolonged stops in each period. It is on the boundaries of existence of these regimes that a sharp change of speed is initiated. Secondly, in each oscillation period of any intensity there are time intervals in which the condition of separation is not realized and the load passes through all the possible subregimes in succession. The presence of velocity gradient along the layer height indirectly indicates that an intensive attenuation of the elastic pulse and its delay is observed in the bulk of the granular body.

By analyzing the results of the experiments, it can be asserted that each particle of the layer is moving according to its distinctive law. The latter is determined by the position occupied by the particle in the layer. Of significant influence are wall and bottom friction, as well as elastic, viscous, and plastic properties of the layer which are manifested in the distortion, attenuation, and phase lag of the force pulse as it passes through the bulk of the granular body. The sharpest changes of the law of motion of the particles located at different levels are observed at the levels corresponding to the stepwise velocity changes that were described earlier. The upper part of the layer, in special conditions, can move substantially slower than the rest of the load, be stationary, or even move backward. If the height of the layer does not exceed the height of such a segment of the layer, the entire load moves practically as a unit without intralayer shears. The particles travelling in accordance with identical laws are located over curvilinear surfaces of equal coupling conditions (see Fig. 1.11).

In the nonseparation regime the steady state motion of the load with stable formation of monolayers commences rapidly and is maintained during the entire period of motion. In the regime with separation, monolayers can from time to time be destroyed due to loosening up and collisions. Durations of the individual phases of motion are changing. This is promoted also by the effect of squeezing out of the deep segments of the layer when the lateral pressure is changing from active to passive. This disruption of stability of the monolayers begins to be observed even in oscillation regimes whose intensities are inadequate for the creation of separation (jolting) conditions. However, at any intensity, the character of motion of separate segments of the layer is repeated after several periods so that the process of vibratory conveyance can be regarded as a steady one.

Thus, the process of vibratory conveyance is accompanied by a complex set of phenomena. Thus, when a layer model is being constructed, one must have a reliable criterion of its quality, i.e., ability to reflect the physical features of the phenomenon. The correlation between theoretical and experimental mean velocities of the model and the layer is a necessary but insufficient condition. Their coincidence in a number of cases might be of a random character. It is another matter if the law of motion of the model were found to be close to the real averaged law of motion of the layer or a segment of the layer for a multimass model. This correlation

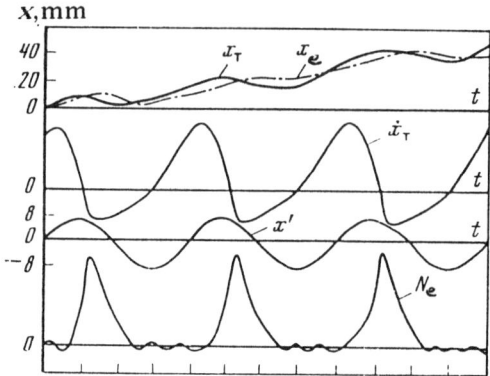

**Figure 1.12** Diagrams of displacement of an elastoviscoplastic model and of the layer.

is the optimum criterion according to which one can judge whether the model is an analog of the real process.

Figure 1.12 shows diagrams of the relative displacement and velocity of a single-mass elastoviscoplastic model of the layer $x_t$, $\dot{x}_t$ which are obtained with the aid of the analog computer EMU-10, displacement of the working element $x'$, normal reaction $N_e$ for $\omega$ = 9.4 Hz, $A$ = 8.5 mm, $\beta$ = 30° and $\alpha$ = 3° which corresponsd to $\Gamma$ = 1.5.

A series of similar oscillograms was obtained during a wide exhaustive search of the characteristics of the medium.

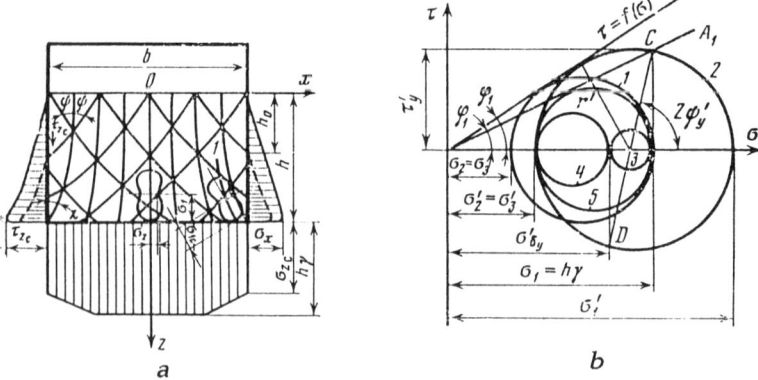

**Figure 1.13** Static stressed condition of a granular body in a trough.

The experimentally obtained averaged law $x_e$ of the displacement of a layer of dry sand is shown in this figure by a dot-dash line. Relatively close correlation of curves $x_t$ and $x_e$ in a number of oscillograms enabled to identify characteristics of the medium.

## 1.2.2 Stressed Condition of a Granular Body under the Action of Vibration

The described experimental techniques enabled one to reveal and partially explain the physical essense of some phenomena that accompany the process of vibratory conveyance. However, a number of problems require further and more detailed investigations. These are the problem of finding the correlations of pressure variation at the bottom and wall of the working element, wall friction, the patterns of loosening and compaction of the load, formation of individual layers, shapes of their boundaries, and so on.

The answers to the raised questions can, to a certain extent, be obtained if the load is considered as a granular body which is formed within the boundaries of the working element. The task of vibratory conveyance of a granular load in a regime without separation, when the continuity condition is not violated, can be ascribed as a special problem of soil mechanics.

Let us first of all consider the case of static stress conditions of a freely poured body in the trough. Figure 1.13 shows the profiles of the normal pressures of the load on to the bottom $\sigma_z$ and walls of the trough $\sigma_x$. In the center of the trough $\sigma_z$ and $\sigma_x$ are, simultaneously, the principal largest $\sigma_1$ and smallest stresses, and $\sigma_x = m\sigma_z$, where $m = \sigma_2/\sigma_1$ is the mobility factor known from soil mechanics. For easily poured loads $m \approx 0.18/f$, and specifically for dry sand $m \approx 0.3$ [4].

The vertical components of tangential stresses $\tau_{zc}$, acting in the plane of the wall partially support the weight of the granular body. As a result, the pressure on the bottom and on the wall near the trough angle is noticeably decreasing.

The convenient means for investigations is the geometrical interpretation of the stress state in the granular body by oval and by circle of stresses (Mohr). The planar stress distribution in the center of the trough is depicted in Fig. 1.13a as an oval with a vertical axis of the largest principal stresses. Fig. 1.13b presents the same picture as a stress circle 1. The stress distribution at the corner of the trough is shown in Fig. 1.13a by a stress oval with an inclined axis of the largest principal stresses $\sigma_{1C}$. The direction of deformations upon disruption of the state of equilibrium is represented by the slip lines 2 and 3, which are deviated from the trajectory of the greatest principal stresses by angle $\psi = 0.5 \, (90° - \varphi)$ where $\varphi = \text{arctg} \, f$ is the angle of internal friction.

If the trough is filled by a coherent load, for example, wet sand, then the pressure pattern will be different, since such loads are capable of forming a vertically standing wall of height.

$$h_0 = \frac{2\tau_0}{\gamma tg \, (45° - 0.5\varphi}$$  (1.1)

where $\tau_0$ is the initial shear resistance of the load. Furthermore, the upper part of the layer up to a depth $z \leq h_0$ does not practically affect pressure on the wall of the trough, and the tangential stress in the wall plane along the whole height $h_0$ is the same and equal to the initial shear resistance $\tau_0$. The profiles of the lateral pressures $\sigma_x$ and tangential stresses $\tau_{zc}$ in Fig. 1.13a are depicted by dashed lines.

It is natural that one should not expect any intralayer shear strains on the boundaries of this segment of the layer, it is being displaced as a solid body. At a large depth the bulk load begins to behave like a granular body.

Vertical oscillations of the trough with acceleration $a_v$ are periodically altering the pattern of the stress distribution in the granular body. The shape of the grid which is formed by trajectories of the principal stresses and of the slip lines is changing. Vertical pressure on the bottom of the trough for a granular body, if its viscous and elastic properties are neglected, can be found from expression.

$$\sigma_x = h\gamma \left(1 \pm \frac{a_v}{g}\right)$$  (1.2)

In Fig. 1.13b the planar static stress distribution is geometrically interpreted by circle 1. The stress distributions for the largest and smallest values of $a_v$ are characterized by circles 2 and 3, respectively. For granular media, not bounded by the rigid boundaries of the working elements, the lateral pressure at any moment of time should differ from the vertical pressure only by a scale factor and repeats its distribution pattern, since $\sigma_x = \sigma_z m$. However, in the trough at the moments corresponding to the greatest $\sigma_{z_{max}}$ and smallest $\sigma_{z_{min}}$, the lateral pressures that are respectively equal to $\sigma_{x_{max}} = \sigma_{z_{max}} m$ and $\sigma_{x_{min}} = \sigma_{z_{min}} m$ are not always detectable. This can apparently be explained by the fact that the stress distribution in the granular body in the trough is largely dependent on the method of its formation. It retains for some time a "memory" about this up to the moment of occurence of a strong pulse. Therefore, the lateral pressure in the nonseparation regime, particularly in the wall region, once it reaches the largest magnitude $\sigma_{x_{max}}$ as a result of load compaction from the first pulse, may remain unchanged and the load avoids becoming loose. The conveyor output noticeably decreases in the regimes where such phenomena are possible, since the value of wall friction increases considerably. The magnitude of this friction becomes commensurable with and, at some moments, can even be in excess of the friction between the load and the bottom.

This circumstance dictates that designers should avoid narrow troughs as far as possible. For very small relative widths of the trough the load might come to a stop even under fairly intensive regimes. This is explained by the fact that in a narrow trough the wedging action of the granular medium is pronounced, and something similar to a hang-up (arch roofing) occurs, thus hindering any loosening or plastic flows. The ultimate height of the trough at a given width can be found from the known relations of soil mechanics by introducing a concept of dynamic poured weight of the load

$$\gamma_g = \gamma \left( 1 \pm \frac{a_v}{g} \right) \qquad\qquad (1.3)$$

By changing the shape of the trough, the described unfavorable phenomenon can be substantially alleviated. A trough which is optimum in shape is one with a rounded bottom, thus copying a surface that is formed by particles with identical dynamic coupling conditions. In this case the layer moves practically without internal shear strains which consume additional energy and reduce the mean translational conveying speed.

A considerable gain in output and energy-intensity of vibratory machines can be ensured by excluding or reducing the phase of reverse slip of the load. This can be achieved either by using a special trough with unidirectional knurling, or by selecting an asymmetric regime of oscillations, or, if possible, by inclining the trough. Under some combinations of vibration parameters, at which the regime of load motion with separation is imminent, one might find out that at some moment the magnitude of wall friction $\tau_{z_C} = \sigma_{xf}$ will be larger than the vertical pressure $\sigma_z$ as a result of compaction. Then, the lateral pressure might squeeze out the load upwards. This is possible if the following condition is observed

$$\sigma_{z_{min}} \leq \sigma_{x_{max}}m = \sigma_{z_{max}}m^2 \tag{1.4}$$

or

$$\frac{A\omega^2\sin\beta}{g} = \Gamma \geq \frac{1 - m^2}{1 + m^2} \tag{1.5}$$

This effect of periodic loosening (squeezing out of the deep monolayers) occurs at values of $\Gamma < 1$ (for dry sand, for example, at $\Gamma \approx 0.85$), i.e., a regime with separation, when vibration intensity is gradually growing, is preceded by a regime with intensive mixing. The effect can be utilized when designing machines for dust-generating materials. Squeezing out of monolayers takes place not only at the top of the trough but also along the trough which is a source of additional shear strains in the layer.

When the concept of the mobility coefficient $m$ is introduced, it becomes possible to define the initial (static) elastic deformation of the layer in the horizontal direction $x_0$, which enters into the equation of motion of the elastoviscoplastic model

$$x_0 = \frac{h\gamma m}{k_x} \tag{1.6}$$

where $k_x$ is the stiffness coefficient of the model layer in the horizontal direction.

Let us now consider the case when the working element of the conveyor oscillates only in the horizontal plane with acceleration $a_h$. The inertial force causes the development of tangential stresses directed along the trough and reaching at the bottom the value

$$\tau'_{g_y} = h\gamma \frac{a_h}{g} \tag{1.7}$$

In Fig. 1.13*b* where the static stress distribution in the trough center at the bottom is represented by the stress circle *1*, we plot point *C* with ordinate $\tau'_{gy}$ on the diagram of $\tau = \varphi$ $(\sigma)$. Thus circle *2* can be obtained which represents considered dynamic stress distribution. By drawing a straight line *CD* through the center, we obtain the value of the lateral pressure $\sigma'_{l_y}$. Simultaneously, circle *2* allows one to obtain magnitude and direction of the new vectors of the principal stresses $\sigma'_1$ and $\sigma'_2$.

The third (the smallest) principal stress $\sigma'_3$ whose vector lies in the transverse plane is also increasing in the presence of acceleration, finally becoming equal to $\sigma'_2$. The lateral pressure in this plane in the presence of only horizontal oscillations is also equal to $\sigma'_3 = \sigma'_2$.

The lateral pressure in the longitudinal plane whose vector is directed along the trough axis is not changing after reaching magnitude $\sigma_{l_y}$ (in the static stressed conditions).

When the sign of the acceleration is changed, magnitudes of vectors of principal stresses $\sigma'_1$ and $\sigma'_2$ reach the same values after changing their direction. However, in passing through the zero value of acceleration the stress state of the layer will be characterized by circle *3*. Such an effect can be explained by the presence of some compaction in the layer: at $a_{max}$ an additional vertical pressure component appears. The principal stresses in three states, when $a = a_{max}$, $a = 0$, $a = a_{min}$, are respectively equal to

$$
\begin{vmatrix}
\sigma'_1; & h\gamma; & \sigma'_1 \\
\sigma'_2; & \sigma_{l_y}; & \sigma'_2 \\
\sigma'_3; & \sigma'_3; & \sigma'_3
\end{vmatrix}
\tag{1.8}
$$

Thus, the spatial stress distribution is represented by three stress circles *3*, *4*, and *5*.

Equilibrium of the granular body reaches a limit when the tangential stress $\tau_y$ attains the value $h \gamma f_1$, i.e., when, as in Fig. 1.13*b*, point *C* reaches curve $OA_1$. Acceleration corresponding to shear of the layer is $a = gf_1$. The angle of inclination of vector $\sigma'_1$ to the vertical axis at the moment of shear is

$$
\psi'_y = 0.5 \ \text{arcsin} \ \frac{\tau'_y}{r'}
\tag{1.9}
$$

where

$$r' = \frac{\sigma'_1 - \sigma'_2}{2} \tag{1.10}$$

The values $\sigma'_1$ and $\sigma'_2$ can be expressed in terms of two known stresses, for example $\sigma_g = h\gamma$ and $\tau'_y = h\gamma \ (a/g)$.

In order to record the true stresses in the layer during experiments, the bottom and the walls of the vibratory trough were equipped with the pressure transducers DD-8.

Figure 1.14 presents diagrams of the dimensionless vertical $\sigma'_v$ and lateral $\sigma_l$ pressures at $A = 7$ mm, $\omega = 43.2$ 1/sec, $\beta = 30°$, and $\alpha = 0°$ ($\Gamma = 0.67$). Also presented here for comparison are theoretical curves of the normal pressure $N$ of a material particle of weight $h\gamma$ and of the normal pressure $N'$ of the elastoviscoplastic model of a layer. It is evident that the diagrams of $\sigma_v'$ and $N$ differ considerably. For example, at moments $\varphi = 0$, $\pi$, and $2\pi$, when the inertial force is absent, the true pressure at the bottom does not become equal to the static pressure $h\gamma$ (in the adopted dimensionless scale $h\gamma$ corresponds to $a_v = g$). A phase shift of the maximum pressure by angle $\delta$ is also observed. Diagrams of $\sigma'_v$ and $N'$ have a satisfactory correlation.

Presented in Fig. 1.15 are analogous diagrams obtained for $\Gamma = 0.98$, i.e., for the upper boundary of the "nonseparation" regime. This experiment confirms the transition of the lateral pressure from active to passive which is manifested in intensive

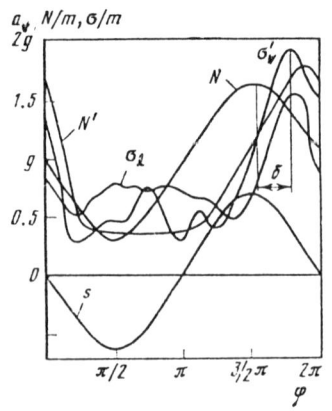

**Figure 1.14**   Pressure diagrams in the layer for $\Gamma = 0.67$.

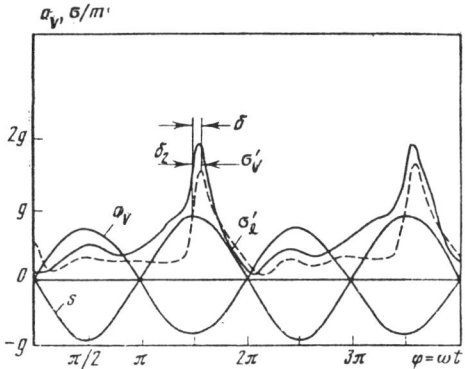

**Figure 1.15** Pressure diagrams in the layer for $\Gamma = 0.98$.

loosening of the material. Sets of pressure diagrams have also been obtained for more intensive regimes with $\Gamma > 1$. Analysis of the results of this series of experiments indicates that peak pressures at the bottom and the walls are developing with phase shifts of $\delta$ and $\delta_2$ relative to the maximum acceleration, indicating viscous properties by the granular medium.

The impact of the load against a plane cannot be called elastic, since an instantaneous rebound of the load is not detected in the diagrams of vertical displacement. Collision of the lower monolayer with the plane and the subsequent impact of the upper monolayers causes a pressure jump exceeding, for example, the static pressure by more than fourfold when $\Gamma = 1.16$ $\sigma_{v_{max}}$.

During the flight phase the pressure does not always drop to zero. The presence of the lateral pressure is explained by the fact that even in flight the layer does not lose contact with the walls of the vibratory trough. The presence of some vertical pressure can be explained by the fact that separation from the bottom of some segments of the layer ought to be regarded not as pure flight, but as loosening.

At any $\Gamma > 1$ the granular body passes through stages that are characteristic for less intensive regimes. For example, up to the moment when the vertical acceleration exceeds the value $a_v \leq 0.8g$ the layer is compacted without periodic loosening. In the interval corresponding to $0.8g < a_v < g$ the conditions are developing for transformation of the lateral pressure into passive pressure which both pushes out and loosens the layer. When $a_v \geq g$ separation of the layer or part of it from the vibration plane becomes possible. The transition from one stress state to another is accompanied by dissipation of energy and attenuation of the elastic pulse.

The investigation of the dynamic stress state by means of photoelasticity is promising. This method is successfully used to obtain the pattern of stress distribution in elements of complex configurations under complex loading.

The effectiveness of application of this method when investigating the stress state of a granular body was explored on the optical-polarization device PPU-5. In order to simulate a granular body, one can use substances based on ethyl cellulose and gelatin-glycerin mixture. The optical and mechanical properties of these substances can be changed within a wide range depending on ratio of the components. Low-modulus jelly-like compositions of water-based glycerin-gelatin mixtures have been considered. Vibration regimes with $\Gamma = 0.8 - 1.4$ were investigated. A changing light interference pattern of the stress state of the model of a layer was observed on the screen as isochore bands. When the walls of the working element were coated with petroleum jelly mixed with resin, the isochores were distorted acquiring a shape outline which was close to the boundaries of monolayers with equal dynamic coupling conditions. The identity between the layer and the model can be verified either by filming marked particles or with the aid of accelerometers placed inside the model and inside the granular medium. By changing characteristics of the model, one can attain the conditions when the signals are identical. If it is taken into account that simple and reliable methods of determining elastic and viscous characteristics for the model substance are known in rheology, then this method can also be used to determine the effective elastic and viscous characteristics of the granular medium.

### 1.2.3  Investigation of Oriented Vibratory Conveyance of Some Lumped Loads with Anisotropic Properties

The process of vibratory conveyance of lumped loads is the simplest and thus is extensively studied. However, the majority of works in this field pertain to investigation of the effect of vibration on a dry, solid, and plane particle. However, frequently there is a need for transporting or some vibratory processing of loads covered with a liquid or a grease, and which are compliant and far from being planar.

Small fish are representative of such products. This biological object has considerable compliance, is capable of changing shape practically without changing its volume, is covered with albumin-water layer and is conveyed over a wetted surface. Furthermore, fish have clearly pronounced anisotropic properties resulting in the different magnitudes of resistance force to motion when sliding with head pointing forward and in the opposite direction. It is also interesting to note that the anisotropic properties, which are explained by the presence of unidirectional scaly cover, become evident not only on a corrugated surface, but also on a smooth surface. In the latter case the greater value of the resistance force in the backward motion (tail forward) is explained by the increase of area of direct frictional contact between the surfaces. The different values of the resistance force to motion depending on orientation elucidate the possibility of vibration conveying of such a load even on horizontal surfaces oscillating according to harmonic laws.

The characteristic feature of such loads is development of the effect of self orientation on the oscillating surfaces, i.e., the tendency of turning until the longitudinal axis coincides with the direction of oscillations. This is explained by the divergence of the position of the centers of gravity of the load and the center of the contact spot and, as a consequence, by the appearance of a force couple which provides the orientation effect. The load comes to a position in which the resistance force to motion becomes minimum. The contact spot of the fish with the oscillating plane is shaped as an elongated oval, and when its directions and direction of the oscillations coincide, rotation on the plane ceases. Later, as it follows from the analysis of the results of high-speed filming, the translational motion of the oriented fish occurs in the direction of oscillations without any noticeable shifts and is practically rectilinear.

In addition to purely theoretical interest, the study of the process of vibration conveyance of such objects also has practical significance. It is with the help of vibratory machines that the problem of directional metered feed of the raw material in fish processing machines is most successfully solved.

To construct reliable feeding systems with stable operation it is necessary to conduct special investigations which would also enable one to make a judicious selection of parameters for a number of other vibratory devices that are

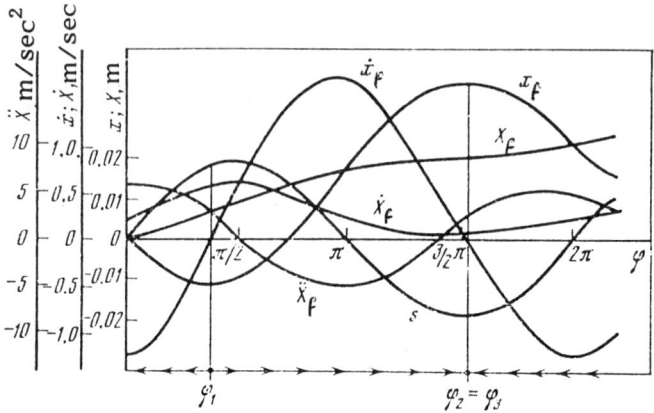

**Figure 1.16**  Diagrams of fish motion along a vibrating surface.

used in the fishing industry. It is natural that consideration of the above enumerated specific properties substantially complicates the problem of vibratory conveying of such objects.

To identify the pattern of vibratory conveyance of oriented fish and the character of resistance forces to motion, which are determined by the biological features of the object, investigations were conducted to determine the law of motion of fish along an oscillating plane. The investigations were carried out on a vibratory test rig of natural dimensions using a device with a continuously moving photographic film according to the technique described above. Diagrams obtained were of fish motion along oscillating surfaces made from stainless steel and rigid PVC under various combinations of vibration parameters.

Figure 1.16 shows the processed diagrams of displacements, velocities, and accelerations of fish when moving along an oscillating stainless steel surface with amplitude $A = 0.02$ m

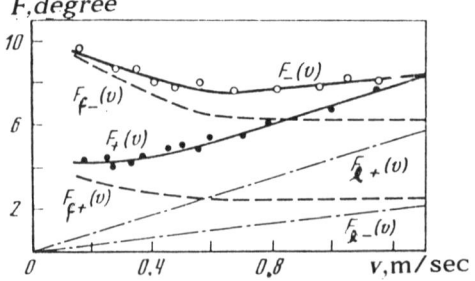

**Figure 1.17**  Resistance force to motion vs. relative velocity for direct and reverse orientation.

and frequency $n = 9.2$ Hz. Similar diagrams of acceleration of the object were obtained using a built-in piezoelectric transducer. The results agree well with the diagrams obtained by filming and enable one to draw the following conclusion.

The character of the curves of the absolute acceleration of the fish $\ddot{X}_f$, which are analogous to the resistance forces to motion, indicates that this force is variable. Moreover, its variation coincides with the change of magnitude of the relative speed of motion of the fish $\dot{x}_f$. Such a correlation of the resistance force with the relative speed can be explained by the presence of viscous resistance. To verify this hypothesis, a second series of experiments was carried out to determine the resistance force to the fish motion along the working surfaces (stainless steel, rigid PVC, rubberized tape) and its dependence on the speed of motion.

Investigations were conducted on an experimental rig comprising a DC motor which drives through a transmission system, a disc with replaceable working surfaces. The force of resistance to the motion of the fish along the surface is transmitted by a flexible filament to a strain-calibrated beam and then recorded on an oscillograph. Experiments were conducted for direct (head forward) and reverse orientation. The results of the investigations were obtained in graphical form and an example of such correlations for fish-stainless steel arrangement is given in Fig. 1.17. These results enable one to conclude that when small fish slide over a hard surface which is wetted with water, semi-viscous lubrication is developing, i.e., there is simultaneously a process of direct contact between the surface elements $F_{f\pm}$ and a process of viscous friction in the intermediate medium $F_{l\pm}$.

In the zone of low speeds (up to 0.6 m/sec), the dominant effect on the resistance force is due to frictional contact of the surface elements $F_{f\pm}$. With increasing slip speed in this region a decrease in the number of frictional contacts takes place, and hence, a decrease of the resistance force. This explains falling characteristics of the experimental curves $F_\pm$. With further increase of the speed, the force of frictional contact in this region remains practically constant, and the resistance force is increasing as a result of increasing force of viscous resistance.

Thus, it can be assumed that the correlation of the resistance force of the intermediate layer with the speed $F_{l\pm}$ in the investigated range is a linear function emerging from the origin of the coordinate frame and equidistant to $F_\pm$ ($v$) at its ascending part.

The tangent of the inclination angle of the straight line $F_{l\pm}$ in the scale of the graph is numerically equal to the coefficient of viscous resistance.

Correlation of the resistance force with speed has the same character for both the direct and reverse orientations. However, the resistance force in the reverse orientation within the studied range of speeds is greater than in the direction orientation. This is explained by the fact that for the reverse orientation the frictional contact force of the surfaces is two times higher than for the direct orientation due to directivity of the scaly skin cover. The less intensive rise of the resistance force, determined by viscous resistances for the reverse orientation, is explained by the increase of thickness of the intermediate albumin-water medium layer.

On the basis of the experimental results for determination of the true law of motion of fish along the oscillating plane and analysis of correlation between the resistance force and the slip speed, it has been found that the resistance force is combined of the force of direct frictional contact and the force of viscous resistance to the motion of the object along the surface caused by the presence of the albumin-water layer on the surface of the conveyed object.

Thus, when solving the problem of vibratory displacement of an object, and also loads that have similar properties, one must take into account the correlation between the resistance force and the relative speed and the difference of the coefficients of resistance to motion for different orientations of the object. The motion of the given object along a vibrating surface can be imagined as the motion of a single-mass viscoplastic model with various values of parameters for the forward and backward motion. Then the differential equation of motion of such a model in case of horizontal harmonic oscillations ($X = A \sin \omega t$) has the form

$$m\ddot{x} = -m\ddot{X} + F_f - C_\pm \dot{x} \qquad (1.11)$$

where

$$F_f = \begin{cases} -mgf_+ & \text{when } \dot{x} > 0 \\ mgf_- & \text{when } \dot{x} < 0 \end{cases} \qquad (1.12)$$

Here, $f_\pm$ is the coefficient of resistance to motion in direct (+) and reverse (-) orientations; $c_\pm$ is a coefficient taking into account viscous properties of the albumin–water layer (from experiments, $c = 0.0435 - 0.017$ N.sec/m).

Since $\ddot{X} = -A\omega^2 \sin\omega t$, then considering (1.12) we obtain

$$\ddot{x} = A(\sin \omega t)\omega^2 \pm gf_\pm - \frac{c_\pm}{m}\dot{x} \tag{1.13}$$

The solution of this differential equation in the various stages of motion has the form

$$\dot{x} = e^{-a_\pm(l-t_{1\pm})}\left(B_\pm\omega \sin \omega t_{1\pm} - E_\pm\omega \cos \omega t_{1\pm} - D_\pm\right)$$
$$+ E_\pm\omega \cos \omega t - B_\pm\omega \sin \omega t + D_\pm \tag{1.14}$$

$$x = \frac{1}{a_\pm}e^{-a_\pm(l-t_{1\pm})}\left(E_\pm\omega \cos \omega t_{1\pm} - B_\pm\omega \sin \omega t_{1\pm} + D_\pm\right)$$
$$+ E_\pm \sin \omega t + B_\pm \cos \omega t + D_\pm(t - t_{1\pm}) + \left(\frac{B_\pm\omega}{a_\pm} - E_\pm\right)$$
$$\times \sin \omega t_{1\pm} - \left(B_\pm + \frac{E_\pm\omega}{a_\pm}\right)\cos \omega t_{1\pm} - \frac{D_\pm}{a_\pm} \tag{1.15}$$

where

$$a_\pm = \frac{c_\pm}{m}; \quad E_\pm = -\frac{A\omega^2}{a_\pm^2 + \omega^2}; \quad B_\pm = -\frac{a_\pm A\omega}{a_\pm^2 + \omega^2};$$
$$D_\pm = \mp\frac{gf_\pm}{a_\pm} \tag{1.16}$$

$t_{1\pm}$ is the moment of time at which $x = 0$, $\dot{x} = 0$, i.e., the moments of commencement of forward (+) and backward (-) motion.

The differential equation of motion of the model is linear only during the various stages of motion. The transition moments from one stage to another could not be obtained in analytical form.

The solution of this problem can be obtained with the aid of computers. The results of the computer-aided computations using a specially developed program agree well with the experimental results. This is a demonstration of the correctness of the mathematical model for the process of vibratory conveyance of an object which possesses specific properties. The criterion of validity of the mathematical model is the coincidence of the real law of motion of the fish over an oscillating surface with the computer-aided solution of the equation of motion.

Experimental investigations were also conducted to determine the mean speed of vibratory conveyance. These were carried out on a vibratory test rig with changeable working surfaces. A frequency range $n$ = 300 - 700 Hz and amplitudes $A$ = 15 - 25 mm, were investigated, which was found to be most advantageous in practice. Less intensive regimes are ineffective from the point of view of attaining output and specific energy consumption. More intensive regimes are not employed due to the increase of dynamic loads and the emergence of flow disorder phenomenon.

The determination of the mean speed of vibratory conveyance was made using time during which fish pass a fixed section of the working surface. During the experiments, conditions similar to production conditions were created. It was established that in the investigated range of parameters, dependence of the mean speed of vibratory conveyance on frequency is well approximated by a linear function.

Special investigations were conducted visually and with the aid of the high-speed camera SKS-1M for determination of the rotation angle of the fish when orientated on an oscillating surface as a function of time and vibration parameters. It was revealed that the preferred rotation of the fish on the oscillating surface was in a direction opposite to the back of the fish. This can be explained by a misalignment between the axis of inertia of the fish and the axis of symmetry of the contact area, by non-uniformity of pressure distribution from fish to the surface, and also by the different values of the

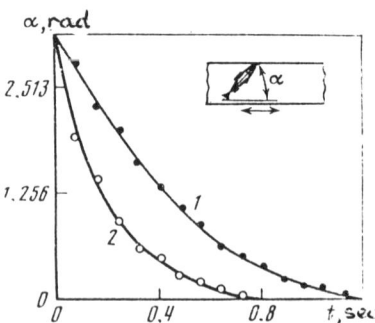

**Figure 1.18** Angle of rotation of fish vs. orientation time. _1)_ $A$ = 0.02 m, $n$ = 6.7 Hz; _2)_ $A$ = 0.02 m, $n$ = 10 Hz.

elementary forces of resistance to shear along the fish length. The dependences of the angle of rotation of the fish during its orientation on the oscillating surface on the time and parameters of vibrations at amplitude $A$ = 0.02 m and frequencies 10 and 6.7 Hz are depicted in Fig. 1.18.

Since two flows of fish leave the orientating surface in opposite directions, a device to turn one of the flows by 180° is provided in actual designs of vibratory machines, i.e., the orientating surface is made as a two-level surface. For the dispersal of the flow following its orientation, inclined surfaces on the conveying sections can be used which leads to an increase in the speed of motion.

The results of the conducted investigations can be used for many similar objects in light manufacturing and food industries.

## 1.3  VIBRATORY BUNKERING (HOPPERS)

In the course of some production processes with the use of vibration, bunkering of the transported product takes place. It can be bunkered in a receptacle where its technological processing takes place, in vibratory loading-delivering machines, mills, cylindrical drums and mixers or directly in vibratory conveying/processing machines. In all cases the vibrationally bunkered product complies with general physical laws and possesses specific motion features which are caused only by the design features of the plant in which the production process is implemented.

Let us consider the mechanism of the process of vibratory bunkering on a rectilinear conveying machine. In this process, front layers of the bulk medium are losing the horizontal component of their speed upon reaching a barrier, while the entire remaining mass continues in motion. Under the pressure, the front layers are compacted and forced upwards while being pressed by the mass of the rest of the product. During this process, not only does the layer height grow, but also a gradual change of its horizontal speed takes place from the speed of vibratory conveyance at the inlet of the bunker (hopper) to zero speed at the barrier. The height of the layer increases with time counting from the commencement of the process. The pressure of the product mass drops with increasing distance from the barrier, and the vibratory bunkering area ends where the pressure force in the layer becomes zero.

The character of the process of vibratory bunkering is determined by such parameters as shape (time history), amplitude and frequency of the working element vibrations, direction of vibrations, and inclination angle of the carrying surface. Investigations show that the process of vibratory bunkering is affected by the structure and configuration of the bunker and the working element, and also by the presence or absence of the overpressure in the bunkered product. Effectiveness of the process is characterized by two principal indicators – rate of bunkering or, which is the same, the rate of rise of the product in the bunker, and the maximum attainable height of the bunkered layer. The first indicator determines the output of the process, the second characterizes its efficiency. The regimes giving high output also ensure the maximum height of the layer.

Analysis of the experimentally established dependence of time of filling the bunker on the frequency and amplitude of vibrations of the carrying element, angle of vibration, and angle of inclination in the regime of rectilinear harmonic oscillations indicates that increasing vibration frequency and amplitude lead to increasing output of the process of vibratory bunkering. A downward inclination of the working element accelerates the filling of the bunker, and its elevation considerably reduces output. At angles of elevation exceeding 10° the process of vibratory bunkering practically comes to a halt. Influence of the vibration angle is more complex. With increasing vibration angle from 30 to 40°, the rise of the product in the bunker is accelerated but the delivery speed decreases. All things considered, effectiveness of the process flow and the output of bunkering are improving. Further increase of the vibration angle accelerates the rise of the medium in the bunker more; however, as a result of reduction of the conveyance speed an insufficient amount of the product enters the bunker and the process output drops.

Dimensions of the bunker and of the working element do not affect the characteristics of vibratory bunkering, provided the required minimum dimensions are maintained. Bunkering is best effected in receptacles with the walls diverging upwards. Efficiency of the process is also enhanced by installation of a partition at the discharge opening, which prevents the product from rolling down from the free surface of the pile.

The height of the bunkered pile is considerably dependent on the value of the induced pressure, i.e., the height of the layer of the load at that part of the working element where bunkering does not take place. The maximum height achieved presently in the regime of rectilinear harmonic oscillations amounts to approximately 800 mm.

A study of distribution of internal dynamic pressures in the pile of the bunkered product created by vibration of the working element enables one to more fully understand the physics of the process of vibratory bunkering. The working element transmits to the medium periodic pulses whose magnitudes are determined by parameters of the vibration regime. However, effects of these pulses in the bulk load are averaged; therefore, it can be assumed that under the action of vibrations some constant dynamic pressure is developing in the pile inducing delivery and bunkering of the product. The pressure component acting along the working element provides transportation of the load, and its vertical component provides its bunkering.

Studies of distribution of the dynamic pressure in the pile in the horizontal and vertical directions indicate that it does not remain constant along the height of the layer (Fig. 1.19). The vertical component of the pressure is diminishing with distance from the working element according to a complex law whose character is largely dependent on the vibratory velocity. However, the direction of its action is always the same – upwards (Fig. 1.19a). The horizontal component of the dynamic pressure decreases along the height of the layer according to a very complicated law which depends on the vibratory velocity (Fig. 1.19b). At some distance from the surface of the working element the dynamic pressure is changing not only by magnitude, but also its direction, acting not only in the direction of but also in opposite direction to displacement. This explains the circulatory motion of the particles in the pile and the presence of countercurrent on its free surface.

The decrease in dynamic pressure along the height is mainly caused by decreasing of vibration amplitude of the monolayers with increasing distance from the vibrating surface and dissipation of pulse energy due to the internal friction in the medium. Distribution of vibration amplitudes of monolayers for a number of products along the height of the pile in the regime of rectilinear harmonic oscillations is depicted in Fig. 1.20. Vibration amplitudes of the monolayers very quickly decreasing with distance from the vibrating surface. The dynamic pressure in the pile drops roughly with the same rate.

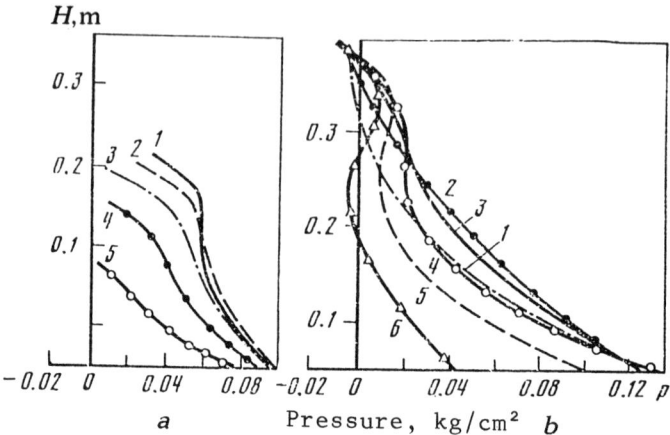

**Figure 1.19** Distribution of dynamic pressures of the bulk load along the height of the layer at different vibratory velocities of the load-carrying element. a - 1) 0.44; 2) 0.40; 3) 0.36; 4) 0.32; 5) 0.28; b - 1) 0.275; 2) 0.300; 3) 0.325; 4) 0.350; 5) 0.375; 6) 0.400 m/sec.

Experiments with vibratory bunkering of a bulk load with 4-10% humidity were conducted. In order to establish parameters of vibratory bunkering with application of elliptical vibrations and find efficiency of the process as compared with the regimes of rectilinear harmonic oscillations.

Study of the height of bunkering as a function of time at different vibration frequencies, and with amplitudes in the vertical direction $A$ = 2.5 mm, in the horizontal direction $B$ = 2.5 mm, and phase shift $\gamma$ = 20° shows that the bunkering height is increasing with increasing vibration frequency.

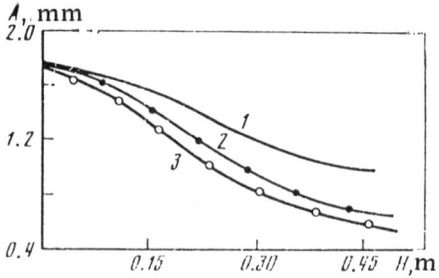

**Figure 1.20** Distribution of vibration amplitudes of monolayers of the bulk load along the height during process of vibratory bunkering. 1) ore; 2) fine sandstone; 3) coarse sandstone.

Furthermore, the higher the frequency, the longer the process. Thus, for example, at 900 osc/min the bunkering height reaches 360 mm, and the bunkering time amounts to 18 - 20 sec; respectively, at 800 osc/min these are 305 mm and 16 - 18 sec; and at 700 osc/min - 220 mm and 14 - 16 sec. In the regime of rectilinear harmonic oscillations of the load-carrying element at a frequency of 860 osc/min and amplitude 2.5 mm, the bunkering height amounts to 230 mm. With increasing amounts of load in the bunker, the height of the pile is increasing to 240 mm and with further addition of the load the height rises to 275 mm. Correspondingly, with amplitude 3 mm and frequency 1100 osc/min, a layer height of 320 mm was obtained.

Thus, with an elliptical trajectory of the bunker vibrations, efficiency of the process increases significantly and the bunkering height rises. In the regime of elliptical oscillations, the angle of slope of the pile in the bunker is also increasing. At frequency 900 osc/min it reaches 50 - 55°, whereas under rectilinear oscillations the angle of slope does not exceed 40° in similar regimes.

Experiments indicate that properties of the poured product greatly affect efficiency of vibratory bunkering. Loads that are characterized by good transportability are worse for bunkering. Hence, it can be concluded that increasing force of internal resistance leads, to a certain extent, to an increase in efficiency of the process of vibratory bunkering.

It can be ascertained that the height of bunkering is mainly affected by the following factors: physical and mechanical properties of the product, amount of the bulk and thickness of the delivered layer to the bunker, the length of the vibrating bottom of the bunker, velocity and direction of vibrations. Increasing the height of vibratory bunkering when using elliptical regimes of vibrations can be explained but the effect of the independent vertical component of vibrations of the working element.

## 1.4  VIBRATORY CRUSHING

Results of investigations of the vibratory crushing process are presented as plots of displacements, velocities and accelerations of the vibratory crusher jaw, phase shifts between position of the unbalanced mass and displacement of the jaw, phase shifts between the exciting force and displacement of the crusher jaw, and energy consumption by the vibratory drive for the crushing process of the rock mass, and the total energy consumption by the crusher motor (Fig. 1.21).

Based on the analysis of high-speed photographic study, mechanisms of the process of vibratory impact crushing of rock mass were established and the parameters of vibrations of the crusher jaw and of motion of the unbalanced vibrator were determined.

The crushing process of rock mass by a vibratory crusher is as follows. An angular chunk of rock after falling into the working chamber of the crusher is gradually chipped, sharp edges on its surface are breaking, the size of the chunk is being reduced, and with each stroke of the jaw it moves downwards, closer to the discharge opening. After 10 - 20 cycles of oscillations, the piece is split into two halves or into several smaller parts. Usually the crack extends from jaw to jaw between the points of contact of the rock piece with the crushing plates. The chipped parts of the chunk, if they are of larger sizes than the discharge opening, continue to be crushed by the described method until they reach the required dimensions. Such a crushing mode of a rock mass in a vibratory crusher is established due to the fact that as a result of the limited oscillation amplitudes during one stroke of the crushing jaws, adequate stresses are not developed in the sharp-edged chunk of rock which are capable of causing its total fracture. Under moderate strains, the stresses arising in the parts of the chunk in contact with the jaw are capable only of breaking the edges. However, with the spalling of the protruding parts the piece acquires a smoother shape, the surface area of contact zones is increasing, the stresses are also increasing and encompassing a larger volume of the crushed rock. When these stresses reach the ultimate breaking magnitudes, the chunk is shattered.

Thus, the most important distinguishing feature of operation of a vibratory jaw crusher is fracture of the rock under the action of repeated two-sided application of impact loads. It is evident that the process of rock crushing in these machines is more fully explained by the hypothesis according to which volumetric crushing of the chunks takes place under the action of tensile stresses accompanied by compression stresses in the zones of contact. In vibratory jaw crushers kernels of compressed crushed rock in which a stress state is created approaching volumetric compression are created in the contact zones with the jaws in the initial period of operation when fracture of the surface edges takes place. With further repeated application of impact loads, volumes of the densified kernels are increasing. This finally leads to the splitting of the rock chunk along the weakest sections under the action of tensile stresses.

Repeated high-frequency strains of the rock piece preceding its fracture induce opening of internal cracks and accumulation of residual strains, which as a result reduce the fracture stress. The high-frequency oscillations of the crushing jaws result in an impact character of the crushing process. The impact occurs due to the fact that the rock chunk does not have enough time to fall when the jaws are moved apart, creating a clearance, and in the process of the closing of this clearance, impact occurs. Impact interaction with the crushed rock enables increasing the acting forces as compared with the forces typical for static crushing methods.

When rock mass is fed into the crushing chamber of an idling vibratory crusher, the oscillations which had a harmonic character at idling are gradually becoming more nonlinear and asymmetric, and phase shift between the jaw travel and rotation of the unbalanced mass of the vibrator is decreasing. If the first impact occurs at a phase angle of the unbalanced mass from 170 to 160°, then the steady-state regime progresses at phase angles from 90 to 60°. Smaller phase angles are set for crushing sturdy rocks and for high degree of filling of the crushing chamber by the rock mass. Duration of the transient regime is short, from ten to several cycles of jaw oscillations. Impacts of the jaws on the chunks of rock mass occur at various displacements of the jaws, but more often this happens in the region of their neutral line of oscillations under load. In the unsteady regime the neutral line of jaw oscillations shifts continuously. When the crusher is working under load, some increase of amplitude of jaw oscillation is observed, often reaching 1.4 - 1.6 of the idling amplitude but can even be considerably higher.

As experiments show, individual rock pieces are crushed in the crusher over 10 - 20 and more working cycles. The principal fraction of the crushing time is wasted in moving the jaws apart. Maximum deviations of the jaws in the crushing regime can exceed the idling amplitude by 2.0 - 2.5 times or more.

In the final cycles of crushing of the rock chunks deviations of the jaws sideways away from the working cavity of the crusher are stabilizing, frequently increasing to

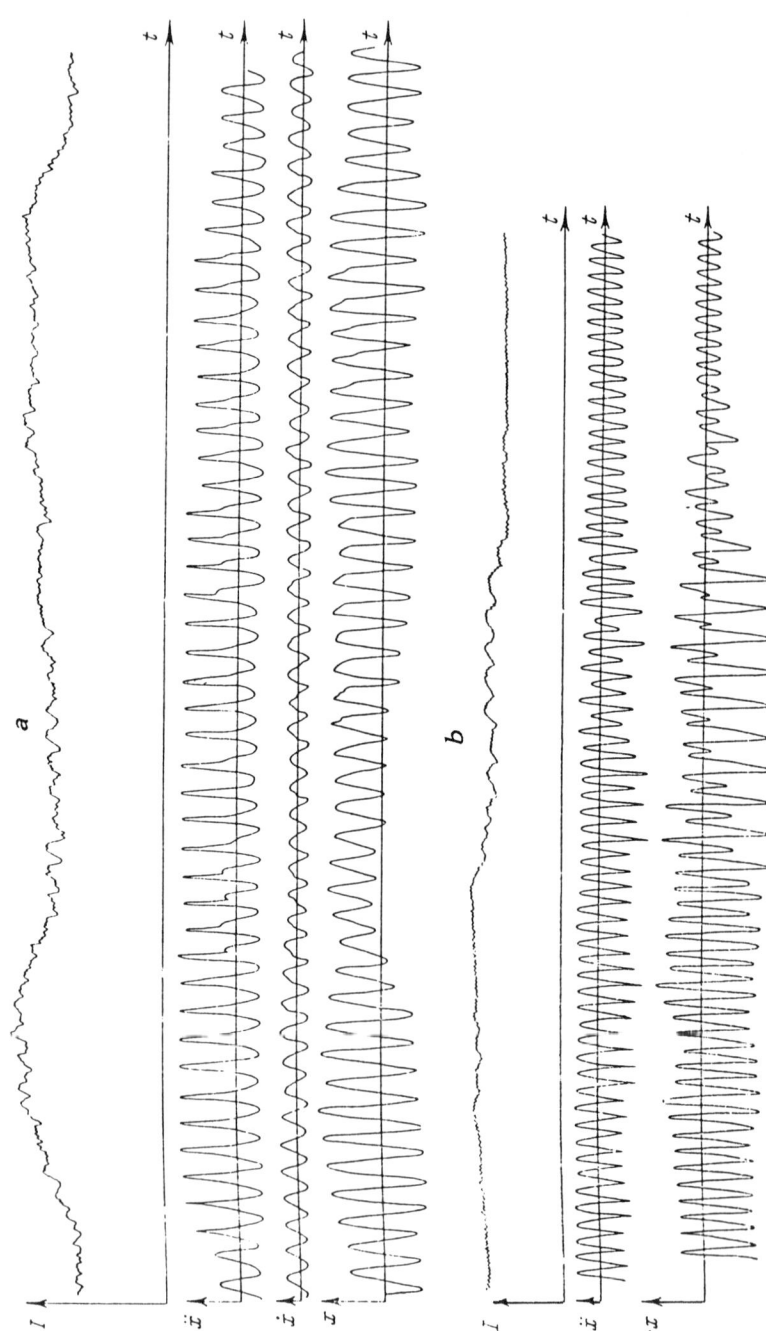

**Figure 1.21** Parameters of a vibratory crusher in the rock crushing process; a) high hardness; b) average hardness.

maximum values which is an indication of the rising resistance of the crushed material to fracture in the final stage of the crushing. Following the break up of the rock, the transition of the crusher into the idling regime lasts 3 - 5 working cycles.

When particularly sturdy pieces are fed into the crusher, very substantial wedging out of the jaws can occur (shifting from the neutral line) and the vibration amplitude increases 3 - 4 times as compared with idling.

As is evident from the presented oscillograms, this is accompanied by a sharp increase of energy consumption by the drive (see Fig. 1.21).

Investigations show that duration of the jaw contact with the crushed rock constitute from 15 to 40% of the period of jaw motion, depending on impact conditions, degree of filling of the crushing chamber, and properties of the rock. The acceleration in the process of collision between the jaw and the rock exceeds the amplitudes of jaw acceleration during idling by 2 - 5 times and, in some regimes, even more. This indicates that the stage of interaction of the jaw with the rock as regards both the duration and the resulting forces (as can be established by the sharp rise of acceleration) is cardinal in shaping the overall pattern of the process of vibration crushing of rocks. Thus, to develop phenomenology of the process, the main attention must be directed toward describing the physical phenomena occurring during this stage. Collision of the jaws with the crushed rock takes place, as a rule, in the range from 0 to 0.3 A from the neutral line of oscillations (where A is the amplitude of the jaw in idling).

Comparison of the operational specifics of the vibratory crusher during breaking of single chunks of rocks and of a bulk rock mass in conditions of high degree of filling of the chamber enables one to note characteristic features that are peculiar to both regimes. When crushing a single chunk of rock, conditions of operation substantially change from the beginning to the end of the process (the latter is characterized by the moment of fracture of the object). During the first cycles, separation of the jaws just begins to show, the oscillations have a quasi-harmonic character, and sharply pronounced peaks are not recorded on the acceleration oscillogram. With the development of the crushing process and dropping of the piece into the crushing chamber, the separation of the jaws is increasing, oscillations become asymmetric and nonharmonic, and at the moments of contacts between the jaw and the rock the accelerations are sharply increasing. Directly prior to the chunk breaking, these parameters approach extremal values – separation of the jaws is sharply increasing, plots of vibratory displacements, velocities, and accelerations of the jaw are greatly distorted. In the steady-state regime of bulk rock-mass crushing, the process runs under slightly and slowly varying characteristics of the operating regime. The opening of the jaws, the shape of the oscillograms of displacements, velocities, and accelerations of the jaw, and the consumed energy all change insignificantly. If a uniform loading of the crusher is provided and rocks of slowly changing composition and properties are processed, variation of the distance by which the jaws are separated does not exceed 10 – 20%.

This analysis shows that the character of crushing of single chunks of rock and rock mass, while having many common features, nevertheless differs significantly. Therefore, to develop the phenomenology of the vibratory crushing process, it is necessary to construct a model which would take into account all the diversities of these phenomena.

## 1.5   VIBRATORY BOILING LAYER AND VIBRATORY COMPACTION

Under vibration acts on a disperse medium, a number of transformations take place whose character depends on the intensity of the acting vibration. As the intensity of treatment of the disperse medium with vibration is increasing within the range of amplitude values of accelerations not exceeding the acceleration of the force of gravity, the disperse medium begins to acquire mobility and pseudoliquidity. Such a state of the disperse medium is referred to as the state of quasiliquefaction. In this state, adhesion between the particles of the disperse medium is decreasing, the particles come closer to each other, the number of pores is diminishing (a more dense packing of the particles is achieved), and the medium is compressed. Its maximum compaction is usually attained at vibratory acceleration amplitudes close to the acceleration of gravity.

Upon further increase of vibration intensity, the disperse medium begins to occasionally lose contact with the vibrating working element, the bonds between the particles weaken and are periodically destroyed, and the medium is experiencing transition to a state similar to boiling. This state, known as vibratory boiling, is characterized by a loosening of the treated medium and by strong circulation of its constituent particles. In the stage of vibratory boiling two characteristic regimes can be identified: the state of segregation of the particles of the medium and the state of strong mixing. The second regime is realized with more intensive vibration.

In a number of cases, to create a boiling state a combined action of vibration and gas blow-thriough (filtering) is used which has substantial advantages as compared with both the vibrational and aerodynamic methods used separately. Thus, for loose products which do not exhibit slugging and have low friction between the particles, both the vibratory and aerodynamic methods provide the required structure of the boiling layer. However, it is rather difficult to develop a uniform boiling layer of a number of finely dispersed products using these methods. Their combined application enables one to attain the required results. Vibrational action reduces particle adhesion, destroys the aggregates being formed, and, thereby, prevents the formation of through channels for gas flow. This creates the conditions for more uniform distribution of the gas flow in the entire layer of the treated product. Vibration also destroys hang-up formations. Under the combined action both the required gas flow velocity and oscillation intensity are considerably reduced.

Let us now consider the processes taking place in the disperse state when treated by vibrations of different intensity: compaction, separation, mixing. Under the action of vibration of a certain intensity on a poured product it undergoes compaction. The latter takes place as a result of reduction of frictional forces between the individual particles of the medium under vibration and under the action of gravitational and inertial forces. It is established experimentally that compaction of a poured medium is mainly determined by the magnitude of acceleration of the transmitted oscillations. The optimum value of acceleration at which the highest degree of compaction is achieved is dependent on the physical and mechanical properties of the medium. For disperse media with weak internal links this acceleration is less than for materials with internal friction forces.

A qualitative picture of the processes taking place under the action of vibration on mass (bulk) loads can be adequately analyzed on the basis of experimental studies. Tests with materials of identical specific weight and grain size (chalk and coal of grain size 30 - 50 mm) poured in a layer into a vibrating bin showed that at the optimum for the compaction process magnitude of acceleration (8.0 m/sec$^2$), a noticeable reduction of the volume occupied by the material is attained. At the same time, its initial layer position in the bin is retained. The same pattern is also noticed under the action of vibration on a fine granular material (0.5 - 3.0 mm) with optimum magnitude acceleration 7.5 m/sec$^2$ in this case.

Experiments on the action of vibration on chunky materials of identical grain size (20 - 30 mm) and different specific weight (coal: 1.2 g/cm$^3$ and ballast: 1.9 kg/cm$^3$) showed that under vibration with optimum acceleration settling of the entire mass of the material takes place, moreover, the heavier pieces move upwards. If the light product was placed at the bottom and the heavy product at the top, then material stratification is not disturbed after being subjected to the same vibration regime.

Tests on the action of vibration on chunky materials of identical specific weight but different grain size (chalk 30 - 45 mm and coal 2 - 10 mm) enabled one to establish that under vibration with optimum acceleration compaction of the entire material takes place. Furthermore, if the chalk layer is placed at the bottom, then following application of vibration the larger chunks were moving upwards; if it was placed at the top, it does not change its position.

Thus, investigations revealed the following qualitative features of the vibratory action on granular loads. Particles with large specific weight and large dimensions under the action of vibration with optimum (from the standpoint of compaction) acceleration rise to the top; in the meantime, the entire mass of the material is compacted. This phenomenon is caused, apparently, by the fact that the values of the optimum vibration accelerations for the larger particles with large specific weight are below the optimum accelerations of the smaller particles with lower specific weight. The vibration regime adopted for compaction was optimized for lighter and smaller particles. Therefore, compaction of these particles occurs more rapidly than compaction of large and heavy particles. As a result, the small particles in the relative motion are travelling downwards more rapidly pushing the larger ones to the surface. It can also be assumed that under some vibration regimes a lifting force is created which is proportional to the specific weight and size of the chunk and which causes the described phenomenon.

When designing vibratory compaction machines the dependence of the degree of compaction on the vibration parameters, namely amplitude, frequency, vibratory velocity, and acceleration, is of principal interest. Experiments indicate that at higher frequencies maximum compaction is attained at lower amplitudes. However, with increasing frequency the degree of maximum compaction is decreasing. Increasing amplitude of oscillations is causing an increasing density only to a certain extent. Further increase of amplitude can even lead to loosening of the material.

To each value of the amplitude corresponds its own most favorable frequency of oscillations at which maximum compaction takes place. When vibrating at frequencies of 200 – 1100 osc/min, some compaction of the material takes place, with increasing frequency the compaction ratio increases from 3 – 4% at 200 osc/min to 15 – 16% at 100 osc/min. With increasing the frequency further (to 1100 osc/min), compaction practically is not changing. If the frequency is increased even further, the compaction process ceases and loosening commences, its intensity increasing with increasing vibration frequency. At different amplitudes the maximum values of the compression ratio correspond to different frequencies. Thus, it can be concluded that the frequency does not uniquely determine the degree of compaction of the material.

From the analysis of the correlation of the compression ratio and the averaged magnitude of the vibratory velocity it follows that different values of the vibratory velocity correspond to the same value of the compression ratio. Thus, the velocity, just as the frequency, does not completely determine the vibration effects. It is well known that each value of the compression ratio corresponds to the practically constant value of the averaged vibratory acceleration. The optimum value of the averaged acceleration, corresponding to the maximum degree of compaction of a material of chunk size 0 – 12 mm, is 6.50 – 7.50 $m/sec^2$.

At frequencies lower than 500 osc/min, the value of the optimum accelerations are unchanged. In the frequency range 2300 – 2500 osc/min the maximum compaction effect somewhat decreases.

Thus, it can be ascertained that the compaction effect for bulk loads is mainly determined by the magnitude of vibratory accelerations. Furthermore, the higher degrees of compaction are reached in the range 500 – 2300 osc/min. In addition to the enumerated main parameters of the vibration regime, the compaction process is also affected by other factors: degree of filling of the bin which is subjected to vibration, physical and mechanical properties of the compacted product, and duration of the vibration process.

The results of studies of a correlation of duration of the vibration process and the degree of compaction show that the compaction process is progressing nonuniformly, with a diminishing speed: most intensively at the initial moment of vibration application, then the speed is decreasing and after vibrating for 60 sec, the process essentially ends. Such nonuniformity is caused by the fact that as the material is compacted, areas of the contact surfaces between individual particles are increasing, causing reduction of effectiveness of the vibratory action. Investigations show that the minimal duration of vibration which is needed to provide maximum compaction is increasing with increasing mass of the vibrating material.

Experimental investigations of effects of the physical and mechanical properties of the compacted material on the characteristics of the vibratory compaction process enabled one to establish the following. Dimensions of the particles exert considerable influence on the process of compaction. With increasing grain size the maximum compression ratio decreases. The maximum achieved degree of compaction of dimensionally homogeneous products corresponds to the smallest size class (0 - 10 mm). For compacting dimensionally diverse products (class 0 - 100 mm), the maximum degree of compaction is reached (17 - 18%). For this case, the averaged value of the optimum vibratory acceleration is 6.50 - 8.00 m/sec$^2$. The granulometric composition of the compacted product also noticeably affects the magnitude of the optimum vibratory acceleration: with increasing grain size the magnitude of the optimum acceleration is increasing.

Effect of humidity of the compacted material on the process of compaction is reflected in the fact that with increase in humidity the optimum acceleration is increasing and maximum compaction decreasing.

In the majority of cases increasing vibratory acceleration to the value of the free fall acceleration does not enable provision of a dense compaction of a disperse material, and increasing acceleration above this critical value causes particle separation and increased porosity. In these cases, to create a state of vibration liquefaction additional cushioned weights are used. The latter are punches which are connected by elastic elements (springs) with the main source of force acting on the formed material (load, press). These weights do not allow the particles to separate from each other and from the bottom of the mold. At the same time, when forming various materials, reduction of internal friction forces under the action of vibration enables achievement of dense packing of the particles at a lower pressure from the weight compared with the pressure required for static compaction.

When acceleration is increased above the critical value, the layer enters the state of vibration boiling and becomes loose. The absolute porosity of the vibratory boiling layer depends on both frequency and amplitude of vibrations. The lower the frequency, the greater the layer porosity that can be attained with similar acceleration magnitudes. In the range from one to four times gravity acceleration, the layer volume increases more intensively than with even higher accelerations. This can be explained to a certain extent by the action of the aerodynamic force which is developing as a result of rarefaction under the layer of the bulk load. The magnitude of this force can exceed by several times the weight of the material itself. For example, aerodynamic pressure on vibratory boiling layer of glass balls of 0.13 mm diameter at frequency 50 Hz and amplitude 1.42 mm exceeds the gravity force by 10 times. The intensity of loosening of the layer is substantially dependent on the physical properties of the medium. Thus, porosity of the layer of some finely dispersed polymer powders is noticeably increasing under oscillation accelerations above 5 $g$.

However, these correlations do not apply to a layer of finely dispersed powders whose particles have high adhesive forces. The vibratory action, by facilitating these particles in drawing together more closely, compresses the layer and does destroy the aggregates being formed. In some cases expansion of a layer of finely dispersed materials can be achieved by the joint action of vibration and gas being blown through the layer or by the use of surface-active agents.

The simultaneous action of vibration and gas, blown through the vibratory boiling layer from the bottom to the top, enables one to obtain a uniform distribution of the gas with a considerable increase of layer porosity. For example, the volume of a layer with 4.3 $\mu$m particles and density 1 $g/cm^3$ under the simultaneous action of vibration and air, blown through the layer from the bottom to the top, is increasing by not less than 30%, which is about 1.3 times greater than increase of the volume of the boiling layer. The volume of a layer of heavier particles increases to a lesser degree. For example, the volume of a powder layer with 63 μm particles and density 7.9 g/cm$^3$ in a state of vibratory boiling increases just by 1.1 times compared with the volume of a layer without vibration and of the same velocity of blown-through air.

Some powders of polymer materials, whose bulk porosity does not increase at low intensities of oscillations, increase the layer volume by 60 - 70% under the action of vibration with gas blow-through. The coefficients of volume increase of a vibratory boiling layer in the case of gas blow-through is higher than for the boiling layer, and, in absolute values, these coefficients are equalized to a considerable degree for materials equally yielding to conversion into a boiling layer.

Application of vibrations is also used for compaction of moulding and core sands when preparing casting moulds. The principal quality criteria of compaction when preparing casting moulds are the density of the mixture and its gas permeability. Determination of the optimum regimes of vibration is guided by these criteria.

Experimental investigations were conducted to determine mechanism of compaction of sodium silicate core sand (100 parts sand and 6 parts sodium silicate) at humidity of 5 - 6%. A mixture in a standard measuring beaker was subjected to vibration in the vertical direction, the height of the mixture layer being equal to 150 mm. Investigations were conducted in the frequency range from 10 to 110 Hz and amplitude range from 1 to 8 mm under harmonic excitations. The duration of vibration treatment was varied from 1 to 10 min.

Observations of behavior of the mixture in the process of vibratory treatment enabled revealing a number of physical laws. At an oscillation frequency of the measuring beaker of 10 Hz and amplitude 1 mm the mixture settles only slightly within 1 - 2 sec following commencement of vibrations, but the degree of mixture compaction is insignificant. After placing the sample into the device for determination of gas permeability, the mixture might even crumble. With increasing frequency from 10 to 40 Hz, effective compaction is detected at amplitudes of 1 - 2 mm, with the most intensive compaction occurring in the amplitude range 3 - 4 mm. Furthermore, sometimes mixture stratification into two unequal parts is observed: the lower (70 - 90% of thickness) is well compacted, and the upper disintegrates into separate lumps of size 1 - 10 mm which are continuously moving and rotating without

**Density and Gas Permeability of Sodium Silicate – Sand Mixture with 5.29% Humidity at Various Frequencies and Amplitudes of Oscillations of the Molding Cylinder and Vibration Duration of 2 min**

| No. of Test | Parameters | | | | | | Characteristics of mixture compaction |
|---|---|---|---|---|---|---|---|
| | Hz | mm | m/sec | m/sec$^2$ | g/cm$^3$ | arbitrary units | |
| 1. | 10 | 1 | 0.063 | 3.95 | 1.1 | not defined | Settling of mixture without effective compaction. |
| 2. | 10 | 2 | 0.126 | 7.9 | 1.13 | 300 | Analogous to test 1. |
| 3. | 10 | 4 | 0.252 | 15.8 | 1.26 | 250 | 20% of volume is loosened with the formation of lumps of 2 – 5 mm size. |
| 4. | 15 | 2 | 0.189 | 17.8 | 1.13 | 320 | Analogous to test 1. |
| 5. | 15 | 3 | 0.283 | 26.7 | 1.24 | 300 | Settling of mixture, formation of small lumps on the surface. |
| 6. | 15 | 4 | 0.38 | 35.6 | 1.26 | 300 | Analogous to test 5 |
| 7. | 20 | 1 | 0.126 | 15.8 | 1.13 | 400 | Analogous to test 1 |
| 8. | 20 | 2 | 0.25 | 31.6 | 1.22 | 300 | Analogous to test 5 |

Table (continued)

| | | | | | | | |
|---|---|---|---|---|---|---|---|
| 9. | 20 | 3 | 0.38 | 47.5 | 1.27 | 250 | Analogous to test 3 |
| 10. | 20 | 4 | 0.5 | 63.0 | 1.35 | 180 | Effective compaction, pellets of size 5 - 10 mm formed on the surface. |
| 11. | 25 | 1 | 0.157 | 24.6 | 1.28 | 230 | Analogous to test 5 |
| 12. | 25 | 2 | 0.314 | 49.4 | 1.24 | 215 | Analogous to test 3 |
| 13. | 25 | 3 | 0.47 | 74 | 1.25 | 200 | Analogous to test 10 |
| 14. | 25 | 4 | 0.67 | 98.6 | 1.14 | 300 | 50% of volume loosened with the formation of pellets of size 5 - 10 mm |
| 15. | 30 | 0.5 | 0.094 | 17.8 | 1.25 | 200 | Analogous to test 5 |
| 16. | 30 | 1 | 0.189 | 35.5 | 1.33 | 185 | Analogous to test 10 |
| 17. | 30 | 2 | 0.38 | 71 | 1.17 | 200 | Analogous to test 14 |
| 18. | 40 | 0.5 | 0.0628 | 15.8 | 1.13 | 400 | Analogous to test 1 |
| 19. | 40 | 0.5 | 0.126 | 31.6 | 1.27 | 110 | Analogous to test 14 |
| 20. | 40 | 1 | 0.25 | 63.2 | 1.2 | 120 | Analogous to test 14 |

adhering to the main mass of the mixture. It should be noted that in all vibration regimes the upper boundary of the sample never becomes smooth. The volume of sand which disintegrates into lumps and is continuously moving is strongly dependent on the amplitude of the vibratory velocity of the measuring beaker. For example, at a frequency 20 Hz, 15 - 25% loosening of the volume occurs in the upper part of the beaker with the formation of lumps for amplitudes above 3 mm. At frequency 40 hz and higher the stated result is obtained for amplitude 1 mm. On the other hand, reducing the oscillation amplitude to 0.25 mm and less with the simultaneous rise in frequency leads to the situation when the mixture compaction practically does not take place. Obviously, at small amplitudes of motion vibration in the mixture is decaying.

After a prolonged vibration compaction (longer than 5 min), some migration of sand and sodium silicate is detected. The bottom of the beaker is always covered by a sticky moist film of sand 1 - 2 mm thick which is easily separated from the remaining part of the mixture. Furthermore, some nonuniformity of the density along the height was observed, which could be easily detected when the molding beaker was cleaned.

The main task of the investigation was to determine gas permeability and density of the sodium silicate sandy mixture. These quantities are inversely related.

The main results of the experiments are presented in the table. At an oscillation frequency 15 Hz of the beaker compaction of the mixture starts at an amplitude of 2 mm, with the most intensive compaction occurring in the amplitude range 3 - 4 mm. In this range the specific weight of the mixture does not undergo significant changes. At amplitudes exceeding 4 mm, mixture stratification into two parts is observed: the lower is well compacted, and the upper moves along a circular trajectory inside the beaker. The amount of the mixture taking part in the circular motion is dependent on the intensity of oscillations. For each frequency there is a corresponding acceleration at which mixture stratification takes place.

At a frequency 20 Hz, compaction starts at an amplitude of 1 mm; in the amplitude range 2 - 3 mm, the specific weight and gas permeability do not undergo significant changes. Compaction in this range is uniform along the entire height. When the amplitude is increased in excess of 3 mm, compaction on the average is increasing, but it is not uniform along the height of the specimen, and at amplitudes higher than 4 mm mixture stratification is observed.

At a frequency 25 Hz there also exists an amplitude range 2 - 3 mm within which gas permeability and the specific weight do not change, but if at an amplitude of 1 mm there is practically no compaction, then at an amplitude of 2 mm stratification of the mixture is already evident, although circular motion is performed by an insignificant part (around 10%). At an amplitude of 3 mm, 30% of the mixture executes circular motion, above 3 mm the mixture is not compacted, since more than half the mixture executes circular motion.

At a frequency 30 Hz and amplitude of 1 mm the mixture was uniformly compacted, although somewhat worse than at lower frequencies. At an amplitude of 2 mm and above mixture stratification takes place, with lumps being formed at the top. At large frequencies, for example 40 Hz, stratification takes place already at an amplitude of 0.5 mm, and at an amplitude of 0.25 mm there is no compaction.

On the basis of the obtained results one can identify several regions of influence of vibrations on compaction of sodium silicate mixtures as a function of frequency and amplitude and find optimum dynamic regimes. Firstly, it is inexpedient to use vibration at low frequencies (below 10 Hz), since for effective compaction higher oscillation amplitudes are required. Secondly, the regime of high frequencies (above 40 Hz) is also ineffective, since the system becomes very sensitive to small variations of the amplitude and instead of compaction of the mixture, transition to the loosening stage is possible. The frequency range 15 - 30 Hz at amplitudes of 4 - 1 mm ought to be regarded as the optimum regimes of the vibratory plant.

When determining the density, a systematic error is always present which is associated with the unevenness and loosening of the upper layer of the mixture. Therefore, in all the regimes the measured density of the mixture, which is subjected to vibration, is less than the reference value. Some scatter of the experimental data is also possible for the same reason. Maximum compaction is attained at accelerations of 50 - 60 m/sec$^2$. For a frequency of 20 Hz, the indicated values correspond to amplitudes of 3.5 - 4 mm.

Despite the large scatter of the values of gas permeability that are obtained for different frequencies, there is a general tendency of its decreasing to a minimum value which is observed for accelerations of 30 - 60 m/sec$^2$, then again increasing for higher accelerations. The minimum gas permeability corresponds roughly to the maximum density.

Of significant interest are the data on the behavior of the compaction process with time. Experiments have shown that compaction of the mixture takes place largely during 5 - 10 sec. Further vibratory treatment is changing its density only insignificantly.

# THEORETICAL PRINCIPLES OF VIBRATORY TECHNOLOGY

## 2.1 PHENOMENOLOGICAL DESCRIPTION OF THE PROCESS LOAD OF VIBRATORY MACHINES

### 2.1.1 General Information on Mechanorheological Phenomenology

The processes of vibratory technology are accompanied not only by displacements of working elements of the vibratory machines and the treated medium but also by their strains. In the process of these strains in the interacting elements, internal stresses are developing which determine, in the final analysis, the performance and reliability of the entire system: vibratory machine-load. In the general case, the vibratory machine-load system can in some time intervals be broken down into its constituent subsystems (load and machine) and then reassembled again. As an example we can cite the sytem: treated medium-working element of a vibratory processing-conveying machine, e.g., a screen. During part of the travel the treated medium and the sieve are in contact. Then the medium is tossed up, after which the sieve and the treated product are moving independently; then they become interactive again. The connection between the sieve and the treated products is non-holding since they can disengage from time to time. Such discontinuous processes are encountered not only in mechanical motions. Deformation processes can also undergo discontinuous phenomena. For example, on the boundary between elastic strain and plastic flow, the law governing the stresses is changing radically. Thus, the mechanorheological model can reproduce not only evolutionary processes, but also discontinuous processes. The efficient operation of mechanorheological phenomenology is achieved by means of development and utilization of the algorithm of the process with a logical control system which continuously analyzes the progress of the process and determines the initial conditions in discontinuous transitions.

Generally, a complex interweaving of the mechanical motions of the elements of the system takes place at some stages and mutual strain at others.

The investigation of such systems cannot be successfully conducted by the methods of mechanics alone. For the description of the process of strain of both machine elements and the treated medium with the purpose of determining the stresses (loads) that arise, the rheology techniques are employed. The overall characteristic of such sytems is attained by the simultaneous application of the methods of mechanics and rheology. The mechanics apparatus describes the motion of the system as a mechanical object, the rheology apparatus is reproducing the dependences between strains and stresses of the elements of the system.

The vibratory machine-process load systems, with predominant concentration on the processed medium, is best investigated using mechanorheological phenomenology. The latter is the science of formulation of complex phenomenological models of objects performing relative mechanical motions and being subjected to mutual strains. For solution of such mechanorheological problems using mathematical techniques, phenomenological concepts of the structure of the investigated objects are being developed. This process is made easier by the construction, at least mentally or on display, of models composed from fundamental rheological bodies and inertial elements. At this stage one important comment must be made. Classical fundamental bodies (elastic, viscous, and plastic) on their own are not sufficient to construct mechanorheological models. In conditions of periodic loadings of the process load, in addition to the stresses created under elastic, viscous, and plastic strains, inertial forces are also generated. Contribution of these forces to the total loading of the system in high-speed installations is significant and commensurate with the other acting loads. In view of this, in vibration rheology the models constructed from elastic, viscous, and plastic bodies are supplemented by inertial elements [8, 1]. These models must be physically reliable to reproduce the main properties of the modeled objects and describe their interaction in terms of force and displacement at some stages of evaluation and in terms of strains and stresses at other stages.

With the aid of the phenomenological model, as is usual in rheology, one can compile a structural formula of the reproduced object which is an abbreviated representation of the model structure and is characterizing specific features of interactions connections between the fundamental rheological bodies in a complex model. The structural formula yields the complete information which is contained in the model itself. It is true that the structural formula is less transparent; however, it provides a more economical form of representation of information on the rheological properties of the object.

Phenomenological models and structural formulas help to qualitatively characterize the mechanorheological properties of the object. They are also regarded as the primary basis for derivation of mechanorheological equations of the various systems that are required for their quantitative description.

At the present time, the procedure of development of phenomenological models is formalized to a high degree on the basis of the techniques of multicriterial design optimization. It is combined with the problem of identification of the model and the real object [43]. The efficient operation of mechanorheological phenomenology in an automatic mode is attained by the development of algorithms for computer-aided computations, and by the sytem of logical conditions which analyze the current process conditions and form its further development.

In a number of cases, the process load can have a random character. In view of this, for investigation of the interactions of a random process load with the working element of the vibratory machine one can use probabilistic mechanorheological phenomenology combining the correlation theory of random processes and mechanorheological methods.

The probabilistic mechanorheological phenomenology can be used not only for description of any process loading, but also for reproduction of characteristics of the vibratory installations.

The system vibratory machine-random process loading will comprise, in the mathematical sense, a set of systems of differential equations. One system will reproduce correlations between forces and displacements, strain and stresses of the process load, the second system will reproduce interrelation between the forces and motions of the moving elements of the vibratory machine and, finally, the third system will reproduce interaction between the load and the moving elements of the vibratory machine.

It is particularly pertinent to underline that such an approach enables one to obtain a complete picture of loading on all the elements of the vibratory plant. Furthermore, it is particularly important that the loads are not specified *a priori* but are developed in the process of interaction of the processed medium with the elements of the vibratory plant, just as in a real object.

In discussing the general principles of constructing phenomenological models, it must be noted that model construction is a typical multicriterial optimization problem. Maximum reliability and accuracy of the characteristcs of the reproduced process must be ensured with minimum complexity of model structure. In other words, the model must be comprised of the least number of basic rheological bodies which must have the simplest interconnections. It is preferable not to use rheological bodies with nonlinear characteristics.

The development of the model goes as follows. The investigator, on the basis of the knowledge of the modeled process, constructs a sufficiently complete initial model establishing connections between the basic rheological bodies comprising the model, and specifies the range of change of the parameters of each element in the model. The model parameters can be treated in a broad sense by questioning whether a given basic rheological body is present in the model or not. If it is present in the model, the parameters are assigned specific values, if it is absent, then the parameter is assumed equal to zero. This procedure is referred to as the establishment of parametric constraints of the model.

Further, the so-called functional constrains are formulated. In order to make it convenient to use a computer, to simplify the program, and to ensure high computational accuracy, it is necessary that the scatter of each parameter, i.e., the deviation of maximum from minimum values, were within the allowable limits. The functional constraints are established from correlations between minimum and maximum values of the parameters.

In accordance with the adopted formulation of the problem, the minimum deviations between the design and experimental values of the most essential parameters of the modeled process and the minimum number of rheological bodies in the model are adopted as the local quality criteria. It must be kept in mind that the phenomenological models for various values of the production regime parameters, for example frequency and amplitude of oscillations, shape of oscillation trajectory and so on, yield sharp deviation from the experimental data. For some values the design data can be acceptable in accuracy, for other values the specified conditions are not met. It is very important that the model ensures the necessary accuracy of computations within the entire range of the production regime parameters. Therefore, as a third quality criterion, the minimum scatter of the design values of the characteristics at the extremal values of the parameters of the production process is adopted. As principal characteristics of the vibratory production process one can adopt: loads on the working organ, energy consumption, trajectories of motion of the medium, and so on, i.e, those parameters which can be accurately measured on a real object.

Quality criteria are arranged according to their importance. As the initial data for the model identification, one can directly use the results of experiments, or the regression equations that are built using these results.

Having developed the structure of the phenomenological model, one must construct its mathematical description and algorithm of the solution with the logical control system. Using the algorithm, a computer, by going through the parameter values in the given range is assessing the variants of the phenomenological model with respect to each quality criterion. The variants of the models are represented in the form of test tables. The first place in the table is given to the best model with respect to the considered criterion. Since each model corresponds to a particular set of parameters which is coded in the number of the test, the investigator, by inspecting the table of model parameters, can see how to change the parameters in order to obtain the best model with respect to one or another quality criterion.

As a rule, there are no models that are optimum with respect to all quality criteria at the same time. Therefore, the results of the analysis of the test tables are used for the substantiation of the compromise solution. Constraints are assumed for each quality criteria. Further, the computer verifies if there is a model meeting the conditions of all quality criteria at the same time. Such variants constitute the allowable set of models. From these models the investigator selects the optimum variant.

The building blocks for various rheological models are the fundamental rheological bodies, namely, viscoelastic and plastic in classical rheology, and elastoinertial, viscoinertial, and plastoinertial in vibration rheology. Thus, the main difference of vibration rheology from classical rheology is the presence of two properties for the fundamental bodies, namely, their own and inertial property.

The need to supplement the classical rheological bodies by adding inertial properties when solving the problems of vibration technology is due to the fact that under loading of the processed medium by periodic actions, the strains are developing with accelerations, thus the stresses are found to be dependent on the developed inertia forces. Inertia loads in some cases can become commensurable and sometimes even exceed elastic, viscous, and plastic stresses.

Accordingly, elastoinertial, viscoinertial, and plastoinertial fundamental rheological bodies are used in vibration rheology. Their behavior in conditions of periodic loading differs from the behavior of the strains in classical rheological bodies.

Let us consider the characteristics of rheological bodies in conditions of periodic loading, starting with an elastoinertial body (Fig. 2.1a).

Upon deformation of an inertial elastic body, characterized by mass $m$ and stiffness $k$, by a periodic force changing according to a harmonic time history, resistance to deformation will be offered not only by the elastic force, but also by the inertial force. If the elastic body is characterized by its stiffness, then the elastoinertial body will be characterized by the relation between its stiffness and mass. This relation determines angular natural frequency of the inertial elastic body, $p = \sqrt{k/m}$. The natural frequency of the elastoinertial body can be determined if it is strained without exceeding the elastic limit and then relaxed. Free oscillations of the body start whose angular frequency is the natural frequency. The elastoinertial body can be strained at this frequency under minimum forces.

When the elastoinertial body is deformed at another frequency, whether it is larger or smaller, additional resistance is always created. Deformation of an elastoinertial body with a frequency not corresponding to its natural frequency requires application of additional forces which, depending on the frequency of deformation, are used to overcome either elastic forces or inertial forces of the body.

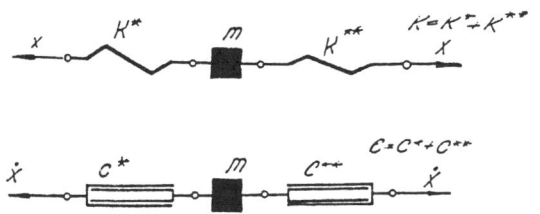

**Figure 2.1** Fundamental rheological bodies. a) inertial elastic body; b) inertial viscous body.

Only when the elastoinertial body is deformed with a frequency equal to the natural frequency does one not have to overcome neither inertial, nor elastic forces. The loading force causes increase of the strain and stress amplitudes in the elastoinertial body. An increase of its internal energy also takes place.

In the elastic body the relation between the strain $x$, stiffness $k$ and unit straining force $p$ under periodic loading $P$ sin $\omega t$ is determined as $x = 1/R\ P$ sin $\omega t$. In the elastoinertial body, whose natural frequency is $p$, the dependence between the strains and the straining force $x = P/R$ sin$(\omega t + \varphi)/1 - \omega^2/p^2$ differs by the term $1/(1 - \omega^2/p^2)$ which is called amplification factor. Furthermore, the straining force and the strain do not always coincide. There is a phase shift $\varphi$ between them.

The physical meaning of the amplitification factor is as follows. It shows how much the magnitude of the strain of the elastoinertial body is different compared with the elastic body under deformation by equal forces. The amplification factor is dependent on the ratio $\omega/p$, which is called the detuning factor, and can be greater or smaller than unity. It is equal to unity when the detuning factor is equal to zero, i.e., when the mass of the elastoinertial body is equal to zero it becomes a simple elastic body. Thus, the elastic body is a special case of the more general elastoinertial body.

Upon deformation of the elastoinertial body with its natural frequency, i.e., at resonance, the detuning factor is equal to unity and the amplification factor is infinite. This corresponds to the infinite rise of strain in the elastoinertial body with time, $x = P/2\sqrt{Km}\ t$ sin $\omega t$. With increasing strain the stresses are also rising and fracture of the body can occur at some moment.

At large values of the detuning factor the amplification factor is less than unity. In this case, the elastoinertial body deforms less than the simple elastic body.

If the detuning factor is less than unity, then the strain and the straining force are in phase. Strain occurs in the force direction. When the detuning factor is greater than unity, the force and the strain are out of phase. Strain occurs in a direction opposite to the force. Under resonance the phase shift is $90^\circ$.

Thus, the laws of deformation of an elastoinertial body by a periodic force are far more complex and their manifestation is much more diverse than the deformation of a simple elastic body. The elastic rheological body is merely a special case of the more general elastoinertial body. The reaction of the elastic body to the periodic action is single-valued (unique). The elastoinertial body reacts to a similar action differently at different frequencies of deformation.

Thus, it can be ascertained that the elastic rheological model does not adequately reproduce properties of real elastic bodies which always have a mass.

Let us consider the rheological properties of viscoinertial rheological bodies (Fig. 2.1$b$). The viscoinertial body is deformed by a harmonic periodic force. Resistance to strain will be comprised of a viscous force and an inertial force. In a viscous rheological body relation between rate of strain $\dot{x}$, effective viscosity coefficient, and straining force $P$ is determined as $\dot{x} = P/2n$ sin $\omega t$, i.e., the rate of strain is proportional to the magnitude of the straining force and inversely proportional to the viscosity coefficient, is independent of the loading frequency, and is in phase with the loading force. In a viscoinertial body the rate of strain is dependent on the frequency $\omega$ of the straining load, $\dot{x} = P/\sqrt{\omega^2 + 4n^2}$ cos $(\omega t + \varphi)$ and has phase angle $\varphi =$ arctg $(-2n/\omega)$ with respect to the straining force.

Comparison of the deformation behavior of viscous and viscoinertial bodies shows that the rate of strain of the viscoinertial body is $n/\sqrt{\omega^2 + 4n^2}$ times less. The difference in rates of strain is increasing with decreasing viscous resistance and increasing application frequency of the straining load. The phase shift is increasing with increasing coefficient of viscous resistance and decreasing frequency.

When a plastoinertial body is strained, the resistances to strain are combined from the force of resistance to plastic shear and the inertial force (Fig. 2.1$c$). In a plastoinertial body, which is characterized by the resistance to plastic shear $F^*$, the relation between strain and the straining load is given by an approximate expression $x = P\sqrt{1 - f^{*2}}$ sin $(\omega t + \gamma)$, where $f^*$

= $F^*/P$. Strain has a phase shift $\gamma$ with the straining load, $\gamma$ = arcsin $(-f^*)$.

The presented dependence indicates that to ensure the deformation of a plastoinertial body by a periodic force it is necessary for the amplitude value of this force to exceed the resistance force to plastic shear. To obtain sufficiently significant strains this excess must be considerable. The phase shfit between the strain and the straining force is defined by the relation between the magnitudes of resistance to plastic shear and the straining force and can be within 0–90$^{\circ}$ interval.

The dependence between the periodic straining force and displacement can be assumed harmonic only as a first approximation. In reality, this dependence has a more complex polyharmonic character. In each special case which requires a specified accuracy of solution, the limit of plastic flow must be expanded in a Fourier series having the number of expansion terms that would enable one to obtain a result with the desired accuracy.

By using fundamental inertial rheological bodies, one can construct more complex phenomenological models such as viscoelastic inertial models, elastoplastic inertial models, and so on, which more accurately reproduce the properties of the processed medium.

Characteristics of the elastoinertial, viscoinertial, and plastoinertial bodies are natural frequency $p$ for the elastoinertial body, effective coefficient of viscous resistances $\sqrt{\omega^2 + 4n^2}$, for the viscoinertial body, and effective coefficient of plastic resistances $f^*$ for the plastoinertial body.

Thus, vibration rheology is the science of deforming a body by periodic loads. Its fundamental feature is the consideration of the inertial properties of the processed body.


### 2.1.2  Combined Inertial Phenomenological Models

The properties of some dispersed products in vibratory processing can be adequately described by viscoelastic, elastoplastic, or viscoelastoplastic rheological models. For reliable description of the deformation features of various media, special phenomenological models have been developed. These models were obtained on the basis of various combinations of inertial fundamental rheological bodies. The main method of arrangement is the series, parallel, or

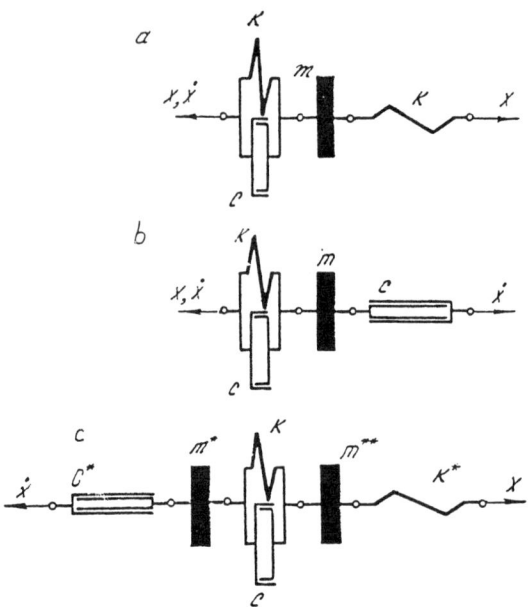

**Figure 2.2** Standard inertial viscoelastic rheological bodies. a) solid; b) liquid; c) universal.

combined series–parallel combination of the fundamental rheological bodies in a specific order. Utilization of these models enables one to solve some important special problems.

The properties of a large group of products which are subjected to vibratory processing can be described by different combinations of elastoinertial and viscoinertial rheological bodies. Products combining elastic and viscous properties are known as viscoelastic. The nomenclature of such products is so large and important for the industry that an independent important trend has developed in rheology devoted to the study of their properties, namely the theory of viscoelasticity. Viscoelastic materials occupy an intermediate position between elastic bodies and viscous fluids.

Among these models, the so-called standard inertial viscoelastic rheological solid and liquid models can be identified. The standard viscoelastic inertial solid rheological model is comprised from an inertial element, two elastic and one viscous bodies. The elastic and viscous bodies are connected in parallel and one more elastic body is connected with them in series (Fig. 2.2a). A standard inertial viscoelastic fluid rheological body includes two viscous and one elastic bodies (Fig. 2.2*b*). One viscous and one elastic bodies are connected in parallel and another viscous rheological body is connected to them. And finally, a standard inertial viscoelastic universal rheological body combines the properties of two previously considered viscoelastic rheological bodies (Fig. 2.2*c*). A standard inertial viscoelastic universal rheological body comprises two elastic and two viscous rheological bodies. The elastic and viscous rheological bodies are arranged in two groups connected in parallel and in series. Then these groups are connected in series.

In a standard inertial viscoelastic solid body the inertial element is inserted between the elastic body and the viscoelastic bodies connected in parallel.

In a standard inertial viscoelastic universal body two inertial elements are used which are inserted between viscoelastic bodies connected in parallel and elastic and viscous bodies, respectively.

Let us consider the laws of deformation by a periodic force of, for example, a standard inertial viscoelastic solid.

Under deformation of the inertial viscoelastic model of a solid body by a force having a harmonic time history, the dependence between the amplitude values of deformation and the unit deforming force is $A_\chi = p^*/\sqrt{(p^{*2} - \omega^2)^2 + 4n^2\omega^2}$ where $p^* = K + K_1/m$. The force and deformation are shifted relative to each other by angle $\varphi = \text{arctg } 2n\omega/p^{*2} - \omega^2$. The amplitude value of deformation of the inertial viscoelastic solid is proportional to the magnitude of the deforming force per unit of the body mass. It is also dependent on the magnitude of the viscous resistances and on the ratio of natural and forced vibration frequencies.

Upon loading of the inertial viscoelastic solid by a periodic force, just as in the case of the elastoinertial body, the magnitude of deformation is changing with a constant force magnitude for different regimes (with varying frequency). However, in the case of inertial viscoelastic body the amplification factor for the deformations is dependent not only on the detuning factor, but also on the magnitude of viscous resistances, $1/\sqrt{(p^{*2} - \omega^2)^2 + 4n^2\omega^2}$. The magnitudes of deformation of the viscoelastic and inertial viscoelastic bodies coincide only when they are being deformed with a very low frequency, in a practically static regime. At very high frequencies they also are close.

When the frequency is increasing, the magnitude of deformation of the solid viscoelastic body is invariably decreasing. The strain of the inertial viscoelastic body is invariably decreasing. The strain of the inertial viscoelastic solid with increasing frequency is increasing, reaching a maximum value at resonance, then it gradually drops again. It should be noted that the deformation magnitude at resonance regimes is limited by the values of viscous resistances. At very large viscous resistances resonance growth of deformation may not even take place. At resonance the phase shift between the force and deformation is $90^\circ$. The process of deformation of the inertial viscoelastic solid is accompanied by energy losses associated with hysteresis. Energy is mainly lost in the resonance regime. This is explained by the fact that deformation amplitude increases sharply at the resonance regime.

Under periodic deformation of inertial viscoelastic rheological bodies, the stresses in the body might not coincide in magnitude and in phase with the deforming force. Thus, for example, in elastoinertial body, in which the stresses are proportional to deformations, the deformation amplification factor is also the stress amplification factor. The shift between the deforming force and the stress is the same as between the force and the deformation.

In an inertial viscoelastic solid the amplification factor of the stresses is equal to the gain factor of strain. However, the phase shift between the deforming force and the deformation in an inertial viscoelastic solid body and in a simple viscoelastic body are different.

The fact of the matter is that under periodic deformation of a common viscoelastic body the deforming force is shifted in phase with respect to deformation. In a simple viscoelastic body the same phase shift is established between the deforming force and the stress. Stress magnitude is equal to the deforming force.

In an inertial viscoelastic solid stress magnitude coincides with the deforming force only at very low deformation frequencies. In this case, the same phase shift is established as in deforming of a simple viscoelastic body.

Specific stresses in an inertial viscoelastic solid, i.e., stresses per unit deformation of the body, are proportional to the dynamic stiffness of $\sqrt{p^4 + 4n^2}$ of the body reduced to the unit mass.

Let us consider the laws of deformation of plastic media.

The deformation of an elastoplastic medium begins from elastic strains. The stresses at which the strains remain elastic are limited by the yield strength. In reaching the yield strength plastic strains set in.

In the zone of plastic deformation the loading and unloading develop in different patterns. Loading is accompanied by an insignificant stress increase. The load is almost constant and the deformation of the medium continues, i.e., as if fluid flow is taking place. The total strain of an elastoplastic medium comprises both elastic and plastic components, with the elastic strain being reversible. When the load is removed from the elastoplastic medium the elastic strains disappear but the plastic strains remain.

Under repeated loading of the elastoplastic medium the process commences again with elastic strains; however, the yield strength is reached now at higher stresses than in the first instance. This points to the fact that the elastoplastic medium has been hardened. In the marjority of cases, the real so-called plastic materials are essentially hardened elastoplastic media. In the region of plastic strains stresses are dependent on the previous history of loading.

For the description of the properties of elastoplastic media in the regime of periodic loadings inertial elastoplastic rheological models are used (Fig. 2.3).

The elastoplastic phenomenological model with series connection of elastic and plastic rheological bodies under quasi-static loading and under stresses not exceeding the yield point is representing the elastic strains (Fig. 2.3a). Upon reaching the yield point, plastic deformation without hardening is simulated. Such a strain of the elastoplastic phenomenological model with series arrangement of the fundamental bodies is comprised of elastic and plastic strains. The work done by the stresses up to the yield point is accumulated as elastic strain energy and is given back without losses when loading is removed. The work done by the stresses above the limits of the elastic strain work is dissipated as a result of internal friction. Such an elastoplastic model has residual strain after removal of loading. They restored elastic strain is equal to the initial elastic strain.

In an elastoplastic phenomenological model with parallel connection of the rheological bodies the deformations of the elastic and plastic bodies are the same (Fig. 2.3b). The strain of the model commences when the stress reaches the yield point. With increasing strain the stresses, which are made up from plastic flow stresses and elastic stresses, are increasing. When the loading is removed from the model, the stresses do

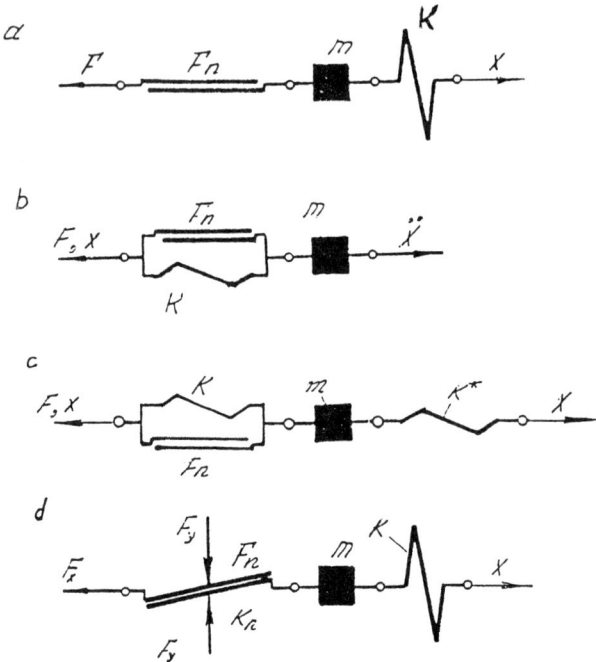

**Figure 2.3** Inertial elastoplastic rheological bodies. *a)* with series connection of elastic and plastic rheological bodies; *b)* with parallel connection of elastic and plastic rheological bodies; *c)* standard elastoplastic rheological body; *d)* hardening elastoplastic body; $K_n$) coefficient of hardening; $F_n$) plastic strain limit

not completely disappear. The model remains in a stress state. The residual stress is equal to the yield strength. As a result of loading and unloading the model receives residual plastic strain. Elastic strain is partially restored under repeated loading, the strains set in after overcoming the yield strength and residual stresses. The elastoplastic model under consideration reproduces the process of hardening and residual stresses.

The standard phenomenological model of an elastoplastic material is comprised of two elastic and one plastic rheological bodies. One elastic and one plastic bodies are connected in parallel and one more elastic rheological body is attached to them in series (Fig. 2.3c). The standard elastoplastic model in the first loading cycle reproduces elastic strains under stresses not exceeding the yield point. In the second cycle elastoplastic strains commence at higher stresses, taking into account the residual stresses following the first loading cycle. Upon reaching the yield point in the first cycle, and residual stresses and yield point in the second, plastic strains commence. As strains are progressing, the stress is rising as a result of adding up the plastic flow stress and residual and elastic stresses. During unloading, the stresses completely disappear in one elastic rheological body, and residual stresses which are equal to the yield strength are retained in the second body. As a result of loading and unloading the standard elastoplastic model undergoes plastic strain. The elastic strain under multicycle loading is completely restored.

Thus, the standard elastoplastic model reproduces the initial elastic strain, the hardening process, and the residual stresses. It enables one to completely describe all the laws of deformation of real elastoplastic media.

The laws of deformation of inertial elastoplastic models in the regime of periodic loading are described by the mathematical relations presented in section 4.3.2.3.

A more comprehensive description of real media is attained within the framework of inertial viscoelastoplastic phenomenology. The standard inertial viscoelastoplastic phenomenological model is comprised of an elastic body connected in series with an inertial element and a viscoelastic body (Fig. 2.4a). An inertial element and a plastic body are connected to the latter in series. The laws of deformation of inertial viscoelastoplastic phenomenological models are described by the mathematical relations presented in section 4.3.2.3.

When investigating a number of processes that are realized by vibratory technology and accompanied by 3-D strains of the processed media, one has to take into account interaction of stresses and strains in mutually perpendicular directions. This is essential in those cases when it is not possible to consider the object in a unidirectional formulation and one has to solve planar or three-dimensional problems. Many important vibration effects cannot be simply explained within the framework of unidirectional systems. For their solution planar or three-dimensional models ought to be solved.

For solution of such problems phenomenological models of interaction of stresses and strains along the principal axes are used.

If in a unidirectional stress state the plastic strains commence when the stresses reach the yield point, then under

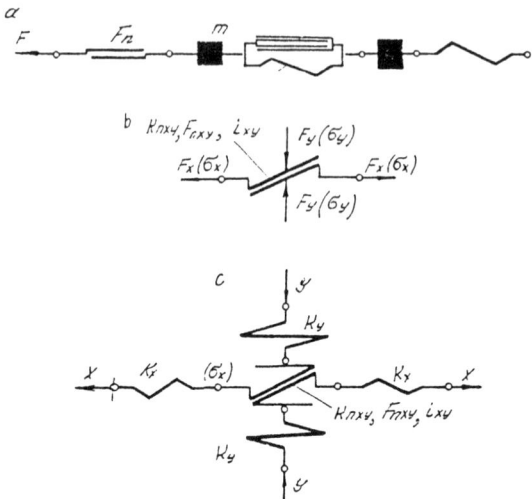

**Figure 2.4** Viscoelastoplastic phenomenological models. a) standard viscoelastoplastic model; b) wedge element of stress interaction; c) elastic wedge element of stress interaction

conditions of planar stress state the beginning of plastic deformation is dependent on the stresses acting in perpendicular directions. In accordance with the theory of plasticity, it is assumed that plastic strains under conditions of stress state commence when the difference of the stresses acting in perpendicular directions reach the yield point. In a more general statement it can be assumed that as applied to the conditions of loading by vibratory forces, a more complex picture of stress interaction is present as a result of periodical interactions. These features can be taken into consideration by some transfer coefficient $i$. This interaction is simulated in the model by a wedge element.

Stress interaction in a plastic material can be represented by a standard phenomenological model comprising a wedge element with two mutually perpendicular plastic rheological bodies (Fig. 2.4b). The standard model of a planar elastoplastic material reproduces elastic strains under stresses in the direction of deformation (Fig. 2.4c) which are not exceeding the sum of the stress acting in a perpendicular direction and the yield strength. Upon reaching the indicated stress limit, plastic strain sets in with hardening. With the pogressing strain, the stress is increasing, being comprised of the stress of plastic flow and the hardening stress, as simulated by a wedge element and a transverse elastic rheological body. When the load is removed, the elastic strain disappears and the model will have residual plastic strains. As already mentioned, the models can reproduce stress interactions in mutually perpendicular directions and by other laws that are characterized by the coefficient $i_{x,y}$.

The random values of the model parameters can be adequately characterized by two parameters, namely the probability distribution and the probability density. In addition to the probabilistic description, one can use the aggregate of random numerical characteristics which might be invariant or might vary with time to characterize the random parameters of the model. This method of estimating the random parameters of the model is preferable, since the numerical characteristics are simple and easier to manipulate in the process of computations. As numerical characteristics of the random parameters of phenomenological models one can use: the mean parameter value which is also known as the expected value, the mean value of its square (mean power), the variance (the mean value of the square of deviations of a parameter from its mean value), the correlation function which characterizes the statistical correlation between the instantaneous values taken at two arbitrary moments of time.

The random parameters of the phenomenol`ogcial model are denoted as $P_0(t)$ (random natural frequency); $N_0(t)$ (random reduced viscosity) and $K_{n0}(t)$ (random reduced plasticity). These parameters of the model can be exactly determined at any moment of time. The specific form $p_0(t)$, $n_0(t)$, $k_{n0}(t)$ which is assumed by the random parameters of the model $P_0(t)$, $N_0(t)$, $K_{n0}(t)$ and established as a result of experimental investigations is a realization of the corresponding random parameters.

A set of realizations of the parameters is obtained as a result of repeated investigations of the model parameters by measuring the characteristics of a real rock mass extracted from mine. The random parameters of the model can be fully characterized by an infinite set of realizations forming an ensemble. The random parameters can be estimated by one more indicator, namely by a cross section which is understood to mean the aggregate of the instantaneous values of the parameters given by the ensemble.

For the cross section of the random process corresponding to moment $t_1$, one can calculate the probability distribution of random parameters $P_0(t)$, $N_0(t)$, $K_{n0}(t)$

$$Q(P_0, t_1) = \lim \frac{i_{p_0}}{I_{p_0}}, \quad Q(n_0, t_1) = \lim \frac{i_{p_0}}{I_{n_0}}, \quad Q(k_{n_0}, t_1) = \lim \frac{ik_{n_0}}{I_{k_{no}}}$$

$i_{p_0}, i_{n_0}, i_{k_{no}}$ is the number of values of parameters $P_0(t)$;

$N_0(t), K_{no}(t)$ satisfies conditions

$$P_0(t_1) \le p_0; \quad N_0(t_1) \le n_0, \quad K_{no}(t_1) \le K_{no};$$

$I_{p_0} \; I_{n_0} \; I_{k_{no}}$ is total number of realizations.

For practical computations for sufficiently large $i_{p_0}, i_{n_0},$ and $i_{k_{no}}$ it can be assumed that

$$Q(p_0, t_1) \cong \frac{i_{p_0}}{I_{p_0}}$$

$$Q(n_0, t_1) \cong \frac{i_{n_0}}{I_{n_0}}$$

$$Q(k_{no}, t_1) = \frac{i_{k_{no}}}{I_{k_{no}}}$$

The probability density is represented by the derivatives of the random parameters of the model $P_0(t_1), N_0(t_1),$ and $K_{no}(t_1)$ with respect to the functions $p_0, n_0,$ and $k_{no}$, respectively and is determined by the formulas

$$q(p_0, t_1) = \lim_{\Delta p_0 \to 0} \frac{Q[p_0 < P_0(t_1) \le p_0 + \Delta p_0]}{\Delta p_0}$$

$$q(n_0, t_1) = \lim_{\Delta n_0 \to 0} \frac{Q[n_0 < N_0(t_1) \le n_0 + \Delta n_0]}{\Delta n_0}$$

$$q(k_{no}, t_1) = \lim_{\Delta k_{no} \to 0} Q[k_{no} < K_{no}(t_1) \le k_{no} + \Delta k_{no}]$$

In addition to the probabilistic characteristics $Q(p_0)$, $q(p_0)$, $Q(n_0)$, $q(n_0)$ and $Q(k_{no})$, $q(k_{no})$ of the random natural frequencies of the reduced viscosities and the reduced plasticity coefficients of the phenomenological model of the medium, the following numerical characteristics can be used:

Mean values of the random parameters

$$\bar{P}_0^{\,2}(t_1) = \int_{-\infty}^{\infty} p_0 q(p_0,\ t_1)\ dp_0$$

$$\bar{N}_0^{\,2}(t_1) = \int_{-\infty}^{\infty} n_0\ q(n_0,\ t_1)\ dn_0$$

$$\bar{K}_{no}(t_1) = \int_{-\infty}^{\infty} k_{no}\ q(k_{no},\ t_1)\ dk_{no}$$

The bars above $P_0(t_1)$, $N_0(t_1)$, $K_{no}(t_1)$ denote averaging operation of the random parameters over the parameter ensemble;

Mean values of the square of random parameters

$$\bar{P}_0^{\,2}(t_1) = \int_{-\infty}^{\infty} p_0^2 q(p_0,\ t_1)\ dp_0$$

$$\bar{N}_0^{\,2}(t_1) = \int_{-\infty}^{\infty} n_0^2 q(n_0,\ t_1)\ dn_0$$

$$\bar{K}_{no}^{\,2}(t_1) = \int_{-\infty}^{\infty} k_{no}\ q(n_0,\ t_1)\ dK_{no}$$

which characterize their intensity.

By subtracting from the random parameters $P_0(t_1)$, $N_0(t_1)$, $K_{no}(t_1)$ their mean values, we obtain the aligned centered random parameters of the model

$$P_0^0(t_1) = P_0(t_1) - P_0(t_1)$$

$$N_0^0(t_1) = N_0(t_1) - N_0(t_1)$$

$$K_{no}(t_1) = K_{no}(t_1) - K_{no}(t_1)$$

One more characteristic of the random parameters of the model, called variance, is obtained by taking the mean value of the square of the aligned random parameters

$$\sigma^2_{p_0}(t_1) = \int_{-\infty}^{\infty} p_0^2 \, q \, (p_0^0, \, t_1) \, dp_0$$

$$\sigma^2_{n_0}(t_1) = \int_{-\infty}^{\infty} n_0^2 \, q \, (n_0^0, \, t_1) \, dn_0$$

$$\sigma^2_{R_{no}}(t_1) = \int_{-\infty}^{\infty} k_{no}^2 \, q \, (k_{no}^0, \, t_1) \, dk_{no}$$

The most important characteristic of the random parameters of the model is the correlation function

$$R_{p_0}(t_1, t_2) = \int_{-\infty}^{\infty} \int_{-\infty}^{\infty} (p_{01} - P_{01})(p_{02} - P_{02}) \, q(p_{01}, p_{02}) \, dp_{01} \, dp_{02}$$

$$R_{n_0}(t_1, t_2) = \int_{-\infty}^{\infty} \int_{-\infty}^{\infty} (n_{01} - N_{01})(n_{02} - N_{02}) \, q(n_{01}, n_{02}) \, dp_{01} \, dp_{02}$$

$$R_{k_{no}}(t_1, t_2) = \int_{-\infty}^{\infty} \int_{-\infty}^{\infty} (k_{n01} - K_{n01})(k_{n02} - K_{n02}) \, q(k_{n01}, K_{n02})$$
$$dk_{n01}, dk_{n02}$$

Correlation functions characterize the rate of change of the random parameters.

The cross correlation function is a particularly important indicator characterizing the interrelation of various parameters of the model (natural frequencies, viscosity, and plasticity)

$$R_{p_0 n_0}(t_1, t_2) = \int_{-\infty}^{\infty} \int_{-\infty}^{\infty} (p_{01} - \bar{P}_{01})(n_{02} - \bar{K}_{n02}) \, q_2(p_{01}, n_{02}) \, dp_{01} \, dn_{02}$$

$$R_{p_0 k_{no}}(t_1, t_2) = \int_{-\infty}^{\infty} \int_{-\infty}^{\infty} (p_{01} - \bar{P}_{01})(k_{n02} - K_{n02}) \, q_2(p_{01}, k_{n02}) \, dp_{01} \, dn_{02}$$

$$R_{n_0 k_{no}}(t_1, t_2) = \int_{-\infty}^{\infty} \int_{-\infty}^{\infty} (n_{01} - \bar{N}_{01})(k_{n02} - \bar{K}_{n02}) \, q_2(n_{01}, k_{n02}) \, dn_{01} \, dk_{n02}$$

Two parameters of the model are noncorrelated if their cross correlation function is equal to zero for any values of the arguments. Noncorrelated parameters do not interact as far as energy is concerned.

## 2.2   VIBRATORY CONVEYANCE OF BULK LOADS

### 2.2.1 Phenomenology of Conveying on a Rigid, Vibrating, Load-carrying Element

The methods of mechanorheological phenomenology enable one to create phenomenological models for study of any types of vibratory conveyance of chunk and bulk loads: along rectilinear horizontal or inclined trough performing oscillations, lifting along a helical trough or vibrating vertical tube, loads bunkered in vibrating receptacle or discharged from a receptacle, and so on.

Figure 2.5 shows a three-dimensional (spatial) two-mass inertial model for study of vibratory conveyance over a trough which enables simulation of viscoelastoplastic properties of a bulk load along axes $x$, $y$, and $z$. Let us consider the general case of vibratory conveyance of a phenomenological model of the transported load along the carrying element of a vibratory machine performing vibrations that are directed at angle $\beta$ to the carrying element which, in turn, in inclined at angle $\alpha$ to the horizontal. Let us introduce a moving coordinate frame $xyz$ which is linked rigidly with the carrying element and two fixed coordinate frames: $\eta o \xi$ whose axis $\eta$ coincides with the direction of vibrations of the carrying element in case of rectilinear oscillations or with direction of the major axis of the ellipse in case of elliptical oscillations, and $x'y'z'$ whose axes are parallel to the axes of the moving coordinate system $xyz$.

The carrying element performs oscillations $\eta = f(\omega t)$ in the plane $\eta o \xi$ ($xoy$); hence, the projections of its displacement on the axis of the fixed coordinate frame will be

$$x' = \cos (\alpha + \beta)\eta, \quad y' = \sin (\alpha + \beta)\eta, \quad z' = 0$$

or, if it performs elliptical or nonharmonic oscillations along two mutually perpendicular axes, projections of its displacement on axis $x'$, $y'$, and $z'$ can be written as

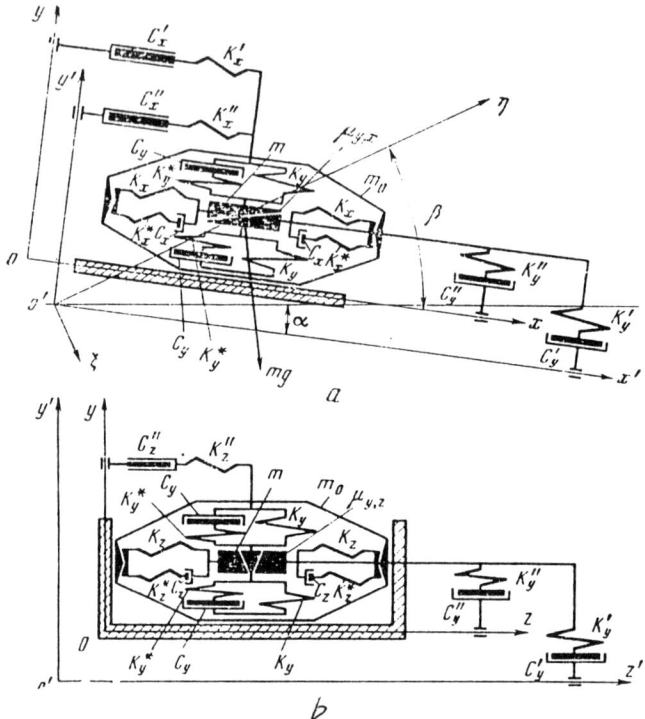

**Figure 2.5** Two-mass viscoelastoplastic phenomenological model of bulk load on the load-carrying element of a vibratory conveying machine. a) section of model by plane $xy$; b) section of model by plane $yz$

$$x' = x'(t), \quad y' = y'(t), \quad z' = 0$$

In the general case of vibratory conveyance there can develop the motion of the model in contact with the carrying element (combined motion) and a free motion. In the direction of axis $ox$, the model is regarded as a two-mass vibratory system with masses $m$ and $m_0$ that are linked together by two elastic elements of stiffnesses $k_x$, $k^*_x$ and two dampers with viscosity coefficient $c_x$. Furthermore, forces of dry friction act in the direction of axis $ox$ in the regime of combined motion, and in the regime of free motion, resistance forces are acting which are proportional to the velocity magnitude (elastic dampers with the coefficients of viscosity $c'_x$, $c''_x$ and stiffnesses $k'_x$, $k''_x$).

In the direction of axis $oy$, the model is a one-mass vibratory system, since mass $m_0$ in the regime of combined motion is stationary and the viscoelastic properties relative to the load-carrying element are modeled by two dampers with coefficients of viscosity $1/2$ $c_y$ and by two elastic elements $1/2$ $k_y$; the irreversible strains are simulated by a wedge element and a dry friction pair. In the regime of free motion masses $m$, $m_0$ are moving together along axis $oy$, they are overcoming viscous resistance forces which are proportional to the relative velocity (elastic damper with coefficients $c''_y$, $k''_y$) and velocity magnitude (elastic damper with coefficients $c'_y$, $k'_y$). The lateral outward pressure $N_z$ acts in the direction of the $z$-axis. It is assumed that $k^*_x = k^*_y = 0$, $N_z = 0$.

In the process of vibratory conveyance the forces acting on the load layer are: on the interval of combined motion – gravitational force $mg$, restoring forces of the elastic links $k_y y$ and $k_x (x - x_0)$, viscous resistance forces which are proportional to the relative velocities $c_y \dot{y}$ and $c_x (\dot{x} - \dot{x}_0)$ and also the forces of dry friction between the load particles $\mu_y k_x (x_i + y \, \text{tg} \, \alpha_0)$ and between the load and the conveying surface $\mu_x N_y^{(1,2)}$. In expression $\mu_y k_x (x_i + y \, \text{tg}\alpha_0)$, $x_i$ is the initial elastic strain of the layer and $\tan \alpha_0$ is a coefficient taking into account the compaction of the layer under compression.

The process of vibratory conveyance of the load is described by a system of differential equations compiled for each stage of motion and by a system of transcendental equations to determine transition moment from one stage of motion or strain to another.

When the model travels along axis $oy$, the following elastoplastic strains are present (in the stage of combined motion mass $m_0$ is stationary) $m\ddot{y} = -m\ddot{y}' - mg \cos \alpha - c_y \dot{y} -$

$k_y \ddot{y}$ (2.19)"

and plastic strains

$$m\ddot{y} = -m\ddot{y}' - mg \cos \alpha - \mu_y k_x (x_i + y \, \text{tg} \, \alpha_0) \qquad (2.20)$$

Here

$$N_y^{(1)} = - c_y \dot{y} - k_y y$$

$$(2.21)$$

$$N_y^{(2)} = -\mu_g k_x (x_i + y \, \text{tg} \, \alpha)$$

are the forces acting on the load during processes of viscoelastic and plastic strain, respectively. Force $N_y^{(1)}$ deforms the transported layer of the load (model), tending to cause irreversible strains (shifting the wedge). The transverse strains of the model will remain elastic up to the moment when the deforming force $N_y^{(1)}$ overcomes resistance to the shift of the wedge $N_y^{(2)}$.

The model is in contact with the carrying element until the normal reaction is reduced to zero

$$- N_y^{(1,2)} = \begin{pmatrix} k_y y + c_y \dot{y} \\ \mu_y k_x (x_i + y \, \mathrm{tg} \, \alpha_0) \end{pmatrix} = 0 \qquad (2.22)$$

At the stage of free motion masses $m$ and $m_0$ are displaced jointly

$$(m + m_0)\ddot{y}$$

$$= - (m + m_0)\ddot{y}{}' - (m + m_0)g \cos \alpha - c(\dot{y} + \dot{y}{}') \qquad (2.23)$$

When the model is displaced along axis $ox$, differential equations for masses $m$ and $m_0$ are compiled.

The motion of mass $m$ in the regime of joint and free motions is described by one equation

$$m\ddot{x} = - m\ddot{x}{}' + mg \cos \alpha - c_x (\dot{x} - \dot{x}_0) - k_x (x - x_0) \qquad (2.24)$$

Force $c_x(\dot{x} - \dot{x}_0) - k_x(x - x_0)$ acts in the plane of conveyance, it deforms the layer of the transported load, and acts on mass $m_0$.

For mass $m_0$, the following regimes are characteristic in the stage of joint motion

relative equilibrium

$$\ddot{x}_0 = 0 \qquad (2.25)$$

slip forward or backward

$$m_0 \ddot{x}_0 = - m_0 \ddot{x}{}' + m_0 g \sin \alpha - N_y^{(1,2)} \mu_x \mathrm{sign}(\dot{x}_0)$$

$$+ c_x (\dot{x} - \dot{x}_0) + k_x (x - x_0) \qquad (2.26)$$

The model on the carrying element stays in a relative rest if magnitude of the force which is moving mass $m_0$ does not exceed the ultimate force of static friction $\mu_{st} \, N_y^{(1,2)}$ (here, $\mu_{st}$

is coefficient of static friction on the carrying element). The force of dry friction changes its direction depending on the motion character of the load

$$\text{sign} \ (\dot{x}_0) \ = \ \begin{cases} + \ 1 \ \text{when} \ \dot{x}_0 \ > \ 0 \\ - \ 1 \ \text{when} \ \dot{x}_0 \ < \ 0 \end{cases}$$

Additionally, the force of dry friction also varies depending on magnitude of the normal reaction $N_y^{(1,2)}$. Therefore, the equation can be written as

$$m_0 \ddot{x}_0 = - \ m_0 \ddot{x}' + m_0 g \ \sin \ \alpha - \left( \frac{k_y y + c_y \dot{y}}{\mu_y k_x (x_H + y \ \text{tg} \ \alpha_0} \right) \mu_x \ \text{sign}(\dot{x}_0)$$

$$+ \ c_x \ (\dot{x} - \dot{x}_0) + k_x \ (x - x_0) \tag{2.27}$$

At the stage of free motion the equation for mass $m_0$ is

$$m_0 \ddot{x}_0 = - \ m_0 \ddot{x}' + m_0 g \ \sin \ \alpha - c_x' \ (\dot{x}_0 + \dot{x}') + c_x \ (\dot{x} - \dot{x}_0)$$

$$+ \ k_x \ (x - x_0) \tag{2.28}$$

If all the terms in the presented differential equations are divided by the coefficient for the highest order derivate and coefficient $\lambda$ is introduced which links masses $m$ and $m_0$, then

$$m = \lambda m_0 \tag{2.29}$$

At the stage of joint motion in the absence of slip in the friction pairs

$$\ddot{y} = - \ \ddot{y}' - g \ \cos \ \alpha - 2n_y \dot{y} - p_y^2 y$$

$$\ddot{x} = - \ \ddot{x}' + g \ \sin \ \alpha - 2n_x \ (\dot{x} - \dot{x}_0) - p_x^2 \ (x - x_0) \tag{2.30}$$

$$\ddot{x}_0 = 0$$

In the presence of slip

$$\ddot{y} = - \ \ddot{y}' - g \ \cos \ \alpha - \mu_y p_x^2 \ (x_H + y \ \text{tg} \ \alpha_0)$$

$$\ddot{x}_0 = - \ \ddot{x}' + g \ \sin \ \alpha + 2n_x \lambda \ (\dot{x} - \dot{x}_0) + p_x^2 \lambda \ (x - x_0) \tag{2.31}$$

$$- \left( \frac{p_y^2 y + 2n_y \dot{y}}{\mu_y p_x^2 (x_i + y \text{ tg } \alpha_0)} \right) \lambda \mu_x \text{ sign } (\dot{x}_0)$$

At the stage of free motion

$$\ddot{y} = - \ddot{y}' - g \cos \alpha - 2n_y' \dot{y} - 2n_y' (\dot{y} + \dot{y}')$$

$$\ddot{x}_0 = - \ddot{x}' + g \sin \alpha - 2n_x' \lambda (\dot{x}_0 + \dot{x}') + 2n_x \lambda (\dot{x} - \dot{x}_0) \qquad (2.32)$$

$$+ p_x^2 \lambda (x - x_0)$$

Here, $p_x$, $p_y$ are natural frequencies of the model of load in the direction of axes $ox$ and $oy$, $p_x = \sqrt{k_x/m}$, $p_y = \sqrt{k_y/m}$; $n_x$ and $n_y$ are damping factors resulting from the internal resistances to the motion of the model of the load-layer in the direction of axes $ox$ and $oy$ in the stage of joint motion, $2n_y = c_y/m$, $2n_x = c_x/m$; $n_y''$, $n_y'$, $n_x'$ are damping factors resulting from the external resistances to the motion of the model of load-layer at the stage of free displacement in the $oy$ and $ox$ directions, $2n_y'' = c_y''/(m + m_0)$, $2n_y' = c_y'/(m + m_0)$, $2n_x' = c_x/m$.

To determine the moment of transition from one regime of motion to another, transcendental equations are used, which determine the choice of the required equation and initial conditions for its solution. This is due to the fact that both analytical and computer-aided solutions are conducted by the fitting technique.

Expression

$$y = g/p_y^2$$

determines the static strain and is regarded as the initial condition for solving the equations.

The moment of transition from an elastic viscous strain of the model to plastic deformation $t_{ep}$ and back $t_{pe}$ are determined as a result of solution of the transcendental equation obtained by equating $N_y^{(1)}$ and $N_y^{(2)}$

$$k_y y + c_y \dot{y} \geq \mu_y k_x (x_i + y \text{ tg } \alpha_0)$$

$$k_y y + c_y \dot{y} < \mu_y k_x (x_i + y \text{ tg } \alpha_0) \qquad (2.33)$$

Transition from slip page of the model to stopping $t \pm 0$ takes place at moment $\dot{x} = 0$ provided that the static friction

friction forces at this moment are greater or equal to the force tending to transfer the load into the slip regime

$$\mu_{st} \ |N^{(1,2)}| \geq | - m_0 \ddot{x}' + m_0 g \sin \alpha + c_x \ (\dot{x} - \dot{x}_0)$$

$$+ k_x \ (x - x_0) \tag{2.34}$$

The reverse transition from stopping to slip takes place provided that the forces of static friction do not exceed the magnitude of the force tending to transfer the model into the slip regime

$$\mu_{st} \ |N^{(1,2)}| < | - m_0 \ddot{x}' + m_0 g \sin \alpha + c_x \ (\dot{x} - \dot{x}_0)$$

$$+ k_x \ (x - x_0) \tag{2.35}$$

The moment of transition from joint motion to free motion (moment of load separation $t_0$) is determined as a result of solving the following transcendental equation

$$N_y^{(1,2)} = 0 \tag{2.36}$$

At moment $t_p$ the model of load drops onto the load-carrying element and the collision phase commences. This moment is determined from the transcendental equation

$$y \ (t) = 0 \tag{2.37}$$

Analysis and solution of the differential equations of motion of an oscillatory system (analog of bulk load) are best conducted with the aid of analog and digital computers. Use of analog computers enables one to obtain the main indicators characterizing the load motion.

Let us study the physical mechanism of the process and establish correlations of parameters of vibratory conveyance with parameters of the vibration regime. Comparative analysis shows that under plastic strains a more intensive rise of flight angles $\delta_p$ is observed with increasing parameter $\Gamma = (A \ \omega^2 \sin \beta)/g$, than in the presence of only viscoelastic strains. Correspondingly, irregular regimes of motion are also commencing earlier (in the range $1.6 < \Gamma < 2.0$ as compared with $1.8 < \Gamma < 2.2$). This is reflected also in the correlation of the mean speed of the model motion and the flight angle. In the first case, maximum of the mean speed corresponds to flight angle 260-280°; in the second, to angle 230-250°. Hence, the speed of model of load motion under plastic strains is reduced.

To identify parameters of the phenomenological model for bulk loads, there were conducted studies of effects of model parameters (natural frequencies, damping factors of internal and external dampers, mass ratio) on the parameters of the conveyance process (flight angles, speed, trajectories of the particles of the transported load, and so on). Results of these studies were compared with the experimental data. The mean speed serves as the integral criterion determining similarity. The coincidence of the slip and flight angles, trajectories of the particles of the transported load and the model, normal reactions, etc., is a sufficient condition of similarity of the model and the real load.

Studies were conducted on the analog machine EMU-10 to determine effects of the model parameters on the vibratory conveyance process in a wide range of regimes – amplitude from 0.5 to 50 mm and frequency from 800 to 1400 osc/min. The dependence of the mean speed of vibration conveying on damping factors $2n_x$ and $2n_y$ at different natural frequencies of the model $p_x^2$ and $p_y^2$ was determined. It was shown that in the range of natural frequencies $p_x$ = 50-200 sec$^{-1}$ at damping factors $n_x$ = 200-500 sec$^{-1}$ and natural frequencies $p_y$ = 200-500 sec$^{-1}$ at damping factors $n_y$ = 200-500 sec$^{-1}$, the model provides good correlation with the experiment within a wide range of amplitudes and frequencies. Changing the natural frequency $p_y$ leads to the change of the mean speed of motion (speed is increasing with decreasing $p_y$). Thus, by changing natural frequency $p_y$, properties of various bulk loads can be modeled. Natural frequencies $p_x$ do not affect the mean speed.

It can be concluded that at damping factors less than $n_y$ = 100-500 sec$^{-1}$ and $n_x$ = 100-400 sec$^{-1}$, motion of the model becomes unstable and at low frequencies irregular regimes of motion develop. Changing the coefficients of friction of the dampers of the model in this range does not cause a variation of the mean speed, but causes a change of energy requirements for conveyance.

The effect of the mass ratio $\lambda$ of the model on the mean speed was studied for the natural frequencies of the model $p_y$ = 50 sec$^{-1}$ and $p_y$ = 200 sec$^{-1}$. Increasing the mass ratio leads to an increase of the conveyance speed. The model mass ratio

$\lambda$ takes into account the thickness of the transported layer of the load and the ratio of the masses directly participating and not participating in vibrations. The load layer transported in a small thickness can be simulated by a single mass model. With increasing load thickness transportability becomes worse and coefficient $\lambda$ must be assigned a lesser value.

The friction coefficient on the carrying element has significant influence on the speed. When this coefficient is increasing, the speed is also increasing.

Viscosity coefficients of the external dampers were found to be the determining factor for the process of vibratory conveyance of the model. Reduction of coefficient $n'_x$ leads to increase of speed; the more intensive is the regime, the more significant is this change. Reduction of damping factors of the internal dampers in the model at low frequencies causes considerable increase of the conveyance speed. This is due to the fact that at small internal viscous resistances of the model $(n_y < p_y)$, intensive separation already takes place at low frequencies, and at frequencies higher than 600 osc/min irregular motion sets in. Increasing the damping factors of the internal dampers shifts the region of initiation of irregular regimes to higher frequencies of the order of 800-900 osc/min which is typical for real bulk loads. The model with natural frequencies $p_y$ = 65-90 and damping factors $n_y$ = 200-500 correspond to bulk loads.

The criterion of identical behavior of the model and the layer of bulk load, which is defined as the equality of the mean speed of the model and the load, does not always express the identity of the true characteristics of the medium and the model. Therefore, as already mentioned, additional criteria are required. The most convenient criterion is the close similarity of the theoretical and experimental time histories of motion, i.e., the similarity of the trajectories of motion of both the model and the particles of the transported bulk load.

Analyzing behavior of vibratory conveyance of a model of a load-layer on an analog computer, the current characteristics of the process can be obtained and recorded on an oscillograph. Figure 1.12 shows oscillograms, obtained on the analog computer EMU-10, in which the displacement diagrams of both the model and the real load are presented.

Analysis of a series of similar diagrams encompassing both discontinuous and continuous regimes enabled one to draw the following conclusions. Time history of motion of the particles is changing with increasing distance from the bottom of the trough as a result of the viscoelastic properties of the layer. At relatively large layer height, attenuation of the pulses and the phase shift can be so substantial that the upper part of the load would travel significantly slower or would stop altogether. In some cases, even a reversed motion is observed. Such a phenomenon is largely due to the fact that the layer has different characteristics in the longitudinal and transverse directions. It is confirmed by the fact that the horizontal components of vibration are attenuated with distance from the bottom of the carrying element more intensively than the vertical components.

The motion of the load in the continuous regime becomes steady rapidly. In the discontinuous regime the particles are grouped in separate layers whose sizes vary in the same way as the time moments of beginning and end of the dropping and separation phases. However, this aperiodicity is relatively small and the law of motion is repeated after several periods so that the process can be regarded as steady. The same pattern is observed with vibratory conveyance of the model.

The method of comparing model motion trajectories obtained with the aid of an analog computer and the experimental laws of motion of the particles is such that in order to reduce the amount of computations, one regime is initially computed to determine the boundaries of the characteristics of the medium (model) giving the law of its displacement which coincides more exactly with the experimental law. This enabled the significant limitation of the range of variation of the coefficients of elastic and viscous resistances of the elements of the model for other regimes. The obtained characteristics of the model are regarded as reliable if relatively close correlation of the indicated laws of motion in the range of parameters observed in practice ($\Gamma$ = 0.8-2.0) is provided and is estimated by standard deviation (3-5%).

Comparison of the experimental and computational laws indicates that at a certain combination of elastic, viscous, and plastic characteristics of the model its mass travels identically with the motion of the real load layer. The selected parameters of the model of load for one regime of oscillations of the carrying element are acceptable also for the other vibration regimes in a wide range. The reliability of phenomenological models and their ability to reproduce the physical picture of the process also corroborates with the relatively close correlation of the theoretical and experimental laws of variation of the normal pressure and trajectory of motion in a direction perpendicular to the load-carrying element.

The conducted investigations to identify the phenomenological model and real bulk loads enable one to ascertain the following.

Conveyance is significantly affected by the mass ratio which takes into account the thickness of the transported load layer, the damping factors of the external viscous elements simulating aerodynamic resistance, and the friction coefficient.

The necessary condition for similarity of conveyance of the model and a certain bulk load is the equality of the mean speeds of their vibratory displacement. The mean speed is an integral and necessary indicator which determines the similarity of vibration displacement of the model and the load. The coincidence of the displacement trajectory of different particles of the load and the model or the reactions of the load on the load-carrying element is a sufficient similarity condition.

For actual bulk loads, damping factors of the internal viscous elements of the model of the order 200–500 $sec^{-1}$ are characteristic. In this range the natural frequencies of the elastic elements of the model can change in the range 60-200 $sec^{-1}$. Increasing the natural frequency above 200 $sec^{-1}$ does not make sense.

Identification of the phenomenological model of a finely dispersed load begins with the determination of the parameters of stiffness $k$ and viscosity $c$ of the load layer at the stage of joint motion as expounded below. By plotting the calculated values on an experimental curve of the angle of separation of the load layer vs. operating regime factor, the conditions under which the real layer of the real load is separated from the surface are determined.

It has been established that for materials with large content of finely dispersed particles at $\Gamma = 1.5$, which corresponds to $A = 4.9$ mm and $\omega = 76.5$ $sec^{-1}$, the parameters of the model are equal to $p_y = 255$ $sec^{-1}$, $n_y = 370$ $sec^{-1}$; at $\Gamma = 2.5$, which corresponds to $A = 5.5$ mm and $\omega = 94.2$ $sec^{-1}$, the parameters of the model are equal to $p_y = 270$ $sec^{-1}$, the parameters of the model are equal to $p_y = 270$ $sec^{-1}$ and $n_y = 370$ $sec^{-1}$; for quartz sand at $\Gamma = 1.9$, which corresponds to $A = 50$ mm and $\omega = 86$ $sec^{-1}$, the parameters of the model are equal to $p_y = 250$ $sec^{-1}$ and $n_y = 370$ $sec^{-1}$; for river sand at $\Gamma = 2.1$, which corresponds to $A = 6.0$ mm and $\omega = 82.5$ $sec^{-1}$, the parameters of the model are equal to $p_y = 260$ $sec^{-1}$ and $n_y$

= 310 $\sec^{-1}$. Comparison of the curves was made with respect to the standard deviation.

The parameters of viscous resistance at the flight stage of the load were found by comparing the experimental dependences of the angles of fall on the vibratory velocity with the data obtained from modeling on the analog computer MN-17M. Best correlation is attained for the curves corresponding to the values $n_y'' = 70$ $\sec^{-1}$ at $\Gamma = 1.5$ for materials with large content of finely dispersed particles. At $\Gamma = 2.5$, $n_y'' = 25$ $\sec^{-1}$ for quartz and river sand. Mathematical modeling on an analog computer was conducted on the assumption that the resistances of load displacements, which are proportional to the absolute speeds of layer motion, are equal to zero. Such simplification is stipulated by the fact that the effect of these resistances at the stage of flight are insignificant.

Based on the physicomechanical conceptions about the character of vibratory conveyance of a load layer, it was assumed that the stiffness parameter $p_x \ll p_y$. The relation between the stiffness parameters in the vertical and horizontal directions was assumed proportional to vibration damping in the layer. A more accurate value for $p_x$ can be determined in the process of identification of the model parameters under plastic strains along a trajectory of the normal reaction of the load layer.

Provided that tan $\alpha$ (parameter of layer porosity) is equal to 0.3-0.5, $x_i$ (parameter of the initial elastic strain which is commensurable with the dimensions of particles) is equal to 0.5-2.0 mm, $\mu_y$ (the coefficient of internal resistance) equal to 0.6, quantity $p_x$ assumes a value in the range 100-130 $\sec^{-1}$. The parameter of the model $\mu_x$ (coefficient of external friction) is assumed equal to the values for the actual granular load. At these values of parameters $p_x$, $x_i$, tan $\alpha$, and $\mu_y$, identification of parameters $\lambda$ and $n_x'$ was carried out.

In the process of vibratory conveyance it is necessary to provide such kinematic parameters of the conveyor at which the motion of the load layer is occurring mainly in the flight stage. The determining resistances in this case are viscous resistances which are proportional to the absolute speed of vibration conveying and are equal to the product of parameters $\lambda$ $n_x'$. By matching the experimental plots of the speed of conveying vs. the oscillation speed with the similar plots obtained by means of mathematical modeling on an analog computer the product $\lambda$ $n_x'$ can be found. For materials with large content of finely dispersed particles this product assumes the following values: at $\Gamma$ = 1.5 $\lambda$ $n_x'$ = 5, at $\Gamma$ = 2.5 $\lambda$ $n_x'$ = 13-15, for quartz and river sand $\lambda$ $n_x'$ are equal to 12-13 and 17, respectively.

The determination of the individual values of parameters $\lambda$ and $n_x'$ is carried out by using the same physical interpretation for parameters $\lambda$ and for criterion Sh. by representing $\lambda$ as 1/Sh we determine that for the studied materials the value of the parameter of energy pulse transmission is equal, respectively, to 0.76, 1.11, 1.16, 1.33.

The parameters of viscous resistances $n_x'$ under these conditions become equal to 6-7, 12-13, 10-12, and 12-13, respectively.

As analysis of the diagrams shows, the displacement of the identified model of the load layer in the regimes of motion with one flight stage over a period of oscillations of the carrying element is determined by its displacement at the flight stage. The displacement of the model at the stage of joint motion in forward direction is compensated for by the displacement in the reverse direction causing only wear of the carrying surface.

In the selection process of the kinematic parameters of vibratory conveying machines for the movement of abrasive granular materials, it is important to account for abrasive action of the load.

For wear assessment, conveying of the model over the slippage sections during the stage of the joint motion was anlayzed. As a result, it is established that with increasing oscillation amplitude at a constant value of the frequency the total path of the granular loads on the slip sections over a period of the carrying plane oscillations is decreasing (for example, 120, 108, 104$^\circ$ at oscillation amplitude of 5, 6, and 7 mm, respectively) and, consequently, abrasion is reduced. Thus, from the standpoint of wear, the regimes with large amplitudes are more beneficial.

## 2.2.2 Phenomenology of Conveying on a Carrying Element Performing Wave Oscillations

A flexible carrying element can perform oscillations with the purpose of load conveying. In this case the oscillations are transmitted by special drive devices or such oscillation can arise independently, for example, during the motion of the carrying element of a belt conveyor along the supporting idlers. In both cases the transported load is subjected to the action of wave perturbations.

Devices of the first category are wave action conveyors and conveyors with a flexible carrying element belong to the second category.

**Wave conveying**. The wave conveying principle is realized most often in devices that are elastic resonators in which longitudinal and transverse travelling waves are excited. Usually, these are miniature devices in which the resonator is made in the form of steel elements whose configuration is selected in accordance with the operating requirements of the wave device (microdisplacement, orientation, and so on) [6]. Oscillations in the resonator are excited by piezoelements or other exciters. Longitudinal and transverse waves can be excited in mutually perpendicular directions which enables displacement of the transported piece in the same direction. Such devices are used for the purpose of accurate orientation of various parts. Excitation of the resonator is effected either kinematically or by impact.

For those branches of industry where it is necessary to move large masses of loads over considerable distances wave conveying devices of different design are being developed.

The principal feature of such a wave conveyor is in the fact that it is an elastic belt (of the type used in ordinary conveyors) without any supports, across which longitudinal and transverse waves are propagating. It is placed directly on the mine. The load on the belt moves, just as on an ordinary vibratory conveyor but on fundamentally different principles, as a result of the vibrations of the surface of the belt. The conveyance speed is close to the speed of wave propagation, if necessary vibration parameters are maintained.

The second operating regime of the wave conveyor is realized in a more complex manner; however, it facilitates the significant increase of output up to the velocity of the travelling wave [7]. Hydrovibrators, magnetostriction elements and built-in miniature electromagnetic vibrators are used as drives for wave conveyors. More advanced principles of excitation are also being developed. The drive system can be vulcanized directly into the belt. The wave conveyor is essentially balanced, it does not require special supports, and allows bending in all directions. Erection of such a conveyor on a mine face is extremely simple and is reduced to spreading the belt and connecting it to the power source.

The configuration of the travelling waves on the conveying surface is determined by the parameters of the longitudinal and transverse travelling waves whose equations have the form

$$y' = A_y \cos (\omega t - kx), \quad x' = A_x \cos (\omega t - kx + \gamma)$$

where $A_y$, $A_x$ are amplitudes of the longitudinal and transverse vibrations; $\omega$ and $\gamma$ are angular frequency and phase shift of vibrations; $k$ is the wave number ($k = \omega/v_{ph}$); $v_{ph}$ is the phase wave velocity ($v_{ph} = \lambda\omega/2\pi$); $\lambda$ is the wave-length.

The configuration of the longitudinal and transverse travelling waves on the conveying surface depends on the parameters of the transverse and longitudinal components of the oscillations and on the phase shift between them. By changing parameters of these components, one can obtain the most diverse configurations of the travelling waves on the conveyor surface.

Not only the parameters, but also the character of the process are substantially dependent on configuration of waves on the conveying surface. This can result in vibratory conveyance regime, conveying with wave velocity and so on. The character of the process is affected by the parameters of both the transverse and longitudinal oscillations; the conveying speed is mainly affected by the magnitude and distribution of the velocity of longitudinal oscillations over the wave surface of the conveying element.

As follows from the loci of the speeds of the conveying wave surface in the absence of phase shift between the components of the oscillations, velocity of the longitudinal oscillations on the outgoing wave surface is directed in the direction opposite to the direction of wave propagation, and on the incoming wave surface the direction is towards propagation. At a phase shift of $90^o$ between the components in the upper part of the wave, the longitudinal vibratory velocity is directed towards wave propagation, and the lower part is directed in the opposite direction. The velocity of longitudinal vibrations reaches maximum on the wave crest.

In using such a regime of longitudinal and transverse oscillations of the travelling wave, when its crest moves with an amplitude velocity, it becomes possible to transport elongated objects with a speed equal to the velocity

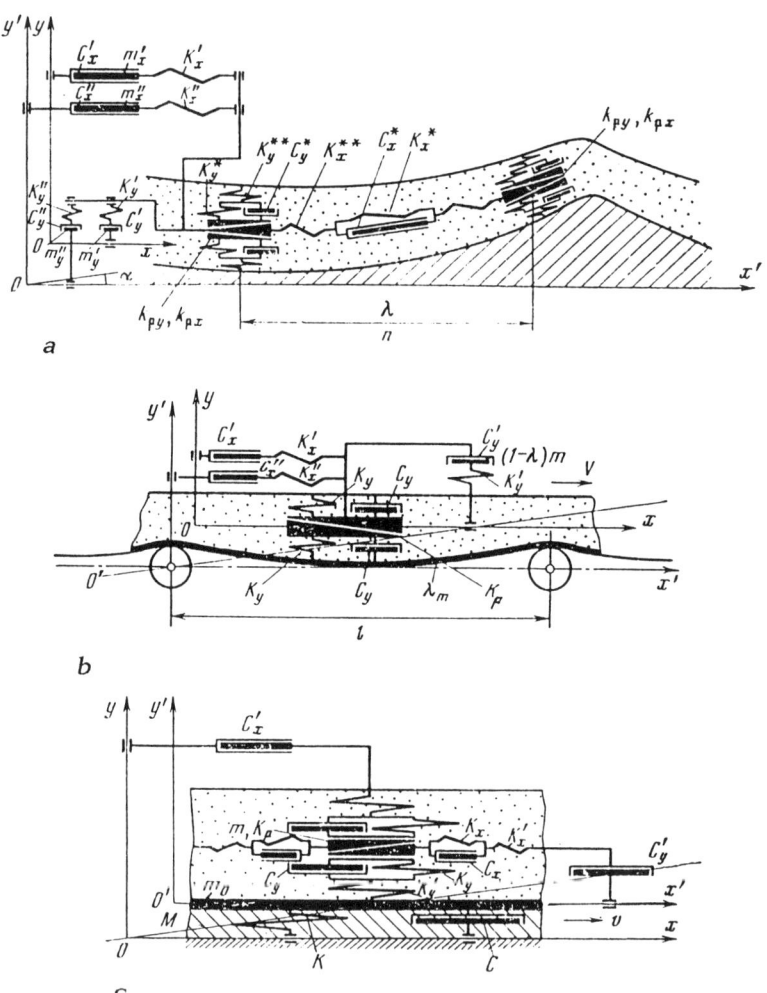

**Figure 2.6** Phenomenological models for investigation of special regimes of vibratory conveyance. a) wave conveying; b) belt conveyor; c) vibrating belt; d) vibrotraction; e) vibropneumatic and vibrohydraulic.

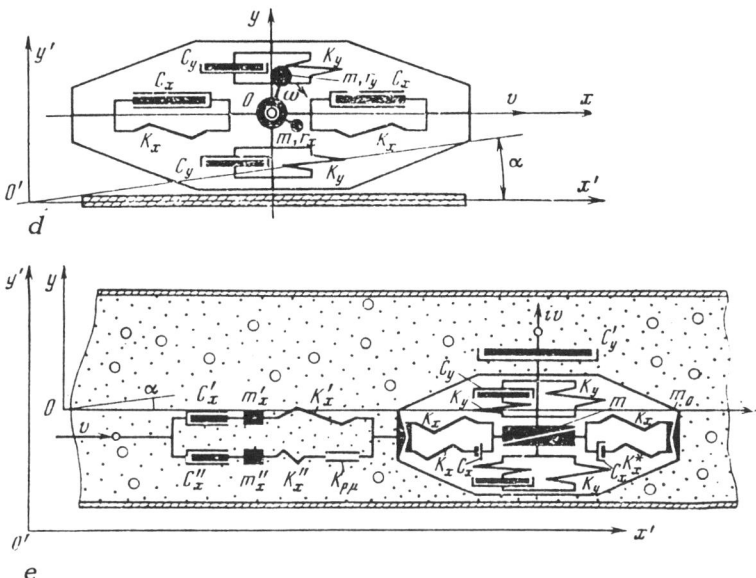

**Figure 2.6** continued.

amplitude of the longitudinal oscillations of the travelling wave.

Generally, the speed of travel of elongated objects (in the presence of some slippage relative to the conveying surface) is determined by expression

$$v = A_x \left(1 - \frac{\dot{x}}{v_{ph}}\right) \tag{2.38}$$

where $\dot{x}$ is the speed of motion of the load relative to the conveying surface.

Let us consider the general case of wave conveyance of a viscoelastoplastic inertial phenomenological model of a monolayer of a transportable load over a carrying surface performing longitudinal and transverse oscillations (Fig. 2.6a). Let us introduce the moving frame of the rectangular coordinates $xoy$ which is rigidly linked with the conveying surface of the carrying element. Along with the moving frame we use the stationary coordinate system $x'o'y'$ parallel to the moving axes.

The transported load is simulated by a two-mass viscoelastoplastic phenomenological model with a concentrated mass $m^*$. The viscoelastic properties of the load are modeled by elastic rheological bodies with the stiffness coefficients $k_y^*$, $k_y^{**}$, $k_x^*$ $k_x^{**}$, by viscous rheological bodies with the viscosity coefficients $c_y^*$, and $c_x^*$, and by plastic rheological bodies with plasticity coefficients $k_{py}$, $k_{px}$. The resistances of the medium caused by displacements of the particles of the load in constrained conditions are simulated by inertial viscoelastic rheological bodies with elasticity coefficients $k_y'$, $k_y''$, $k_x'$, $k_x''$, viscosity coefficients $c_y'$, $c_y''$, $c_x'$, $c_x''$, and masses $m_y'$, $m_y''$, $m_x'$, and $m_x''$. The viscoelastic link between the concentrated load masses located at a distance of $1/n\ \lambda$ is introduced to account for the longitudinal strains of the load on the conveying element of the wave conveyor. If the aim is to investigate these strains, then one can limit oneself to considering a single-mass phenomenological model of load by assuming $k_x^* = k_x^{**} = c_x^* = 0$.

In the investigations conducted below a number of coefficients are assumed to be equal to zero, particularly $m_y' = m_y'' = m_x' = m_x'' = 0$.

In the process of displacement of the load by a wave conveyor, the load will be subjected to viscoelastic and viscoelastoplastic strains (for the simplification of the problem we shall assume that instead of viscoelastoplastic strains only purely plastic strains are present), slip over the surface of the carrying element, and perform free motion (flight).

Let us analyze the behavior of wave-induced conveyance, starting with some special regimes.

We rearrange the expressions of the time history of oscillations of the carrying element. Assuming that $x = \dot{x}t$ and $k = \omega/v_{ph}$, we obtain $kx = (\dot{x}/v_{ph})\ \omega t$.

Substituting the value $kx$ from the expression for the vibratory velocity, we obtain a new expression for the law of propagation of the travelling waves across the carrying element

$$y' = A_y \cos \left(1 - \frac{\dot{x}}{v_{ph}}\right) \omega t$$

$$x' = A_x \cos \left[\left(1 - \frac{\dot{x}}{v_{ph}}\right) \omega t + \gamma\right]$$

If the wave propagates across the carrying element with a speed significantly exceeding the speed of load conveyance, then it can be assumed, without significant losses of accuracy, that $\dot{x}/v_{ph} = 0$. Then, from the obtained expressions we determine the time history of oscillations of the carrying element.

$$y' = A_y \cos \omega t$$

$$x' = A_x \cos (\omega t + \gamma)$$

Thus, when $\dot{x}/v_{ph} = 0$, we obtain the usual harmonic oscillations which do not take into account the presence of the travelling wave. In this case, the calculation of the wave conveying installation is reduced to the calculation of a vibratory installation and can be conducted by the method expounded in section 2.2.1 of the present chapter.

For approximate calculations it can be assumed that $\dot{x} = v_{ph}$. In this case the time history of oscillations of the working element can be written in the following form

$$y' = A_y \cos (1 - \rho) \omega t$$

$$x' = A_x \cos [(1 - \rho) \omega t + \gamma]$$

where $\rho$ is the coefficient of velocity ratio, $\rho = v_m/v_{ph}$. Coefficient $\rho$, depending on the signs of velocities $v_m$ and $v_{ph}$, can have both positive and negative values (if the load velocity and wave travel are in different directions).

The obtained expressions show that the presence of the travelling wave over the surface of the carrying element causes a change of the effective oscillation frequency. If the wave and the load are travelling in the same direction, the effective frequency is decreasing. If they travel in opposite directions, the effective frequency is increasing. At commensurable speeds of travel of the load and the wave, the effect of factor $\rho$ can be substantial. For example, in the case of the wave and the load moving in opposite directions with close speeds at low frequencies of oscillations of the carrying element, the frequencies acting on the load can be practically doubled. In some cases, such a conveying regime can be of practical significance.

Thus, if the carrying element oscillates according to a time history characterized by the last formulas, the calculation of wave conveyance can be conducted using the method of section 2.2.1 of the present chapter.

Let us now develop an exact solution for the relationships of conveying load which is identified with the viscoelastoplastic inertial phenomenological model.

The above enumerated characteristic regimes of strain and motion of the load on a wave conveying plant are described by the following system of differential equations.

Viscoelastic strains of the load

$$m\ddot{y} = A_y \left(1 - \frac{\dot{x}}{v_{ph}}\right)^2 \omega^2 \cos\left(1 - \frac{\dot{x}}{v_{ph}}\right)\omega t - mg \cos \alpha$$

$$- k_y^* y - k_y^{**} e^{-\frac{k_y^{**}}{c_y^*}t} \int \dot{y} e^{\frac{k_y^{**}}{c_y^*}t} dt - k_y'' e^{-\frac{k_y'}{c_y''}t} \int \dot{y} e^{\frac{k_y'}{c_y''}t} dt$$

$$m\ddot{x} = A_x \left(1 - \frac{\dot{x}}{v_m}\right)^2 \omega^2 \cos\left[\left(1 - \frac{\dot{x}}{v_m}\right)\omega t + \quad + mg \right. \qquad (2.39)$$

$$- k_x^* (x - x^*) - k_x^{**} e^{-\frac{k_x^{**}}{c_x^*}} \int (\dot{x} - \dot{x}^*) e^{\frac{k_x^{**}}{c_x^*}} dt$$

$$- k_x' e^{-\frac{k_x'}{c_x'}t} \int \left\{ \dot{x} - \left(1 - \frac{\dot{x}}{v_m}\right) \omega \sin\left[\left(1 - \frac{\dot{x}}{v_m}\right)\omega t + \gamma\right]\right\} e^{\frac{k_x'}{c_x'}t}$$

where $x^*$ and $\dot{x}^*$ are the displacement and speed of the second mass in the phenomenological model of the load.

During the process of the viscoelastic strain the following force acts on the carrying element from the displaced load

$$F_y\text{ve} = - k_y^* y - k_y^{**} e^{-\frac{k_y^{**}}{c_y^*}t} \int \dot{y} e^{\frac{k_y^{**}}{c_y^*}t} dt$$

$$\qquad\qquad\qquad\qquad\qquad\qquad\qquad\qquad\qquad (2.40)$$

$$F_x\text{ve} = - k_x^* (x - x^*) - k_x^{**} e^{-\frac{k_x^{**}}{c_x^*}t} \int (\dot{x} - \dot{x}^*) e^{\frac{k_x^{**}}{c_x^*}t} dt$$

Viscoelastic strain of the displaced load will continue until the following condition is satisfied

$$F_y\text{ve} \geq k_{py} x_i$$

$$F_x\text{ve} \geq k_{px} y_i$$

After the first condition is complied with, the plastic strains of the phenomenological model of the load set in and they are described by the differential equations

$$m\ddot{y} = A_y\left(1 - \frac{\dot{x}}{v_m}\right)^2 \omega^2 \cos\left(1 - \frac{\dot{x}}{v_m}\right)\omega t - mg\cos\alpha - k_{py}(x_i + y)$$

$$- k_y'' e^{-\frac{k_y''}{c_y''}t} \int \dot{y}e^{\frac{k_y''}{c_y''}t}\, dt$$

$$m\ddot{x} = A_x\left(1 - \frac{\dot{x}}{v_m}\right)^2 \omega^2 \cos\left[\left(1 - \frac{\dot{x}}{v_m}\right)\omega t + \gamma\right] + mg\sin\alpha$$

$$- k_{px}(y_i + x) - k_x'' e^{-\frac{k_x''}{c_x''}t}\ \int \left\{\dot{x} - A_x\left(1 - \frac{\dot{x}}{v_m}\right)\omega\sin\right.$$

$$\left. \times\left(1 - \frac{\dot{x}}{v_m}\right)\omega t\right\}e^{\frac{k_x''}{c_x''}t}\, dt$$

(2.41)

During the process of plastic strain, the following force acts on the carrying element from the displaced load

$$F_{py} \geq -k_{py}(x_i + y)$$

$$F_{px} \geq -k_{px}(y_i + x)$$

The moment of transition from plastic strain to viscoelastic strain is determined by the condition

$$F_{py} \leq F_{vey}$$

$$F_{px} \leq F_{vex}$$

The load is in contact with the carrying element until the following condition is satisfied

$$F_{(ve, p)y} = 0$$

After this condition is satisfied, free motion (flight) of the load commences, which is described by equation

$$m\ddot{y} = mA_y\left(1 - \frac{\dot{x}}{v_m}\right)^2 \omega^2 \cos\left(1 - \frac{\dot{x}}{v_m}\right)\omega t - mg\cos\alpha$$

$$- k_y'' e^{-\frac{k_y''}{c_y''}t} \int \dot{y}e^{\frac{k_y''}{c_y''}t}\, dt$$

$$m\ddot{x} = mA_x\left(1 - \frac{\dot{x}}{v_m}\right)^2 \omega^2 \cos\left[\left(1 - \frac{\dot{x}}{v_m}\right)\omega t + \gamma\right] + mg\sin\alpha$$

$$- k''e^{-\frac{k''_x}{c''_x}t} \int \left\{ \dot{x} - A\left(1 - \frac{\dot{x}}{v_m}\right) \omega \sin\left[\left(1 - \frac{\dot{x}}{v_m}\right)\omega t + y\right]\right\} e^{\frac{k''_x}{c''_x}t} dt \qquad (2.42)$$

The free motion of the load comes to an end and joint motion commences when condition $y = 0$ is satisfied.

The presented equations and the system of logical conditions enable one to investigate conveyance in the regime of wave oscillations.

**Conveying by flexible carrying element.** In the process of transporting bulk loads by means of a flexible carrying element the load is continuously "stirred" and displaced relative to the conveying surface. The energy expended in the process of these motions is used up in abrading the belt, crushing of the transported load, and in heating of both belt and load.

At the same time redistribution of the transported load by grain size takes place. It is established by experimental investigations that in the process of motion over the idlers, the flexible carrying element performs transverse and longitudinal oscillations. The main generating mechanism for these oscillations is interaction of the load and the carrying element when they move in wave trajectory formed as a result of the sag of the loaded carrying element between idlers. Thus, the trajectory of motion of the load is the line of sag of the carrying element on the idlers to which oscillations of different forms are superimposed. These oscillations are caused by the interaction of the load with the driving element and the idlers.

The motion of the carrying element can be represented as a translational displacement with constant speed $v$ to which longitudinal and transverse periodic oscillations are superimposed. Their time histories are formed from the displacements along a wavy trajectory between the rollers $x'$ and $y'$ and oscillations $x''$ and $y''$ resulting from interaction between the transported load with the carrying element. It ought to be mentioned that these interactions are very significant, since the mass of the load can exceed the mass of the carrying element. Assuming that in the given case the principle of superposition is valid, we find that the carrying element performs periodic oscillations $x^* = x'(t) + x''(t)$ and $y^* = y'(t) + y''(t)$ where functions $x'(t)$ and $y'(t)$ are known.

Experiments indicate that a wide range of low- and relatively high-frequency harmonics is contained in the vibration spectrum of the carrying element.

For the simplification of the solution of the problem, and taking into account that the mechanics of the conveying process is of principal interest to us, the part of the carrying element under consideration is modeled as a lumped parameter system with mass $M$, elastic elements $K_x$, $K_y$, and viscous elements $C_x$, $C_y$. The transported load is simulated by a two-mass viscoelastoplastic model with masses $m$ and $m_1$, elastic elements $k_y'$, $k_y$, $k_x$ and viscous elements $c_y$, $c_x$. The resistances to the layer motion resulting from the presence of aerodynamic effects are modeled by viscous elements with the coefficients $c_x'$ and $c_y'$.

In accordance with the above statement the differential equations of load motion along the section where viscoelastic strains occur, are

$$m\ddot{y} + c_y\dot{y} + (k_y + k_y') y = k_y'y^* + c_y\dot{y}_1 + k_yy_1$$
$$m_1\ddot{y}_1 + (c_y + c_y') \dot{y}_1 + k_yy_1 = (c_y + c_y') y + k_yy \qquad (2.43)$$
$$m_1\ddot{x}_1 + (c_x + c_x') \dot{x}_1 + k_xx_1 = c_x\dot{x} + k_xx - c_x'v$$

It is assumed here that the near-wall layer of the load of mass $m$ moves jointly with the carrying element in the direction of conveying. However, according to the method expounded in section 2.2.1, the slip over the carrying element can be considered.

The dynamic equations of the load-carrying element have the form (Fig. 2.6b, c)

$$M\ddot{y}^* + C_y\dot{y}^* + (K_y + k_y') y^* = K_yy' + C_y\dot{y}' - k_y'y$$
$$M\ddot{x}^* + (C_x + c_x) \dot{x}^* + (K_x + k_x) x^* = C_xv - K_xvt - c_x\dot{x}_1 \qquad (2.44)$$
$$- k_xx_1 - \mu_rN_y$$

where $y'$ is the displacement of the lower surface of the carrying element determined by the location of the idlers and belt deflection (sag) lines between them; $N_y$ is the pressure on the supporting idlers during the travel along them of the carrying element with a load; $\mu_r$ is the effective coefficient of friction of the idlers.

The pressure on the idlers is determined from expression

$$N_y = C_y (\dot{y}^* - \dot{y}') + K_y (y^* - y')$$

the pressure of the load on the carrying element is equal to

$$N'_y = k'_y (y + y^*)$$

the pressure inside the layer of the transported load is

$$N''_y = c_y (\dot{y}_1 - \dot{y}) + k_y (y_1 - y)$$

$$N''_x = c_x (\dot{x}_1 - x) + k_x (\dot{x}_1 - x)$$

Energy expenditure on overcoming resistance in the idlers when the loaded carrying element moves over them can be determined from expression

$$W_r = \frac{1}{T} \int_0^T [C_x (v - \dot{x}^*) + K_x (vt - x^*)] (v - \dot{x})^* \, dt$$

Energy expenditures on "stirring" the load in the process of conveying are determined by expression

$$W_9 = \frac{1}{T} \left[ \int_0^T N'_y (\dot{y} - \dot{y}^*) \, dt + \int_0^T N''_y (\dot{y}_1 - \dot{y}) \, dt + \int_0^T N''_x (\dot{x}_1 - \dot{x}) \, dt \right]$$

It is also not difficult to determine energy losses associated with the aerodynamic resistances of the transported bulk load. It must be born in mind that these resistances in high-speed units can be considerable.

Thus, the developed computational method enables one to estimate energy expenditures on "stirring" the load and the level of its crushing during the conveying process. By analyzing the effect of various factors on their intensity, one can solve the problem of selection of the optimum (from the standpoint of the minimum energy losses) regimes of conveying bulk loads by belt conveyor installations.

### 2.2.3 Phenomenology of Conveying Using Combination Transport Means

In order to expand the sphere of application and increase the efficiency of vibratory conveying machines while preserving such advantages as the absence of wear of the carrying element and crushing (reduction of grain size) of the transported load, low resistances to motion, and so on, a number of combination transport means have been developed.

In combination conveying machines, among which the present section will consider vibropneumatic, vibrohydraulic, vibrotraction, and vibromagnetic machines, vibrations are used mainly to reduce the resistance to conveying. Displacement is performed by additional means, namely by aerodynamic or hydrodynamic jet pressure, special driving weight, or forces of magnetic attraction.

In all such combination conveyors, a considerable increase of the conveyance speed is feasible, in excess of the attainable speed with purely vibratory techniques, due to the presence of considerable moving forces.

The specific feature of vibropneumatic conveying is the dependence of the aerodynamic head of the air jet on the speed of motion of the transported load. With vibromagnetic conveying the moving force is independent of the speed of load motion and is determined by its position relative to the electromagnet and the operating regime of the latter. The traction force of vibrotraction plants is dependent on the speed and elasticity of the tractive element.

In vibropneumatic conveying, the motion of the transported load is realized in the general case by the dynamic and static pressures of the air jet. In the process of motion over the load-carrying element the air expands and the velocity of the air jet increases. However, due to a limited length of the vibratory conveyors, the pressure drop is small and the increase of air jet velocity can, as a first approximation, be disregarded along the length of the load-carrying element. Thus, at low velocities of air motion, encountered in the process of vibropneumatic conveying, the lift force of the air jet is negligible and its static head (pressure) can be expressed in terms of the dynamic pressure in conditions of constrained motion. In vibrohydraulic conveying the jet velocity along the length of the load-carrying element remains unchanged.

Taking the aforementioned into account, let us consider characteristics of aero- and hydrodynamic conveying in the presence of vibration field, especially of vibropneumatic and vibrohydraulic conveyance. The phenomenological process of pneumatic (hydraulic) conveying in a vibration field is presented in Fig 2.6$d$. Let the transported load be acted upon by the fluctuating flow of the carrying medium moving with a speed of $v = V [1 + \theta f (\omega t + \varphi)]$ [8].

The equations of motion of the load in the pulsating flow of the medium in the presence of a vibration field have the form

viscoelastic strains

$$\ddot{y}' + 2n_y\dot{y}' + (p_y^2 + p_y'^2)y' = -\ddot{\bar{y}}' - g \cos a$$

$$\ddot{y}'' + 2n_y\dot{y}'' + (p_y^2 + p_y'^2)y'' = -\ddot{\bar{y}}'' - g \cos a \qquad (2.45)$$

plastic strains

$$\ddot{y}' = -\ddot{\bar{y}}' - f_p - g \cos \alpha$$

$$\ddot{y}'' = -\ddot{\bar{y}}'' - f_p - g \cos \alpha \qquad (2.46)$$

on the motion segments with loss of contact

$$\ddot{y}' + 2n_y\dot{y}' + p_y^2 y' = \ddot{\bar{y}}' - g \cos \alpha$$

$$\ddot{y}'' + 2n_y\dot{y}'' + p_y^2 y'' = \ddot{\bar{y}}'' - g \cos \alpha \qquad (2.47)$$

slippage

$$\ddot{x} + (2n_x + 2n'_x)\dot{x} + p_x^2 x = -\ddot{\bar{X}} + p_x^2 V [1 + \theta f(\omega t + \varphi)]t$$

$$+ 2n_x V [1 + \theta f(\omega t + \varphi)] - \text{sign}(\dot{x})\mu N_y - g \cos \alpha \qquad (2.48)$$

Here, the following notations are adopted (Fig. 2.6d)

$$p_y^2 = \frac{k_y}{m_y}, \quad p_y'^2 = \frac{k_y'}{m_y}, \quad 2n_y = \frac{c_y}{m_y}$$

$$p_x^2 = \frac{k_x}{m_x}, \quad p_x'^2 = \frac{k_x'}{m_x}, \quad 2n_x = \frac{c_x}{m_x}, \quad 2n'_x = \frac{c_x'}{m_x}$$

$N_y$ is the normal reaction of the load on the walls of the carrying element; $\bar{Y}'$ and $\bar{Y}''$ are the displacement of the lower and upper walls of the carrying element in the direction of axis $y$; $\bar{X}$ is the displacement of the carrying element in the direction of axis $x$.

Calculations carried out using presented equations when $\bar{Y}' = A_y \sin \omega t$, $\bar{Y}'' = -A_y \sin \omega t$, $\bar{X} = 0$, $v = -V$, $\alpha = 90°$ enable one to obtain all the characteristics of the process of conveying.

Load conveying can also be effected by the combination vibratory-pulling method. Between the transported load of mass $m_1$ in the transport device and the ground there is an elastic system characterized by the stiffness coefficient $K_y$ and coefficient of viscous resistances $C_y$. Mounted on the working section of the machine is an inertial vibrator having unbalanced masses $m'$ and $m''$ with eccentricity $r$. The masses rotate in opposite directions with an angular velocity $\omega$ (Fig. 2.6$d$).

The transport system has a pulling drive system characterized by stiffness $K_X$ and viscous resistances $C_X$. Displacement occurs with a mean speed $v$ and acceleration $a$. In the general case, the transport system can travel with a periodic loss of contact with the ground. The motion of the machine is described by a system of differential equations

$$\ddot{y} + 2n_y\dot{y} + p_y^2 y = -g\cos\alpha + q_y r\omega^2 \sin\omega t$$
$$\ddot{y}^* + 2n_y'\dot{y}^* = -g\cos\alpha + q_y r\omega^2 \sin\omega t$$

(2.49)

The asterisk corresponds to the motion of the machine while separated from the ground.

$$\ddot{x} + 2n_x\dot{x} + p_x^2 x = -g\sin\alpha + q_x r\omega^2 \sin(\omega t + \theta)$$
$$+ p_x^2\left(\frac{1}{2}\,at^2 + vt\right) + 2n_x(at + v) - \mu(p_y^2 y^* + 2n_y^*\dot{y}^*)$$

(2.50)

$$\ddot{x}^* + 2n_x^*\dot{x}^* = -g\sin\alpha + q_x r\omega^2 \sin(\omega t + \theta) + p_x^2\left(\frac{1}{2}\,at^2 + vt\right)$$
$$+ 2n_x(at + v)$$

The following notations are adopted here: $q_y$ and $q_X$ are the rations of the rotating and the total mass of the machine,

$$q_y = \frac{m' - m''}{m_1 + m' + m''}, \qquad q_x = \frac{m' + m''}{m_1 + m' + m''}$$

$p_y$ and $p_X$ are the natural frequencies of the vibratory machine along axes $y$ and $x$

$$p_x^2 = \frac{K_x}{m_1 + m' + m''}, \qquad p_y^2 = \frac{K_y}{m_1 + m' + m''}$$

$n_X$ and $n_y$ are effective coefficients of viscous resistances of the elastic links of the vibratory machine

$$2n_x = \frac{c_x}{m_1 + m' + m''}, \qquad 2n_y = \frac{c_y}{m_1 + m' + m''}$$

$n_x^*$ and $n_y^*$ are effective coefficients of viscous resistances of the vibratory machine at the free motion segment,

$$2n_x^* = \frac{c_x'}{m_1 + m' + m''}; \qquad 2n_y^* = \frac{c_y'}{m_1 + m' + m''}$$

θ is the phase shift between the components of the disturbing vibrator force in the direction of axes $y$ and $x$; $\mu$ is the coefficient of friction between the vibratory machine and the ground.

For the most characteristic regime of motion without separations from the ground, the displacements and speeds are determined by the following expressions

$$y = -\frac{g \cos \alpha}{p_y^2} + q_y r \frac{\omega^2}{\sqrt{4n_y^2\omega^2 + (p_y^2 - \omega^2)^2}} \sin(\omega t + \varphi_y)$$

$$\dot{y} = q_y r \omega \frac{\omega^2}{\sqrt{4n_y^2\omega^2 + (p_y^2 - \omega^2)^2}} \cos(\omega t + \varphi_y)$$

$$x = -\frac{g(\sin \alpha + \mu \cos \alpha) + a}{p_x^2} + \frac{1}{2} at^2 + vt$$

$$+ \mu q_x r \frac{\omega^2 \sqrt{p_y^2 + 4n_y^2\omega^2}}{\sqrt{[(p_x^2 - \omega^2)^2 + 4n_x^2\omega^2][4n_y^2\omega^2 + (p_y^2 - \omega^2)^2]}} \sin(\omega t + \varphi_x)$$

$$\dot{x} = at + v +$$

$$+ \mu q_x r \omega \frac{\omega^2 \sqrt{p_y^2 + 4n_y^2\omega^2}}{\sqrt{[4n_x^2\omega^2 + (p_x^2 - \omega^2)^2][4n_y^2\omega^2 + (p_y^2 - \omega^2)^2]}} \cos(\omega t + \varphi_x)$$

where $\varphi_y$ and $\varphi_x$ are phase shifts between the components of the disturbing force and the displacements of the vibratory machine. The force on the pulling device is equal to

$$F = (m_1 + m' + m'') [g(\sin \alpha + \mu \cos \alpha) -$$

$$- \mu q_y r \omega^2 \sqrt{\frac{(4n_x^2\omega^2 + p_x^4)(4n_y^2\omega^2 + p_y^4)}{[4n_x^2\omega^2 + (p_x^2 - \omega^2)^2][4n_y^2\omega^2 + (p_y^2 - \omega^2)^2]}}$$

$$\times \sin(\omega t + \varphi_x + \xi_x)]$$

where $\xi_x$ is the phase shift between the disturbing force and the displacements of the vibratory machine in the direction of axis $x$.

The power needed for the conveyance

$$N = (m_1 + m' + m'') \Big\{ g (\sin \alpha + \mu \cos \alpha) (v + at)$$

$$+ (m' + m'') r^2 \omega^3 \frac{n_y [4n_x^2 \omega^2 + (p_x^2 - \omega^2)^2] - \mu q_x n_x (4n_y^2 \omega^2 + p_y^4)}{[4n_x^2 \omega^2 + (p_x^{2'} - \omega^2)^2] [4n_y^2 \omega^2 + (p_y^2 - \omega^2)^2]}$$

The combination vibratory-pulling method of conveying, as shown by computations and experimental investigations, is characterized by lower resistances than the displacement by a pure pulling drive.

The vibromagnetic conveyor is a vibrating load-carrying element along which driving magnets are mounted. The displaced loads are magnetic. In vibromagnetic conveying, different regimes of motion of the load can be realized depending on the character of operation of the electromagnet. Let us consider the case of vibration conveying of most practical interest when the load moves in a discontinuous or in a pulsating flow. The packaged load traffic is formed by virtue of the specific operating features of the magnetic traction system.

If there is a continuous layer of magnetic load on the carrying element, then when the electromagnets are switched on, the load particles located in the domain of attraction of the electromagnet are pulled inside the magnet. Since the electromagnets are mounted along the carrying element at such a distance that their domains of attraction were close to each other without overlapping, the continuous load becomes discontinuous and forms separate portions (packages) when the electromangets are switchedc on. The magnets are placed in such a way that the air gaps between the individual portions of the load would be minimum, thus excluding the elementary portions from ending up in the domain of attraction of the field directed against the direction of conveying. In order to exclude the decelerating influence of the magnetic field, the electromagnet operates in a regime of interrupted power supply, i.e., the commutation system is tuned as follows: when the core of the load element is in the center of the electromagnet, i.e., when the motive force becomes equal to zero, the elecromagnet is switched off. Another regime is also possible, in which the electromagnet is switched off with some delay, thus the load which falls on the carrying element is somewhat decelerated. The electromagnet is again energized when the load reaches the domain of attraction of the next electromagnet in the direction of motion. In such a regime of operation the attractive force acts on the load only in the direction of conveying. Thus, the magnetic attractive force can be written as

$$F_{i-1} = k_1 (x - iL) \qquad\qquad \text{when } iL \le x \le iL + L_1$$

$$F_{i-1} = -k_2 x - k_2 iL + L_1 (k_1 + k_2) \qquad \text{when } iL + L_1 \le x \le (i + \tfrac{1}{2}) L$$

$$F_{i-1} = 0 \qquad\qquad \text{when } (i + \tfrac{1}{2})L \le x \le (i + 1)L$$

where, $i$ is the successive number of the electromagnet; $L$ is the sphere of attraction of the electromagnetic field on the load layer element under consideration; $L_1$ is the distance from the beginning to the maximum of the function of the attractive force; $k_1$ and $k_2$ are the characteristics of the field of the electromagnet.

The force of magnetic attraction as a function of the distance to the center of the electromagnet is a curvilinear function which can be approximated by a nonequilateral triangle. It has been established by experimental investigations that the attractive force of the load layer element is determined by the ratio of the length of the load package and the length of the electromagnet. Within some limits, the magnitude of the attractive force increases with the length increase of the element of the transported load for a given length of the electromagnet. The attractive force is considerably dependent also on the location of the load element with respect to the electromagnet. As the load element moves into the electromagnet the attractive force is increasing. It reaches a maximum when the load element moves into the electromagnet by 0.5–0.7 of its length; with further advance of the load the attractive force drops, becoming equal to zero at the moment when the center of the load element coincides with the center of the electromagnet and the load is situated symmetrically relative to electromagnet. Further advance leads to development of a force in the opposite direction, i.e., against the direction of the motion of the load.

Let us consider the regime of vibromagnetic conveying in which the action of the magnetic and vibratory systems are synchronized so that the force of magnetic attraction acts on the element at the moment when it has no contact with the carrying element. Furthermopre, the equations of motion of the load model in the direction of conveying relative to the carrying element will have the form

on interval $0 \leq x \leq L_1$

$$\ddot{x} + 2n_x\dot{x} - p_1^2 x = -\ddot{x}' - 2n_x'\dot{x}' + g \sin \alpha \qquad (2.51)$$

on interval $L_1 \leq x \leq 1/2 \ L$

$$\ddot{x} + 2n_x\dot{x} + p_2^2 x = -\ddot{x}' - 2n_x'\dot{x}' + (p_1^2 + p_2^2)L_1 + g \sin \alpha \qquad (2.52)$$

On the interval of slippage along the carrying element, equations of motion of the load are

$$\ddot{x} + 2n_x^*\dot{x} = -\ddot{x}' - 2n_x'^*\dot{x}' + g \sin \alpha + \text{sign} \ (\dot{x})\mu \ (p_y^2$$

$$+ \ 2n_y'^*\dot{y}) \qquad (2.53)$$

Here, $m$ is the mass of the load element; $n_x$, $n_x'$ are the resistances to the motion of the load element which are proportional to the relative and absolute velocities of load motion in the direction of the $x$-axis; $\dot{x}$, $\ddot{x}$ are the components of vibratory velocity and acceleration of the load-carrying element in the direction of axis $x$; $\alpha$ is the angle of inclination of the load carrying element; $p_y$, $n_y$, and $\mu$-rheological parameters of the load and the coefficient of friction against the carrying element; $p_1$, $p_2$ are the parameters of the electromagnet, $p_1^2 = k_1/m$, $p_2^2 = k_2/m$.

Since the force of the electromagnetic attraction in the case under consideration acts only in the direction of conveying, the equations of motion of the load in the direction of axis $y$ will be determined only by vibration parameters of the carrying element.

When investigating the laws of vibromagnetic conveying using these equations, the moments of transition from the joint motion to free motion, from free motion to joint motion, from elastic strain of the layer to slippage and vice versa are determined from correlations obtained for the regimes of vibratory conveyance.

Energy expenditure on the displacement of the load in the process of vibromagnetic conveying is comprised from the expenditures of the vibratory and magnetic systems.

## 2.3  VIBRATORY PROCESSING OF DISPERSED MEDIA AND BULK LOADS

Processing large masses of small-chunk and dispersed media is carried out when executing many technological operations with the use of vibration processing in enrichment, in conveying, and others.

Vibration processing of large masses of rocks in blocks or loose state is realized in vibration discharge of the ore from the blocks, massif working with the aim of weakening, vibration diagnostics and so on.

The features of the action of vibration on large masses of dispersed media or rocks are based on general physical processes and linked mainly with the damping of oscillations in the massif which occurs as a result of dispersion of the energy of oscillations, filtering properties of the medium, and the formation of exponential waves.

The common physical basis of these processes calls for considering the laws of processing of the massif and various dispersed media in one section of the book. The problem is reduced to developing phenomenological models that might take into account the fact that different parts of the massif vibrate with different amplitudes changing from maximum near the source of vibration to zero at the boundary of the zone of vibration propagation.

### 2.3.1  Phenomenology of Vibration Action on Dispersed Media

When vibrations act on a dispersed medium, a number of transformations take place whose specifics depend on oscillation intensity. As the intensity increases with values of the acceleration amplitudes not exceeding the free fall acceleration, the dispersed medium begins to acquire mobility, and pseudoliquidity (state of quasiliquefaction). In this state, adhesion between the particles becomes weak, they approach each other more closely, the number of pores decreases (a more dense arrangement of the particles takes place), and the medium is compacted. Maximum compaction is attained at values of acceleration amplitudes close to free fall acceleration.

With further increase of oscillation intensity the particles of the dispersed medium begin to lose contact with the vibrating working element, the links between the particles are reduced and periodically destroyed, i.e., the medium seems to change into a state of boiling. This state, called vibration boiling, is characterized by the loosening of the medium and by strong circulation of the particles making up the medium. Vibration boiling can be divided into two characteristic stages - a stage of segregation of the particles of the medium and the stage of intensive mixing. The second regime of vibration boiling is realized at more intensive vibration regimes.

Investigation of dispersed systems indicates that the transition from such characteristic states as quasiliquefaction to vibratory boiling takes place either upon reaching a specified level of vibratory acceleration by the medium, or upon imparting a specific energy level to the medium. the first criterion is more applicable to coarsely dispersed systems, the second is applicable to microheterogeneous systems. It is pertinent to note that even in coarsely dispersed systems transition from the state of quasiliquefaction to the state of vibratory boiling occurs, as a rule, at accelerations exceeding free fall acceleration. The critical accelerations become close to the free fall acceleration only when processing very thin layers of coarsely dispersed systems having low internal resistances. In the general case, critical accelerations are dependent on the properties of the medium, thickness of its layer, adhesion forces between the particles, and a number of other factors.

Similarly, the energy criterion is also a function of many factors characterizing the medium. In this case, since only a part of the energy generated by the vibration exciter is absorbed by the medium, the absorbed energy ought to be adopted as the energy criterion of transition of the dispersed medium from one characteristic state to another, instead of the energy dissipated by the vibration exciter as is currently the practice. It is this part of energy that causes the transition of the medium from one state to another.

This section considers fundamental mechanics of vibration action on coarsely dispersed and microheterogeneous media. The goal is to understand formation of the criteria of transition of the medium from the state of quasiliquefaction to vibration boiling, both energy and acceleration criteria. The energy balance of the vibration exciter-disperse medium system is specifically analyzed and the generated energy which is absorbed by the medium is determined.

The real dispersed medium comprising an enormous number of particles represents a volumetric multi-mass system. Depending on disperse composition of the medium and the number of phases (presence, apart from the solid phase, also of gaseous and liquid phases) it can be represented as a multi-mass elastoinertial system in which hysteretic resistances of different types are generated causing damping of oscillations applied to it.

Damping in the considered dispersed media is a very complex phenomenon. It can arise for various reasons among which one should first note the friction of the surfaces, dry or wet, of the particles against each other; resistance to the motion of solid particles in the liquid or gaseous phase; passage of these phases through the pores of the solid phase; irreversible strains of insufficiently elastic phases, presence of various forces of adhesion, etc. Generally, when energy dissipation takes place, damping forces are present in all cases. Usually several of its forms are in action simultaneously. The presence of damping forces causes the manifestation of nonlinear effects in the system which is subjected to vibratory processing. This, in its turn, extremely complicates the investigation of such systems. In practice, usually different methods of approximation of the real forms of damping by viscous damping are used. Such a method enables one to reduce complex nonlinear systems to simpler linear systems, in particular by using the method of energy balance which is to the effect that the dissipated energy during a cycle, caused by real resistances, is equated to the dissipation of energy from viscous resistances. Complex forms of resistances can be reduced to viscous and dry resistances with sufficient accuracy for practical purposes.

Thus, for the solution of the majority of practical problems, the dispersed medium can be represented by a three-dimensional multi-mass viscoelastoplastoinertial system (Fig. 2.7). Each rheological cell of such a system is characterized by inertial properties (masses $m$ and $m'$), by elastic properties (elastic elements $k'$, $k''$), and also by hysteresis properties - viscous $c$ and plastic $k_p$ elements. The rheological cells interact via elastic elements and viscous and dry friction connections. Thus, by using the developed standard rheological elements, and in some cases by supplementing them with special elements, all the diversity of acting in real systems adhesion forces between the particles, which include atomic coagulation and true phase contacts, can be reduced to equivalent viscous and dry resistances by the techniques of equivalent modeling. The majority of dispersed media, including multi-phase media, can be modeled with the aid of standard viscoelastoplastoinertial phenomenological models.

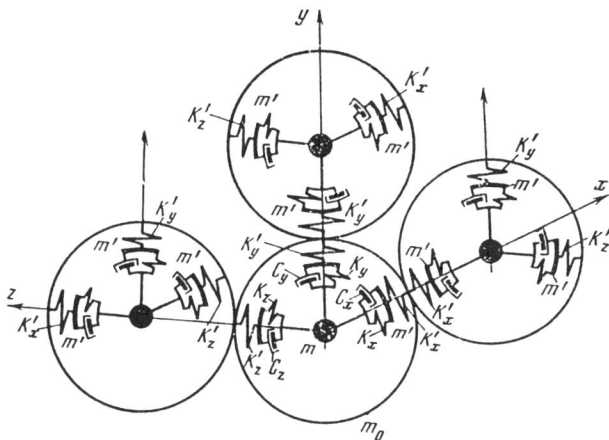

**Figure 2.7** Phenomenological model of a three-dimensional dispersed medium

The laws of behavior of dispersed media under the influence of vibration and the characteristic properties of the mass-transfer processes are extremely complex, since they are dependent on a large number of various factors. The theory of these processes is built on the basis of a thorough study of the physical principles of the phenomenon. Certain successes in this direction have been achieved by using the techniques of phenomenological rheology.

It should be noted that mass-transfer processes in the vibratory field are changing somewhat upon transition from coarsely dispersed systems to highly concentrated microheterogeneous systems in which the decisive role is played by surface phenomena on the interphase boundary and by the adhesion forces between the particles. This is caused by the fact that as a result of increase of concentration and degree of dispersion of the solid phase in liquid and gaseous dispersed media the active interphase surface increases substantially, and the free surface energy is increasing accordingly on the boundary of phase separation. Therefore, the role of intermolecular forces of adhesion between the particles increases. As a result of the appearance of the molecular adhesion forces, particle aggregates and spatial structures are formed in highly concentrated microheterogeneous systems. In large dispersed systems, in which there is no adhesion between the particles, only forces of friction are acting. Spatial structures are not formed and, therefore, mass-transfer processes take place, as a rule, under less intensive vibrations.

The properties of spatial structures of highly concentrated microheterogeneous systems are determined by surface phenomena on the interphase boundary and by the type of contacts between the particles of the solid phase. All structures are usually divided into three types with respect to the strength of contacts between the particles of the solid phase [9]. Structures with coagulation contacts are characterized by the least simple contacts. These are formed between particles of the solid phase separated by layers of liquid dispersion medium. Structures with direct point (atomic) interactions which are usually formed in highly dispersed powders are characterized by strong contacts. The strongest contacts are encountered in condensed (crystallized) structures with true phase contacts that are created following solidification of the layer between the particles of the solid phase. The behavior of finely dispersed loads are largely influenced by aero- and hydrodynamic resistances of the gaseous and liquid phases.

When operating a vibratory plant its working element performs oscillatory motions as a result of which force pulses are imparted onto the medium it is carrying, thus deformation waves propagate through the medium. It is established that during the vibration processing of bulk media, the process is largely affected not only by the parameters of the source of vibration, but also by the features of interaction between the constituent monolayers and particles. When processing poured media, the monolayer coming in contact with the vibration source receives force impulses from it. The impulses are transmitted from the lower monolayer to the upper monolayers. As a result of inertia, and the presence of frictional forces and irreversible strains, the impulses are gradually attenuated as they are transmitted from one monolayer to another. The degree of attenuation is determined by the properties of the medium and also by the character and magnitude of the force pulses. The energy of the oscillatory motion of the working element during the passage of the waves is spent on accelerating the processed medium and in making up for the losses during the irreversible strains.

Investigations show that when dealing with dispersed media, a phase shift is observed in displacements of contiguous monolayers. Also, if vibratory processing is combined with the conveying process, the mean speed of motion is decreasing with increasing distance from the source of vibration. The indicated trend (reduction of speed of the upper monolayers) is predominant; however, for some media, layer thicknesses, and operating regimes, the upper layers can lead the lower layers. This is explained by the elastic properties of the medium.

Thus, a travelling attenuating wave propagates in the dispersed medium. The presence of a phase shift in the displacement of adjacent particles weakens the contacts between them imparting to them the property of quasiliquefaction, and under intensive vibrations the particles are transferred to a state of pseudoboiling.

As a result of the different velocities of the upper and lower layers of the medium (in magnitude and in some cases also in direction), periodic strains of the whole mass are constantly created. In some regimes the upper monolayers in their travel do not uniformly lag from the lower, but perform complex motions. There are also cases when some monolayers have stepwise phase shifts. This is witness to the fact that complex spatial displacement of the constituent particles and monolayers takes place in the mass of the processed medium. Such internal circulation in the layer defines the character of the process of mass transfer and energy losses associated with it.

In the steady-state regimes phase shift between the displacement of a layer of the medium and the working organ can be very significant. It must be born in mind that the phase shift is found in the motion of both the upper and the lower surfaces of the layer, the latter being in contact with the source of vibration. Energy consumption in the process of vibratory working of the massif and the delay of the moments of separation of the medium from the vibrating surface are determined by the phase shift. The process of interaction of the source of vibration with the medium is accompanied by constant change of the value of the forces acting between them (reaction of the medium on the vibration exciter). Under sufficiently intensive oscillations, the normal reaction of the medium becomes periodically equal to zero, the lower layer loses contact with the working element and, as a result of the accumulated kinetic energy, performs free motion. In such toss-up regime the lower monolayer, after imparting all its kinetic energy to the layers lying on the top, in time begins a reverse motion, although the upper monolayers can continue to move upwards. At this moment loosening of the processed medium begins, i.e., boiling. There could exist such a regime when the upper monolayers are still in a suspended state, and the lower, coming into contact with the vibration exciter, again receives an impulse upwards. Thus, the upper and lower monolayers meet in the suspended state and are dispalced in the process. The upper monolayer receives an impulse from the lower layer in an upward direction and continues its motion, and the lower uses up its kinetic energy and returns once more to the surface of the vibration exciter. In meeting with the surface the layer of the medium is decelerated and compacted. Upon dropping, the medium can partially regain its kinetic energy or, conversely, cause additional expenditure of energy by the source of the vibration. This energy will be spent on crushing and reducing the grain size of the processed medium and also on wear of the working organ.

The described regime of vibratory processing of a dispersed medium, which is realized only at sufficient layer thickness, is characterized by the fact that two zones can be identified. In the zone closest to the vibration exciter, intensive oscillations of the medium take place. In the second zone oscillations are smoothed and the medium is as if in a state of levitation.

Depending on the paraemters of the source of vibration, the medium can be displaced steadily, reaching an identical height with identical speed when being tossed in each cycle of the oscillations, or unsteadily when large and small tossings alternate. Unsteady regimes are established when the acceleration of the oscillation exciter as the medium falls is close to the values required by the conditions of tossing or exceeds them. In this case, the falling medium loses contact again with the vibration exciter before it succeeds in acquiring the speed of motion. With this, a small toss occurs and the medium falls already at a more favorable time considering the subsequent tossing. Therefore, the next event of medium separation progresses normally, but the moment of drop is again unfavorable for the subsequent tossing. Such vibratory processing regimes are quite frequently encountered in practice. The stochastic character of change of the dispersed medium properties affects to a certain extent the character of vibratory processing.

Vibratory processing of finely dispersed loads is largely formed under the influence of the gaseous or liquid phases. As a result of the poor air permeability, the medium becomes subject to large aerodynamic loads. The time of free motion (flight) of finely dispersed media is insignificant even at very intensive oscillations of the working organ as a result of the action of the aerodynamic resistances.

Aerodynamic resistances arise as a result of the phase shift between the pressure fluctuations of the gas medium and the motion of the solid particles. Therefore, aerodynamic forces arise which hinder the motion of the solid phase. Thus, in the space between the surface of the working organ and the lower monolayer of the processed medium rarefaction is created upon tossing up and the pressure rises relative to the atmospheric pressure upon falling. The balancing of these periodic oscillations in the space between the layer of the microdispersed medium and surface of the vibration exciter is attained due to the periodic outflow of the excess and inflow of the deficit amount of air which passes through the pores in the layer of the medium. Therefore, the particles of the finely dispersed medium are acted upon by the fluctuating aerodynamic pressure (head), oriented mainly with some phase shift in a direction opposite to their displacement. The values of the aerodynamic forces acting on the particles of the medium are essentially functions of the bulk weight and specific gas permeability, and are also dependent on the regime of vibrations.

When vibration acts on microheterogeneous, highly filled dispersed media with the aim of their quasiliquefaction or the creation of a vibration boiling layer as a result of the presence of substantial forces of adhesion between the particles, one has to apply significantly more intensive regimes of vibrations than when processing coarsely dispersed media. Thus, for the conversion of finely dispersed media to a boiling state, the vibratory accelerations must exceed the gravity acceleration by several times, and energy influx must also be significantly increased. When vibratory processing media with considerable contact interactions, energy consumption can become so significant that the realization of the tchnological process becomes ineffective. In such cases, a complex combination of surface-active substances and vibratory processing is used. Surface-active substances create the conditions for a substantial decrease of the level of input energy by decreasing contact interaction between the particles of the solid phase.

Many technological processes are realized in flow-line installations with the simultaneous displacement of the processed medium. In view of this, we are going to consider the main laws of the process of vibration conveying.

The humidity of the loads with tendency to sticking has substantial influence on the process of vibratory conveyance. The dry friction of these loads on the surface of the carrying element is replaced by viscous friction. Furthermore, the magnitude of adhesion with the carrying element is only slightly, if at all, dependent on the pressure against the surface (the magnitude of the normal reaction). As a result, the principle of vibratory conveyance is violated – the value of the force moving the load when the carrying element moves forwards or backwards is found to be the same; hence, the directional displacement of the load is not realized. Conveyance of sticky loads is feasible only by providing asymmetric oscillations of the carrying element, which can be attained by using, for example, biharmonic oscillations.

For the elucidation of the physics of the process of vibration conveying of mass loads, the study of the energy dependences is essential. The resistances, irreversible strains, and friction of the particles during compression and loosening of the load layer result in energy consumption when the carrying element moves in a vertical direction. Energy expenditures also take place in the process of collision of the load with the carrying element. They are associatd with irreversible strains, grain-size reduction, and friction between the particles, and also with hysteresis losses in the carrying element and elastic elements of the vibratory machine that are caused by their deformation upon collision. A certain amount of energy is lost in the process of free motion, in particular under the action of air resistance. Compression and rarefaction of the air cushion under the load layer also cause damping of vibrations of the carrying element.

Upon displacement in the horizontal direction, energy is spent on friction of the load during its slippage over the carrying element and relative slippage between layers. The latter occurs since the layers of the load coming into contact with the conveying surface move, as a rule, more rapidly than the particles on the surface of the layer.

The energy for the acceleration of the transported load and overcoming the resistances encountered in the process of vibratory conveyance is imparted by the carrying elements via the friction forces of the load on the conveying surface. Energy transfer in this case from the carrying element to the load takes place on the interval of joint motion in the absence of relative slippage or under (relative) slippage of the load in the direction opposite to conveying, i.e., when the load lags behind the motion of the carrying element. Such a method of transferring energy to the transported load is associated with unproductive energy expenditures whose measure is the relative speed of the load. The magnitude of energy imparted to the load is dependent on the duration of the joint motion. In particular, at identical regimes of vibrations a load which is not easily transportable moves for a longer time with the carrying element. It is established that with increasing duration of the phase of the joint motion the stability of the process of vibration conveying increases.

Energy dissipation in the process of vibratory conveyance occurs on all stages of motion of the load, namely in free motion, during collision, and in the period of joint motion with the carrying element. However, the intensity of these processes is different and is characterized by different energy dissipation factors.

The phase shift between displacements of the load and the carrying element is the integral criterion that takes into account all the different factors of energy dissipation.

Let us now consider the characteristic properties of the other processes associated with the vibratory handling and displacement of a medium.

The characteristic property of vibratory processing in large receptacles with simultaneous vibratory discharge is the presence of large masses of the medium on the surface of the vibrating element. Pulses created by the working element may not reach the free surface of the bulk of the load, but rather may be damped inside the core. In this case, the monolayers of the load coming into contact with the vibrating surface perform periodic motion with practically the same amplitude as excitation. As the distance from the vibrating element increases, the vibration amplitude of the monolayers decreases and, finally, becomes equal to zero. The layers lying at the top are not affected by vibration. As a result of the constant pressure from the mass above, it is sometimes not feasible to realize regimes involving tossing up. In such conditions of vibration propagation, the motion of the medium does not take place over the entire mass but just near the working organ.

Mixing of the processed medium occurs in vibratory mixers, vibratory mills, during the process of vibratory bunkering, etc., as a result of the periodic contacts of the medium with the walls of the container in which it is being handled. Vibratory processing with simultaneous mixing of the medium differs substantially from the process of ordinary vibratory processing. When the layer of the medium comes into contact with the wall of the working organ, it comes to rest, and the constituent particles acquire high mobility as a result of exposure to vibratory action and begin to rise upwards over the walls of the working organ under pressure from the newly delivered mass. Furthermore, the first monolayer forms a small slope at the wall. As more of the medium is delivered, the lower monolayer forming a slope remains, as a rule, stationary and the newly delivered portions of the medium are displaced over the lower monolayer. Once sufficiently close to the wall, they begin to rise practically vertically. The medium located at some distance from the wall rises at an angle. The further from the wall, the larger the deviation of the moving medium from the vertical. In the initial stages of the process, the angle of slope of the free surface of the massif is usually larger than at the end of the process when the bunker is full to a greater degree. In addition to the translational motion, the medium performs oscillatory motion with the oscillation frequency of the working organ. In this case, just as in the processes of ordinary vibratory processing, the amplitude of the monolayers decreases as the distance increases from the surface of the vibration exciter.

When the height of the layer reaches a maximum at the wall of the working organ (for a given operating regime), in addition to the main fast process - vibratory conveyance and vibratory bunkering - a number of slowly flowing processes occur. The latter include down rolling (motion in a direction opposite to the direction of conveying) of the particles along the upper inclined surface of the bunkered medium. This phenomenon is explained by the fact that as a result of attenuation vibration, the pressure forces in the upper layers are small or are altogether absent. Therefore, the particles on the free surface of the massif, whose mutual adhesion is small owing to the action of vibration, roll down the inclined surface under the action of the gravitational force. Since part of the bunkered medium in its motion over the free surface is retarded on the slope, the free surface is gradually moved away from the barrier forming a horizontal surface on the upper part of the pile. The particles rising from the lower layers are detained on this surface which leads to an increase in the height of the bunkered pile.

As the height of the pile increases and its inclined surface is moved back, the process of arrival of new batches at the bunker is made more difficult and is finally terminated altogether. Thus, upon reaching a specific height in the bunkered mass, circulatory motion of the particles commences along elliptical trajectories. In the lower section of the trajectory the particles rise to the surface of the mass, shift back and join the motion again in the direction of displacement.

In the process of circulatory motion of the medium in the pile, segregation of the constituent particles by sizes takes place. Small particles accumulate at the bottom, and the large particles are constantly carried away to its surface. The accumulated small particles at the bottom form a layer which is impermeable to air. Therefore, in the process of vibrations, fluctuation of the air pressure under the layer takes place, where the peak values can exceed the atmospheric pressure. Occasionally, the compressed air breaks through the layer of the bunkered medium and is accompanied by local ejections of the constituent particles. The accumulation of the small fractions at the bottom of the working organ reduces efficiency of vibratory processing. It is interesting to note that in the process of vibratory processing, the medium is in a loose state, and when vibrations are terminated, compaction of the medium takes place.

The indicated laws of the behavior of dispersed media under the action of vibrations can be adequately reproduced with the aid of multi-mass rheological models. However, in many cases simpler methods are required which simultaneously yield qualitatively reliable and sufficiently accurate description of the investigated processes. When developing a simplified model of the medium, we shall proceed from the fact that it must reproduce the elastodissipative and inertial properties of the medium. The above described experimental investigations of the action of vibration on dispersed media show that when handling large masses, the propagating oscillations in the medium are gradually dissipated and damped. Oscillation amplitude of the particles of the medium decreases with increasing distance from the vibration source and tends to zero at some distance, farther than which the oscillations do not propagate. Thus, when processing large masses, the medium vibrates with different amplitudes and at a significant distance from the vibration source it might not be excited at all. The phenomenon of attenuation of oscillations in dispersed media is associated mainly with energy dissipation of the wave propagating in the medium under the influence of hysteresis losses. However, attenuation can also occur in the media with small hysteresis losses as a result of their filtering properties if excitation takes place at frequencies lower or higher than the system transmission range. In this case, the waves propagating in the medium from the vibration source attenuate rapidly with increasing penetration into the bulk. If the dispersed system is acted upon with frequencies less than the lower limit of transmission, then in this case exponential waves propagate in the medium. Amplitude of these waves is decreasing exponentially as the distance increases from the source of vibrations. In order to provide for a deeper vibration processing of dispersed media one must use excitation frequencies lying in the range of natural frequencies of the system.

Especially effective processing of the massif is achieved when it is executed with varying frequency but within the transmission range; in this case, various forms of oscillations are excited in the massif at natural frequencies and the massif is processed uniformly in all its parts. Processing can be conducted at several frequencies simultaneously, i.e., polyfrequency vibrations are used which are also useful for effecting deep uniform working of the massif.

In the general case, in order to investigate the laws of processing dispersed media and determine the energy consumption, one must use a rheological system which would enable one to take into account the fact that the medium oscillates with a variable amplitude and some part of it cannot even participate at all in the vibrations. Such possibilities are revealed in full when using multi-mass phenomenological models; however, difficulties of computational character limit the scope of their practical application. Presented below is a phenomenological model with a limited number of masses which, nonetheless, enables one to satisfactorily cope with this problem.

Figure 2.8a depicts the rheological model of a disperse medium which takes into account the requirements set above and which is intended for the investigation of the mechanisms of vibration processing when the vibration source is in the center of the medium and when it acts on its periphery.

The total mass of the massif is equal to $m_x + m_y$. This mass in the rheological model is divided into two parts. The first part is the effective mass $q_x m_x = q_y m_y$ participating in the oscillations. The second mass $(1 - q_x)m_x + (1 - q_y)m_y$ does not participate in the oscillations but exerts on the oscillating mass the static pressure $[(1 - q_x)m_x + (1 - q_y)m_y]$. The mass

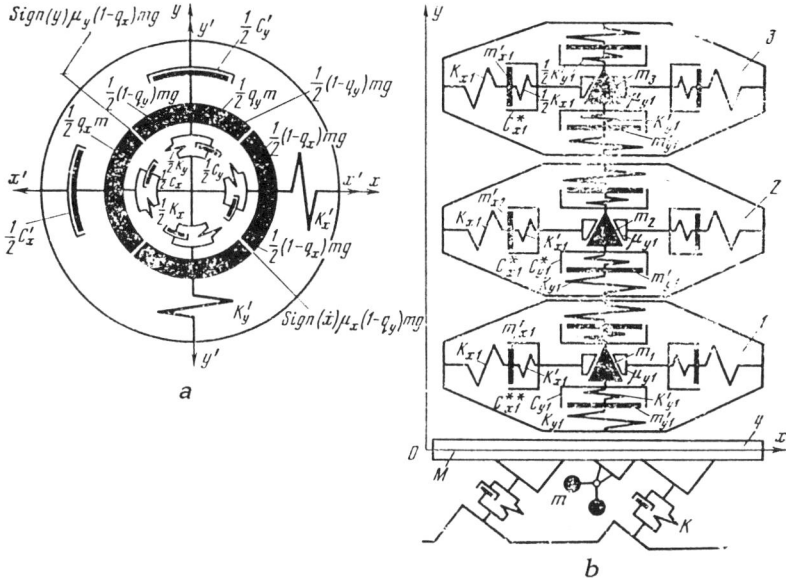

**Figure 2.8** Phenomenological model of a dispersed medium. *a)* finely dispersed medium, *b)* caved ore during vibration discharge

of the massif which is subjected to oscillations is characterized by the viscoelastic properties $k_x$, $c_x$, and $k_y$, $c_y$. The surrounding massif exerts on the oscillating mass viscous and dry resistances

$$c'_x \dot{x}, \ c'_y \dot{y} \text{ and } \text{sign}(\dot{x}) \mu_x (1 - q_y) m_y g \text{ and } \text{sign}(\dot{y}) \mu_y (1 - q_x) m_x g$$

The equations of motion of the dispersed medium which is subjected to vibration $A_x \sin \omega t$ in the direction of axis $x$ and $A_y \sin(\omega t + \gamma)$ in the direction of axis $y$ has the form

$$\ddot{x} + 2n_x \dot{x} + p_x^2 x = A_x \omega^2 \sin \omega t - 2n'_x \omega A_x \cos \omega t$$

$$- \frac{4}{\pi} q_{yx} \mu_x g \cos(\omega t + \varphi_x) - q_x^* g$$

$$\ddot{y} + 2n_y \dot{y} + p_y^2 y = A_y \omega^2 \sin(\omega t + \gamma) - 2n_y \omega A_y \cos(\omega t + \gamma)$$

$$- \frac{4}{\pi} q_{xy} \mu_y g \cos(\omega t + \varphi_y) - q_y^* g \tag{2.54}$$

where $n_x$ and $n_y$ are the total damping factors due to internal resistances of the massif participating in the oscillations and to external resistances of the nondeformable massif

$$2n_x = \frac{c_x + c_x'}{q_x m_x}, \qquad 2n_y = \frac{c_y + c_y'}{q_y m_y}$$

$n_x^*$ and $n_y^*$ are the damping coefficients due to internal resistances of the deformable massif

$$\left(2n_x^* = \frac{c_x}{q_x m_x}, \qquad 2n_y^* = \frac{c_y}{q_y m_y}\right)$$

$n_x'$ and $n_y'$ are the damping coefficients due to external resistances of nondeformable massif

$$\left(2n_x' = \frac{c_x'}{q_x m_x}, \qquad 2n_y' = \frac{c_y'}{q_y m_y}\right)$$

$p_x$ and $p_y$ are natural frequencies of the massif in the directions of axes $x$ and $y$

$$\left(p_x^2 = \frac{k_x}{q_x m_x}, \qquad p_y^2 = \frac{y}{q_y m_y}\right)$$

$q_x$ and $q_y$ are the parts of the massif taking part in the oscillations in the directions of axes $x$ and $y$; $q_x^*$, $q_y^*$, and $q_{xy}$, $q_{yx}$ are ratios between the masses of the medium

$$\left(q_x^* = \frac{1-q_x}{q_x}; \quad q_y^* = \frac{1-q_y}{q_y}; \quad q_{xy} = \frac{1-q_x}{q_y}; \quad q_{yx} = \frac{1-q_y}{q_x}\right)$$

In equations (2.50) and (2.51), the following substitutions are made for the purpose of their linearization

$$\text{sign}\,(\dot{x})\,q_{yx}\mu_x g = \frac{4}{\pi}\,q_{yx}\mu_x g\,[\dot{x}]$$

$$\text{sign}\,(\dot{y})\,q_{xy}\mu_y g = \frac{4}{\pi}\,q_{xy}\mu_y g\,[\dot{y}]$$

where $[\dot{x}]$ and $[\dot{y}]$ are dimensionless velocities of the medium in the directions of axes $x$ and $y$

$$([\dot{x}] = \cos\,(\omega t + \varphi_x), \qquad [\dot{y}] = \cos\,(\omega t = \gamma + \varphi_y))$$

$\varphi_x$ and $\varphi_y$ are phase shift angles between the exciting force created by the vibratory machine and the displacements of the processed medium in the direction of axes $x$ and $y$, respectively; $f_x$ and $f_y$ are factors defining the ratios of the dry friction forces and the exciting force

$$\left(f_x = \frac{4q_{yx}\mu_x g}{\pi A_x \omega^2}, \qquad f_y = \frac{4q_{xy}\mu_y g}{\pi A_y \omega^2}\right)$$

$\psi_x$ and $\psi_y$ are phase shift angles

$$\left( \psi_x = \operatorname{arctg} \frac{\dfrac{2n_x}{\omega} + f_x \cos \varphi_x}{1 - f_x \sin \varphi_x}; \quad \psi_y = \operatorname{arctg} - \frac{\dfrac{2n_y}{\omega} + f_y \cos \varphi_y}{1 - f_y \sin \varphi_y} \right)$$

By solving equations (2.50) and (2.51), we determine time history of strain in the dispersed medium

$$x = -\frac{q_x^{\ast}}{p_x^2} g + \overline{X} \sin(\omega t + \varphi_x)$$

$$y = -\frac{q_y^{\ast}}{p_y^2} g + \overline{Y} \sin(\omega t + \gamma + \varphi_y) \qquad (2.55)$$

where the amplitudes of forced oscillation are

$$\overline{X} = A_x \frac{\sqrt{(2n_x^{'}\omega + f_x\omega^2 \cos \varphi_x)^2 + \omega^4 (1 - f_x \sin \varphi_x)^2}}{\sqrt{(p_x^2 + \omega^2)^2 + 4n_x^2\omega^2}}$$

and the phase shifts between the displacements of the processed medium and oscillations of the exciter are

$$\varphi_x = \operatorname{arctg} \frac{2n_x\omega}{p_x^2 - \omega^2} - \psi_x$$

Expressions for $\bar{y}$ and $\varphi_y$ are similar.

As follows from these expressions, the magnitude of the phase shifts is dependent on the combined resistances acting in the processed medium, namely dry friction $f_x$ and $f_y$, viscous friction in the oscillating layers $n_x$ and $n_y$, and viscous friction in the surrounding massif $n_x^{'}$ and $n_y^{'}$.

Let us analyze contributions made by each of these resistances.

The amplitude and phase shift can be determined from the following formulas:

in the presence of only viscous resistances of the deformable part of the massif

$$\overline{X} = A_x \frac{z_x^2}{\sqrt{(1 - z_x^2)^2 + 4v_x^{\ast'}z_x^2}} \qquad (2.56)$$

$$\varphi_x = \operatorname{arctg} \frac{2v_x^{\ast}z_x}{1 - z_x^2} \qquad (2.57)$$

in the presence of only viscous resistances in the nondeformable part of the massif

$$\underline{X} = A_x \frac{2v'_x z_x}{\sqrt{(1 - z_x^2)^2 + 4v'^2_x z_x^2}} \tag{2.58}$$

$$\varphi_x = \operatorname{arctg} \frac{2v'_x z_x}{1 - z_x^2} - \operatorname{arctg} \frac{2v'_x}{z_x} \tag{2.59}$$

in the presence of only dry resistances

$$\underline{X} = A_x \frac{z_x^2 \sqrt{1 - f_\infty^2}}{\sqrt{(1 - z_x^2)^2}} \tag{2.60}$$

$$\varphi_x = \operatorname{arctg} f_x \tag{2.61}$$

Here, the following notations are assumed: $v^*_x$ and $v^*_y$ are damping coefficients of the deformable part of the massif ($v^*_x = n^*_x/p_x$, $v^*_y = n^*_y/p_y$); $v'_x$ and $v'_y$ are damping coefficients of the nondeformable part of the massif ($v'_x = n'_x/p_x$, $v'_y = n'_y/p_y$); $z_x$ and $z_y$ are the detuning factors of the worked medium and vibration exciter

$$(z_x = \omega/p_x, \quad z_y = \omega/p_y)$$

Resistances, exerted by a unit mass of the massif which participates in the oscillations, to the motion of the vibratory exciter are

$$\frac{F_x}{q_x m_x} = - q^*_x g + \sqrt{p^4_x + 4n^{*2}_x \omega^2} \; \tilde{X} \sin(\omega t + \varphi_x + \theta_x) \tag{2.62}$$

where the phase shifts between the displacements of the massif and the reactions on the vibratory exciter are

$$\theta_x = \operatorname{arctg} \frac{2n^*_x \omega}{p^2_x} \tag{2.63}$$

Vibratory processing transcends from a nonseparating regime (quasi-liquefaction state) to a separating regime (vibration boiling state) in the direction of axis $x$ when $(\omega t + \varphi_x + \theta_x) = \pi/2$. Therefore, the conditions of commencing of the separating regime will be

$$\frac{q_x g}{\sqrt{p^4_x + 4n^{*2}_x \omega^2}} = 1 \tag{2.64}$$

As is evident from expression (2.64), the vibration parameters at which the separating regime commences are dependent on the properties of the processed massif which are defined by parameters $v^*_x$, $v^*_y$, $v'_x$, $v'_y$, $z_x$, $z_y$, $f_x$, $f_y$ and on the ratio of the masses participating in the oscillations and creating the static head.

The work done by the vibratory machine per cycle of oscillations on bringing a unit mass of the massif into an oscillatory motion is determined by formulas

$$\frac{W}{qm} = \Sigma \left[ \frac{W_x z}{q_x m_x} + \frac{W_y}{q_y m_y} \right] \tag{2.65}$$

$$\frac{W_x}{q_x m_x} = \pi A^2_x \omega^2 z^2_x \sqrt{\frac{(1 + 4v^{*2}_x z^2_x)\left[ \left( \frac{2v'_x}{z_x} + f_x \cos \varphi_x \right)^2 + (1 - f_x \sin \varphi_x)^2 \right]}{(1 - z^2_x)^2 + 4(v^*_x + v'_x)^2 z^2_x}} \times \sin(\varphi_x + \theta_x) \tag{2.66}$$

The processed massif does not absorb all the energy spent by the vibratory machine but just part of it. The other part of the energy is unproductively dissipated.

The energy received by the processed massif during a cycle of oscillations and spent on imparting motion to it is determined by formulas

$$\frac{W_{us.}}{qm} = \Sigma \left[ \frac{W_{xus.}}{q_x m_x} + \frac{W_{yus.}}{q_y m_y} \right] \tag{2.67}$$

$$\frac{W_{uus}}{q_x m_x} = \pi A^2_x \omega^2 \frac{z^4_x \sqrt{1 + 4v^{*2}_x z^2_x} \left[ \left( \frac{2v'_x}{z_x} + f_x \cos \varphi_x \right)^2 + (1 - f_x \sin \varphi_x)^2 \right] \sin \theta_x}{(1 - z^2_x)^2 + 4(v^*_x + v'_x)^2 z^2_x} \tag{2.68}$$

Using relations (2.66), (2.68), we determine the efficiency of energy utilization in imparting oscillations to the massif

$$\eta = \frac{W_{us}}{W} = \sqrt{\frac{\left( \frac{2v'_x}{z_x} + f_x \cos \varphi_x \right)^2 + (1 - f_x \sin \varphi_x)^2}{(1 - z^2_x)^2 + 4(v^*_x + v'_x)^2 z^2_x}} \frac{\sin \theta_x}{\sin(\varphi_x + \theta_x)}$$

$$+\sqrt{\frac{\left(\dfrac{2v_x'}{z_y}+f_y\cos\varphi_y\right)^2+(1-f_y\sin\varphi_y)^2}{(1-z_y^2)^2+4(v_y^*+v_y')^2 z_y^2}}\ \frac{\sin\theta_y}{\sin(\varphi_y+\theta_y)} \tag{2.69}$$

It is evident that the efficiency of energy transmission to the massif from the vibration source is determined mainly by its properties but it is also dependent on the tuning of the system, i.e., on the ratio of the natural and forced frequencies. In particular, for a given massif, its part which is entrained into vibratory process $(q_x,\ q_y)$ is considerably dependent on the range of frequencies within which the vibratory exciter operates.

## 2.3.2 Phenomenology of Vibration Action on Caved Ore During the Process of Vibratory Discharge

The conducted investigations show the following character of interaction of the working organ of the vibratory plant for the discharge of ore with caved rock mass. Vibrations are transmitted from the load carrying element to the adjacent layer and farther to the rock mass above. The vibration frequency of the vibrating ore mass is the same and the amplitude decreases with increasing distance from the load-carrying element, until it is damped altogether. Thus, the characteristic feature of operation of vibratory plants operating while buried under rock mass is that the caved mass is subjected to vibration only in a limited zone near the working organ. The dimensions of this zone are defined by the regime of vibrations and location of the vibratory plant and also by the properties of the caved mass.

For simulation of the characteristic features of the action of vibration on the caved mass a multiple-mass phenomenological model with non-restraining contacts between the masses is used.

The schematic diagram of a vibratory plant buried under a caved rock mass and represented by a phenomenological model is shown in Fig. 2.8b. The exciting force created by the inertial self-balancing vibrator has the form

$$F_y = 2\ mr\omega^2 \sin\beta\cos\omega t \tag{2.70}$$

$$F_x = 2\ mr\omega^2 \sin\beta\cos\omega t \tag{2.71}$$

where $m$ is the unbalanced mass; $r$ is the eccentricity of the unbalanced mass; $\beta$ is the vibration angle; $\omega$ is the angular velocity of the unbalanced mass.

When all four masses $1$-$4$ simulating the caved ore mass and the load-carrying element are in contact with each other, the motion of the system: load-carrying element – core mass is described by the system of equations[1]

$$\begin{cases} M_y = F_y - ky - N_{1,4}(t) \\ m_1 y_1 = - m_1 \ddot{y}' - k_1(y_1 - y_2) - c^* y_1 - c^*(\dot{y}_1 - \dot{y}_2) \\ m_2 \ddot{y}_2 = - m_2 \ddot{y}' - k_1(y_2 - y_1) - c^*(\dot{y}_2 - \dot{y}_1) \\ \qquad\quad - k_1(y_2 - y_3) - c^*(\dot{y}_2 - \dot{y}_2) \\ m_3 \ddot{y}_3 = - m_3 \ddot{y}' - k_1(y_3 - y_2) - c^*(\dot{y}_3 - \dot{y}_2) \\ M\ddot{x} = F_x - kx + \text{sign}(\dot{x}) N_{1,4}(t)\,\mu \end{cases} \qquad (2.72)$$

where $M$ is the mass of the load-carrying element and vibrator; $N_{1,4}$ is the normal reaction of the moving load on the carrying element; $(N_{1,4}(t) = k_1 y_1 + c^* \dot{y}_1)$; $m_1$, $m_2$, $m_3$ are masses of the first, second and third load layers, $c^*$ is the equivalent resistance coefficient to the displacement of the phenomenological model in the direction of axis $Y$ during the interval of joint motion, which is proportional to the relative velocity; $k_1$ is the equivalent stiffness of the mass layers; $k$ is the stiffness of the elastic system of the vibratory conveying machine; $\mu$ is the coefficent of dry friction of the load on the carrying element

$$\text{sign}(\dot{x}) = \begin{cases} -1 \text{ when } \dot{x} > 0 \\ +1 \text{ when } \dot{x} < 0 \end{cases} \qquad (2.73)$$

The system of equations (2.72) describes the motion of the load-carrying element and the ore mass from the reference moment to a certain moment of time which is characterized by loss of contact between any two layers out of four under consideration. Relative positions of these layers depend on the vibration parameters of the vibratory plant and properties of the caved mass. In the process of motion the layers can be in the following eight different positions relative to each other (see Fig. 2.8b). In the initial stage (position $1$) the layers with masses $m_1$, $m_2$, $m_3$ and the load-carrying element $M$ travel in contact with each other. In this case, the motion of the load-carrying element – mass system is described by equations (2.72). After position $1$, only positions $2$, $3$, and $4$ can be practically realized. In order to determine which of these three

---

[1] In this case, for simplification it is assumed that $F^* = \text{sign}(x^*)\mu N_{1,4}(t)$, and the relative motion of the ore layers in the direction of the x-axis is not considered.

positions will take place, one must find the smallest positive roots of the equations

$$N_{1,4}(t) = k_1 y_1 + c^* \dot{y}_1 = 0 \qquad \text{(root 2)} \qquad (2.74)$$

$$N_{1,2}^*(t) = k_1 (y_1 - y_2) + c^* (\dot{y}_1 - \dot{y}_2) = 0 \quad \text{(root 3)} \qquad (2.75)$$

$$N_{2,3}^*(t) = k_1 (y_2 - y_3) + c^* (\dot{y}_2 - \dot{y}_3) = 0 \quad \text{(root 4)} \qquad (2.76)$$

(here $N_{1,2}(t)$ and $N_{2,3}(t)$ are initial reactions of the first layer on the second and the second on the third) and select the smallest among them. The sign under the root indicates the number of the next position. Depending on which position comes after the first, the motions of the layers with masses $m_1$, $m_2$, $m_3$ and the load-carrying element at the corresponding positions will be described by the following equations.

   Position 2 (masses 1, 2, and 3 are in contact, masses 3 and 4 do not come into contact)

$$\left\{ \begin{aligned} & M\ddot{y} = F_y - ky \\ & m_1\ddot{y}_1 = -m_1\ddot{y}' - c\dot{y}_1 - k_1(y_1 - y_2) - c^*(\dot{y}_1 - \dot{y}_2) \\ & m_2\ddot{y}_2 = -m_2\ddot{y}' - k_1(y_2 - y_1) - c^*(\dot{y}_2 - \dot{y}_1) \\ & \qquad\quad - k(y_2 - y_3) - c^*(\dot{y}_2 - \dot{y}_3) \\ & m_3\ddot{y}_3 = -m_3\ddot{y}' - k_1(y_3 - y_2) - c^*(\dot{y}_3 - \dot{y}_2) \\ & M\ddot{x} = F_x - kx \end{aligned} \right. \qquad (2.77)$$

   Position 3 (masses 1, 4, and 2, 3 are in contact, masses 1 and 2 do not come into contact)

$$\left\{ \begin{aligned} & M\ddot{y} = F_y - ky - N_{1,4}(t) \\ & m_1\ddot{y} = -m_1\ddot{y}' - k_1 y_1 - c^* y_1 \\ & m_2\ddot{y}_2 = -m_2\ddot{y}' - c(\dot{y}_2 - \dot{y}_1) - k_1(y_2 - y_3) - c^*(\dot{y}_2 - \dot{y}_3) \\ & m_3\ddot{y}_3 = m_3\ddot{y}' - k_1(y_3 - y_2) - c^*(\dot{y}_3 - \dot{y}_2) \\ & M\ddot{x} = F_x - kx + \text{sign}(\dot{x})\,\mu N_{1,4}^*(t) \end{aligned} \right. \qquad (2.78)$$

   Position 4 (masses 4, 1, and 2 are in contact, masses 4 and 3 do not come into contact)

$$\left\{ \begin{aligned} & M\ddot{y} = F_y - ky - N_{1,4}^*(t) \\ & m_1\ddot{y}_1 = -m_1\ddot{y}' - k_1 y_1 - c^* y_1 - k_1(y_1 - y_2) - c^*(\dot{y}_1 - \dot{y}_2) \\ & m_2\ddot{y}_2 = -m_2\ddot{y}' - k_1(y_2 - y_1) - c^*(\dot{y}_2 - \dot{y}_1) - c(\dot{y}_2 - \dot{y}_3) \\ & m_3\ddot{y}_3 = -m_3\ddot{y}' - c(\dot{y}_3 - \dot{y}_2) \\ & M\ddot{x} = F_x - kx + \text{sign}(\dot{x})\,\mu N_{1,4}^*(t) \end{aligned} \right. \qquad (2.79)$$

where $c$ is the equivalent resistance coefficient to the displacement of the load in the direction of the $y$-axis in the interval of free motion, which is proportional to the relative velocity.

Position 5 (root 5) can practically develop only from positions 2 or 3. If position 5 develops following position 2, then it happens when

$$N^*_{1,2}(t) = 0$$

where $y_1$, $y_2$, $\dot{y}_1$, $\dot{y}_2$ are obtained from the solution of system (2.77).

If position 5 develops from position 3, then it happens when for system (2.78)

$$N^*_{1,4}(t) = 0$$

at the value $t = t_s$.

The equations of motion to the system "carrying element-load" in position 5 will have the form (masses 3 and 2 in contact, masses 4, 1, and 2 do not come into contact)

$$
\begin{cases}
M\ddot{y} = F_y - ky \\
m_1\ddot{y} = -m_1\ddot{y}' - c\dot{y}_1 - c(\dot{y}_1 - \dot{y}_2) \\
m_2\ddot{y}_2 = -m_2\ddot{y}' - c(\dot{y}_2 - \dot{y}_1) - k_1(y_2 - y_1) - c^*(y_2 - y_1) \\
m_3\ddot{y}_3 = -m_3\ddot{y}' - k_1(y_3 - y_2) - c^*(\dot{y}_3 - \dot{y}_2) \\
M\ddot{x} = F_x - kx
\end{cases}
\tag{2.80}
$$

From position 5 only position 6 or positions 2 and 3, discussed above, can practically develop. Position 6 develops if $N^*_{2,3}(t) = 0$ for system (2.80).

The equations of motion of the system "carrying element-load" in position 6 will have the form (masses 4, 1, 2, and 3 do not come into contact)

$$
\begin{cases}
M\ddot{y} = F_x - ky \\
m_1\ddot{y}_1 = -m_1\ddot{y}' - c\dot{y}_1 - c(\dot{y}_1 - \dot{y}_2) \\
m_2\ddot{y}_2 = -m_2\ddot{y}' - c(\dot{y}_2 - \dot{y}_1) - c(\dot{y}_2 - \dot{y}_3) \\
m_3\ddot{y}_3 = -m_3\ddot{y}' - c(\dot{y}_3 - \dot{y}_2) \\
M\ddot{x} = F_x - kx
\end{cases}
\tag{2.81}
$$

Positions 7 and 8 and position 5, considered above, can practically emerge from position 6. At the moment of transition from position 6 to position 7, the following conditions apply

$$y_1 = 0$$

The equations of motion of the system "carrying element-massif" in position 7 will have the form (masses 4 and 1 are in contact, masses 1, 2, and 3 do not come into contact)

$$\begin{cases} M\ddot{y} = F_y - ky - N_{1,4}^*(t) \\ m_1\ddot{y}_1 = -m_1\ddot{y}' - k_1y_1 - c^*\dot{y}_1 - c(\dot{y}_1 - \dot{y}_2) \\ m_2\ddot{y}_2 = m_2\ddot{y}' - c(\dot{y}_2 - \dot{y}) - c(\dot{y}_2 - \dot{y}_3) \\ m_3\ddot{y}_3 = -m_3\ddot{y}_3' - c(\dot{y}_3 - \dot{y}_2) \\ M\ddot{x} = F_x - kx + \text{sign}(\dot{x})\,\mu N_{1,4}(t) \end{cases} \qquad (2.82)$$

If position 6 is followed by position 8, the following conditions, obtained from system (2.81), are satisfied at the moment of transition

$$y_2 - y_1 = 0$$

The equations of motion of the system "carrying element-massif" in position 8 will have the form (masses 1 and 2 are in contact, masses 4 and 1, 2, and 3 do not come into contact)

$$\begin{cases} M\ddot{y} = F_y - ky \\ m_1\ddot{y}_1 = -m_1\ddot{y}' - c\dot{y}_1 - c^*(\dot{y}_1 - \dot{y}_2) \\ m_2\ddot{y}_2 = -m_2\ddot{y}' - c^*(\dot{y}_2 - \dot{y}_1) - k_1(y_2 - y_1) - c(\dot{y}_2 - \dot{y}_3) \\ m_3\ddot{y}_3 = -m_3\ddot{y}' - c(\dot{y}_3 - \dot{y}_2) \\ M\ddot{x} = F_x - kx \end{cases} \qquad (2.83)$$

Let us consider the solution of the presented systems on the example of system (2.72) describing the motion of the massif layers in the direction of axis $y$ and the load-carrying element of the vibratory conveying machine in the direction of axes $x$ and $y$ in position 1.

By dividing the equations of system (2.72) by the coefficients at the highest derivative and introducing new constants, we obtain

$$\begin{cases} \ddot{y} + p^2 y + 2n^* \dot{y} + p^{*\varrho} y_1 = qr\omega^2 \sin\beta \cos\omega t \\ \ddot{y} + \ddot{y}_1 + 4n_1^* \dot{y}_1 + 2p_1^2 y_1 - 2n_1^* \dot{y}_2 - p_1^2 y_2 = 0 \\ \ddot{y} - 2n_2^* \dot{y}_1 - p_2^2 y_1 + \ddot{y}_2 + 4n_2^* \dot{y}_2 + 2p_2^2 y_2 - 2n_3^* \dot{y}_3 \\ \qquad - p_3^2 y_3 = 0 \\ \ddot{y} - 2n_3^* \dot{y}_2 - p_3^2 y_2 + \ddot{y}_3 + 2n_3^* \dot{y}_3 + p_3^2 y_3 = 0 \\ \mathrm{sign}\,(\dot{x})\,\mu 2n^* \dot{y}_1 + \mathrm{sign}\,(\dot{x})\,\mu p^{*\varrho} y_1 + \ddot{x} + p^2 x \\ \qquad = qr\omega^2 \cos\beta \cos\omega t \end{cases} \tag{2.84}$$

where $q$ is the factor of the ratio of the masses of the vibrator and vibratory conveying machine ($q = 2m/M$); $p$, $p^*$, $p_1$, $p_2$, $p_3$ are the natural frequencies of the vibratory conveying machine and the layers of the massif model, ($p^2 = k/M$; $p^{*2} = k_1/M$; $p_1^2 = k_1/m_1$; $p_2^2 = k_2/m_2$; $p_3^2 = k_3/m_3$); $n^*$, $n_1^*$, $n_2^*$, and $n_3^*$ are the effective coefficients of viscous resistance to displacement of the layers of the massif model during the interval of joint motion ($2n^* = c^*/M$; $2n_1^* = c^*/m_1$; $2n_2^* = c^*/m_2$; $2n_3^* = c^*/m_3$); $n$, $n_1$, $n_2$, and $n_3$ are the effective coefficients of viscous resistance to the displacement of the layers of the massif model on the interval of free motion ($2n = c/M$; $2n_1 = c/m_1$; $2n_2 = c/m_2$; $2n_3 = c/m_3$).

Denoting the operator of differentiation with respect to time by the letter $D$, we obtain

$$\begin{cases} (D^2 + p^2)y + (2n^*D + p^{*2})y_1 = qr\omega^2 \sin\beta \cos\omega t \\ D^2 y + (D^2 + 4n_1^*D + 2p_1^2)y_1 - (2n_1^*D + p_1^2)y_2 = 0 \\ D^2 y - (2n_2^*D + p_2^{*2})y_1 + (D^2 + 4n_2^*D + 2p_2^2)y_2 \\ \qquad - (2n_3^*D + p_3^2)y_3 = 0 \\ D^2 y - (2n_3^*D + p_3^2)y_2 + (D^2 + 2n_3^*D + p_3^2)y_3 = 0 \\ [\mathrm{sign}\,(\dot{x})\,\mu 2n^*D + \mathrm{sign}\,(\dot{x})\,\mu p^{*2}]y_1 + (D^2 + p^2)x \\ \qquad = qr\omega^2 \cos\beta \cos\omega t \end{cases} \tag{2.85}$$

System (2.85) in a matrix form is

$$f(D)\,y = B \cos\omega t \tag{2.86}$$

where $f(D)$ is the system matrix; $y$ is the column matrix of the unknown functions.

$$\begin{vmatrix} D^2 + p^2 2nD + p^{*2} & & 0 \\ D^2 D^2 + 4n_1^*D + 2p_1^2 & 2n_1^*D + p_1^2 & 0 \\ D^2\ 2n_2^*D + p_2^2 & D^2 + 2n_3^*D + p_3^2 2n_3^*D + p_3^2 & 0 \\ D^2 & 2n_3^*D + p_3^2 D^2 + 2n_3^*D + p_3^2 & 0 \\ \mathrm{sign}\,(\dot{x})\mu 2nD + \mathrm{sign}\,(\dot{x})\mu p^2 & & D^2 + p^2 \end{vmatrix}$$

$$
Y = \begin{Vmatrix} y \\ y_1 \\ y_2 \\ y_3 \\ x \end{Vmatrix} \qquad B = \begin{Vmatrix} qr\omega^2 \sin\beta \\ 0 \\ 0 \\ 0 \\ qr\omega^2 \cos\beta \end{Vmatrix}
$$

We calculate the roots of the characteristic equation

$$
\Delta D_i \begin{vmatrix} = 0 \\ D = x \end{vmatrix}
$$

For the sake of definiteness we shall assume that the order of the characteristic equation is equal to the sum of the orders of the highest derivatives, i.e., 10. Let all the roots of the characteristic equation be real and simple $x_1, x_2, \ldots, x_{10}$.
We calculate the adjoint matrix

$$
F(D) = \begin{Vmatrix} A_{11}A_{21} \ldots A_{51} \\ \cdot\ \cdot\ \cdot\ \cdot\ \cdot\ \cdot \\ \cdot\ \cdot\ \cdot\ \cdot\ \cdot\ \cdot \\ A_{15} \ldots A_{55} \end{Vmatrix}
\tag{2.87}
$$

where $A_{i\gamma}$ is the algebraic cofactor of the element with number $i, \gamma$ of matrix $f(D)$.

Substituting the values of the roots in the adjoint matrix we compute matrices $F(x_1)$, $F(x_2)$, ..., $F(x_{10})$ and write their modal columns $v_1, v_2, \ldots, v_{10}$. Then, the general solution of the homogeneous system - system (2.86) - is represented as follows

$$
y_0 = C_1 v_1 e^{x_1 t} + C_2 v_2 e^{x_2 t} + \ldots + C_{10} v_{10} e^{x_{10} t}
\tag{2.88}
$$

where constants $C_1, C_2, \ldots, C_{10}$ are found from the matrix of the initial conditions.

$$
\begin{Vmatrix} \dot{y}(0) \\ \dot{y}_1(0) \\ \dot{y}_2(0) \\ y_3(0) \\ x(0) \end{Vmatrix} = \begin{Vmatrix} 0 \\ 0 \\ 0 \\ 0 \\ 0 \end{Vmatrix} \qquad \begin{Vmatrix} y(0) \\ y_1(0) \\ y_2(0) \\ \dot{y}_3(0) \\ \dot{x}(0) \end{Vmatrix} = \begin{Vmatrix} 0 \\ 0 \\ 0 \\ 0 \\ 0 \end{Vmatrix}
$$

The particular solution of the nonhomogeneous system (2.83) is represented in matrix form as

$$Y_1 = Re\Phi(i\omega) e^{i\omega t} \tag{2.89}$$

where $\Phi(i\omega) = [F(D)B/\Delta(D)]_{D=i\omega}$; $Re$ is the real part; $\Delta(D)$ is the determinant of matrix $f(D)$.

The general solution of system (2.85) has the form

$$Y(t) = C_1 v_1 e^{x_1 t} + \dots + C_{10} v_{10} e^{x_{10} t} + Re\ \Phi(i\omega)\ e^{i\omega t} \tag{2.90}$$

The solution of the system of equations describing the motion of the caved massif and load-carrying element of the vibratory conveying machine is carried on by a similar method, and using a computer.

The presented correlations enable the investigation of the specific features of the operation of vibratory conveying machines under large loads in conditions of rock burst and the study of the laws of motion of the layers of the caved massif.

## 2.4  VIBRATION CRUSHING OF ROCK MASSES

### 2.4.1  Main Requirements for the Phenomenological Model of the Rock Mass, Model Parameters

When designing a vibratory crusher, one must know the dynamic reaction of the rock inside the crusher chamber on the jaws. The problem of describing the laws of formation of the dynamic reaction of the rock mass in the process of vibratory crushing is fairly complex, since its magnitude and character are dependent on a number of factors such as the physicomechanical properties of the rock, extent of cracking, homogeneity, strength, hardness, viscosity, density, humidity, size, shape, and mutual location of different parts of the rock in the crusher chamber, and so on. The degree of filling of the crusher chamber by the rock mass also plays a significant role. Furthermore, these factors change to some degree in the process of crusher exploitation.

For the development of a model of crushed rock methods of statistical dynamics which enable determination of the probabilistic characteristics of the rock mass can be used. One can also use methods of phenomenological rheology which yield the opportunity to describe deformation-dissipation properties of the medium as applied to the probabilistic parameters of the object. The phenomenological approach in developing the rock mass model enables one to reproduce fundamentally important features of the process of vibratory crushing and provides the necessary accuracy of the calculations by selecting the model parameters by means of its identification with the statistically processed experimental data.

Analysis of the experimental results, described in the previous sections of the chapter, of the crushing process of lump rock masses in vibratory jaw crushers enabled revealing of the principal mechanisms of the phenomenon under consideration and formulating of the essential requirements of the phenomenological model of the crushed rock mass.

The model must reproduce the elastoplastic strains, hysteretic dissipation of energy under cyclic loading, and brittle fracture of real rock masses on the one hand, and describe the motion of rock mass in the crushing chamber in the presence of internal friction and pressure of the rock in the bunker, on the other hand.

In accordance with the current understanding of the mechanics of the fracture process of rock masses and the characteristic properties of the crushing process in vibratory jaw crushers, one must take into account the phenomena associated with the formation of compacted core, chipping of the protuberances of the rock pieces and the resultant rise in resistance to deformation, gradual energy consumption on the development of cracks and formation of new free surfaces.

Taking these requirements into account, a rock mass model and the phenomenology of vibration crushing by jaw crushers have been developed. For a double-jaw crusher the adopted model is a three-mass inertial viscoelastoplastic rheological body (Fig. 2.9). It is assumed within the framework of lump parameter representations that the total mass of the rock is concentrated in three elements of the model: the central core of the mass $(1-2\,\xi)m$ which does not participate in oscillations under symmetrical applications of the loads encountered in double-jaw vibratory crushers, and two oscillating masses $\xi m$. In reality, chunk of the rock mass represents a system with distributed inertial, elastic, and plastic properties. During the crushing process attenuating strain waves propagate across the chunk from the periphery to the center from opposite sides of both jaws. Furthermore, the particles of the chunk perform oscillations with variable amplitudes with the maximum value at the boundaries and zero at the center. For the description of such a complex phenomenon within the bounds of the lumped parameter model,

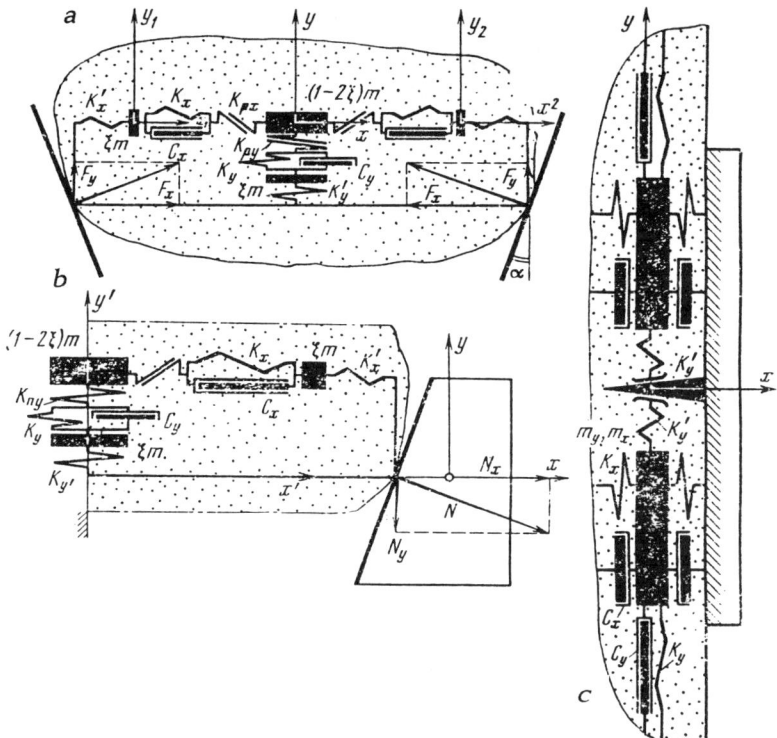

**Figure 2.9** Phenomenological models of crushed rock mass for the investigation of oscillations of crushing jaws. a) non-synchronized; b) synchronized antiphase; c) polyfrequency

the well-known technique of introducing a reduced or equivalent oscillating mass is used. In the developed model a reduced mass $\xi m$ is assumed which takes part in the oscillations and represents only a part of the total rock mass; moreover, the reduction factor $\xi$ is determined by the Rayleigh method or by more accurate methods that account for the attenuation of strain waves in the rock. The elastic properties of the rock mass are reproduced by the elastic elements of the model with stiffness coefficients $k'$ ($k'_x$, $k'_y$) and $k$ ($k_x$, $k_y$). Energy dissipation (hysteretic losses) in the region of the elastic strains of the model are realized by dampers with the coefficients of viscous resistances $c$ ($c_x$, $c_y$) included in parallel with the elastic elements $k$ and in series with the elastic elements $k^1$. Plastic strains with hardening are modeled by wedge elements of dry friction which are characterized by the coefficient of plastic strain $k_p$ ($k_{px}$, $k_{py}$). The friction processes of the model on the crushing jaws are estimated by the coefficients of static $\mu_{st}$ and dynamic $\mu$ friction. The resistances offered by the rock mass in the working space of the crusher to the motion of a separate rock chunk from the loading receptacle to the discharge opening are modeled by dampers with the coefficients of resistance $c_y$ (proportional to the relative speed) and $c'_y$ (proportional to the absolute speed). Since the crushing process is considered in the coordinate system $x$, $y$, the radial viscoelastic elements $k$, $c$ of the model can be replaced, for the sake of convenience of investigations, by a combination of the viscoelastic elements $k'_x$, $k_x$, $c_x$, and $k'_y$, $k_y$, $c_y$ positioned in the directions of the corresponding axes.

## 2.4.2  Phenomenology of the Process of Vibration Crushing of the Rock Mass

Let us consider the phenomenology of the process of vibration crushing (cyclic loading of the model) using force-strain and energy consumption-strain diagrams as applied to a single rock chunk and the whole rock mass inside the crushing chamber. The force-strain diagram in the direction of axis $x$ of the viscoelastoplastic phenomenological model of a chunk of the crushed rock mass in the loading regime with the application of cyclic pulsations is presented in Fig. 2.10. Upon loading in the first cycle, viscoelastic strains take place initially and they are described by the rheological equation

$$F_{ve_1} = k'_x x \tag{2.91}$$

where $x$ is the strain (displacement of mass $\xi m$) of the rock relative to the jaw.

The curvilinear character of function $F(x)$ is explained by the hysteresis losses in the rock during the deformation process.

Viscoelastic strains in the first loading cycle continue until the stresses reach the plastic limit $\sigma_p$ and the strain forces become equal $F_{p1l} = \sigma_p f_1$ (where $f_1$ is the area of the rock mass engulfed by the strains in the first cycle); from this moment onward, plastic strains with hardening set in and they are described by the rheological equation

$$F_{p1} - f_{ve_1} = k_p (x - x_{vec1}) \tag{2.92}$$

where $x_{vee1}$ is viscoelastic strain of the model of rock chunk in the first loading cycle up to the moment when plastic strains set in, and $F_{vee1}$ is the strain force at the end of the first cycle of elastic loading.

Plastic strains develop simultaneously with viscoelastic strains as long as model deformation continues unless, of course, fracture stresses are reached.

As soon as unloading of the phenomenological model of the rock mass commences, elastic will be removed and the rheological equation acquires the following form (in Fig. 2.10 function $F(x)$ in the stage of unloading is denoted by the dashed line)

$$F_{ve\ u1} = F_{e1} - k'_x\ (x_{C1} - x) \tag{2.93}$$

where $x_{C1}$ is strain of the model of chunk at the end of the first loading cycle; $F_{e1}$ is the force of its strain at the end of the first cycle.

As a result of the first loading cycle the specimen receives residual strain $x_{01}$. In order for the following loading cycle to start, clearance $x_{01}$ between the jaw and the model of chunk which is formed as a result of its irreversible plastic strains, should be closed. Thus, to realize a crushing process in all the loading cycles the relative displacement between the jaw and the chunk must be comprised of a constant component to which pulsations are superposed. The clearance forming in the process of irreversible strains in a vibratory jaw crusher is totally or partially taken in when the rock mass is descending between the crusher jaws. Thus, each subsequent loading cycle starts following the taking up of the clearance appearing as a result of the residual strain of the chunk.

Figure 2.10 also shows the dependence of the absorbed energy by the crushed rock mass on the strain of the rock $W(x)$. This energy is spent on overcoming hysteresis losses and irreversible strains associated with the chunk being chipped and with the formation of free surfaces in it as a result of development of a system of microcracks. It is evident from the graph that in the process of viscoelastic strains, the crushed rock absorbs energy. This process continues with the onset of plastic strains; however, the intensity of absorption varies (either increases or decreases depending on the physicomechanical properties of the rock). With the start of elastic unloading of the rock mass, part of the energy accumulated in it as elastic strain energy is returned to the

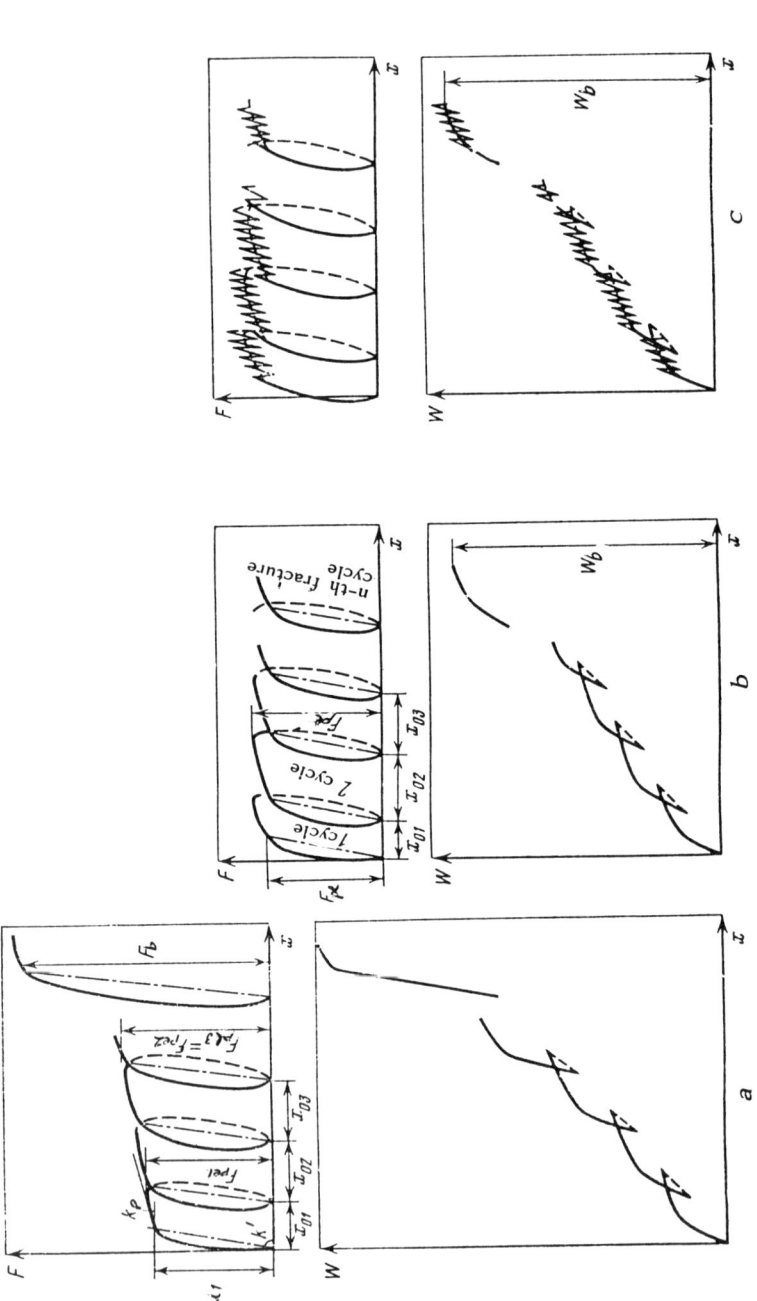

**Figure 2.10** Stress–strain and energy consumption–strain diagrams of a rheological model of rock mass. a) individual viscoelastoplastic chunks of rock mass; b) viscoelastoplastic rock mass; c) brittle rock mass

system (the process of energy recovery is indicated in the figure by the dashed line). Energy consumption in the process of vibration crushing of rock mass is determined from expression

$$W = F\dot{x} \tag{2.94}$$

The second loading cycle also starts with viscoelastic strains which will develop until the stresses reach the plastic limit and the strain forces become equal to $F_{pl2} = \sigma_p \, f_2$ (where $f_2$ is the area of the piece engulfed by strains in the second cycle). Since the process of strain is accompanied by an increase of the loaded area of the piece following the elimination of surface roughness, the plastic strains commence upon reaching larger loading than in the first cycle. We assume that the latter corresponds to the maximum load occurring in the preceding cycle: $F_{pl\,i} = F_{pe(i-1)}$. Such an approach to the determination of the initial loads of plastic deformations of the chunk in each subsequent loading cycle reflects actual correlations established in the practice of operating vibratory crushers. It allows one to assume as constants of the model, in addition to the viscoelastic characteristics, only the plastic limit and yield point (the loads corresponding to its fracture) or the energy expenditures on crushing to a specified grain size. The strain forces and energy expenditures in the intermediate cycles are formed automatically and are determined by the presented rheological equations. Thus, when investigating the crushing process with the aid of a model of a single rock chunk, we establish that during the course of the crushing the dynamic reaction rises until the piece is fractured. This same phenomenon was identified when analyzing experimental results when the maximum separation of jaws was recorded prior to breaking up the single rock chunk (see Chapter 1). The same correlation is observed by analyzing the character of change of the energy expenditure at various stages of the crushing process of a single rock chunk. In both cases, energy consumption per cycle increases towards the end of the crushing process. All this is a witness to the fact that the model under consideration adequately reflects, in a qualitative sense, the physical features of interaction of the crushed rock mass with the jaw of the vibratory crusher.

All intermediate cycles are formed similarly to the above mentioned. Specific features are present only for the last $n$-th breaking cycle where the process comes to an end when the ultimate strength is reached (the corresponding breaking forces of the rock mass chunk $F_b$) or if the crushed chunk absorbs enough energy to be crushed to the specified size. In the first case, the end of the crushing process is determined by the force criterion; in the second case, by the energy criterion. We assume that upon reaching the given criterion the model breaks and the stresses are removed (the residual strains and stresses in the newly formed elements of the model are not considered in this case). Henceforth, only the accumulated energy in the generated pieces, associated with the formation of free surfaces in the microcracks, is taken into account. This is the situation when an individual rock piece is crushed. When rock mass is crushed while the crushing chamber is completely filled, the process (as it was indicated when analyzing the experimental data) has another pattern and is more stable.

When rock mass is crushed the process is substantially averaged. This occurs due to the fact that the crushing chamber contains a large number of chunks of various sizes and properties; in addition, their crushing is at different phases: one chunk is in the stage when protuberances are initially chipped, others have already acquired a round shape, still others have totally been broken, and so on. Thus, when crushing rock mass in the crusher working chamber, the chunks of rocks are at all possible stages of breaking. The consequence of this is the averaging and stabilization of the dynamic reaction of the rock on the crusher jaw.

This circumstance considerably simplifies the phenomenological model of the rock mass compared with the model of a rock chunk. The phenomenology of the whole process of vibration crushing is easier to describe. In this case, each subsequent loading cycle of the rock progresses analogously to the preceding one (see Fig. 2.10b). Unlike the single chunks, the mean energy expenditure on crushing rises linearly. The developed model can reproduce the laws of strain not only of viscoelastoplastic rocks, but also of brittle rocks. In this case, the stage of irreversible strains simulates the brittle fracture of the protuberances on the surface of the rock chunks. For deformation of a brittle rock, the $F - x$ diagram will look as shown in Fig. 2.10c. At the start of the loading process the viscoelastic deformation of the rock takes place. When the ultimate strength $\sigma$ is reached at one of the protuberances of the chunk, it breaks, and the load on the rock instantly decreases and the strain is redistributed in the unfractured parts. When the rock is further loaded, the forces and strains are increasing again until they reach the breaking limit at any other protuberance of the chunk, and so on. If a medium line is drawn on the section of the $F - x$ diagram corresponding to the irreversible strains, the force will increase as the strain increases. The averaged diagram will resemble the loading diagram of a viscoelastoplastic rock. The energy consumption–strain diagram is also exactly similar to the diagram obtained for a viscoelastoplastic rock. The magnitude of the equivalent factor of proportionality $k_{pe}$, which determines the angle fo inclination of the curve $F(x)$ at the stage of irreversible strains (fracture) for brittle rocks, must be determined by identifying the theoretical dependences with the experimental data. Thus, the developed phenomenological model of the crushed rock is equally applicable for the description of the process of fracture of both viscoelastoplastic and brittle rocks.

### 2.4.3  Rheological Equations of Rock Mass Deformation in the Process of Vibratory Crushing

Let us consider the general case of interaction of the viscoelastoplastic inertial model of rock mass with the vibratory crusher. Since for the phenomenology of the crushing process only laws of strain of the rock mass are of interest, and not the displacements in the direction of the $x$-axis which are possible only at non-synchronous oscillations of the jaws, the rock model can be simplified. Let us consider its interaction with just one jaw, which is quite admissible, since it interacts with the second jaw at the same time and in a similar manner. Since the loads from the crushing jaws acting on the rock mass are equal and directed in opposite directions, they are balanced and the center of gravity of the chunk does not move in the direction of the $x$-axis. In this case, only one part of the chunk with mass $\xi m$, interacting with one jaw, takes part in the oscillations. A more general case was discussed when the processes in the crusher were considered (see Fig. 2.9). Here, collisions of the crushing jaws with the rock mass do not occur simultaneously and the central part of the mass of the rock $(1 - 2\,\xi)m$ also performs translational motion in the working chamber of the crusher.

Let us introduce a moving rectangular coordinate frame $xoy$, which is rigidly linked with the crushing jaw of the vibratory crusher and a stationary frame $x'o'y'$, which is parallel to the moving axes and linked with the frame of the crusher. Let us consider the mechanism of the process of vibratory crushing of rock mass for the general case of jaw oscillation in the direction of the $x$-axis and having time history $x'$. In the direction of the $y$-axis there are no jaw oscillations. The gravitational force acts constantly on the rock mass in the working chamber of the crusher. Furthermore, if the chunk of rock mass lies on the crushing jaw, a reaction in a direction perpendicular to the jaw surface is exerted at the point of contact with the jaw. Using the schematic diagram in Fig. 2.9, we can write the projection of the reaction on the $x$ and $y$ axes

$$F_x = - mg \ \frac{1}{tg \ \alpha} \tag{2.95}$$

$$F_y = mg$$

where $\alpha$ is the angle of inclination of the surface of the crushing jaw to the vertical.

The forces $N_x$, $N_y$ resulting from the weight of the piece and acting on the jaw at the spot it touches the chunk are equal to the reaction of the jaw on the piece but oriented in opposite directions. In addition to the gravitational forces, the part of the chunk participating in the oscillations during the crushing process is acted upon by the elastic forces $k'_x x'$, $k_x(x + x')$ and $k'_y y$, $k_y(y + y')$; forces of viscous resistances simulating hysteretic losses $c_x(\dot{x} + \dot{x}')$, $c_y(\dot{y} + \dot{y}')$; forces of resistance proportional to the plastic strains of the material $k_{px}(x + x' + x_{ve})$ and $k_{py}(y + y' + y_{ve})$, or equivalent forces of brittle fracture and also dry friction on the surface of the jaw sign $(\dot{x}) \ \mu N_y$; sign $(\dot{y}) \ \mu N_x$.

Since motion of the chunk in the working chamber takes place in tight conditions (in a midst of crushed rock mass), additional resistances to its displacement arise, namely medium environmental resistances. We assume that the latter are proportional to the relative $\dot{y}$ and absolute $(\dot{y} + \dot{y}')$ velocities of the chunk in the working chamber of the crusher and proportional to the coefficient of viscous resistances $c_y$, $c_y'$. The resistances to the chunk motion in the crusher working chamber which are proportional to the relative velocity are equal to $c_y \dot{y}$, and the resistances which are proportional to the absolute velocity are equal to $c_y$ $(\dot{y} + \dot{y}')$. It is assumed here that $x'$ and $y'$ are the displacements of the vibratory crusher jaw in the directions of axes $x$ and $y$ (in the case under study the jaw performs only horizontal oscillation, therefore, $y' = 0$).

Thus, the viscoelastic deformation process of the crushed rock mass model in a vibratory jaw crusher in projections on axes $x$ and $y$ in relative coordinates will be described by the following system of differential equations (coordinates $x$ and $y$ counted from the static equilibrium position of the system)

$$\xi m\ddot{x} + (k'_x + k_x)x + c_x\dot{x} = -\xi m\ddot{x}' - c_x\dot{x}' - k_x x' + F^*_x \tag{2.97}$$

$$\xi m\ddot{y} + c_y\dot{y} + (k'_y + k_y)y = -\xi mg - c_y\dot{y}' - k_y\dot{y}' + F^*_y \tag{2.98}$$

where $F^*_x$, $F^*_y$ are the pressure of the rock mass in the feed hopper on the rock mass in the crushing chamber acting in the direction of axes $x$, $y$.

Viscoelastic straining of the rock mass in the crushing chamber takes place until stresses corresponding to the onset of plastic strains (or brittle fracture) are reached. Plastic strains of the rock mass commence when the following conditions are fulfilled

$$k_x (x + x') + C_x (\dot{x} + \dot{x}') = F_{pb} \tag{2.99}$$

$$k_y (y + y') + C_y (\dot{y} + \dot{y}') = F_{pb} \tag{2.100}$$

where $F_{pb}$ is the load corresponding to the beginning of plastic deformations (brittle fracture) of the rock mass in the crushing chamber.

The plastic strains (brittle fracture) of the rock mass are described by the equations

$$\xi m\ddot{x} + k_{px} (x - x_{vee}) = -\xi m\ddot{x}' + F^*_x \tag{2.101}$$

$$\xi m\ddot{y} + k_{py} (y - y_{vee}) = -\xi mg + F^*_y \tag{2.102}$$

where $x_{vee}$, $y_{vee}$ are the displacements of the rock mass in the direction of axes $x$ and $y$ at the end of the viscoelastic deformation.

Slippage of the rock mass along the crusher jaw can begin in the case when the viscoelastic or plastic strain forces exceed the frictional force on the surface of the jaw crushing plate. Forces $F_x$ sin$\alpha$ and $F_y$ cos$\alpha$ act in the plane of the jaw, deform the rock mass, and tend to shift it along the jaw surface. Forces $F_x$ cos$\alpha$ and $F_y$ sin$\alpha$ act in the perpendicular direction, also deforming the rock mass and pressing it against the jaw surface. The rock mass is in contact with the crushing jaw without slipping only if the total shearing force $F_x$ sin$\alpha$ + $F_y$ cos$\alpha$ does not exceed in magnitude the limit value of the static friction force $\mu_{st}$ ($F_x$ cos$\alpha$ + $F_y$ sin$\alpha$) of the rock mass on the jaw surface. When the shearing force exceeds the static friction force, the rock mass begins to slip over the jaw. The pressure of the crushing jaw in the direction of axes $x$ and $y$, which is exerted on the rock mass in the working chamber, is determined by the following expressions at the stages of both viscoelastic and plastic deformations

$$F_x = k'_x x \tag{2.103}$$

$$F_y = k'_y y \tag{2.104}$$

The condition of the start of slippage of the rock mass at the stages of both viscoelastic and plastic strains (brittle fracture) is written as follows

$$/k'_x x \, \sin\alpha - k'_y y \, \cos\alpha/ \geq \mu_{st} / (k'_x x \, \cos\alpha + k'_y y \, \sin\alpha)/ \tag{2.105}$$

The equations of slippage of the rock mass along the jaw surface at the stages of both viscoelastic and plastic strains (brittle fracture) are

$$\xi m \ddot{x} + c_x \dot{x} + k_x x = -m \ddot{x}' - c_x \dot{x}' - k_x x' - \text{sign}\,(\dot{x})\, \mu\,(k'_x x \, \cos\,\alpha$$

$$+ \, k'_y y \, \sin\,\alpha)\, \cos\,\alpha + F_x^* \tag{2.106}$$

$$\xi m \ddot{y} + c_y \dot{y} + k_y y = - \, \xi m g - \text{sign}\,(\dot{y})\, \mu\,(k'_x x \, \cos\,\alpha + k'_y y \, \sin\,\alpha)$$

$$\times \, \sin\,\alpha + F_y^* \tag{2.107}$$

Equations (2.106) and (2.107) are nonlinear since they include a dry friction force. Depending on the sign of velocity of the rock mass relative to the jaw the friction force changes its direction. This is taken into account by the dependences

$$\text{sign}\,(\dot{x}) = \begin{cases} + \, 1 \text{ when } \dot{x} > 0, \\ - \, 1 \text{ when } \dot{x} < 0, \end{cases} \quad \text{sign}\,(\dot{y}) = \begin{cases} + \, 1 \text{ when } \dot{x} > 0 \\ - \, 1 \text{ when } \dot{x} < 0 \end{cases}$$

The dry friction force also varies depending on the value of the rock mass reaction on the crushing jaw. Therefore, the values of reactions $F_x$ and $F_y$ of the rock mass in the direction of axes $x$ and $y$ in the presented equations must be used in accordance with the character of the strains taking place at the moment of slip, namely viscoelastic or plastic strains (brittle fracture).

The crushed rock mass will be in contact with the jaw until its reaction on the jaw becomes equal to zero:

$$F_x = F_y = 0 \qquad (2.108)$$

If this condition is satisfied, the crushed rock mass loses contact with the jaw and begins to fall in conditions of constrained motion in the crushing chamber.

Let us consider the laws of motion of the rock mass in the direction of the $y$-axis (from the feed receptacle to the discharge opening). In this case, interaction of the rock mass in the crushing chamber with the mass in the feed hopper is of considerable significance. Since we are mainly interested in the phenomenology of the crushing process and the displacements of the rock mass must be known only to establish the moments of start of the rock deformation, the effect of the rock mass in the feed hopper is accounted for by the force of the pressure acting on the rock in the crushing chamber $F_y$.

Thus, the differential equation of the constrained motion of rock mass in the working chamber of the crusher in the direction of the $y$-axis without contact with the crushing jaws is written as

$$m_y \ddot{y} + (c_y + c_y') \dot{y} = -mg + c_y' y' + F_y^* \qquad (2.109)$$

The free fall of the piece in the crushing chamber comes to an end at the moment of coincidence of the coordinates of the rock mass and the jaw which takes place when the following condition is satisfied

$$x' = x - y \operatorname{tg} \alpha \qquad (2.110)$$

The second term on the right-hand side of the presented expression takes into account descending of the rock mass in the working chamber and angle of inclination of the crushing plate.

During the crushing process forces from the rock mass being crushed act on the jaw:

in the absence of slippage of the rock mass along the jaw

$$N_x = -F_x \tag{2.111}$$

$$N_y = -F_y \tag{2.112}$$

and when the rock mass slips over the jaw

$$N_x = \text{sign } (\dot{x})\, \mu\, (k'_x x \cos \alpha + k'_y y \sin \alpha) \cos \alpha \tag{2.113}$$

$$N_y = \text{sign } (\dot{y})\, \mu\, (k'_x x \cos \alpha + k'_y y \sin \alpha) \sin \alpha \tag{2.114}$$

Equal in magnitude but opposed in sign, forces act on the rock mass during the process of slippage.

The described viscoelastoplastic strains (brittle fracture) are sustained due to the drive energy imparted to the crushed rock by the jaw of the vibratory crusher. As already mentioned above, in each jaw stroke the rock receives a specific amount of energy which is spent on viscoelastic and plastic strains, the formation of new surfaces (development of cracks and fracture), overcoming the internal and external friction, and so on. Usually, the energy imparted to the crushed rock per a jaw stroke due to limited amplitude of the jaw oscillations is insufficient for a total fracture. Rock chunks undergo an increase of the number and expansion of the existing cracks, sharp protuberances are chipped, and residual strains are generated. Only after the rock receives the energy which is required for the formation of new free surfaces corresponding to the necessary degree of grain-size reduction does its crushing to the specified grain size commence.

When investigating the process of crushing in a vibratory crusher according to the phenomenological dependences developed above, the completion of the crushing process is defined either by the moment the stresses in the rock reach the ultimate strength or by the magnitude of the specific energy imparted to the rock during the process of irreversible straining. The first criterion is used when investigating the crushing process of a single rock chunk. In this case, the force criterion is more applicable, since the obtained phenomenological dependences enable determination of the growth of the internal stresses in the chunks during development of the crushing process. Therefore, when computations are carried out with the aid of analog or digital computers, comparison is conducted between the internal stresses in the model of rock mass chunk and the ultimate strength of the investigated rock. Matching of these values indicates conclusion of the crushing process.

When investigating the crushing of rock mass the energy criterion is adopted. In crushing large volumes of rock mass filling the chamber, different chunks are at different stages of fracture. Therefore, the criterion of attainment of a specified stress level is found to be unacceptable. The energy criterion is more accurate, since it is easily applicable and simple to determine from the experimental data. When conducting computations using the energy criterion, the program provides for comparison of the values of the irreversible expenditures upon crushing of the rock mass with the reference values. Matching of these indexes indicates the conclusion of the crushing process of the studied volume of the rock mass.

The energy criterion of the crushing process was adopted on the basis of the improved Kirpichev-Kick energy volume hypothesis which gives the best correlation with the experimental data. Also used were studies of the crushing process in jaw crushers which were conducted by V. A. Bauman and then by other investigators. According to these studies, the crushing work is proportional to the magnitude of the elastic strain in the crushed rock and depends on its properties, volume, and the degree of grain-size reduction. In the method developed the crushing process is considered as dependent on the irreversible viscoplastic strains and brittle fracture. Such an approach enables more fully reflecting the physical laws of the crushing process; however, its application is possible only on the basis of the phenomenological viscoelastoplastic inertial models of the crushed rock mass which are developed in the present method and also enable simulating the brittle fracture of the rock during the crushing process.

### 2.4.4 Rheological Equations for Rock Mass Deformation in Crushers with Inertial-eccentric Drive

In the process of crushing in the inertial-eccentric crusher, the rock is subjected to a complex force action of the crushing jaw comprising two components: one of low frequency and large amplitude, and another of high frequency and small amplitude. In this case, the crushing takes place mainly under the action of the low-frequency forces, and the high-frequency component of the pressure change between the jaw and the rock mainly causes a decrease in the friction on the contact surface. The presence of the high-frequency components of the crushing force also leads to opening of cracks and weakening of the internal bonds in the rock mass.

Let us consider the vibration crushing process of the rock under biharmonic oscillations of the jaw. To study this process, the phenomenological model of the crushed rock developed in section 2.4.2 is supplemented by introducing the compaction core and considering its wedging action on the crushed chunk of rock. The improved model of the rock-crusher jaw system is shown in Fig. 2.9c. The phenomenological model of the system is comprised of the crushing jaw of mass $M$ and the rock which is represented by a two-component inertial viscoelastic system characterized in the direction of the $x$-axis (jaw stroke) by the reduced mass $m_x$, stiffness $k_x$, and viscosity $c_x$; in the direction of the $y$-axis by mass $m_y$, stiffness $k_y$, and viscosity $c_y$. In the presence of compaction wedge, interaction of deformations in the direction of axes $y$ and $x$ takes place. The interaction coefficient of deformations along these axes is denoted by the letter $\xi$. The mechanism of this interaction can be pictured as follows: when the rock is compressed by the crushing jaw (rock strain in the direction of the $x$-axis), a compaction core is formed, which under the action of the jaw is implanted deep into the rock chunk and tending to cleave it. The implantation of the core and fracture of the chunk are hindered by the internal bonds in the rock and by the force of rock friction on the jaw surface. The presence of the latter force calls for the need to increase the crushing forces, this causing increasing unproductive energy expenditures associated with the crushing process, and increasing jaw wear. In further investigation the possibility of reducing the forces of friction of the rock on the crushing jaw by applying high-frequency vibrations will be considered, which facilitates the crushing process. It is assumed that the displacement, velocity, and acceleration of the crushing jaw are generally described by the laws of vibration $x'$, $\dot{x}'$, $\ddot{x}''$. By assuming specific time histories describing jaw motion, we determine the most effective operating regime.

The force of resistance to the penetration of the compaction core into the rock is due to the internal strain forces of the rock in the directions of axes $x$ and $y$ at the place of core penetration and to the frictional forces along on its surface. Moreover, the principal forces are associated with the wedging of the piece and moving its halves in the direction of the $y$-axis. Taking into account the interaction of deformations along axes $x$ and $y$ by coefficient $\xi$, the forces of core penetration into the rock can be written as $F'_x = k'_y (\xi \bar{x} - y)$.

The total deforming force of the rock in the direction of the $x$-axis (force of resistance to jaw motion) is comprised from the force of resistance to penetration of the compaction core and the strain force of the rock in the direction of the $x$-axis. The latter force, taking into consideration that jaw mass $M$ is several times larger than the oscillating mass of rock chunk being crushed (comprising only a part of its total mass), is expressed in terms of its inertial properties in the direction of the $x$-axis. Taking this into account, the force of resistance to the strain of the rock in the direction of the $x$-axis can be expressed in terms of displacement and velocity of jaw motion $F''_x = kx' + c\dot{x}'$. Since the parameters of jaw motion are known, it is not difficult to determine the interaction force between the jaw and the rock

$$F_x = k'_y (\xi x' - y) + k_x x' + c_x \dot{x}'$$

To determine strain of the rock in a direction $y$ parallel to the plane of the jaw, differential equations are written. The strains of the rock along the $y$-axis are described by three equations for the cases of absence and presence of slippage along the jaw.

The force created by the compaction core $F' = k_y (\xi x' - y)$, which is tending to cleave the rock chunk, is opposed by the friction force of the rock on the jaw surface $\mu (k_x x' + c_x \dot{x}')$ and by forces of internal bonds $F''_y = (k_y y' + c_y \dot{y}')$.

Accordingly, the generalized equation of rock strain considering the compaction core can be written as follows

$$m_y \ddot{y} + c_y \dot{y} + (k_y' + k_y) y = k_y \xi x' - \mu \, \text{sign}(\ddot{y}) (k_x x' + c_x \dot{x}')$$

$$(2.115)$$

$$\text{where sign}(\dot{y}) = \begin{cases} +1 & \text{when } \dot{y} > 0 \\ 0 & \text{when } \dot{y} = 0 \\ -1 & \text{when } \dot{y} < 0 \end{cases}$$

Equation (2.115) is a nonlinear equation which is broken down into three independent equations for the cases $\dot{y} = 0$ (no slip), $\dot{y} > 0$ (slip during crack development), and $\dot{y} < 0$ (slip during crack compression). The third equation (for case $\dot{y} < 0$), in the case under consideration, with the positive jaw velocity ($\dot{x} > 0$) is not practically realized. Therefore, two regimes are considered: elastic strain and slip. The condition of transition of the process from the elastic strain is the equality of shearing elastic forces to frictional forces and forces of internal resistances

$$k'_y \left( \xi x' - y \right) \geq \mu \left( k_y y + c_y \dot{y} \right) + k_y y + c_y \dot{y} \tag{2.116}$$

Given the oscillation time history of the jaw of the investigated inertial-eccentric crusher as $x' = A(\sin \omega t + 1/a \sin i \omega t)$, the equation of rock strain can be written, after transformations, in dimensionless variables in a form more convenient to solve and analyze

$$\ddot{x}^* + \frac{\zeta i \omega}{z_x^2} \dot{x}^* + \frac{1}{z_x^2} x^* = \xi \frac{1}{z_y} x' - \mu \, \text{sign}\,(\dot{y}) \left( \frac{1}{z_y} x^* + \frac{\zeta i \omega}{z_y} \dot{x}^* \right) \tag{2.117}$$

Here $\xi$ is the proportionality coefficient between viscous resistances and rock stiffness ($c = \xi k$); $z_x$ and $z_y$ are detuning factors of the crusher; $i$ is the frequency ratio of the eccentric and unbalanced mass of the crusher; $a$ is the ratio of the amplitudes of low-frequency and high-frequency oscillations; $A$ is the amplitude of low-frequency oscillations of the jaw; $x^*$, $\dot{x}^*$, and $\ddot{x}^*$ are the dimensionless deformation velocity, and acceleration of the rock deformation, $x^* = x/A$, $\dot{x}^* = \dot{x}/(Ai\omega)$, $\ddot{x}^* = \ddot{x}/(Ai^2\omega^2)$; $x'$ and $\dot{x}'$ are the dimensionless displacement and velocity of jaw

$$x' = \sin \omega t + \frac{1}{a} \sin i \omega t, \qquad \dot{x}' = \frac{1}{i} \cos \omega t + \frac{1}{a} \cos i \omega t$$

As a result of solving equation (2.117) on a computer, determined were the dimensionless strain $x^*$, velocity and acceleration $\dot{x}^*$ and $\ddot{x}^*$ of rock strains, and forces $F_x'$, $F_x''$, $F_x$, $F_y'$, $F_y''$, $F_y$ created in the process of rock crushing.

The evaluation of the efficiency of the crushing process is carried out on the bais of mean energy expenditures on crushing which are determined from the expression

$$W = \frac{1}{\omega T} \int_0^{\omega T} F_x \dot{x} \, d\omega t \tag{2.118}$$

and energy expenditures on overcoming the forces of friction on the jaw

$$W_m = \frac{1}{\omega T} \int_0^{\omega T} (\mu x' + c_x \dot{x}') \, \dot{y} \, d\omega t \tag{2.119}$$

with the superposition of high-frequency oscillations or without it. Here, $T$ is the interval (in time) of the realization of the crushing process.

## 2.5  CASTING AND WORKING OF METALS IN VIBRATORY ENVIRONMENT

### 2.5.1  Formulation of the Problem.  The Basic Laws of Crystallization of Metals under Vibrations

Casting-rolling aggregates create real prospects for the transition to in-line production in metallurgy and for the sharp increase in productivity and quality of castings and rolled products. The main elements of casting-rolling assemblies are continuous-casting machines (CCM). Continuous casting shortens the metallurgical production cycle and sharply increases the rolling quality. The most versatile currently are radial machines which have a number of advantages, including smaller dimensions and cost, as well as product high-quality.

The technology of continuous casting is widely applied in the developed industrial countries. VNIImetmash has developed and successfully applied various CCM modifications in a number of metallurgical plants. It must be noted that despite the very high production efficiency of the existing CCM there are needs for their further development. In particular, it would be important to increase the rate of casting, improve structure and surface of the ingot for a number of difficult-to-roll steels and others. Work is being carried out for application of vibratory technology with the aim of increasing the efficiency of operation of the individual units of casting rolling assemblies and increasing the product quality. The accumulated experience shows that the combination of the techniques of continuous casting with vibratory technology can be very fruitful. In particular, imparting high-frequency mechanical oscillations to the mold of a radial CCM at the Rustavy metallurgical works enabled increasing the casting speed and improving the quality of the ingot. In the meantime, the obtained results show that the continuous casting-rolling technology in metallurgy can be significantly improved with the proper use of the methods of vibration engineering.

Careful analysis of the available, albeit uncoordinated, experiences of application of the methods of vibration engineering in continuous-casting technology enables one to systematically survey the achievements in this area [10].

The following main aspects of application of vibratory technology in casting rolling assemblies can be noted. When the casting passes through the crystallizer, fairly significant resistances to motion are encountered which are caused by the friction between its surface and the mold walls. Consequently, in the relatively thin crystallized section of the ingot, significant stresses are created by the tractive system which frequently cause rupture of the crust of the casting and spill out of the liquid metal contained inside it. The significant resistances to motion impose a limitation on the rate of casting which limits the output and makes matching of the casting machine with the rolling mill more difficult. Consequently, difficulties arise in the realization of advanced in-line operations in metallurgical production. Thus, one of the problems that needs to be solved is the increase of casting speed which is linked with the decrease of the resistances to ingot motion through the crystallizer (mold) and the acting stresses in it.

The second problem which can be simultaneously solved by the developed vibratory mold is the acceleration of heat- and mass-transfer processes and equalization of temperature distribution in the ingot. The consequence of this must be intensification of the process and more uniform cooling of the ingot in the mold, improvement of the structure of the metal both with respect to decreasing the grain size and in ensuring more uniform distribution of inclusions.

The process of crystallization begins with the formation in the liquid phase of solid-phase inclusions, which are called crystallization centers, their subsequent development, and increase in volume. The process of formation of a new phase in this case is linked with the expenditure of energy on the creation of interphase boundaries. The kinetics of conversion from the liquid to solid phase is determined by two main factors: the number of crystallization centers developing per unit time per unit volume, and the speed of crystal growth. With the increasing rate of initiation of crystallization centers and decreasing rate of their growth, the grain sizes decrease.

Crystallization centers are formed either in the volume of the liquid phase or on foreign solid particles present in the liquid. The most effective catalytic action is displayed by the solid particles in the melt, which have physical, structural, and dimensional affinity with the crystallizing substance. The reduction of the work required for the embryo formation is facilitated by a decrease in surface tension at the crystal-liquid boundary and by a reduction of the edge wetting angle. In the final analysis, the indicated factors cause an increase in the speed of initiation of crystallization and the formation of small-grain structure of the casting.

In the absence of vibratory action in the crystallizing metal contained in the mold, crystallization centers are formed near the walls as a result of overcooling. In the first stage crystal growth occurs mainly at the surface in different directions; with time the crystals growing from various centers come close to each other. Furthermore, a region of fine equiaxial crystals are formed near the wall of the casting. Then the crystals begin to grow deep into the melt and elongated column-shaped crystals are formed. The crystals are oriented in the direction of heat transfer and develop in the opposite direction to the heat flux. With time the developing crystals come into contact with each other. Thus, the column-shaped structure of the ingot is created. Usually, it is composed of an external layer of fine equiaxial crystals, regions of column-shaped crystals, and a central zone of equiaxial crystals.

In castings with the column-shaped structure, the central part of the ingot is usually enriched by contaminants which cause deterioration of its mechanical properties. The region of column-shaped crystals is also characterized by low strength. From the viewpoint of raising the mechanical properties of the metal the equiaxial structure is desirable with the grains having equal dimensions in different directions and random orientation in space.

The role of vibratory (at the present time mainly ultrasound) excitation on the casting must be in the enhancement of the factors which lead to obtaining fine-grain, randomly oriented metal structure. Generalization of the available experience of treating melts by oscillations in the ultrasound range [11] enabled one to identify the following changes of the ingot structure: decrease of the average grain size; replacement of the column-shaped structure with equiaxial grains; increase of ingot homogeneity; more uniform distribution of nonmetallic inclusions.

According to existing knowledge, the obtained improvement of the ingot structure is attained as a result of the following phenomena being introduced into the melt: cavitation; fluctuations of viscous friction forces; increase of frequency of initiation of crystallization centers, and dispersion of the formed crystals. The effectiveness of the process of ultrasound treatment is largely affected by the conditions of crystallization and by the nature of the metal. It has been established that the effectiveness of the vibratory treatment increases in conjunction with other treatments, for example, by the introduction of inclusions into the melt which serve as centers of crystallization. Under the joint action of these factors the specified level of grain-size reduction of the structure is reached at lower intensities of the exciter.

The accumulated experience and conducted investigations on the crystallization of metals under vibrations (mainly in the range of ultrasound frequencies) do not adequately reveal the physical laws of the processes involved. This makes it difficult to develop a phenomenology of the crystallization process of metals in conditions of vibratory actions.

It is currently feasible to develop a preliminary phenomenological approach and rheological models of the melt which enable one to estimate the level of energy input into the melts and the laws of motion of the latter for specified parameters of the external vibration actions.

## 2.5.2 Continuous Casting of Metals with Vibrating Crystallizer

It is pertinent to note that current CCM, despite their general technological effectiveness, suffer from a number of shortcomings. The main shortcoming is the limited casting rate (speed of passage of the workpiece through the mold). The ultimate pulling speed increase is limited by the strength of the skin which is formed during passage through the mold. The quicker the ingot passes through a mold of specified length, the thinner the skin as a result of inadequate heat transfer. Increasing the mold length does not quite solve the problem, since the resistances to the displacement of the ingot and, consequently, stresses in the ingot are increasing in this case in the outlet region from the mold. Under large pulling forces skin rupture is possible with the resulting spill out of the liquid metal.

Work is under way on imparting vibration to the mold of a CCM with the aim of reducing the pulling forces and elimination of sticking and skin rupture. The cast metal inside the mold is a physicochemical structure having complex rheological properties – part of the metal is in liquid state and part in various crystallization phases. In order to reproduce the fundamental properties of the metal in the mold in its complex two-phase S – L state, a viscoelastoplastic inertial model has been developed. The parameters of the model are determined by means of identification of the characteristics of motion and deformations of the model with the actual system. Figure 2.11a shows a section of the model in the plane $yx$, the section in plane $yz$ is analogous (not shown in the figure). All possible motions and deformations of the model in the vibrating mold are described by a nonlinear system of differential equations.

Axis $y$ coincides with the direction of ingot motion, and axes $x$ and $z$ are perpendicular. Henceforth, we shall limit ourselves to considering the model parameter along axis $x$, since the parameters along axis $z$ are essentially analogous.

The elastic properties of the model in the direction of $x$-axis are simulated by elements with stiffness coefficients $k_x$ and $k_x'$, viscous elements – by dampers with coefficients of viscosity $c_x$, plastic elements by a wedge pair of dry friction with coefficient $k_{px}$; mass of the skin is simulated by inertial elements $m_x$. The walls of the mold are under ferrostatic pressure $F_f$ from the contained metal.

The model in the direction of axis $y$ is characterized by the elastic properties $k_y$ and $k_y'$, viscous properties $c_y$ and $c_y'$; the ingot mass taking part in the oscillations in the direction of axis $y$ is denoted $m_y$. The gravitational force of the metal in the mold $mg$ acts downwards along axis $y$. The pulling speed of the ingot is $v$.

In the general case, the following strains and motions of the ingot are feasible: viscoelastic and plastic strains, and also free motion in the direction of axis $x$; viscoelastic and plastic strains, and also free motion or motion with dry friction in the direction of axis $y$.

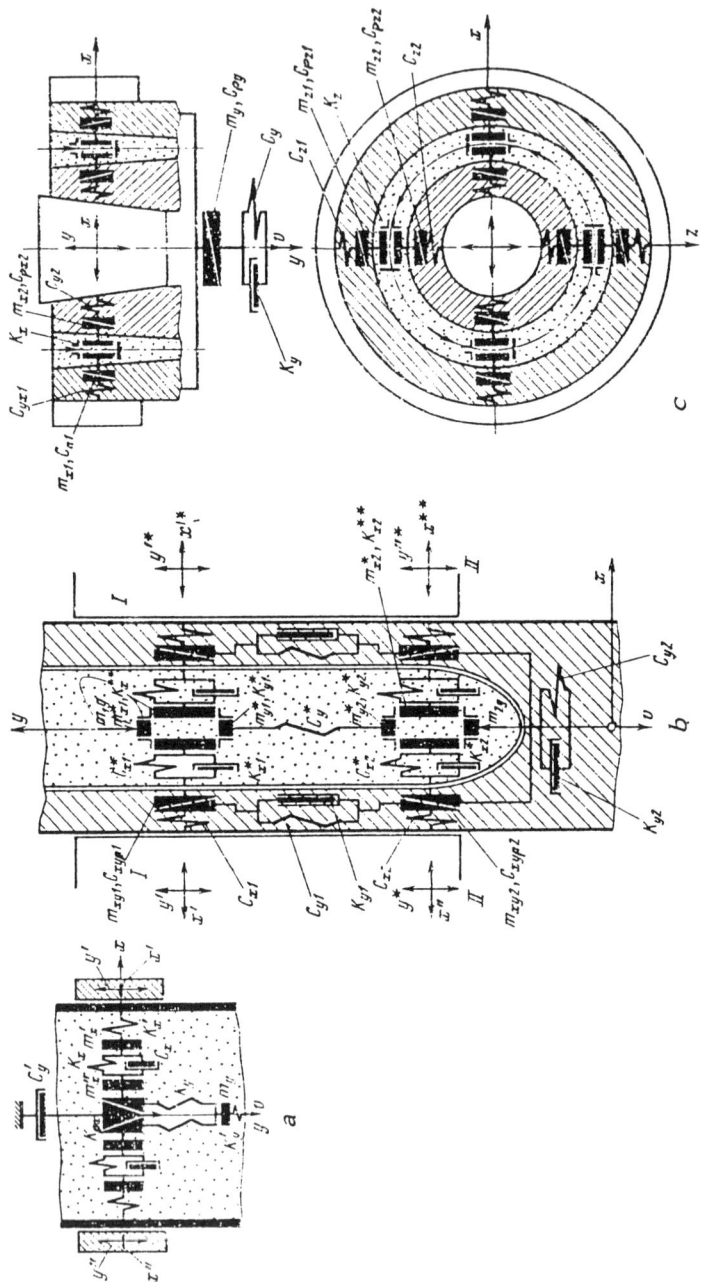

**Figure 2.11** Phenomenological models of the melt for investigation of the continous-casting process through vibrating mold. a) two-component; b) multi-component; c) mold with vibrating core

The viscoelastic strains of the ingot in the direction of axis $x$ are described by the system of differential equations

$$m_x \ddot{x}' + c_x \dot{x}' + (k_x' + k_x)x' = - m_x \bar{\ddot{X}}' + F_f \tag{2.120}$$

$$m_x \ddot{x}'' + c_x \dot{x}'' + (k_x' + k_x)x'' = - m_x \bar{\ddot{X}}'' + F_f \tag{2.121}$$

where $\ddot{X}'$ and $\ddot{X}''$ are the vibratory accelerations of the mold walls; $F_f$ is the ferrostatic pressure on the mold walls.

Viscoelastic strains of the ingot continue until the yield point is exceeded.

Upon compliance with the conditions

$$\left| k_x (x' + \bar{X}') + c_x (\dot{x}' + \bar{\dot{X}}') \right| \geq F_p \tag{2.122}$$

$$\left| k_x (x'' + \bar{X}'') + c_x (\dot{x}'' + \bar{\dot{X}}'') \right| \geq F_p \tag{2.123}$$

viscoelastic strains are transformed into plastic strains which are described by the equations

$$m_x \ddot{x}' + k_{xp} x' = - m_x \bar{\ddot{X}}' + F_f \tag{2.124}$$

$$m_x \ddot{x}'' + k_{xp} x'' = - m_x \bar{\ddot{X}}'' + F_f \tag{2.125}$$

where $F_p$ is the force corresponding to the onset of plastic deformations.

When the load is removed, the plastic deformations (strains) can be transformed into viscoelastic strains.

When the following conditions are satisfied

$$\bar{\dot{X}}' + \dot{x}' = 0 \tag{2.126}$$

$$\bar{\dot{X}}'' + \dot{x}'' = 0 \tag{2.127}$$

plastic strains are transformed into viscoelastic strains, which are described by the equations presented above at corresponding initial conditions. In the general case, when the load is removed (tensile strains), the coefficients characterizing the viscoelastic properties can have different values from those under compression strains, $k_x^*$, $k_x^{'*}$, and $c_x^*$. The skin may periodically lose contact with the walls of the mold under conditions of high-intensity oscillations.

The conditions of loss of contact of the skin with the mold wall are written as follows

$$x' \leq 0 \qquad\qquad (2.128)$$

$$x'' \geq 0 \qquad\qquad (2.129)$$

The motion of the skin during the interval of contact loss with the mold walls is described by equations

$$m_x \ddot{x}' + c_x \dot{x}' + k_x x' = - m_x \ddot{X}' + F_f \qquad\qquad (2.130)$$

$$m_x \ddot{x}'' + c_x \dot{x}'' + k_x x'' = - m_x \ddot{X}'' + F_f \qquad\qquad (2.131)$$

The conditions of contact reestablishment have the form

$$x' \geq 0 \qquad\qquad (2.132)$$

$$x'' \leq 0 \qquad\qquad (2.133)$$

The motion of the ingot in the direction of axis $y$ is described by the nonlinear differential equations

$$m_y \ddot{y} + (c_y + c'_y)\dot{y} + k_y y$$
$$= - m_y \ddot{Y} + C_y v + k_y vt - \text{sing}(\dot{y}) F_{fr} + gm \qquad\qquad (2.134)$$

where $\ddot{Y}$ is the vibratory acceleration of the mold in the pulling direction (axis $y$)

$$\text{sign}(\dot{y}) = \begin{cases} 1 & \text{when } \dot{y} > 0 \\ -1 & \text{when } \dot{y} < 0 \end{cases} \qquad\qquad (2.135)$$

Magnitude of the friction force $F_{fr}$ featured in the above presented equation can be assumed either independent from the speed of ingot

$$F_{fr} = \mu N \qquad\qquad (2.136)$$

or dependent on the speed of

$$F_{fr} = (\text{sign } \dot{y} - \alpha_1 \dot{y} + \alpha_3 \dot{y}^3) \mu N \qquad\qquad (2.137)$$

where $\alpha_1$ and $\alpha_3$ are positive constants.

The ingot pressure on the mold walls is determined from expression

$$N = (k'x' + k'x'')\qquad(2.138)$$

As a result of solving the system of equations (2.140)-(2.141), (2.144)-(2.145), (2.150)-(2.151) and (2.154), one can determine deformations of the ingot and displacements of its skin $x'$, $x''$, $y$, and their velocities and accelerations $\dot{x}'$, $\dot{x}''$, $\dot{y}$, $\ddot{x}'$, $\ddot{x}''$, $\ddot{y}$ at all the characteristic stages of the process.

The tractive force $F_{tr}$ required for pulling the ingot through the mold is determined from expression

$$F_{tr} = k\,(vt - y) + c\,(v - \dot{y})\qquad(2.139)$$

Energy expenditures are:
   on ingot broaching:
      total

$$W_y = \int F_{tr} v\ dt\qquad(2.140)$$

      useful

$$W_{y\,us.} = \int F_{tr}\ \dot{y}\ dt\qquad(2.141)$$

   on vibratory treatment of the ingot:
      total

$$W_x = \int N'\,(\dot{x}' + \ddot{X}')dt + \int N''\,(\dot{x}'' + \ddot{X}'')dt\qquad(2.142)$$

      useful (absorbed by the ingot)

$$W_{x\,us.} = \int N'\dot{x}'dt + \int N''\dot{x}''dt\qquad(2.143)$$

Features of vibratory treatment of the ingot under various time histories of mold oscillations both in the direction of perpendicular to ingot motion are considered.

In the marjority of cases, direct solution is preceded by a reduction of the given system of equations into a form which is convenient for modeling on an analog computer. Furthermore, this form takes into account the method of solution, selected from known methods or specially developed; the imposed limitations and additional conditions; the character of external excitations; and also the methods of read-out and recording to information, which is obtained as a result of the solution [12].

The equations which need to be realized on the analog computer, i.e., the so-called machine equations, are compiled and scaled with respect to the specific type of analog computer. Furthermore, the technical characteristics of the analog machine must be considered. These are the capability to solve the equations given only in an explicit or implicit form, the methods of approximation of nonlinear dependences, the capability of solving equations with constant and variable coefficients, logical capabilities of the analog computer, and so on. The formulated problem was solved on an ac analog computer A-110.

Since the equations from which the coordinates $X'$ and $X''$ are sought have similar structure, all discussions henceforward will be based on one of them.

The equations of ingot motion through the mold corresponding to the forms of strains indicated above are reduced into one equation which is written as

$$m_x \ddot{x} + \Sigma = -m_x \ddot{X} + F_f \qquad (2.144)$$

Here $\Sigma (x, \dot{x})$ is nonlinear function characterizing the force under viscoelastic or plastic strains vs. displacement and velocity of the ingot, and $\ddot{X}$ is acceleration of the external harmonic excitation ($X = a \sin \omega t$). From the standpoint of the specific features of modeling, only the realization of function $\Sigma$ on the computer is of interest.

Let us consider three possible cases of representing function $\Sigma (\dot{x}, x)$, which can characterize the process of straining the ingot in different degrees of approximation.

Firstly, we assume the basic piece-wise linear character of $\Sigma (x, \dot{x})$ as shown in Fig. 2.12a. Sections I, III, IV, VI generally correspond to the viscoelastic strains, and II and V to the plastic strains; moreover, the value of $\Sigma$ during their entire duration remains unchanged.

If in the range of variation of $x$ corresponding to plastic strains the rectilinear sections are replaced by curvilinear sections, we arrive at the case of representation of the functions $\Sigma$ depicted in Fig. 2.12b. And, finally, if magnitude of the force $\Sigma$ on the sections of plastic strains is rising by a linear law, we arrive at a third case whose qualitative pattern is depicted in Fig. 2.12c by the solid line.

As the figures show, the inclination angles of the characteristics in the negative region of the values of strain forces differ from the angles of inclination in the positive region. Thereby, the different characteristics are accounted for under tensile and compressions strains of the workpiece.

Since the load-deflection characteristics of the types presented above are often encountered in the practice of computer-aided modeling when investigating physically diverse processes, we shall dwell on their realization in more detail.

The expression for function $\Sigma$ $(x, \dot{x})$ for the first case is written as

$$\Sigma = (x - \Delta x)(k' + \alpha k) \tag{2.145}$$

where $\Delta x$ are current values of coordinate $x$ on the interval corresponding to the start and end of plastic strain; $\alpha$ is a factor acquiring values 0 or 1 depending on whether the system is operating under tension ($\alpha = 0$) or compression ($\alpha = 1$); $k'$ and $k$ are the stiffness coefficients determining the inclination angle of the characteristic $\Sigma$ relative to the abscissa on intervals of viscoelastic strains. Switching over from one value of $\alpha$ to another is effected by an instruction from the comparator with $\Sigma$ applied to one input and zero - to the other.

Quantity $\Delta x$ is operated upon by the integration block with $\dot{x}$ as an input. The block operates in the following regimes: on segment I of characteristic $\Sigma$ (see Fig. 2.12) location of zero initial values, on segment II integration, on segments III and IV location of the attained values on segment II, on segment V again integration and, finally, on VI - location of the values attained on segment V. It has to be noted that the last segment of the characteristic does not terminate at the origin of the coordinates 0 as shown in the figure, but at some other point on the abscissa as a result of the occurred strains. The closure of the loop at the origin is done to shorten discussions of the descriptive character without violating the generality of the picture.

Let us dwell briefly on the operation of the scheme which provides automatic solution of the equations with minimum interference by the operator of the analog computer. Such a regime of computer operation is obtained by supplying the necessary logical conditions and their realization circuitry. When $t = 0$ all integrators, with the exception of the one operating on $\Delta x$, are changed over on instruction from the

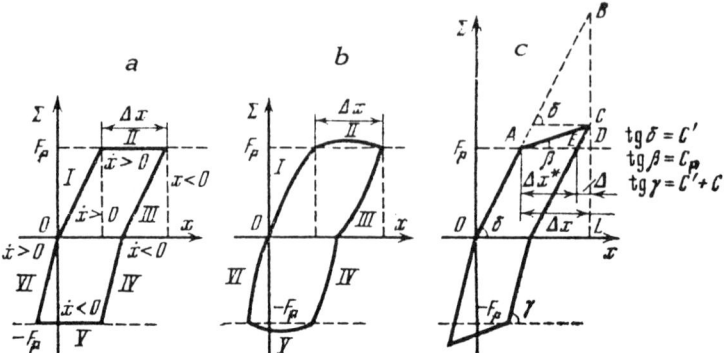

**Figure 2.12** Loading diagrams of viscoelastoplastic strains of an ingot. *a)* for constant loads; *b)* for variable loads; *c)* with hardening

regime of assigning initial conditions to the integration regime. The external harmonic excitation causes a change of $\dot{x}$ and $x$, and value of $\Sigma$ is simultaneously computed. On the first segment $\Sigma$ continuously rises ($\Sigma = xk'$) since $\Delta x = 0$ and $\alpha = 0$ when $\Sigma > 0$. Just as $\Sigma$ reaches the value $F_p$, the $\Delta x$ integrator is switching over into the integration regime of $\dot{x}$. Due to the fact that $x = \int \dot{x} dt$ and $\Delta x = \int \dot{x} dt$, the difference $(x - \Delta x) =$ const (increment in $x$ is equal to the increment in $\Delta x$) and $\Sigma$ will retain a constant value equal to $F_p$ until the sign of speed $\dot{x}$ changes from plus to minus.

The moment the sign of $\dot{x}$ changes, the comparator ($\dot{x}$ and 0) receives an instruction to index the value of $\Delta x$; with this, $x$ begins to decrease and, therefore, $\Sigma = (x - \Delta x)k'$ will decrease with the same angle of inclination of the characteristic as on the first segment. At the moment when $x - \Delta x$ becomes equal to zero (and $\Sigma = 0$), coefficient $\alpha$ on instruction from the comparator ($\Sigma$ and 0), changes its value from 0 to 1, the angle of inclination of the characterristic would increase, and the equation of the viscoelastic strains acquires the form

$$\Sigma = (x - \Delta x_{ind})(k' + 1k) \tag{2.146}$$

When $\Sigma$ in the process of decreasing reaches the level $-F_p$, the integrator operating on $\Delta x$, on instruction from the comparator ($\Sigma_p$, $-F_p$), changes over to the integrating regime. During this process, by virtue of the reasons considered above $\Sigma$ will retain a constant value equal to $-F_p$ until the sign of velocity changes (from minus to plus). From this moment on the value of the integrator $\Delta x$ is again indexed and $\Sigma$, computed from (2.146), will decrease in magnitude. Following the passage of the characteristic $\Sigma$ through zero, the angle of inclination of the next segment would decrease ($\alpha$ becomes equal to zero). Further on, operation of the scheme will be analogous to that described above.

Realization of the second case (Fig. 2.12b) on an analog computer is not fundamentally different, although the expression for the function will have another form

$$\Sigma = (x - \Delta x)(k' + \alpha k) + c\dot{x} \tag{2.147}$$

By introducing into the equation another term $c\dot{x}$, curvilinearity of the viscoelastic and plastic segments is achieved.

In modeling of the third case of variation of the strain forces $\Sigma$, all the logical constructions and algorithms considered for the first case are valid. However, the equation for function $\Sigma$ becomes more complex. Let us derive this equation using the notations adopted earlier. For the sake of clarity and elucidation of the discussion we shall make use of Fig. 2.12c.

We assume as a basis that on segment $OA$ ($\Sigma < F_p$)

$$\Sigma = (x - \Delta x)(k' + \alpha k) = xk' \tag{2.148}$$

($\Delta x = 0$, since integrator $\Delta x$ is on segment $OA$ in the regime of assignment of zero initial conditions, $\alpha = 0$ since $\Sigma > 0$). Had the computation of $\Sigma$, after it becomes greater than $F_p$, taken place using (2.148), the value of $\Sigma$ would have equalled $BL$ after an interval of change of $x$ equal to $\Delta x$. But, on the segment of plastic strain $AD = \Delta x$, the inclination angle of the characteristic must decrease in accordance with the given value of the stiffness coefficient $k_p$, i.e., $\Sigma$ must be equal to $CL$

$$\Sigma = CL = BL - BC; \quad BL = x\mathrm{tg}\delta = x(k' + \alpha k), \quad \alpha = 0$$

From the figure $BC = \Delta x^* \mathrm{tg}\ \delta = \Delta x^*(k' + \alpha k)$, $\alpha = 0$ where $\Delta x^* = \Delta x - \Delta$. From the similarity of triangles $ABC$ and $ECD$ ($AB \parallel EC$) we find

$$\frac{CD}{\Delta} = \frac{BD}{\Delta x} \text{ or } \Delta = \Delta \dot{x} \cdot \frac{CD}{BD} = \Delta x \frac{\dfrac{CD}{\Delta x}}{\dfrac{BD}{\Delta x}} = \Delta x \frac{\mathrm{tg}\beta}{\mathrm{tg}\delta} = \Delta x \frac{k_p}{k' + \alpha k}$$

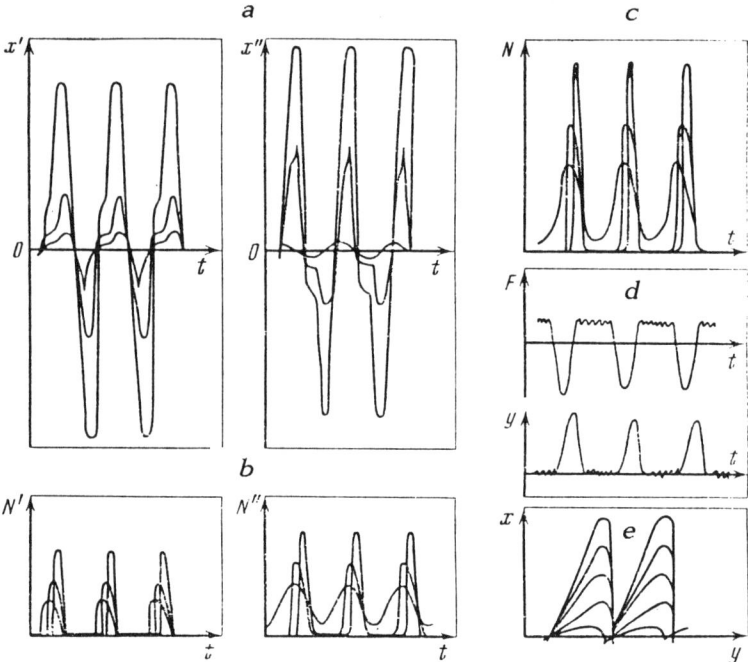

**Figure 2.13** Characteristics of the process of workpiece passage through a vibrating mold. a) vibratory velocities of the solidified phase of the workpiece; b) workpiece pressure on mold walls; c) total pressure of workpiece on mold walls; d) velocity and resistance to the motion of the workpiece in the mold; e) trajectory of motion of the solidified phase of the workpiece

Thus

$$\Delta x^* = \Delta x \left( 1 - \frac{k_p}{k' + \alpha k} \right) \text{ and } BC = \Delta x \left( 1 - \frac{k_p}{k' + \alpha k} \right) (k' + \alpha k)$$

Consequently, for the third case the expression for $\Sigma$ is finally written as

$$\Sigma = BL - BC = \left[ x - \Delta x \left( 1 - \frac{k_p}{k' + \alpha k} \right) \right] (k' + \alpha k) \tag{2.149}$$

When solving the problem, certain elements are introduced into the structural circuits which realize the equations of ingot motion along the $x$-axis on analog computers as applied to three discussed cases of presentation of function $\Sigma$.

Modeling of the equations of the ingot motion in the direction of the $y$-axis does not require further elucidation since known methods and standard schemes are used in this case.

As a result of the investigations on an analog computer of the ingot motion through the mold performing oscillations along the axis and perpendicular to it, a set of parameters fully characterizing behavior of the investigated process has been obtained. As an example, Fig. 2.13 shows the current values of the velocities of the skin $\dot{x}'$ and $\dot{x}''$ and their trajectories, the pressures on the walls of the mold $N'$, $N''$, and $N$, workpiece velocity $\dot{y}$, and the pulling force $F_{tr}$ at various parameters of mold vibrations.

Analysis and generalization of these results allowed one to establish effective oscillation regimes of the mold providing a sharp decrease of the forces of workpiece pulled through the mold $F_v/F$ (where $F_v$ are the resistances of workpiece being pulled through the vibrating mold; $F$ are the resistances in the absence of vibration) depending on some parameters of the workpiece and on the regime of vibrator oscillations.

### 2.5.3 Formation of Solid and Hollow Workpieces under Special Regimes of Mold Oscillation

The method of continuous casting is presently used to produce not only solid but also hollow workpieces for manufacturing seamless tubes.

Conducted investigations have established [13] that the most promising trend for obtaining quality continuously-cast hollow round billets is casting with an internal cooled core. To increase the effectiveness of the production process of hollow billets, a number of new techniques is being employed, one of which is to impart ultrasound oscillations to the core.

Work is currently under way [14] on providing oscillations to the core and to the mold in a wide range of frequencies, amplitudes, trajectories of motion, and so on.

When solving the problem of increasing effectiveness of the production of hollow billets by the continuous-casting method with the application of vibration, one goal is reduction of wall-thickness variation, in addition to the usual requirements relating to reducing resistance to workpiece motion through the mold, improving the structure of the metal, and upgrading surface quality.

According to the mentioned studies, the main cause of variation of wall thickness in the continuous hollow billet is the longitudinal strain of the formed internal skin of the hollow ingot. The internal skin of the billet comes into contact with the core at its upper part (along 100-150 mm from the level of the liquid metal), and then a clearance is formed between them. As a result of such interaction of the billet with the core, the crystallization process develops extremely nonuniformly - intensive cooling and rapid rise of the skin take place at the upper part of the core, then the process sharply decelerates to accelerate again only near the joint of the crystallization fronts.

To study the mechanism of formation of a hollow ingot with the application of vibration to the mold and core simultaneously or only to one of the elements of the system, a phenomenological model of the billet has been developed enabling reproduction of the behavior of its strain in the longitudinal and transverse directions (see Fig. 2.11$b$).

The hollow ingot located between the walls of the mold and the core is a complex one- or two-phase rheological system. At the meniscus level this is a completely liquid metal; after coming into contact with the walls of the mold and core it crystallizes and forms a solidified shell which contains a liquid phase. The thickness of the internal and external shells are growing gradually and the liquid metal layer is reduced. At some distance from the meniscus of the metal the crystallization fronts of the external and internal shells meet and the liquid phase is completely transformed into a solid phase.

Under the action of vibration which can be excited from the core or mold sides and also simultaneously from the internal and external sides, the billet shells, to which viscoelastoplastic properties are ascribed, will behave like two annular concentrated masses separated by a viscoelastic liquid medium. Furthermore, the properties and parameters of the solid and liquid phases change along the length of the ingot. In other words, as a rheological body, the hollow ingot is a two-phase multiple-mass system possessing inertial, viscous, elastic, and plastic properties in the direction of the axes.

The model reproduces two phases of the hollow casting: solid phase with masses $m_{x1}$, $m_{z1}$, and $m_{x2}$, $m_{z2}$ located from the internal and external sides (in the figure the solid phase is shaded), liquid phase of mass $m$ located at the center (designated by dots).

The viscoelastoplastic properties of the hollow billet are reproduced by elastic, viscous, and plastic rheological bodies with the coefficients $k_{x1}$, $k_{xp1}$, $c_x^*$, $k_{x2}$, $k_{xp2}$, $k_{z1}$, $k_{zp1}$, $c_z^*$, $k_{z2}$, $k_{zp2}$, $k_y$, $c_y$, $k_{yp}$.

The inertial properties of the elements of the workpiece are modeled by masses $m_{x1}$, $m_{x2}$, $m_{z1}$, $m_{z2}$, $m_y$. At the contact of the solid and liquid phase complex, mainly viscous, resistances are generated. By taking into consideration that the thickness of the liquid phases in the hollow casting is small and is surrounded from two sides by solid phases, and also taking into account that the resistances at the phase contact are large, particularly due to the presence of dendrites on the surface of the solid phase, we assume that all phases move in the direction of motion without lag and phase shift.

Dry friction exists in the contact area of the solid phase with the walls of the mold and core, which we assume to be velocity dependent as

$$F_{fr} \ (\text{sing} \ \dot{y} - \alpha_1 \dot{y} + \alpha_3 \dot{y}^3)$$

where $\alpha_1$, $\alpha_3$ are constant coefficients; $y$, $\dot{y}$ are displacement and velocity of workpiece relative to the mold wall

$$\text{sign} \ \dot{y} = \begin{cases} + \ 1 \ \text{when} \ \dot{y} > 0 \\ - \ 1 \ \text{when} \ \dot{y} < 0 \end{cases}$$

In order to find effective regimes of vibratory treatment of a hollow casting, the developed phenomenology envisages the possibility of conducting investigations under longitudinal and transverse oscillations of the mold and the core.

The strains and displacements of the hollow casting in a mold with the core vibrating in the transverse direction, modeled by the presented phenomenological model, are described by the following system of differential equations

$$m_{x1}\ddot{r}_1 = - \begin{vmatrix} k_{x1} \\ k_{xp1} \\ 0 \end{vmatrix} x_1 - \begin{vmatrix} c_{x1} \\ 0 \\ c_{x1} \end{vmatrix} \dot{r}_1 + c_{x1}\dot{r}_2 + fmg$$

$$m_{x2}\ddot{r}_2 = - m_{x2}\ddot{X} - \begin{vmatrix} k_{x2} \\ k_{xp2} \\ 0 \end{vmatrix} x_2 - \begin{vmatrix} c_{x2} \\ 0 \\ c_{x2} \end{vmatrix} (\dot{r}_2 + \dot{X}) + c_{x2}\dot{r}_1 - fmg$$

$$(2.150)$$

where $fmg$ is the ferrostatic pressure on the walls of the mold; $\ddot{X}$ and $\dot{X}$ are vibratory acceleration and velocity of the core in the direction of the x-axis.

The last term in the first equation of the system is the perturbation which is transmitted from the core through the liquid phase to the external solidified part of the ingot. Thus, even in the absence of oscillations of the mold, all parts of the workpiece are more or less subjected to vibratory treatment.

In the presence of components of core oscillations in the direction of the z-axis the equations of workpiece oscillations have a similar form.

In accordance with the character of strain and displacement of the hollow casting, the coefficients of the variables and their derivatives in equations (2.150) acquire the following values: under viscoelastic strains $k_{x1}$, $c_{x1}$, $k_{x2}$, $c_{x2}$; under plastic strains $k_{xp1}$, 0, $k_{xp2}$, 0; upon separation from the wall of the mold or core 0, $c_{x1}$, 0, $c_{x2}$.

The moments of transition between certain types of strains and displacements of the hollow casting are determined as a result of solving a system of transcendental equations.

The moment of transition from viscoelastic strains to plastic strains $t_{ve,p}$ is determined from the following relations:
in the external crystallizer layer

$$x_1 \ (t_{ve,p}) = \frac{F_p}{k_{x1}} \qquad\qquad (2.151)$$

in the internal layer

$$x_2 \ (t_{ve,p}) = \frac{F_p}{k_{x2}} \qquad\qquad (2.152)$$

where $F_p$ is the force corresponding to the onset of plastic strains.

The moment of transition from plastic strains to viscoelastic strains $t_{p,ve}$ is determined from the relations:
for the external layer

$$\dot{x}_1 \ (t_{p,ve}) = 0 \qquad\qquad (2.153)$$

for the internal layer

$$\dot{x}_2 \left( t_{p,ve} \right) = 0 \tag{2.154}$$

The moments of transition from joint motion (in contact with the walls) to free motion $t_0$ are found from relations:
for the mold

$$x_1(t_0) \geq 0 \tag{2.155}$$

for the core

$$x_2(t_0) \geq 0 \tag{2.156}$$

and the moments of transition from free motion to joint motion $t_p$ are found from the expressions:
for the mold

$$x_1 \left( t_p \right) \leq 0 \tag{2.157}$$

for the core

$$x_2 \left( t_p \right) \leq 0 \tag{2.158}$$

The motion through the mold and the strains of the hollow ingot in the direction of pulling (axis $y$) are described by the following differential equations:
under viscoelastic strains

$$m_y \ddot{y} + c_y \dot{y} + k_y y = - m_y \ddot{Y} + m_y g + c_y v + k_y vt - F_{fr} \tag{2.159}$$

under plastic strains

$$m_y \ddot{y} + k_y y = - m_y \ddot{y} + m_y g + c_y v + k_y vt - F_{fr} \tag{2.160}$$

where $\ddot{Y}$ is acceleration of mold oscillations in the direction of pulling; $f_{fr}$ is the frictional force between the hollow ingot and the walls of mold and core.

Study of the process of formation of a hollow ingot with the aid of analog computers and determination of all its parameters are carried out by the method expounded in section 2.5.2. Imparting to the mold simple rectilinear harmonic oscillations directed along or across the workpiece motion with judiciously selected parameters provide, as was shown in section 2.5.2, some decrease of resistances to the motion and accompanying stresses in the crystallized surfaces of the ingot layer.

Meanwhile, experience with the application of vibration techniques in many production processes is witness to the fact that special regimes of oscillations of the actuators of the machines are frequently found to be many times more effective than simple harmonic oscillations (see chapter 1). In special designs of the mold the latter can perform complex spatial polyfrequency oscillations causing spatial strains of the ingot including phase shift along its longitudinal axis. Similar oscillation regimes can cause not only the decrease of resistances to the motion of the ingot through the mold, but also the equalization and intensification of heat-transfer processes in the thick of the metal which helps to improve its structure.

To investigate the special regimes of vibratory treatment, one must use more refined phenomenological workpiece models. In developing such a model of the ingot, the task was to provide adequate physical reliability for a minimal number of rheological elements and simplicity of the calculation algorithm.

The metal in the mold, from the rheological standpoint, is a complex two-phase system. The melt enters the mold by gravity from the ladle through a metering device in a continuous regulated flow. Upon coming into contact with the cooled walls of the mold, the melt begins to crystallize from the surface and a solid crust is formed in which liquid phase is contained. As the ingot moves in the mold, the thickness of the crystallized crust grows. The workpiece comes out of the mold with a fairly thick layer of crystallized metal.

Thus, as a rheological body, the workpiece is a two-phase (S - L) system which possesses inertial, viscous, elastic, and plastic properties. Moreover, these properties change gradually along the length of the workpiece.

When developing a phenomenological model to study its motion in the mold under conditions of multiple-component polyfrequency vibratory field, the aim was to have the model, without unduly complicating the system, satisfactorily reflect the main physical properties of the investigated object in the specific conditions of vibratory action under consideration. The phenomenological model of the workpiece in the mold is presented in Fig. 2.11c.

To reproduce the deformation properties of the workpiece along its longitudinal axis (axis $y$), the model has two masses

$m_{y1}$ and $m_{y2}$, which enables study of the longitudinal wave strains in the ingot. The model reproduces two phases of the workpiece: liquid (denoted in the figure by dots and located at the center of the model) and solid (denoted by the shaded area and located on the periphery of the model). The viscoelastic properties of the liquid phase are reproduced by elastic and plastic rheological bodies with the coefficients $k_x^*$, $k_y^*$, $c_x^*$, $c_x^{**}$, $c_y^*$. These coefficients pertaining to the first mass have subscript 1, and those pertaining to the second mass have subscript 2. The ferrostatic pressure at different levels of the ingot is modeled by ascribing initial pressure head $m_1g$ and $m_2g$ to the damper. The inertial properties of the liquid phase are taken into consideration by masses $m_x^*$, $m_y^*$ with the corresponding subscripts.

The viscoelastic, plastic, and inertial properties of the solid phase of the ingot are represented by elastic, viscous, hardened (wedge) plastic, and inertial rheological bodies with the coefficients $k_x$, $k_y$, $c_y$, $k_{px}$, $m_{zy}$.

The contact area between the solid and liquid phases is acted upon by viscous resistances which are represented by a viscous rheological body with coefficient $c_y'$ (not shown in the figure). The contour of the solid phase with the mold is acted upon by a dry friction with coefficient $\mu$ (sign $\dot{y} - \alpha_1\dot{y} + \alpha_3\dot{y}^3$), (where $\alpha_1$ and $\alpha_3$ are positive constants, in the simplest case $\alpha_1$ and $\alpha_3$ can be assumed equal to zero).

In the process of investigations, the capability to impart various oscillations to the mold is provided: linear, planar, and spatial oscillations with any harmonic, polyharmonic, and wave time histories.

When writing the equations of motion and strains of the right and left parts of the solid phase of the workpiece, one ought to keep in mind that they are analogous and, in some cases, can differ only by the character of perturbation. If the perturbation is adequate, one can limit oneself to considering the system of equations compiled for any (right or left) part of the solid phase of the workpiece. Since the structure of the equations describing the strains and motions of individual phases of the workpiece in sections I-I and II-II is analogous, all discussions henceforth will be conducted with respect to one of the sections.

The strains and displacements of the workpiece modeled by the above discussed rheological system in the mold, both in contact with its walls or with loss of contact, are described by the following system of differential equations:

liquid phase

$$m_x^* \ddot{x}^* = m_x^* \underline{\ddot{X}} + fmg - c_x^{**}(\dot{x}^* + \underline{\dot{X}}) - k_x^*(x^* - x) - c_x^*(\dot{x}^* + \dot{x})$$
(2.161)

solid phase

$$m_x \ddot{x} = -m_x \underline{\ddot{X}} - \begin{vmatrix} k_x \\ k_{px} \\ 0 \end{vmatrix} x + \begin{vmatrix} k_x^* \\ 0 \\ k_x^* \end{vmatrix} (x^* - x) + \begin{vmatrix} c_x^* \\ 0 \\ c_x^* \end{vmatrix} (\dot{x}^* - \dot{x})$$
(2.162)

where *fmg* is the ferrostatic pressure on the walls of the mold.

Depending on the character of strain of the solid-phase workpiece the coefficients at the variables and derivatives of the integration in equation (2.162) acquire the following values: for viscoelastic strains fo the workpiece in contact with the mold walls $k_x$, $k_x^*$, $c_x^*$, for plastic strains both in contact with the mold walls and separated from them $k_{px}$, 0, 0; on the section of free motion (upon separation from the mold walls) and viscoelastic strains 0, $k_x^*$, $c_x^*$.

The moments of transition from one form of workpiece strains to another, and from motion in contact with the mold walls are determined as a result of solving the system of transcendental equations.

The moment of transition from viscoelastic strains to plastic strains $f_{\text{ve,p}}$ is determined from condition

$$x(t_{\text{ve,p}}) = \frac{F_{\text{p}}}{k_x}$$
(2.163)

where $F_{\text{p}}$ is the force corresponding to the beginning of plastic strains.

The moment of transition from plastic strains to viscoelastic strains is determined from the condition

$$x(t_{\text{p,ve}}) + \underline{\ddot{X}}(t_{\text{p,ve}}) = 0$$
(2.164)

The moment of transition from joint motion (in contact with the mold wall) to free motion $t_0$ is found from the relation

$$x(t_0) \geq 0$$
(2.165)

and the moment of transition from free motion to joint motion $t_p$ is found from

$$x(t_p) \leq 0 \tag{2.166}$$

The motion through the mold and the strains of solid and liquid phases of the workpiece in the direction of pulling (axis $y$) is described y the system of equations:

solid phase

$$m_{y1}\ddot{y}_1 = -m_{y1}\ddot{Y}_1 + m_{y1}g - k_{y1}(y_1 - y_2) - c_{y1}(\dot{y}_1 - \dot{y}_2)$$
$$- c_y(\dot{y}_1 - \dot{y}_1^*) - a\,\text{sign}\,(\dot{y}_1)\,F_{fr1}$$
$$m_{y2}\ddot{y}_2 = -m_{y2}\ddot{Y}_2 + m_{y2}g + k_{y2}(y_1 - y_2) + c_{y2}(\dot{y}_1 - \dot{y}_2)$$
$$- c_y(\dot{y}_2 - \dot{y}_2^*) + k_{y2}(vt - y_2) + c_{y2}(v - \dot{y}_2) \tag{2.167}$$
$$- a\,\text{sign}\,(\dot{y}_2)\,F_{fr2}$$

liquid phase

$$m_{y1}\ddot{y}_1^* = m_{y1}g - c_{y1}^*\dot{y}_1^* - k_{y1}^{**}(y_1^* - y_2^*) - c_{y1}^*(\dot{y}_1^* - \dot{y}_2^*)$$
$$m_{y2}\ddot{y}_2^* = m_{y2}g - c_{y2}^*\dot{y}_2^* - k_{y2}^{**}(y_1^* - y_2^*) - c_{y2}^*(\dot{y}_1^* - \dot{y}_2^*) \tag{2.168}$$

where $F_{fr1}$ and $F_{fr2}$ are forces of dry friction of masses $m_1$ and $m_2$ on the walls of the mold ($F_{fr1} = \mu N_1$, $F_{fr2} = \mu N_2$); $c_y$ is the coefficient of viscous friction of the solid phase on the liquid phase;

$$\alpha = \begin{cases} 1 \text{ for the motion of workpiece in contact with the mold} \\ \quad \text{walls} \\ \\ 0 \text{ for the free motion of the mold} \end{cases}$$

$$\text{sign}\,(\dot{y}) = \begin{cases} +\ 1 \text{ when } \dot{y} > 0 \\ -\ 1 \text{ when } \dot{y} < 0 \end{cases}$$

$\ddot{Y}_1$, $\ddot{X}_1$, $\ddot{Y}_2$, and $\ddot{X}_2$ are vibratory acceleration of mold in the directions of axes $y$, $x$ at the beginning (mass $m_1$) and end (mass $m_2$) of casting

$$\ddot{Y}_1 = -A_y\left(1 - \frac{\dot{y}_1}{\lambda\omega}\right)^2 \omega^2 \cos\left(1 - \frac{\dot{y}_1}{\lambda\omega}\right)t$$

$$\ddot{X}_1 = -A_x\left(1 - \frac{\dot{y}_1}{\lambda\omega}\right)\omega^2 \cos\left[\left(1 - \frac{\dot{y}_1}{\lambda\omega}\right)t + \gamma\right]$$

$$\ddot{Y}_2 = -A_y \left(1 - \frac{\dot{y}_2}{\lambda\omega}\right)^2 \omega^2 \cos\left[\left(1 - \frac{\dot{y}_2}{\lambda\omega}\right)t + \frac{2\pi}{n}\right]$$

$$\underline{\ddot{X}}_2 = -A_x \left(1 - \frac{\dot{y}_2}{\lambda\omega}\right)^2 \omega^2 \cos\left[\left(1 - \frac{\dot{y}_2}{\lambda\omega}\right)t + \frac{2\pi}{n} + \gamma\right]$$

(2.169)

$\lambda$ is the wavelength; $\omega$ is the angular frequency of oscillations; $2\pi/n$ is the phase shift of oscillations of masses $m_1$ and $m_2$; $N_1$ and $N_2$ are the normal reactions of masses $m_1$ and $m_2$ on the mold walls ($N_1 = k_{x1}x_1$, $N_2 = k_{x2}x_2$).

The resistance to workpiece pulling through the mold is determined from expression

$$F = k_{y2}(vt - y_2) + c_{y2}(v - \dot{y}_2)$$

(2.170)

Total energy expenditures on workpiece pulling through the mold

$$W = \int Fvdt$$

(2.171)

Energy spent on overcoming resistances to workpiece motion is

$$W_p = \int F\dot{y}_2 dt$$

(2.172)

Energy spent on vibratory treatment of the workpiece is:
total

$$W = \int (N_1\underline{\ddot{X}}_1 + N_2\underline{\ddot{X}}_2)dt$$

(2.173)

useful

$$W_{us.} = \int (N_1\dot{x}_1 + N_2\dot{x}_2)dt$$

(2.174)

Study on an analog computer of the process of workpiece forming under special regimes of mold vibration is performed by the technique described above.

## 2.5.4 Plastic Deformation of Metals under Vibrations (Upsetting, Extrusion, Rolling, Drawing)

Diverse technological processes, including pressure working in vibration field, such as vibratory rolling, vibratory drawing, vibratory hardening, have many things in common with respect to their physical essence. This allows the consideration of these processes from general methodological positions in various regimes of vibration excitation.

Development of conditions of effective application of vibration with the purpose of intensification of metal working processes requires a thorough study of the mechanics of the phenomenon and the development of the theory and the computational methods for defining the optimum working regimes. A significant role in solution of this problem is played by the methods of modern rheology, particularly of vibration rheology. Application of the methods of phenomenological rheology with the use of classical rheological bodies does not always yield the desired result, since in this case one cannot reproduce the inertial phenomena which significantly contribute to such dynamic processes.

For the solution of such problems, mechanorheological methods and viscoelastoplastic and inertial models of the worked metals are presently available. Considered below are the generalized systems of nonlinear differential equations describing the strains of the product in a vibratory field.

Depending on the specific application conditions, the regimes of vibration actions can be very diverse. In practice, oscillations used can range from infrasound to ultrasound. Low-frequency or high-frequency regimes of vibratory working are found to be preferable in various cases. In some special cases it might be expedient to use vibration regimes beyond this range.

When using high-frequency oscillations (ultrasound), the worked metal and the tool are subjected to elastic oscillations. Usually, elastic waves are generated in the product. For the generation of a standing wave all the elements of the oscillatory system or of its individual parts must have the dimensions which ensure the creation of resonance of the desired form at the given frequency. The dimensions of the elements of the oscillatory system and the wavelength must be coordinated. To increase effectiveness of vibratory working it is expedient to excite the oscillations at the strain area through the working tool. Oscillations can also be transmitted through the deformed metal. In the second case, the introduction of oscillations to the strain area becomes more difficult if the mass of the worked piece is less than the mass of the tool. Longitudinal, transverse, radial, torsional, and flexural oscillations of the tool are used.

Despite the fact that the mechanism of action of vibration in different processes of metal working are not fully understood up to now, the effect of vibration on the processes under consideration has, however, been convincingly demonstrated. It has been established that vibrations influence both the structure and properties of metals, as well as the contact friction between the worked metal and the tool.

Experimental investigations have established that during vibratory working the static stresses needed for the realization of plastic strain are decreased, the plasticity of the metal is increased, and the opportunity is created for increased workpiece reduction per pass and for working high-strength materials [2]. In the meantime, hardening of the test pieces as a result of pressure working coupled with vibration has been revealed. It is also known that superposition of vibrations on the process of plastic strain affects not only the structure and properties of the worked metal, but also the boundary conditions on the contact surface. Under the influence of vibration contact interaction changes considerably and, as a rule, a decrease of the effective coefficients of friction takes place.

For the plastic straining it is necessary to have motion of dislocations, if such are present, or initiation of new ones, if there are none or they have little mobility. Thus it is natural to link the change of plasticity of the specimens with the characteristic features of the behavior of dislocations in vibratory conditions.

The decrease of resistancees to plastic strain is attributed to the activation of dislocations and increase of their mobility. Activation occurs as a result of absorption of acoustic energy at the locations of the defects of the crystal lattice.

The characteristic properties of pressure working of metals under vibrations are determined by both the processes taking place in the body of the worked piece and by the character of contact interaction with the tool.

Let us consider the application of phenomenological methods of vibration rheology to the description of the behavior of the worked metal in the realization of several characteristic production processes that are performed with superposition of vibration such as, for example, vibratory upsetting (or vibratory extrusion, vibratory rolling, vibratory drawing, and vibratory hardening). The first two processes are carried out with preliminary heating of the worked metal, whereas the last process is carried out at a normal temperature.

For adequate description of these processes, methods of mechanics and rheology must be used. To conduct investigations mechanorheological phenomenological models of the worked metal, machine elements, and the object as a whole are constructed. Let us consider the methods of construction and description of mechanorheological phenomenological models on the example of the processes of vibratory extrusion, vibratory rolling, and vibratory drawing.

Phenomenological models of worked media for the investigation of the processes of pressure working in a vibration field, unlike purely classical rheological models, take into account the inertia of the simulated objects. Mechanorheological phenomenological models representing the object as a whole, i.e., the worked medium and the production machine, take into account, in addition to viscoelastoplastic parameters, statistical properties of the medium, interaction of stresses and strains in various directions. Introduced also are unilateral constraints which enable reproducing the loss of contact and the break up of the complex model into a number of simpler ones for some time, for example, when describing the process of upsetting and separation of the punch from the worked piece. In the intervals when the system is in the state of separation, its elements are characterized by individual "behavior", the reference points of which are the initial conditions at the moment of separation. Upon liquidation of discontinuities and return of the system to the united state, it again behaves as a single object; however, it "remembers" the independent behavior of its elements owing to the initial conditions of the moment of uniting the system.

The developed method of construction of the mechanorheological models enables reproducing, with the required reliability, any real objects. Furthermore, only the condition of attaining the required accuracy using the simplest possible model is stipulated.

For the development of the phenomenological model of the product which is subjected to upsetting, we consider the main laws of this process under vibrations. Detailed experimental investigations [2] have revealed the fundamental features of the process of free metal upsetting when the product is subjected to high-frequency (in the ultrasound range) oscillations and its differences with the process of normal upsetting.

In the conducted experiments the punch periodically lost contact with the product; therefore, a vibroimpact working regime was developed. Briefly, the important process features are as follows. The maximum degree of strain of the specimen in normal free upsetting occurs at the most distant zones from the end face, whereas in vibroimpact upsetting everything is reversed - maximum strain occurs at the end face of the specimen. Under normal free upsetting, the strain is minimum as a result of the large resistances on the contact surfaces in the region of end-face surfaces. Under upsetting in vibroimpact conditions the character of formation of end-face surfaces is drastically changing. In this case, intensive slip of the metal over the end faces occurs with the least degree of product strain at the center of the end face. The material flows from the central parts of the specimen to the end faces. The indicated features of specimen strain are caused by the manifestation of its inertial properties in conditions of pulse load application. The effect of the reduction of resistances due to friction forces on the punch-specimen contact surface is also noticeable.

It has been established that under free upsetting of the specimen in vibratory and vibroimpact conditions the forces of plastic strain are reduced considerably. In vibroimpact regimes strain occurs without static loading and the acting loading ought to be determined as the impact pulse averaged over the cycle of tool oscillations. Under vibration and vibroimpact actions an increase of strain level is noticed and the plasticity of the material is improved. Magnitude of plastic flow stresses is determined by the extent of strain and intensity of vibratory actions.

The reduction of static stresses under free upsetting in conditions of vibration actions is explained by the decrease of contact friction forces and internal resistances, in view of which the slip of specimen elements relative to each other becomes easier, and possibly by the change in the relation between the main directions. Vibration conditions prevent development of medium discontinuity.

Presented in Fig. 2.14 is the schematic of free upsetting of metal and three-component viscoelastoplastic phenomenological model of the worked piece. The elastic strains in the part of mass $m_y + m_x + m_z$, which reflects the inertial properties, are modeled in the process of vibration compaction by elastic bodies with stiffness coefficients $k_x$, $k'_y$, $k_y$, $k_z$. Energy dissipation (hysteresis losses) in the deformation process is

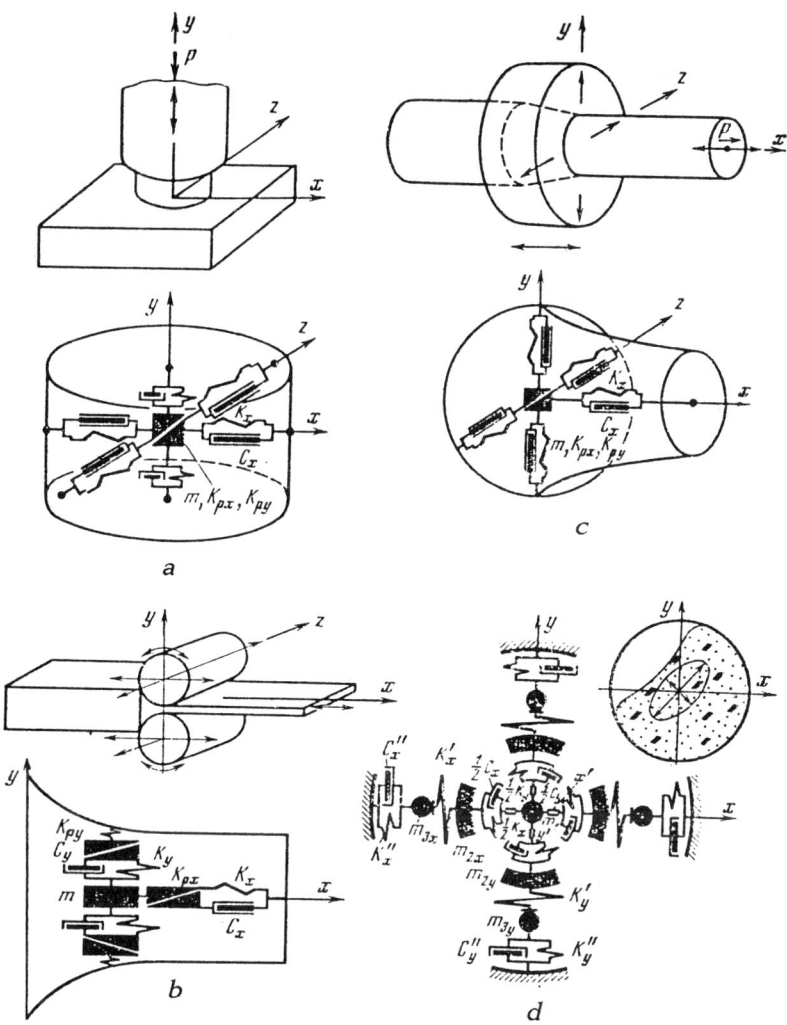

**Figure 2.14** Phenomenological models for study of metal working processes. a) vibratory upsetting; b) vibratory rolling; c) vibratory drawing; d) vibratory hardening of a round element

reproduced by viscous bodies with viscosity coefficients $c_x$, $c_y$, $c_z$; plastic strains are modeled by wedge elements with coefficients $k_{px}$, $k_{py}$, $k_{pz}$; interaction of strains and stresses in mutually perpendicular directions can be accounted for by coefficients $i_{yx}$, $i_{yz}$, $i_{xz}$.

Let us consider the process of free vibroimpact upsetting as the most general case, due to occurrence of periodic losses of contact between the worked element and the punch. In this case the following forms of strains and motions of the worked element are possible: viscoelastic strain, plastic strain, and also free viscoelastic unloading (when the punch losses contact with the element) and slippage over the surfaces in contact.

Without specifying the character of motion, we assume that the punch travels according to some prescribed time history and projections of its motion on the coordinate axes are $y'$, $x'$, $z'$. For example, in the direction of axis $y$ the punch can perform a translational motion with the superposed oscillations and in the direction of axes $x$, $z$ - vibrations. The viscoelastic and plastic strains of the worked element are described in this case by the equations

$$m_y \ddot{y} + \begin{vmatrix} c_y \\ 0 \\ c_y \end{vmatrix} \dot{y} + \begin{vmatrix} k_y + k_y' \\ k_{py} \\ k_y \end{vmatrix} y = m\ddot{y} - {}'F_x - F_z$$

$$m_y \ddot{x} + \begin{vmatrix} c_x \\ 0 \end{vmatrix} \dot{x} + \begin{vmatrix} k_x \\ k_{px} \end{vmatrix} x = -F_y - F_z \qquad (2.175)$$

$$m_z \ddot{z} + \begin{vmatrix} c_z \\ 0 \end{vmatrix} \dot{z} + \begin{vmatrix} k_z \\ k_{pz} \end{vmatrix} z = -F_y - F_x$$

where $F_x$, $F_y$, $F_z$ are the reactions of the worked medium in the directions of axes $x$, $y$, $z$.

In accordance with the character of strain and displacement of the workpiece the coefficients at the variables and their derivatives acquire the following values: for viscoelastic strains $c_y$, $(k_y + k_y')$; $c_x$, $k_x$; $c_z$, $k_z$; for plastic strains 0, $k_{py}$; 0, $k_{px}$; 0, $k_{pz}$; and when separation of the punch occurs - $c_y$, $k_y$.

The moments of transition from viscoelastic strains to plastic strains $t_{ve,p}$ in the directions of respective axes are determined from the relations

$$c_y \left[ \dot{y}(t_{ve,p}) + \dot{y}'(t_{ve,p}) \right] + k_y \left[ y(t_{ve,p}) + y'(t_{ve,p}) \right] = F_{pb}$$

$$c_x \dot{x}(t_{ve,p}) + k_x \dot{y}(t_{ve,p}) = F_{pb} \qquad (2.176)$$

$$c_z \dot{z}(t_{ve,p}) + k_z \dot{z}(t_{ve,p}) = F_{pb}$$

where $F_{pb}$ are forces at the beginning of plastic deformation of the specimen.

The moments of transition from plastic strains to viscoelastic strains are found from equations

$$\dot{y}(t_{ve,p}) + \dot{y}'(t_{ve,p}) = 0$$

$$\dot{x}(t_{ve,p}- = 0 \qquad\qquad (2.177)$$

$$\dot{z}(t_{ve,p}) = 0$$

The moment of the punch separation from the worked part $t_0$ and the moment when it comes back into contact with the part $t_p$ are determined from the relations

$$y(t_0) \leq 0$$

$$\qquad\qquad (2.178)$$

$$y(t_p) \geq 0$$

The forces of specimen strain under viscoelastic and plastic deformations are, respectively,

$$F_{ve} = k_y' y$$

$$\qquad\qquad (2.179)$$

$$F_p = F_{pb} + k_{py} y$$

Figure 2.14*b* and *c* show the schematic diagrams of the processes of vibratory rolling and vibratory drawing, as well as two-component and three-component viscoelastoplastic phenomenological models of a rolled strip and a drawn rod. In this case, more complex phenomenological models of the worked parts are used which enable adequately reproducing properties of the worked metal. The methodical approach to the investigation of the process of vibratory rolling and vibratory drawing is similar to the approach discussed above.

### 2.5.5 Vibratory Hardening

Figure 2.14*d* shows the schematic diagram and the mechanorheological model of the process of vibratory hardening. In this process, the worked elements are placed in a vibrating container filled with hard-alloy balls. As a result of container vibration, the worked elements are subjected to high-frequency impacts of the hard-alloy balls which causes surface cold working and remove internal stresses in the crystal lattice.

The worked piece is represented by a multiple-mass ($m$, $m'_x$, $m_y$) viscoelastoplastic phenomenological model with elastic and viscous elements $k'_x$, $k_x$, $c_x$, and $k'_y$, $k_y$, $c_y$, and with plastic elements $k_{px}$, $k_{py}$. The worked medium is simulated by the inertial viscoelastic system $m^*_x$, $m^*_y$, $k^*_x$, $c^*_x$, $k^*_y$, $c^*_y$ and can be in contact with the piece or be separated from it.

The motion of the worked medium is described by the equations

$$m^*_x \ddot{x}^* + \begin{vmatrix} c^*_x \\ c^{*\prime}_x \end{vmatrix} \dot{x}^* + \begin{vmatrix} k_x \\ 0 \end{vmatrix} x^* = \alpha k'_x x - m^*_x \ddot{\overline{X}}$$

$$m^*_y \ddot{y}^* + \begin{vmatrix} c^*_y \\ c^{*\prime}_y \end{vmatrix} \dot{y}^* + \begin{vmatrix} k_y \\ 0 \end{vmatrix} y^* = \alpha k'_y y - m^*_y \ddot{\overline{Y}} \qquad (2.180)$$

where $c^{*\prime}_x$, $c^{*\prime}_y$ are coefficients of resistance to the motion of the particles of the medium in the bulk; $\alpha$ is the discontinuity factor ($\alpha = 1$ when moving in contact with the piece; $\alpha = 0$ when moving separately from the piece); $\overline{X}$ and $\overline{Y}$ are the displacements of the container.

The motions and strains of the worked piece in the direction of axis $x$ are described by the equations

$$m_x \ddot{x}' + \begin{vmatrix} c_x \\ 0 \end{vmatrix} (\dot{x}' + \dot{x}^* - x) + \begin{vmatrix} k_x \\ k_{px} \end{vmatrix} (x' + x^* + \Delta x - x)$$

$$+ \begin{vmatrix} k'_x \\ 0 \end{vmatrix} x' = 0$$

$$m \ddot{x} + \begin{vmatrix} c_x \\ 0 \end{vmatrix} (\dot{x} - \dot{x}' - \dot{x}^*) + \begin{vmatrix} k_x \\ k_{px} \end{vmatrix} (x - x' - x^* - \Delta x) = 0 \qquad (2.181)$$

where $\Delta x$, $\Delta y$ are the residual plastic strains in the piece. In the direction of axis $y$ the formulas are analogous.

In accordance with the form of the displacements and with the character of the strains, the coefficients at the variables and their derivatives have the following values: for viscoelastic strains $k_x$, $c_x$; $k_y$, $c_y$; for plastic strains $k_{px}$, 0; $k_{py}$, 0; when moving in contact with the working medium $k'_x$, $k'_y$; when moving separately from it 0, 0.

Analysis of the process of vibratory hardening of the workpiece is conducted using methods discussed above with the utilization of the presented equations.

## 2.6  PARAMETER IDENTIFICATION OF THE RHEOLOGICAL MODELS

### 2.6.1 Fundamentals of the Identification Method

For identification of the parameters of phenomenological models and real media which are processed in vibratory machines, one can use the results of experimental investigations on actual installations and analog and digital models for studying the effect of model parameters (natural frequencies, resistance coefficients of the model itself and the surrounding medium, limits of commencing of plastic strains, etc. fracture, ratio of the oscillating and stationary masses, etc.) on parameters of the beginning and end of collision, displacements, velocities, and accelerations of deformations and motions of the medium, forces created in the process, trajectories of motion, energy expenditure, and so on. The obtained results of experimental and modeling investigations are compared with each other. The model parameters determined in the course of this comparison render the model as an analog of the real medium with specific physical properties which are exhibited in the process of its vibratory working.

The developed method of identification of calculated and experimental data and determination of parameters of the rheological models of the worked media is based on information theory and on the method of dynamic testing of objects. The essence of the identification process is as follows: reaction of the medium on the working element of the vibratory machine, determined experimentally by dynamic tests, is represented in accordance with the adopted rheological model as a multiple-valued function of displacement and velocity. Accuracy (credibility) of the adopted rheological model is estimated by the width of the graph-layer of the multiple-valued function of the medium reaction on the working element of the vibratory machine. As a working value of the reaction, one can assume the graph-surface, which is equidistant from the upper and lower surfaces of the graph-layer of the multiple-valued function of the constraint reaction. The reaction of the medium which is obtained experimentally is compared with the calculated reaction, namely the function of displacement and velocity of the working element of the vibratory machine; the values yielding the best correlation are selected. Parameters of the rheological model of the medium are assumed according to this case. For verification of the characteristic features of the vibratory process, one can make comparison between the computational and experimental values of such parameters as phase angles of the motion stages, travel velocities, energy expenditure in the vibratory process, and so on.

## 2.6.2 Theoretical Background of the Dynamic Test Method

Let us consider the dynamic test method which underwrites identification of rheological models on the example of a simple vibratory machine.

The dynamic equation of the working element of a vibratory machine of mass $M$ set into oscillating motion by a limited external force $F(t)$ is

$$M\ddot{x} = R\,(x,\dot{x}) + F\,(t) \tag{2.182}$$

Here $x$ and $\dot{x}$ are displacement and velocity of the working element (mass $M$); $R(x,\dot{x})$ is the function characterizing the dependence of the reaction of the worked medium (crushed rock, conveyed load, and so on) on displacement $x$ and velocity $\dot{x}$ of the working element of the vibratory machine.

Let the displacement of the working element of the vibratory machine under a given disturbance $F(t)$ be described by function $x(t)$. This means that function $x(t)$ identically satisfies (for all the moments of time $t$) equation (2.182). We introduce notation $y = M\ddot{x} + F(t)$, then equation (2.182) can be written as

$$Y = R\,(x,\,\dot{x}) \tag{2.183}$$

Geometrically, the presented equation in a three-dimensional space with coordinates $y,\ x,\ \dot{x}$ can be represented by the graph of function $R$ which is a nonlinear surface.

For solution of equation (2.183) and determination of function $x(t)$, let us consider the three-dimensional trajectory $\bar{\omega}\,(t)$ defined as $\bar{\omega}\,(t) = (y(t),\ x(t),\ \dot{x}(t))$.

Investigations show that the graph of trajectory $\bar{\omega}\,(t)$ for all moments of time $t$ will belong to the graph of function $R(x,\ \dot{x})$ if only function $\bar{\omega}(t)$ is computed for solution $x(t)$ of equation (2.183)

Let us now fix some moments in time $t_1$, $t_2$, ..., $t_N$ ... and note at these moments points $\bar{\omega}\,(t_i)$ = $(y(t_i),\ x(t_i),\ \dot{x}\,(t_i))$ of the three dimensional space through which trajectory $\bar{\omega}\,(y)$ is passing. The coordinates of points $\bar{\omega}\,(t_i)$ are arranged in a table which is called the dynamic test table. This table enables one to present function $R|x,\ \dot{x}|$ in a tabular form.

Using the obtained dynamic test table, the function of the dependence of the constraint reaction $R$ on the displacement $x$ and velocity $\dot{x}$ of the vibratory machine working element (mass $M$) can be restored, i.e., the dynamic equation describing the observed motion of the vibratory machine is restored.

Thus, having the experimental data on the dependences of $x$, $\dot{x}$, $\ddot{x}$, and $F$ on time $t$, the model of dynamic system can be constructed in the form of equation (2.183).

It is particularly important that the knowledge of the dependence of the external force $F$ on time $t$ is absolutely necessary. This is due to the fact that in the absence of such information there is an arbitrariness in selection of function $R$. Thus, for any bounded function $R^*(x,\ \dot{x})$ under arbitrarily observed law of motion $x^*(t)$ the bounded function $F^*(t)$ can be selected for which the dynamic equation will be identically satisfied. In this case, function $F^*(t)$ is computed as

$$F^*\,(t) = M\ddot{x}^*\,(t) - R^*\,(x^*(t),\ \dot{x}^*\,(t))$$

In this case the problem of constructing the dynamic model in the form of equation (2.183) becomes indeterminate, since a model with an arbitrary bounded function $R$ has been assumed.

Within the framework of the presented geometrical constructions we shall assume the restoration (or localization) of the dependence of the constraint reaction $R$ of the worked medium on displacement $x$ and velocity $\dot{x}$ of the vibratory machine working element of mass $M$ to be the solution of the problem of constructing the dynamic model. As initial data we adopt the information on the experimental data which is listed in the dynamic test table. Furthermore, a satisfactory solution of the problem would be one in which function $R(x,\ \dot{x})$ is restored with a specified accuracy and the accuracy is weakly dependent on the change of the initial data.

If information is filed into the dynamic test table, for example, about three points $\omega(t_1)$, $\omega(t_2)$, and $\omega(t_3)$, then it is evident that one plane or unlimited number of nonlinear surfaces can be drawn through these three points.

By filing into the table information on all points $\omega(t_i)$, where $i$ = 1, 2, ..., $n$, ... and also on all the intermediate points between points $\omega(t_i)$ and $\omega(t_{i+1})$ and so forth, the volume of the tabulated information can be increased infinitely. However, this does not lead to the desired result, since an infinite number of nonlinear surfaces can be drawn through the curve defined on the time interval $t_i$ $t_{i+1}$. This circumstance is an indication that the number of "bits" of information, i.e., the measure of information volume, cannot serve as a measure of completeness of information about the initial data of the problem. Conducted experiments show that if points $\omega(t_i)$ are "scattered" over the surface so that the distance between all neighboring points is small, then the shape of the analyzed surface can be traced in sufficient detail. Furthermore, the more points are given and the less is the distance between the neighboring points, the more information we obtain about the surface.

Thus, the problem is reduced to the determination of the number of points and of the method of their arrangement on some region of the analyzed surface so that it would be possible to determine the surface shape with the specified accuracy.

It is necessary that the reaction being restored of the worked medium on the vibratory machine working element, namely function $R|x, \dot{x}|$, at the possible region of its location $D$ should have a limited rate of change, i.e., the function must belong to the Lipschitz category.

In region $D$, in which the reactions on the worked medium of the vibratory machine can practically be located, points $s_i$ are defined ($i$ = 1, 2, ..., $n$) with coordinates $x_j$, $x_j$, which possess the common property that $(x, \dot{x}) = \vec{x}$, which are belonging to region $D(\vec{x} \in D)$. Among these points there ought to be at least one for which the distance between it and point $s_i$ will be less or equal to $\varepsilon$. The points possessing this property are called points of $\varepsilon$-network of region $D$. It is significant that if region $D$ is bounded, as it is in our case, then the number of $\varepsilon$-network points can be made finite. This finite set of $\varepsilon$-network points of region $D$ is denoted as $s_\varepsilon$ ($D$).

If we now enter into the dynamic test table points $\omega_i$ which, being projected on region $D$, yield $\varepsilon$-network, the function $R(x, \dot{x})$ being restored will be given by the table with accuracy $\delta$. The numbers $\varepsilon$ and $\delta$ are linked by inequality $\delta \leq L_\varepsilon$, where $L$ is the Lipschitz constant characterizing the maximum rate of change of function $R(x, \dot{x})$ in region $D$.

As a result, the following scheme of the problem analysis can be outlined. If constant $L$ is known, then from $\delta$ (the desired accuracy of the function restoration which is specified in advance) value of $\varepsilon$ can be assumed. The value of $\varepsilon$ for region $D$, in its turn, determines the number of $\varepsilon$-network elements which will be denoted as $N_\varepsilon(D)$. This number determines the length of the dynamic test table (the number of columns), i.e., it characterizes the "memory volume" required for the initial data of the problem.

This scheme of the analysis of the problem of dynamic model construction which is based on the geometrical properties of a finite number of $\varepsilon$-network points, formulates the quantitative requirements for organization, planning, and accuracy of dynamic experimentation with the analyzed system.

If now, using inequality $\delta \le L_\varepsilon$, we assume $\varepsilon = \delta L^{-1}$, then the number $N_\varepsilon(D)$ in the representation $N_\varepsilon(D) = N\delta\ L^{-1}(D)$ would determine both quantitative and qualitative content of the initial data required for the determination of a reliable value of the reaction of the worked medium on the working element of the vibratory machine.

Thus the number $N_\varepsilon(D)$ when $\varepsilon = \delta L^{-1}$ is the main characteristic of the problem of determination of the reaction of the worked medium on the machine working organ. It should be noted that for further estimates and analysis it is more convenient to use the number $\log N_\varepsilon(D) = H_\varepsilon(D)$. The number $H_\varepsilon(D)$ is called $\varepsilon$-entropy of region $D$.

It is important to note that the model cannot be constructed with accuracy $\delta$ by using points of the region $D$ if they do not form an $\varepsilon$-network ($\varepsilon = \delta L^{-1}$) and if their number is less than $N_\varepsilon(D)$. This is explained by the fact that what is known about the restored function $R(x, \dot{x})$ is only that it is bounded in the region and that the rate of its change in this region does not exceed $L$. However, with such properties one can construct in region $D$ a continuum of various functions of two variables. Therefore, in order to localize the true function from this set with accuracy $\delta$, one must consider all parts of region $D$. Using the fundamental concept of analysis associated with the compactness of the metric space, it was established that there is a possibility to construct a finite set of elements in them forming a finite $\varepsilon$-network.

The above considered scheme of reducing the problem of constructing the dynamic model of the process of vibratory machine operation to the task of tabulating of the unknown reaction of the worked medium on the vibratory machine working organ $R(x, \dot{x})$, with the subsequent restoration of this function using the dynamic test table is the dynamic test method.

The information on the initial parameters of the process, for which the restored model satisfies the specified accuracy condition $\delta$, is regarded as complete information for the construction of the dynamic model of the process of a vibratory machine operation. The measure of this completeness is expedient to estimate by the introduced above number $N_\delta$ ($D$), or by any appropriate function of this number.

The analytical representation of the reaction of the worked medium on the operating organ of the vibratory machine $R(x, \dot{x})$, which is given by the dynamic test table, is formulated as an approximation problem.

The construction of the dynamic model of the process of a vibratory machine operation for complete information on the character of its operation has been considered above. Practically, it is important that the concept of complete information indicates the content and volume of the initial experimental data for which the dynamic model can be represented by ordinary differential equations.

In the problem under consideration, it was assumed that the vibratory machine–load system can be completely described by the dynamic equation of a single-mass oscillatory system in which the reaction of the worked medium is introduced, namely the function of displacement and velocity of the working organ of the vibratory machine. Furthermore, the worked medium is a rheological system which is completely characterized by viscoelastic properties.

When handling large masses of the process medium such an approach is not sufficiently accurate; for the reliable description of the behavior of the worked medium a generating viscoelastic, rheological system must be used. Then the vibratory machine–load system acquires an additional degree of mobility and one must know the coordinates of displacement of the worked medium to describe its motion. However, in the process of the experiments one cannot always measure displacement of the medium and thus the experiments can be limited by recording of the motion parameters of the working organ. This constitutes the problem of determination of the reaction of the medium with incomplete information on the operating regime of the vibratory machine.

With incomplete information on the operating regime of the vibratory machine, it becomes impractical to represent the dynamic model in the form of ordinary differential equations. Let us illustrate this by the example of a vibratory machine under load, which is modeled by a mechanical system with two degrees of freedom of the following particular form

$$M\ddot{x}_1 = R_1$$

$$m\ddot{x}_2 = R\ (x_1,\ x,\ \dot{x}_1,\ \dot{x}_2) \tag{2.184}$$

where $M$ and $m$ are the oscillating masses of the working organ of the vibratory machine and of the worked medium.

In this case, the first equation describes the motion of the working organ, and the second describes deformations of the worked medium. $R(x_1,\ x_2,\ \dot{x}_1,\ \dot{x}_2)$ is the dynamic reaction of the worked medium; $F(t)$ is the force acting on the working organ.

Since in the course of the experiment only the parameters of motion of the working organ are recorded, the initial data of the problem, just as in the first case, are formed exclusively from the points of trajectory $\bar{\omega}\ (y_1(t),\ x_1(t),\ \dot{x}_1(t))$, where $y_1(t) = m_1\ddot{x}\ (t) - F(t)$. It is necessary to verify whether it is possible, with these initial data, to restore function $R_2(x_1,\ x_2,\ \dot{x}_1,\ \dot{x}_2)$ with the accuracy specified in advance and determined by the number $\delta$. Let us consider the information which can be obtained from function $R(x_1,\ x_2,\ \dot{x}_1,\ \dot{x}_2)$ with these initial data. Let us perform the following construction.

Function $R(x_1,\ x_2,\ \dot{x}_1,\ \dot{x}_2)$ is determined in region $D$ and its graph is plotted. The region of the graph of the function denoted as $G$ is projected on plane $y_1,\ x_1$ and the projection is denoted by $\Pi G$. The geometrical region $\Pi G$ is represented as a graph-layer of some multi-valued function $F(x_1)$. This multiple-valued function is denoted as $y \in F(x_1)$, i.e., at each fixed value of argument $x_1^*$ this function can assume a set of values $y$.

Let the trajectory of motion $\bar{x}(t) = (x_1(t),\ x_2(t))$ of system (2.184) belong to region $D$ and densely fill it. Then the trajectory $\bar{\omega}(t) = (y(t),\ x_1(t),\ x_2(t))$ will densely cover the region $G$ on the graph of function $R(x_1,\ x_2)$. Function $x_2(t)$ is unknown, therefore at moment $t$ of time point $\omega(t)$ in the three-dimensional space cannot be marked, although the projection of this point on plane $(y_1,\ x_1)$ can be marked. By marking all the remaining projections of the points of trajectory $\bar{\omega}(t)$, the graph-layer of the multiple-valued function $F(x)$ can be plotted. If these points are entered into the dynamic test table, we obtain the tabular method of representation of the multi-valued function $y \in F(x_1)$.

We construct a cylindrical graph-layer $y \in F(x_1, x_2)$. We identify a Lipschitz function $R^*(x_1, x_2)$ whose graph on the points of region $D$ belongs to the graph of the cylindrical layer, and the projection of region $G^*$ on plane $(y, x)$ coincides with the region filled by the multi-valued function $F(x_1)$. It is evident that unlimited number of similar functions $R_1^*(x_1, x_2)$ can be constructed. For each of functions $R_1^*$, one can select such an unobservable function $x_2^*(t)$ for which the following identity occurs

$$M\ddot{x}_1(t) \equiv R_1^*(x_1(t), x_2^*(t), \dot{x}_1(t), \dot{x}_2^*(t)) + F(t)$$

The noted circumstances lead to the fact that the dynamic model of the vibratory machine operation under load can be represented as a differential equation, also called differential inclusion, with a multi-valued right-hand side of the following form

$$M\ddot{x}_1 \in F(x_1) + F(t)$$

The differential inclusion is the form of representation of the dynamic model with incomplete information about the initial experimental data.

### 2.6.3 Rheological Equations of the Dynamic System: Vibratory Crusher Jaw-Rock Mass

Let us consider the computational model of a vibratory crusher under load (Fig. 2.9). As was shown above, one can limit oneself to the consideration of the operation of one jaw of the vibratory crusher without loss of generality. A jaw of mass $M$ is driven with oscillatory motion by an inertial vibrator with $n$ unbalanced masses $m_j$. The rotations of the unbalanced masses with angular velocities $\varphi_j$ cause the jaw to oscillate in the direction of axis $x$. Jaw oscillations excite complex motions of the rock chunks inside the crushing chamber. Motion of each chunk with mass $m_k$ ($k = 1, 2, ..., n$) is determined by three linear components $q_{k1}, q_{k2}, q_{k3}$. In the general case, polyharmonic oscillations of the vibratory crusher jaw are generated, which are described by the following differential equation

$$\left( M + \sum_{i=1}^{n} m_i \right) \ddot{x} + T(x, \dot{x}) + R(x, \dot{x}, \vec{\alpha}, \vec{\lambda})$$

$$= \sum_{i=1}^{n} m_i r_i (\cos \varphi_i)^{**} \qquad\qquad (2.185)$$

Function $T(x, \dot{x})$ in the equation characterizes the nonlinear restoring forces of the elastic system and the structural friction generated due to interaction of the jaw with the other elements of the structure. Function $R(x, \dot{x}, \vec{\alpha}, \vec{\lambda})$ determines the dynamic reaction of the crushed rock on the jaw arising under the strain and the displacement of the chunks with masses $m_k$. Vector $\vec{\lambda}$ is determined by the generalized coordinates, velocities, and accelerations of the rock chunks which interact with the jaw, i.e., we can write $\vec{\lambda}$ = $(\vec{q}, \dot{\vec{q}}, \ddot{\vec{q}})$. The vector takes into account the change of physicomechanical properties fo the rock from the moment of delivery to the moment of discharge from the crushing chamber. The action of the vibrating jaw on the rock mass inside the chamber is accompanied by changes in its physicomechanical properties, such as granulometric composition, density, elastic deformation properties, and so on. These changes are reflected on the character of dynamic interaction of the elements of the jaw-rock mass system. The latter is expressed in the change of the reaction of the rock mass on the jaw, namely in the function in equation (2.185). Thus, the vector of the parameters characterizes the sufficiently slow change of the physicomechanical properties of the rock mass inside the crushing chamber.

For complete description of the dynamics of the system: crushing jaw-rock mass equation (2.185) must be supplemented by equations characterizing changes of vectors $\vec{\lambda}$, $\vec{\alpha}$ and of cyclic coordinates $\varphi_i$ ($i$ = 1, 2, 3, ..., $n$).

It must be born in mind that in the steady operation regime of vibratory crusher with the uniform filling of the crushing chamber with rock mass, the rock chunks at different stages of crushing interact with the jaw at a given time. Thus it can be assumed that the change of vector $\vec{\alpha}$ for the averaged rock mass which is subjected to the action of the oscillating jaws is insignificant. When considering the crushing process of single chunks or small batches of rock, it is necessary to take into account the change of the properties on the interval of motion beween the feeding and discharge

openings. As a first approximation, the behavior of vector $\vec{\lambda}$
can be characterized by rheological equations of the strain of
the rock mass (see section 1.4), which can be written as

$$m\ddot{x} = R\ (x - x',\ \dot{x} - \dot{x}',\ \vec{\lambda},\ \vec{\alpha}) = 0 \qquad\qquad (2.186)$$

Thus, the behavior of the vibratory crusher under load is
described, as a first approximation, by equation (2.183); in a
more accurate formulation (second approximation), it can be
described by the system of equations (2.184). However, in both
the first and second formulations adequate information on the
behavior of the system is not obtainable. In both cases, one
must use the methods of the dynamic model with incomplete
information on the parameters of the system; however, in the
second case the accuracy is higher than in the first. The
accuracy of determination of the reaction of the rock mass $R$
considerably increases when the equation of jaw motion is
supplemented by a rheological strain equation of the rock
mass.

Let us consider the method of determination of the main
rheological trends of the rock mass reaction $R(x,\ \dot{x},\ \vec{\lambda},\ \vec{\alpha})$ on
the basis of experimental investigations. We rewrite equation
(2.185) as follows

$$\left(M + \sum_{i=1}^{n} m_i\right)\ddot{x} + T\ (x,\ \dot{x}) - \sum_{i=1}^{n} m_i\, r_i\ (\cos\ \varphi_i)'' - R\ (x,\ \dot{x},\ \vec{\lambda},\ \vec{\alpha})$$

$$(2.187)$$

The situation is handled as follows: without analytically
investigating the laws of change of vector $\vec{\lambda}$, we assume that
$\bar{\Lambda}$ is some region from which vector $\vec{\lambda}$ adopts its values, i.e.,
$\vec{\lambda} \in \bar{\Lambda}$. We introduce the multi-valued function

$$R(x,\ \dot{x},\ \vec{\lambda},\ \vec{\alpha}) = \underset{\lambda \in \bar{\Lambda}}{U}\ (-\ R\ (x,\ \dot{x},\ \vec{\lambda},\ \vec{\alpha}))$$

which reflects the fact that for some fixed values of $x,\ \dot{x},\ \vec{\alpha}$
vector $\vec{\lambda}$ can assume arbitrary values from the region of
dynamic states $\bar{\Lambda}$ and, therefore, the function can also assume
the corresponding set of values. As a result, equation (2.187) is
replaced by a multi-valued differential equation of the type

$$\left(M + \sum_{i=1}^{n} m_i\right)\ddot{x} + T(x,\,\dot{x}) - \sum_{i=1}^{n} m_i\, r_i\,(\cos\varphi_i)'' \in R(x,\,\dot{x},\,\vec{\alpha})$$

(2.188)

The differential inclusion (2.188) is a first approximation rheological model of the jaw-rock mass system. The multi-valued function $R(x,\,\dot{x}\,\,\vec{\alpha})$ ought to be determined experimentally.

The second approximation of the rheological model of the system is obtained by using an additional rheological strain equation of the rock mass. The differential inclusion of the second approximation has the form

$$\left(M + \sum_{i=1}^{n} m_i\right)\ddot{x} + T(x,\,\dot{x}) - \sum_{i=1}^{n} m_i\, r_i\,(\cos\varphi_i)''$$

$$\in R(x - x',\,\dot{x} - \dot{x}',\,\vec{\alpha})$$

(2.189)

## 2.6.4 Methods of Determination of the Rheological Properties of the System: Rock Mass-Jaw of Vibratory Crusher

In accordance with the dynamic test method the following information on the motions of the jaw of a vibratory crusher as functions of time $t$ can be obtained in the experiment: $\ddot{x}(t)$, $\dot{x}(t)$, $x(t)$. Furthermore, from the design data the following constant parameters are known: mass of the working organ $M$, masses and eccentricities of the unbalance vibrators $m_i$ and $r_i$, and from the preliminary experiment function $T(x,\,\dot{x})$ characterizing the force of structural friction has been determined.

Let us turun to equation (2.188). In accordance with the aforesaid, we introduce the notation

$$y = \left(M + \sum_{i=1}^{n} m_i\right)\ddot{x} + T(x,\,\dot{x}) - \sum_{i=1}^{n} m_i\, r_i\,(\cos\varphi_i)''$$

then the differential inclusion (2.188) can be replaced by a scalar multi-valued function

$$y \in R (x, \dot{x}, \vec{\alpha})$$                                      (2.190)

Since the multi-valued relation (2.190) is obtained from equation (2.188), then any observed motion of the jaw of the vibratory crusher, which is characterized by functions $\ddot{x}(t)$, $\dot{x}(t)$, $x(t)$, $\varphi_i(t)$ will satisfy relation (2.190). This means that for each moment of time the following condition will be satisfied

$$y (t) \in R (x(t), \dot{x}(t), \vec{\alpha})$$

Further, we select the moments of time $t_1$, $t_2$, ...., $t_i$, ...., $t_n$ and compile a dynamic test table of the form

$$y (t_1) \, ... \, y (t_2) \, ... \, y (t_i) \, ... \, y (t_n)$$

$$x (t_1) \, ... \, x (t_2) \, ... \, x (t_i) \, ... \, x (t_n)$$

$$\dot{x} (t_1) \, ... \, \dot{x} (t_2) \, ... \, \dot{x} (t_i) \, ... \, \dot{x} (t_n)$$

which yields a tabular representation of the reaction of the rock mass on the crushing jaw which is a multi-valued function $R(x, \dot{x}, \vec{\alpha})$.

One of the decisive conditions for achieving the necessary accuracy of determination of the worked medium reaction on the working organ by the dynamic test method is the dense filling of the region of the phase trajectory $\vec{x} (t) = x(t), \dot{x}(t)$ in which can be practically located values of the medium reactions on the working organ. In a number of cases, to provide dense filling of the phase trajectory the use of multi-debalance vibrator with flexible links between the unbalanced masses or tri-debalance vibrator with fractional transmission ratios between the shafts can be recommended. This vibrator generates an exciting force which is changing along a multi-lobe trajectory.

When investigating vibratory crushers there is no need to use a special drive, since the phase trajectory densely fills the phase plane even when a standard vibrator is used. Figure 2.15 shows, as an example, the phase trajectories of jaw oscillations of a vibratory crusher.

The region of the phase plane $(x, \dot{x})$, which is densely filled with the phase trajectory $\vec{x} (t) = x(t)$, is divided into rectangular cells with sides $\Delta x$ and $\Delta \dot{x}$ as seen in Fig. 2.16. The quantities $\Delta x$ and $\Delta \dot{x}$ characterizing the division steps of $x$ and $\dot{x}$ are specified from the condition of accuracy of the analysis and are in accordance with the accuracy of the experiment.

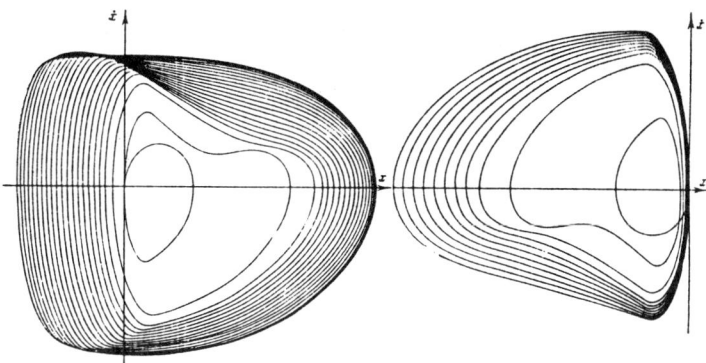

**Figure 2.15** Phase trajectories of the jaw of a vibratory crusher and of the crushed rock

Further, the data entered into the dynamic test table are grouped on the basis of their belonging to the rectangular cell of region $D$ with the center at point $(x_i, x_j)$. Coordinates of the central point $(x_i, x_j)$ are ascribed to all the experimental points ending up in the cell with number $i, j$. Subscripts $i$ and $j$ here are defined as

$$i = 1, 2, ..., \frac{x_{max} - x_{min}}{\Delta x}; \quad j = 1, 2, ..., \frac{\dot{x}_{max} - \dot{x}_{min}}{\Delta \dot{x}}$$

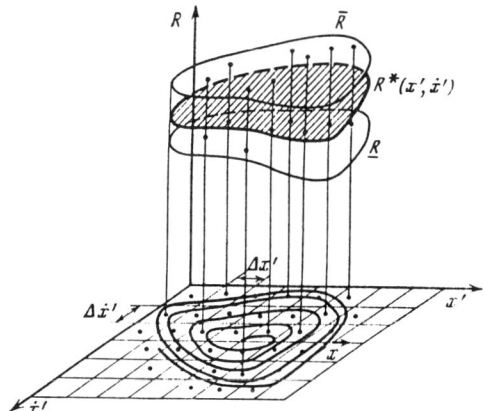

**Figure 2.16** Multi-valued function of reaction of rock mass on the jaw

Thus, all experimental points corresponding to region $D$ and which are filed into the dynamic test table are described by the coordinates $x$, $\dot{x}$ with an accuracy up to quantities $\Delta x$ and $\Delta \dot{x}$.

Due to the ambiguity of function $R(x, \dot{x}, \vec{\alpha})$ which represents reaction of the rock mass on the crushing jaw, several values of function $y(t_{ij})$ which correspond to the moments of time $t_{i,j}$ when the trajectory intersected the rectangular cell with number $i$, $j$, will correspond to each point $x_i$, $\dot{x}_j$.

The construction of the multi-valued function $R(x, \dot{x}, \vec{\alpha})$ for some fixed value of the vector of parameters $\vec{\alpha}$ is performed as follows. All the points with coordiantes $R_{ij}$, $x_i$, $\dot{x}_j$, are identified and surfaces are drawn through them which graphically specify the upper and lower bounds of function $y = R(x, \dot{x}, \vec{\alpha})$, i.e., of the function which bounds the multi-valued function from above and from below (see Fig. 2.16).

The rheological function of the reaction force of the crushed rock mass on the jaw of the vibratory crusher will be characterized, on the average, as

$$y = R^* (x, \dot{x}, \vec{\alpha})$$

which corresponds geometrically to the graph of the surface which is equidistant from the upper and lower surfaces bounding the graph-layer. The computation of function $R(x, \dot{x}, \vec{\alpha})$ is performed according to the rule

$$Y = R^* (x, \dot{x}, \vec{\alpha}) = \tfrac{1}{2} [\bar{R} (x, \dot{x}, \vec{\alpha}) + R(x, \dot{x}, \vec{\alpha})]$$

These constructions can be conducted graphically. Should there be a need for obtaining an analytical dependence, the approximation problems must be solved. With this in mind, the restored reaction $R(x, \dot{x}, \vec{\alpha})$ is represented as a finite sum in the form

$$R (x, \dot{x}, \vec{\alpha}) \cong a_0 + a_1 F_1 (x, \dot{x}) + a_2 F_2 (x, \dot{x}) + \dots$$

$$+ a_n F_n (x, \dot{x}) = a\vec{F} (x, \dot{x})$$

Here $\vec{a} \ F(x, \dot{x})$ is the scalar product (compact representation of the sum) of a vector of unknown coefficients $a_0$, $a_1$, ..., $a_n$ and a vector of specified functions 1, $F_1$, ..., $F_n$. Functions $F_i$ $(x, \dot{x})$ must correspond to the physical content of the investigated system and satisfy a number of mathematical conditions at which the sum $[a, \; F(x, \; \dot{x})]$ would replace function $R(x, \; \dot{x}, \; \alpha)$ with accuracy $\delta$. For the determination of the unknown coefficients the least squares method must be used.

The accuracy (credibility) of the obtained rheological function of the reaction of the rock mass on the jaw of the vibratory machine will be characterized by the width of the graph-layer of the multi-valued function which is computed as $\bar{R}R = \Delta R(x, \; \dot{x}, \; \vec{\alpha})$. By employing various approximations in the solution of the problem with respect to the width of the graph-layer of the multi-valued function, the credibility of the adopted model can be estimated.

### 2.6.5 Methods of Organization, Planning, and Conducting Automated Experiments for Identification of Dynamics of the System: Jaw of Vibratory Crusher-rock Mass

**1. Recording of the dynamic variables.** For the realization of the above-stated methods of analysis of dynamics of the system: jaw of vibratory crusher-rock mass, one must, as a first approximation, conduct simultaneous recording of the following variables: accelerations of the jaw $\ddot{x}(t)$ and phases $\varphi_i(t)$ of the unbalanced masses of the vibrator. The permissible error in measuring momentary acceleration magnitude of the jaw is estimated at 5% of its maximum value. The phase recording must be performed by a discrete method, using 100 positions per revolution of the rotating unbalanced mass. The integral accuracy of the experiment in this case will not exceed 10%.

**2. Investigation of function $R(x, \; \dot{x}, \; \vec{\alpha}, \; \vec{\lambda})$ in the second approximation.** In this case, information must be obtained on the motion of the jaw and of the rock mass. Two accelerometers are used; one is mounted on the jaw and the other in the rock mass. Recording of angular velocity of the unbalanced mass in this case is not required. It is assumed here that function $T(x, \; \dot{x})$ has been determined in preliminary experiments.

**3. Scheme of formation processing.** Processing the experimental information is carried out using standard techniques. Automation of the scheme is carried out either by using microprocessors intended for execution of specified operations, or by automatic imput of information on the functions $\ddot{x}(t)$, $\varphi_i(t)$ into a computer and the subsequent processing according to the described algorithms. The information on the dynamic test tables must be stored in the external memory of the computer.

**4. Determination of friction forces in the vibratory crusher.** The determination of the friction force $T(x, \dot{x})$ acting in the structural elements of the vibratory crusher must be carried out using equation of motion of the jaw

$$\left( M + \sum_{i=1}^{n} \right) \ddot{x} + T(x, \dot{x}) = \sum_{i=1}^{n} m_i r_i (\cos \varphi)"$$

In accordance with the dynamic test method, we obtain

$$y = T(x, \dot{x})$$

where

$$y = \sum_{i=1}^{n} m_i r_i (\cos \varphi_i)" - \left( M - \sum_{i=1}^{n} m_i \right) \ddot{x}$$

Thus, to determine function $T(x, \dot{x})$, simultaneous recording of the jaw acceleration $\ddot{y}$ and of $\varphi_i$ of the unbalanced vibrator must be made. The required function $y = T(x, \dot{x})$ can be obtained by using the automated experiment scheme considered above.

**5. Investigation of function $R(x, \dot{x}, \vec{\lambda}, \vec{\alpha})$ in the absence of information on phase $\varphi_I(t)$.** In this case, the vibrator is connected with the jaw through an elastic system (for example, by mounting on rubber pads). The dynamic equations of the reaction of the rock mass on the jaw then are

$$\left( M' + \sum_{i=1}^{n} m_i \right) \ddot{x}' + T(x', \dot{x}) = \sum_{i=1}^{n} m_i r_i (\cos \varphi_i)"$$

$$M\ddot{x} + T(x, \dot{x}) + R(x, \dot{x}, \vec{\lambda}, \vec{\alpha}) = 0$$

where $M'$ is the mass of the vibrator with the elements associated with it; $x'$, $\dot{x}'$, and $\ddot{x}'$ are displacement, velocity, and acceleration of the vibrator.

Phase variables $\varphi_i$ do not enter in the second equation. For determination of reaction of the rock mass on the jaw $R(x, \dot{x}, \vec{\alpha})$ it suffices to obtain information on motions of masses $M$ and $M'$ by simultaneously recording their acceleration. It is assumed here that function $T(x, \dot{x})$ is determined beforehand.

**6. Control of filling of the phase trajectory of the region.** The visual monitoring of the time when the trajectory $\vec{x} = (x(t), \dot{x}(t))$ of region $D$ of the phase plane with coordinates $x$, $\dot{x}$ is sufficiently densely filled must be performed using two-coordinate oscillographs.

**7. Monitoring of dynamics of the vibratory crusher when physicomechanical properties of the rock mass are changing.** Fluctuations of the physicomechanical properties of the crushed rock mass causes a change of the vector of parameters $\vec{\alpha}$. This is reflected in function $R(x, \dot{x}, \vec{\lambda}, \vec{\alpha})$, which determines the dynamic reaction of the rock mass on the jaw. Vector $\vec{\alpha}$ is changing slightly as compared with the change of the phase variables $x$, $\dot{x}$. Therefore, the experiment must be planned so that vector $\vec{\alpha}$ would be changing only slightly during the time of collection of the necessary information for construction of function $R(x, \dot{x}, \vec{\alpha})$. Implementation of this condition must be verified experimentally. As a result of this step-by-step experiment the dependence of function $R(x, \dot{x}, \vec{\alpha})$ on the vector of parameters $\vec{\alpha}$ can be established. In reality such a dependence will be determined by the time $\tau$ during which the rock mass stays in the crushing chamber.

# BASIC SCHEMATICS AND DESIGN OF VIBRATORY MACHINES

## 3.1 PRINCIPAL TYPES OF VIBRATORY MACHINES

Vibration technology is a relatively new and rapidly developing discipline, and vibratory machines are successfully used in various branches of industry. The design of these machines was preceded by thorough analysis of the theory and computational techniques along with experimental investigations. It is pertinent to mention that despite the relative structural simplicity of vibratory conveying machines, their theory is complex and its development requires application of the techniques of nonlinear mechanics and the use of modern analog and computer technology.

The following operations performed by these machines are of the same type with respect to the character of the external resistances encountered by the working organ: imparting vibration onto various bulk media and dispersed systems in loose conditions; penetration of the vibrating elements into soil or rock; vibratory cutting and vibratory fracture of rocks, bounded soil, metals; vibratory crushing and grain-size reduction of various materials.

Vibratory machines with respect to their intended industrial use can be divided into conveying, process conveying, loading, bunkering, as well as machines for crushing, reducing grain size, separation of bulk and multi-phase media (screens, centrifuges, sieves, and so on), for transporting multi-phase media such as cement, food products, and so on, machines for metal cutting and rock fracture, devices for increasing effectiveness of pressure metal working and improving cast structures, and also auxiliary equipment.

Vibratory conveying machines include conveyors, feeders and screen feeders, elevators, and hoppers - dosators. Conveyors and feeders of universal design for the transportation of poured and lump loads are the most widely used. They differ by the diversity of their fundamental and structural design and are produced for different productivities. For the displacement of the loads in the vertical direction, conveyers with screw-like load-carrying elements are used. For the delivery of small single workpieces and parts in transfer lines of metal working machines, vibratory machines and hoppers-dosators, are used. Vibratory devices are widely used in the mining industry.

On the basis of conveying machines a number of process-conveying machines have been developed such as screens, dryers, dehumidifiers, granulators, classifiers, dosators, hopper-dosators, and many others. To provide continuous operation of powerful conveying installations comprising belt conveyers, high-output screen-feeders are successfully used as loading and load transfer devices which ensure the optimum conditions of load transfer from one conveyer to another.

When handling underground ore deposits vibratory conveying machines are used for automating of such basic operations as unloading (discharge) of the ore, and delivery and loading of the blasted ore. As a result, efficiency of these processes increases considerably, operation becomes safer, quality improves, and the disintegration of oversize ore (chunks) is made easier.

The group of loading machines comprises continuous-operation machines with vibratory working organ, machines with auxiliary vibrating devices such as bucket-type devices with heaping arms ("claws"), combers, and also loading/delivering machines and greifers (clamshells).

Devices used in metallurgy and in metal working are relatively new. They are used to improve quality of castings in pressure metal working (vibratory upsetting drawing and rolling) and in cold-work hardening, during welding, to relieve welding stresses, etc.

Auxiliary vibratory devices include dosators (batchers), exciters for bunkers and troughs (chutes), compactors of bulk loads and various process mixtures, unloaders of railway freight cars, devices for cleaning castings from the molding sand, rippers of frozen or compressed loads, cleaners of freight cars and mine cars and devices for cleaning conveyer belts. The use of auxiliary vibratory equipment yields certain operational and economic effects, since it increases the level of mechanization of labor-intensive auxiliary operations.

Among the auxiliary devices one must, first of all, underline an entire class of equipment for unloading and cleaning the conveying means and compaction of transported loads in them. Intensification of unloading and cleaning of the railway rolling stock and truck bodies, and also compaction of the loads in them leads to an increase of transport productivity. Thus, when compacting some transported granular loads the useful capacity of the rolling stock can be increased by 15 - 20% in the process of loading and subsequent vibratory cleaning following unloading. The use of vibratory cleaning devices in conveyers increases their operational reliability, particularly in conditions of transportation of sticking and freezing loads and thus enables the increase of the turnover of the transportation hardware.

Vibratory unloading, cleaning, and compacting devices have a great deal in common with respect to their fundamental design; unloaders and cleaners are usually arranged in one assembly.

The essence of the vibratory method of intensification of unloading of poured loads is application of vibration from special exciter to the body of the car (buggy). By using special devices, vibrations can also be imparted directly to the poured load and not to the body. Under the action of vibration the forces of adhesion between the stuck load and the walls of the body, and also the internal friction and adhesion of individual particles of the poured load become significantly weaker, its flow is facilitated, and the load is discharged by gravity. In the case of a monolithic frozen load, it breaks along the freezing planes under the action of the vibratory ripper. The unloading of frozen materials requires, as a rule, powerful equipment and intensive oscillations. The most effective here are vibro-impact devices.

Another type of auxiliary equipment is vibratory compactors. Under their influence poured loads in cargo vehicles, casting molds, etc., are compacted as a result of reduction of the forces of internal friction under vibration and the simultaneous action of the force of gravity. Due to presence in the transported load of different chunks in both size and shape and due to their arbitrary arrangement, the volume of the carrying body is not fully utilized. When vibratory compactors are used, the fine particles fill the space between the large chunks, the load is arranged more uniformly and is compacted, and, hence, the volume utilization is increasing. Thus, the possibility of reducing the size of the rolling stock or increasing the carrying capacity of the vehicle is created. At the present time, vibratory compactors are used in some cases in underground loading mine stations; however, there are no fundamental difficulties for their implementation in railway transportation.

## 3.2  VIBRATORY MILLS AND CRUSHERS

Preliminary fine grinding of the product is a prerequisite for effective realization of many production processes in a number of industries. Fine grinding contributes to increasing the rate of interphase processes and to bettering the structure of the end product. This is due to the fact that with fine grinding of the product, the sizes of its particles decrease, their specific surface area increases, and homogeneity is improved. Hence, the reactions are accelerated, the temperatures required to carry out the production process are usually decreased, the consumption of rare elements, and in many cases, the consumption of energy, are reduced.

The most effective and currently used method is vibratory grinding. Vibration grinding is carried out in vibratory mills in a grinding medium. Figure 3.1 shows the basic arrangement of the simplest single-drum vibratory mill with unbalanced inertial drive. The vibratory mill is comprised from a container 1 filled with the grinding elements (balls are often used) and an inertial vibrator which is made as an eccentric shaft with unbalanced masses mounted on its ends 2 and linked with the container by means of bearing blocks. The shaft of the inertial vibrator is rotated through an elastic coupling 3 by an electric motor and the container is mounted on plate 4 by means of elastic mounts 5 in the form of helical springs or rubber pads. The capacity of one container of modern vibratory mills can range from fractions of a cubic decameter to 1000 $dm^3$. Calibrated sieves 6 are mounted inside the working chamber. The processed product is fed into the mill through the feed opening A and discharged through opening B.

Rotation of the shaft of the inertial vibrator generates centrifugal forces from the unbalanced masses, which bring the working chamber into oscillatory motion. The form of the

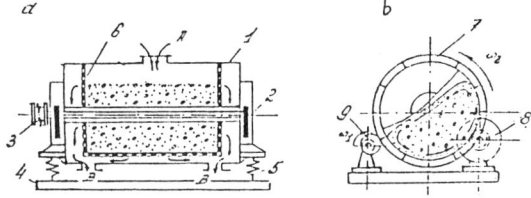

**Figure 3.1** The principle of operation and design of a vibratory mill. a) with a vibrating chamber;  b) with a vibrating and rotating chamber.

trajectory of container oscillation is considerably dependent in this case on parameters of the elastic system, and is varying from an elongated ellipse to a circle. The elliptical form is the most typical trajectory. Under the influence of oscillations each grinding element in the working chamber is brought into rapid oscillatory motion repeating the form of chamber trajectory and, simultaneously, all the masses of the grinding elements and the product being ground perform slow circulatory motions. The motion of the mass filling the working chamber of the vibratory mill usually takes place in a direction opposite to the direction of rotation of the unbalanced shaft. Fracturing the processed product during vibration grinding is achieved due to the relative displacement of the grinding elements and the medium as a result of impacts, squeezing, and abrasion. The main parameters of the vibratory mill are the frequency and amplitude of oscillations, form, dimensions, material of the grinding elements, and parameters of the container. The quality of grinding in the vibratory mill is dependent on the degree of filling by the product, relation between the amounts of the grinding elements and the ground product, type of the worked product and its granulometric composition, and the required grain size at the end. The effectiveness of the grinding process is also dependent on the design features and dimensions of the container, the size, form, and density of the grinding elements, the conditions of delivery and discharge of the product from the mill. The determining factors of the intensity of operation of the vibratory mill are the parameters of container oscillations (shape of trajectory, frequency and amplitude of oscillations).

Depending on the type of the product, its granulometric composition, and the desired degree of grinding, three principal operating schematics of vibratory mills are used and, correspondingly three design layouts of the arrangement of the working chambers of the containers: for very fine, fine, and medium grinding.

For very fine product grain size the working chamber is equipped at the ends by two calibrated sieves. The ground product is loaded from above through one loading door in the center of the container and discharged from the sides of the container through two discharge doors. With such an arrangement of the working chamber of the container, maximum path of travel of the product to the discharge doors is achieved.

For a coarser product grain size, the working chamber has calibrated sieves at the ends and at the bottom. The rapidly ground part of the product in this case spills through the sieve at the bottom of the chamber and the less fragmented part spills through the end sieves after travelling a longer path in the working chamber. Loading is also carried out through one central door and discharged through two side doors.

For an even coarser grinding, the working chamber is provided with a calibrated sieve only at the bottom. The particles, in this arrangement, rapidly cross the layer of the grinding elements reaching the discharge door without being subjected to sufficiently intensive rubbing. Loading of the working chamber is carried out from the top through two doors at the sides and the product is discharged through three doors at the bottom of the chamber. The directions of motion of the ground product are clearly seen in the schematic diagram (Fig. 3.1).

The high efficacy of using vibratory mills for fine grinding and homogenization of a wide range of products has led to development of a large variety of designs and sizes of vibratory mills both in the USSR and abroad.

Vibratory grinders of other designs have also been developed for the handling of special problems. More uniform distribution of the ground product amongst the mass of the grinding elements and, hence, an increase of grinding quality can be achieved in a vibratory mill with a rotating chamber. The chamber 7 of such a mill is cylindrical. The conical ends of the chamber rest on pneumatic cylindrical rollers 8 and the cylindrical part on the driving eccentric roller 9. The diameter of the roller is significantly less than the diameter of the cylindrical part of the chamber: therefore, it provides high oscillation frequency together with slow rotation (Fig. 3.1b).

The fundamental arrangements of traditional vibratory mills of various modifications are depicted in Fig. 3.2.

Figure 3.2a shows a vibratory mill with one working chamber *1* which rests by means of bracket *2* on frame *4* through rubber-metal elastic elements *3*. Loading of the mill is effected through door *5* and discharge through door *6*. Oscillations of the working element are excited by vibrator *7*, whose axis coincides with the center of gravity of the oscillating mass of the vibratory mill.

Figure 3.2b and 3.2c depict two versions of a two-chamber vibratory mill with working chambers arranged at the same level and at different levels. The second version requires less operating area for the installation. The vibratory mills with the containers arranged at the same level have the working chambers *8* and *9* which are joined by a common bracket *10* with the inertial vibrator *11*. The oscillatory system is supported on frame *12* by rubber-metal elastic elements *13*. Loading of the working chambers is effected through doors *14* and *15* and discharge through doors *16* and *17*. In the second version, container *18* is mounted higher than container *19*, and the unbalanced vibrator *21* is mounted on a common bracket *20* between the two. Loading of the upper working chamber is effected through door *22*, and the lower through the same door and ducting *23*; discharge takes place through ductings *24* and *25*. The oscillatory system is supported on pads *27* and *28* by means of rubber-metal elastic elements 26.

The three-chamber vibratory mill is shown in Fig. 3.2d. Chambers *29, 30,* and *31* are connected by a common bracket *32* with the inertial vibratory *33*. The chambers are loaded through doors *34, 35,* and *36* and unloaded through doors, *37, 38,* and *39*. The oscillatory system is supported on frame *41* by means of rubber-metal elastic elements *40*.

The schematic diagram of a five-chamber vibratory mill is shown in Fig. 3.2e. The mill has two containers *42* and *43* mounted above, two containers *44* and *45* mounted below, and one chamber *46* in between. All five containers are linked with two inertial vibrators *48* and *49* by bracket *47*. The two left-hand side working chambers are loaded through door *51*, the right-hand side chambers are loaded through door *52*, and the middle chamber through door *53*. The chambers are unloaded through doors *54, 55,* and *56*, respectively. The

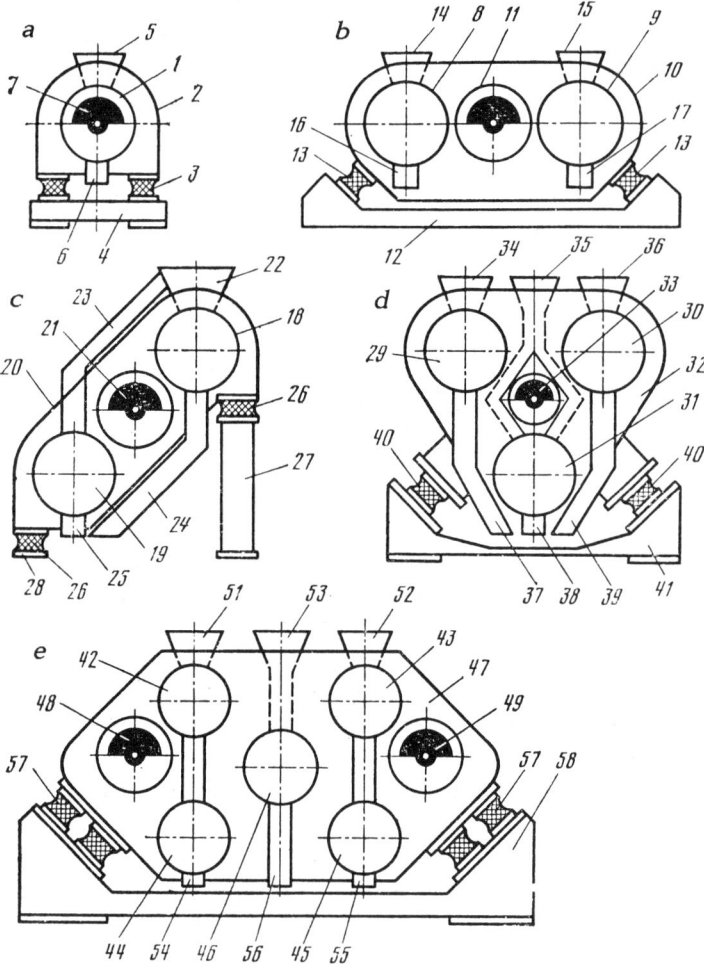

**Figure 3.2** Fundamental arrangements of vibratory mills. a) single-chamber, b) two-chamber with the chambers at one level; c) two-chamber with the chambers at different levels; d) three-chamber; e) five-chamer.

oscillatory system is supported on frame 58 through the rubber-metal elastic elements 57.

In the described types of vibratory mill, the chambers are assembled in one mobile frame. However, there are designs in which the working chambers are assembled in two frames that are movable relative to each other. Such designs enable one to construct balanced dynamic systems; however, these are distinguished by their greater complexity. A system of four-chamber vibratory mill has been developed in which the containers are mounted in pairs on two independent frames that are connected with each other by eccentric vibrator. The frames are supported on the carrying structure by means of elastic elements.

The working chambers of the vibratory mill can be arranged not only horizontally, but also vertically. The oscillations of such chambers usually take place in the horizontal plane, while rocking of the vertical axis is simultaneously excited. There are also vibratory mills in which the chamber is made as a torus. Combined torsional and translational oscillations are imparted to the chamber causing the filler to perform circular motions in the cross section of the torus and simultaneously move slowly in the horizontal plane along the ring.

In order to improve traditional solutions, fundamentally new systems have been designed. In all the designs of vibratory mills under consideration the oscillations are imparted directly to the working chamber. This poses certain inconveniences: on the one hand, one has to bring into vibration heavy components, on the other hand, to protect structural elements of the building from vibration. This problem is not encountered in a vibratory mill with loaded vibrator. This mill comprises a high cylindrical vertical chamber inside which a tube of a significantly smaller diameter is suspended on elastic elements. The unbalanced vibrator inside the tube generates circular oscillations along the longitudinal axis of the chamber. The free space in the chamber is filled with the product being ground. Grinding occurs under the action of oscillations of the inner tube. In such a vibratory mill, there are significantly less oscillating masses. The frame, which is mounted on the carrying structures by means of springs, practically does not oscillate. However, in some designs both the frame and the inner tube are oscillating.

There also exist balanced vibratory mills with vertical working chambers. Vertical vibratory mills of this type consist of two identical, mutually interchangeable grinding chambers mounted one above the other and on a frame by means of shock absorbers. The drive is a two-shaft eccentric vibration exciter mounted on the frame symmetrically with respect to the axis of the mill between the grinding chambers.

Grinding chambers are divided heightwise by perforated partitions into a number of sections, each of which contains the grinding elements with a clearance between the upper layer and the perforated partition located above. During the operation of the mill the grinding elements interact with the perforated partitions twice per cycle: once with the lower and then with the upper partitions. The frame is mounted on the foundation with the aid of shock absorbers. When the mill is working, the grinding chambers perform rectilinear oscillations in antiphase, and the inertial forces generated during motion are mutually balanced.

The rectilinear trajectory of oscillations of the grinding chamber in the vertical plane and the double impact interaction of the grinding elements with the perforated partitions at optimum height of their layer enable ensuring high power intensity and, hence, effectively carrying out the grinding of hard materials.

The initial material is fed into the crushing chamber at the top, is uniformly distributed over the area of its cross section, and fragmented as a result of their collisions as it passes progressively through the layers of the grinding elements of the working sections of the grinding chambers.

There are vibratory mills with one grinding element. Such a vibratory mill is a cylindrical chamber in which there is a cylinder of somewhat smaller diameter. Product fragmentation takes place between the chamber wall and the surface of the cylinder whose rotation is imparted by the oscillations of the frame on which the entire system is mounted. Inertial unbalanced vibrators are mounted on the frame on both sides of the working chamber. The frame is mounted on elastic elements. The synchronous rotation of both vibrators, which is needed to ensure the specified law of motion of the working chamber, is provided by the action of the vibratory moment of the excited oscillations of the common frame on the unbalanced masses.

Jaw vibratory crushers are mechanical systems with incomplete constraints. Therefore, the laws of motion of the crushing jaws are not determined by the kinematic parameters of the crusher but are formulated as a result of interaction of the drive and operating resistances (forces created during rock crushing). A substantial role in the formulation of the law of motion of the crushing jaws is played by the tuning of the crusher oscillating system (natural frequency, ratio of the natural and forced frequencies, resistances in the elastic system). In vibratory jaw crushers, unlike ordinary jaw crushers, crushing is not effected by static breaking, but by vibration impact. Vibratory impact crushing is realized as a consequence of periodic generation of a clearance between the jaw and the rock. The clearance is generated during the crushing process as a consequence of the fact that at high-frequency jaw oscillations, the rock does not have the time to fall with the required speed in the working space of the crusher and loses contact with the crushing jaw.

Presently existing vibratory jaw crushers can be divided with respect to the fundamental arrangement into two main categories: with a kinematic link of the jaw with the frame and with an elastic link of the jaw with the frame (Fig. 3.3). The kinematic connection of the jaw with the frame is by means of suspending it on a knuckle pin and also by using a crank or eccentric drive. The elastic connection of the jaw with the frame is used with the inertial drive. When the kinematic attachment of the jaw is used, practically all the forces generated during crushing of the rock mass are transmitted to the frame. With the use of elastic suspension, transmission of these forces to the frame of the crusher can be considerably reduced.

The single-jaw vibratory crusher with inertial drive (Fig. 3.3a) has moving *1* and stationary *2* jaws. The stationary jaw is rigidly mounted in the frame *3*, and the moving jaw is suspended on a pivot *4*. The moving jaw is driven by an inertial vibration exciter of the unbalanced-mass type *5*. The free end of the moving jaw is supported by the elastic system *6*.

During rotation of the unbalanced mass, the moving jaw performs oscillations relative to the pivot with a sufficiently high frequency. At a constant angle of oscillation the amplitude increases from the suspension axis to the end of the jaw. The rock mass is crushed between the moving and stationary jaws, and the forces created in the process of crushing are transmitted from the jaw to the frame of the crusher mainly through the stationary jaw and partially through the joint. Magnitudes and directions of the forces, transmitted to the frame by the stationary and oscillating jaws, are different and do not balance each other.

Crushers of this type are simple in design and reliable in operation. However, due to the fact that the forces generated in the process of crushing are transmitted to the frame, the latter must be made heavier in order to reduce oscillations of the structure. Nonetheless, as a result of the imbalance of the

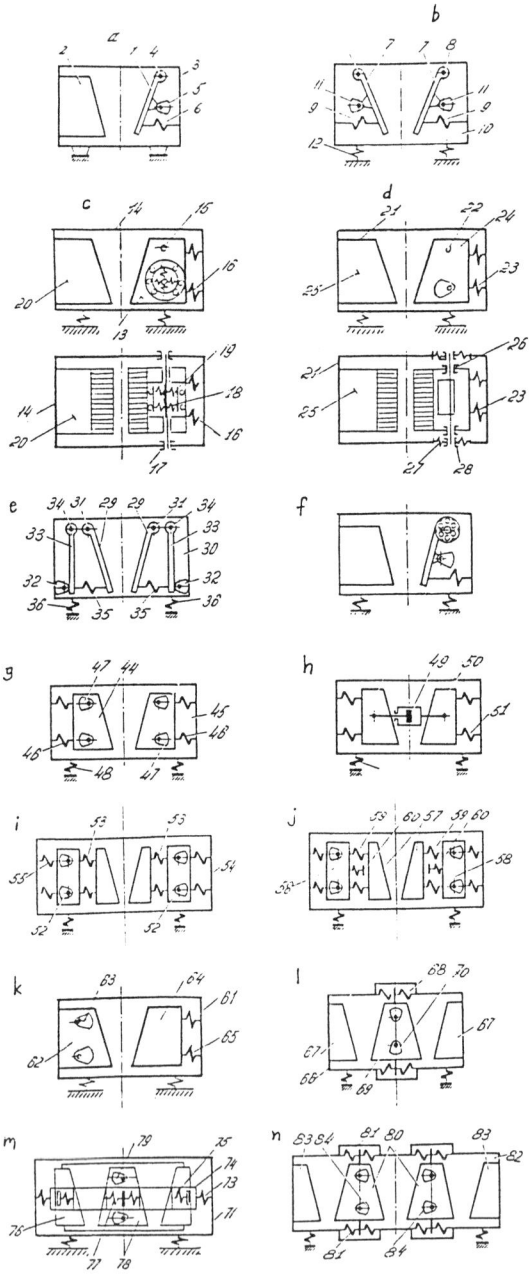

**Figure 3.3** Fundamental arrangements of vibratory jaw crushers.

forces the frame vibrates during the crushing process and thus has to be mounted on a foundation.

Double-jaw vibratory crushers (Fig. 3.3*b*) have moving jaws 7 suspended by means of pivots *8* and elastic elements *9* in frame *10*. The jaws are driven by unbalanced vibration exciters *11*. In a crusher with two moving jaws, the crushing forces are also transmitted to a significant degree to the frame; however, due to the symmetry of construction, the horizontal components are balanced to a certain degree which reduces the vibration of the frame during the crushing process. The crusher is usually rigidly mounted on the foundation.

To reduce transmission of vertical components of the load to the carrying structure, the two–jaw crusher is mounted sometimes on the foundation by means of elastic mounts *12*. Increased vertical oscillations of the suspended frame contribute to the increase of crusher output. The process of acceleration of the rock-mass develops as follows. At the uppermost location the rock is in a compressed state between the jaws and a motion is imparted to it downwards with the velocity of the frame, since there is no rock motion relative to the jaw. By an appropriate adjustment, the jaw starts opening at the moment when the downward speed of frame motion, and hence of the rock mass, acquires maximum value. Therefore, the free motion of the material begins with an initial speed corresponding to the frame velocity at the moment the jaws are freed.

The single-jaw vibratory crusher with parametric eccentric drive (Fig. 3.3*c*) consists of moving jaw *13* suspended in frame *14* on axis *15* and elastic connections *16*. The crusher is driven by an eccentric vibration exciter whose shaft is mounted in bearings *17* in frame *14*. Mounted on the shaft is an elastic eccentric *18* whose yoke *19* is installed in the moving jaw *13*. When the elastic element is rotated, both parametric and kinematic excitation of oscillations is occurring simultaneously since the jaw is not only displaced by the amount of eccentricity, but stiffness of the elastic element in the drive is also changing simultaneously in the direction of oscillations. The stiffness is changing due to eccentricity in the elastic element.

The combined kinematic and parametric oscillations of the jaw are brought about by the inertial exciter whose drive shaft is installed in the bearings of the jaw by means of rotating elastic eccentric bushings. The single-jaw crusher with such a drive (Fig. 3.3*d*) consists of a frame *21* in which a moving *24* and stationary 25 jaws are mounted on axis 22 and on the elastic system 23. It is driven by an unbalanced mass whose shaft is installed in the jaw in bearings 26 and in frame 21 by means of elastic bushings 27 of variable stiffness and bearings 28. When the unbalanced mass is rotating a rotary exciting force is created and the stiffness of the elastic system linking the jaw with the frame is simultaneously changing.

The most heavily loaded elements of vibratory crushers are the bearing supports of the vibrators. These are acted upon by the centrifugal forces of the rotating unbalanced masses and also by impact impulses from the crushed rock. The vibratory crusher with shock-absorbing bearings (Fig. 3.3e) consists of moving jaws 29 suspended on axes 31 in frame 30. Inertial vibration exciters 32 of the unbalanced-mass type are, in their turn, mounted on levers 33 that are suspended with pins 34 on the crusher frame 30 and connected with the jaw by elastic links 35. The frame of the crusher is mounted on the foundation on shock-absorbing elastic connectors 36.

In the considered design, the crushing forces are transmitted from the jaw to the vibrator via elastic elements, as a result of which the loads on the bearings of the elements are significantly reduced. However, in this case, the crushing effectiveness becomes dependent on the tuning of the system and can decrease if the crushed material is fed nonuniformly.

In vibratory crushers with hinged suspension and rocking motions of the jaws, the amplitude of oscillations is small, particularly in the upper section of the jaws. Because of this, some materials are not efficiently crushed.

Vibratory crushers with two-frequency jaw oscillations have been developed: high frequency (rotation frequency of the unbalanced mass of the vibration exciter) with small amplitude, and low frequency with large amplitude. Such oscillations are generated using only one inertial vibration exciter of the self-balanced type by suspending the jaw on an eccentric roller fitted with overrunning clutch.

A vibratory crusher with one moving jaw and two-frequency oscillatory motion driven by an inertial-eccentric system with overrunning clutch (Fig. 3.3f) consists of a frame 37 in which a stationary jaw 38 is mounted, and a moving jaw 41 is suspended on an eccentric roller 39 with overrunning clutch 40. The lower section of the moving jaw is connected with the frame by an elastic system 42. The high-frequency oscillatory motions are transmitted to the jaw by an inertial vibration exciter 43 of the unbalanced-mass type. When the jaw oscillates, the overrunning clutch turns the eccentric roller *39*, as a result of which the upper end of the jaw performs slow-speed oscillation of sufficiently large amplitude along circular trajectories. The action of two-frequency oscillations on the crushed rock mass reduces the duration of crushing, and in a number of cases it contributes to reduction of energy consumption on the crushing process.

Vibratory crushers with three and four moving jaws have been designed, though not used, in which the jaws are mounted at an angle of $120^\circ$ or $90^\circ$ relative to each other. In three-jaw crushers the moving jaws are suspended by pivots in the frame by means of axes, the lower ends of the jaws being supported on elastic elements. The oscillatory motion is imparted to the jaws by hydraulic vibration exciters. During the crushing process the rock is subjected to three-sided compression. It is assumed that this leads to an increase in efficiency and quality of crushing. The crushing forces are transmitted to the frame via the pivots, the elastic system, and the vibration exciter. Since the jaws are mounted on the sides of a triangle, these forces must be balanced to a significant degree. Since such crushers have not been tested in practice, they cannot be given comprehensive evaluation. However, it can be conjectured that their design would be characterized by significant complexity.

Two-jaw vibratory crusher (Fig. 3.3*g*) can be of interest. The moving jaws *44* in these crushers are joined with frame *45* by elastic elements *46*. The jaws are driven by self-balanced vibration exciters *47*. The crusher frame is isolated from the load-carrying structure by shock absorbers *48*. Vibration exciters receive synchronous rotation in opposite directions as a result of which the jaws oscillate in antiphase, thus crushing the rock that is fed between them. The synchronous jaw rotation is obtained either by a kinematic connection of the vibration exciters shafts or by establishing a dynamic synchronization regime.

The merit of the considered vibratory crusher with two moving jaws is the closure of the crushing forces on the crushed material. Connecting the jaws with the frame by the elastic system prevents the direct transmission of the crushing forces to the frame. The horizontal components of the reaction forces from the elastic system to the frame are small as a result of transresonant tuning of the system and are directed from each jaw into opposite directions, which practically assures their total balance. In order to prevent transmission of the vertical force components as well as of random oscillations, the frame is isolated from the load-carrying structure by shock absorbers. The vertical components of the reaction forces in the vibratory crusher under study remain unbalanced, as a result of which they contribute to the oscillations of the crusher frame and increase the speed of material passage [32, 33, 34].

One of the deficiencies of this type of crusher is the large jaw opening in some loading regimes. In this case excessive overloading of the elastic system takes place, which causes its accelerated breakdown. To prevent inacceptable opening, a pneumatic compensating elastic system is installed between the jaws and the frame. This system, which has low stiffness, is capable, without practically changing the total stiffness of the crusher elastic system, of creating significant static forces that compensate for the constant component of the crushing forces. The pneumatic compensating system can be equipped with a system of automatic pressure control in the pneumatic cylinders depending on the magnitude of the outward thrust. In this case, optimum jaw opening is maintained, which provides normal operating conditions of the elastic system.

A vibratory crusher of a similar design (Fig. 3.3h) can be brought into action by a hydraulic vibration exciter 49 located between the moving jaws which are mounted in frame 50 by means of elastic elements 51. An advantage of the vibratory crusher with hydraulic drive is the possibility of a simple adjustment of frequency and amplitude of oscillations of the crushing jaws which enables making adjustments for optimum operation when the properties of the crushed rock change. Furthermore, the absence of bearings increases the reliability of the vibrator. This arrangement, just as the previous one, is characterized by high degree of equilibrium of the dynamic loads.

A number of modifications of the vibratory crushers with two moving jaws has been developed. In vibratory crushers with two moving jaws and inertial vibration exciters of the self-balanced type (Fig. 3.3i), the vibration exciters 52 are linked with the crusher jaws via the elastic elements 53, unlike the basic design considered above, in order to prevent the transmission of impact loads to the bearings. The jaws are connected to the frame 54 by additional elastic links 55. The crusher is vibration-isolated from the load-carrying structure by the elastic elements 56.

In a two-jaw vibratory crusher of such a design (Fig. 3.3j) impact buffers 60 are installed between the jaws 57 and the vibration exciters *58* that are connected by the resonant elastic system *59*.

When crushing a material of limited hardness, the resonant elastic system is deformed insignificantly and the buffers do not come into contact with each other, i.e., crushing takes place in the normal regime. When a particularly hard material is encountered the jaws come to a stop, and as a result of using an elastic system with resonant tuning the amplitude of the oscillations of the vibrator increases until the buffers collide. In the process of buffer collision, large crushing forces are generated and crushing of particularly hard materials is assured.

A crusher of the considered design can be recommended for use in special conditions when rocks with sharply varying strength are crushed. Such design is distinguished by its complexity. The bearing units of the vibration exciters are subjected to considerable impact loads.

In order to increase the oscillation amplitude of one of the crushing jaws, to increase effectiveness of the crushing process, and to provide symmetrical loading of the oscillatory system, a two-jaw crusher has been developed whereby an inertial vibration exciter of the self-balanced type is mounted in the crushing jaw which is stationary relative to the frame (Fig. 3.3*k*). A stationary jaw *62* is installed on frame *61*, and an inertial vibration exciter *63* of the self-balanced type is mounted inside the stationary jaw. The moving jaw *64* is connected with the frame by elastic elements *65*. The frame is vibration-isolated by elastic elements *66*. When the vibration exciter is started, the stationary jaw starts to oscillate together with the frame. The displacements of the frame via the elastic elements *65* induce oscillations of the moving jaw. Stiffnesses of the elastic elements of the moving jaw are so selected that the jaws would oscillate in antiphase. In such a system the amplitudes of jaw oscillation are distributed inversely proportionally to their masses. Consequently, the nondrive movable jaw which is considerably lighter than the system comprising the frame, stationary jaw, and vibration exciter would perform oscillations with substantially larger amplitude.

The merit of this arrangement, in addition to the possibility of creating large oscillation amplitudes of the movable jaw, is use of only one vibration exciter which considerably simplifies the design of the vibratory crusher.

To create a possibility of simultaneous crushing of several products or crushing of one product to different grain sizes, and also ensuring the symmetrical loading of the oscillatory system, jaw crushers with several crushing chambers have been developed. In the vibratory crusher with two crushing chambers, two stationary jaws 67 are mounted on frame 66 (Fig. 3.3*l*). One swinging jaw 69 with two crushing surfaces is also mounted on the same frame via the elastic system 68. Oscillations are imparted to the movable jaw by the inertial self-balanced vibration exciter 70.

Such a crusher design eliminates idling stroke of the crushing jaw, since with each stroke of the jaw in any direction the material in the corresponding working space of the crusher is crushed. However, this crusher design is unbalanced as a result of which the frame performs significant oscillations. The symmetrical loading of the crushing jaw increases the stability of the crushing process as well as a more uniform loading of the drive and of the load-bearing structural elements. The other advantage of the crusher is a need for only one vibration exciter. As a result, its design is simplified, since there is no longer a need for a device to synchronize vibration exciters.

However, unbalance of the loads on the frame causes the transmission of the dynamic loads to the load-carrying structure of the installation in which the crusher is being used.

A reduction of the level of dynamic loads transmitted to the load-carrying structure is achieved in a vibration-isolated two-chamber crusher. The crusher (Fig. 3.3*m*) consists of a vibration isolating frame 71 mounted on a load-carrying structure on shock-absorbing elastic connectors 72. The working frame of the crusher 74 is mounted on the vibration isolating frame by means of the elastic elements 73. Two thin jaws 76 (and a central jaw 78 between them on elastic elements 77) are mounted on the working frame by means of elastic elements 75. A self-balance type inertial vibrator 79 imparts oscillations to the central jaw in the horizontal plane. Oscillations to the side paired jaws are transmitted from the working frame 74 via elastic elements 75.

Just as in the previous design, loads on the working frame are found to be asymmetrical due to the presence of two crushing chambers. However, due to the fact that the working frame is connected with the vibration isolating frame by elastic elements, only insignificant dynamic forces are transmitted to the load-carrying structure.

The vibratory crusher with three crushing chambers (Fig. 3.3*n*) has two movable jaws *80* mounted with the aid of elastic elements *81* in frame *82*, and two stationary jaws *83*. The jaws are driven by self-balanced inertial vibrators *84*. The three-chamber vibratory crusher is more stable than a two-chamber crusher; however, structurally it is much more complex.

The crushers under consideration operate in the material crushing mode under vibration and vibro-impact actions (with the exception of the vibratory crusher in Fig. 3.3*d* operating in the regime of two-frequency oscillations of the jaws). The quest for rationalization of the process of rock-mass crushing gave birth to a series of fundamentally new approaches in the development of vibratory crusher designs. In the already mentioned two-frequency crusher, one can reduce the losses on overcoming friction forces. Reduction of the frictional losses is also a goal in crushers in which elastic materials are incorporated in the design of the working surfaces. A design of a vibratory crusher with segmented crushing jaws has been developed in which bending stresses are generated in the crushed rock mass.

The vibratory crusher jaw with elastic layers consists of a frame in which layers from elastic material with metallic vulcanized inserts are fixed. Oscillations of the crushing surfaces of the jaws are effected by means of periodic pumping of the working medium under them or applying compressed air with the aid of a regulator. The merit of elastic jaws is their light weight and, hence, lower dynamic loads, the possibility of executing large strokes, and the replacement of the friction forces of the jaw against the rock at the site of formation of destructive cracks by forces of elastic strain in the elastic jaw material. Elimination of the frictional forces reduces the forces created in the process of fracture of the crushed material, thus increasing, in the final analysis, the crushing effectiveness.

For the elimination of the friction forces on the jaw with the formation of destructive cracks, composite elastic coatings of ordinary crushing jaws can be used. The coating comprises elastic material in which spheres are vulcanized. The elastic coating is fixed to the crushing surface of the jaw. The spheres can be vulcanized in such a way that either they come into contact with the jaw surface or a gap remains between them. In the presence of a gap, the spheres impact on the jaw during the process of crushing, the impact impulse is transmitted to the material, and the effectiveness of the crushing process is increased.

The proposed vibratory crusher designs are advanced designs; however, in order to make elastic jaws and elastic coatings with the necessary strength, formidable technical problems must be overcome. The lack of experience does not enable one to evaluate possibilities for application of this design in the near future.

In a number of designs of vibratory crushers there are various additional devices which are excited by the vibration of the main crusher elements. Such devices ensure, for example, a uniform supply of the crushed rock mass to the crushing chamber, delivery of fines and loading of the crushed material onto conveyer belts, recirculation of uncrushed pieces, and screening out of the quality product.

For the uniform delivery of the rock to the working crushing chamber, and also to provide preliminary screening of the fines, special designs of vibratory crushers have been proposed. A vibratory crusher with feed chutes for uniform material supply to the crushing chamber has been built using the basic model of Fig. 3.3. Loading chutes are mounted on the frame of the crusher via elastic elements. Under the frame vibration, the grates oscillate, thus facilitating the uniform delivery of the material to the crushing chamber. By varying inclination of the chutes, the loading capacity of the crusher can be regulated.

A single-jaw vibratory crusher-screen has been proposed which is intended for crushing off side rocks and preliminary screening of the fines upon loading the crushed material onto the conveyer belt. The vibratory crusher-screen consists of a grate screen which is mounted with elastic elements. The crushing jaw with an inertial vibration exciter is attached to the top of the grate screen by means of an elastic system. The frame of the vibratory crusher-screen is mounted on the metal structure of a line conveyer by shock absorbers. Upon delivery of a load of rock mass to the grate screen, the fines seep through the openings and the oversize rocks are crushed by the jaw and loaded in crushed form from the grate screen onto the conveyer belt.

As a drive for jaw vibratory crushers one can use inertial, hydraulic, and eccentric vibration exciters, and also vibration exciters of a new type for the generation of parametric oscillations of jaws.

The other types of vibratory crushers suitable for crushing in stationary and mobile plants include rotary and cone crushers.

The rotary crusher with eccentric drive (Fig. 3.4a) consists of frame 1 on which stationary or spring-supported with elastic elements 2 jaws 3 are installed. A rotor 4 is mounted between the jaws on elastic eccentrics 5. The crushing element is a cylindrical roller 6 which can rotate freely on bearings. Two crushing chambers are formed between the rotor and side jaws. With the rotation of the driving eccentric shaft, the rotor performs circular oscillations and crushes the rock mass alternately in one chamber then in the other. Since the stiffness of the elastic eccentrics is different in different directions, parametric oscillations which are distinguished by high stability under the action of substantial loads are induced in addition to forced oscillations.

The cone vibratory crusher with an inertial drive (Fig. 3.4b) consists of three main units: bowls 7 rigidly connected to frame 8; a crushing cone 9, and an unbalanced mass 10. An electrical motor spins the unbalanced mass via a drive shaft with two Cardan joints and a spline connection. The shaft of the unbalanced mass is mounted in two bearings. The cone is connected with the bowl by a spherical joint 11. The bowl and the frame are vibration-isolated with the aid of tie-rods or springs.

The centrifugal force of the unbalanced mass presses the crushing cone to the surface of the bowl imparting to it rotational motion. With this, the axis of the crushing cone deviates from the axis of the frame and performs circular oscillations as a result of which inertial forces are generated, differing from the inertial forces of the unbalanced mass on the stationary axis. Under the action of the inertial forces, the cone slowly rotates in the opposite direction around its axis and rolls along the surface of the bowl crushing the products in the region of approach between the cone and the bowl. The clearance between the cone and the bowl can be equal to zero in the case of direct contact (idling) and can reach a certain value depending on the properties of the crushed rock mass. The changes in the crushing force are larger the more substantial the deviation of the cone axis from the crusher axis during rotation. As a result of this, the increase of the initial clearance to a certain value leads to an increase of crusher capacity and to a decrease in grain size of the final product. However, this is true only up to a certain limit. For

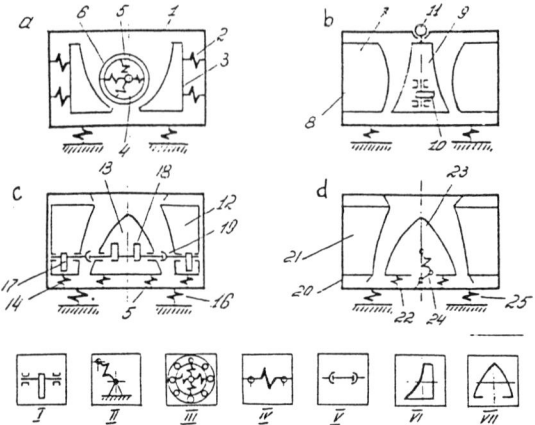

**Figure 3.4** Schematic diagram of rotary and cone vibratory crushers: *I)* unbalanced vibration exciter; *II)* eccentric vibration exciter; *III)* parametric vibration exciter; *IV)* elastic element; *V)* Cardan shaft; *VI)* crushing jaw; *VII)* crushing cone.

very large initial clearances the regime of spinning motion is distorted and crushing of the material might be halted altogether. If the crushing force which is generated by the unbalanced mass and the frame is not adequate for crushing, the cone is jammed and stops rolling, although the unbalanced mass will continue to freely rotate inside the cone. Furthermore, the centrifugal force of the unbalanced mass will cause small oscillations of the cone as a result of which some jamming chunks are crushed, freeing the cone. During this period, crushing results in large grain-size product. A similar phenomena is observed under excessive feeding of the crusher. Therefore, the crusher must not be overloaded above the admissible capacity limit, otherwise the normal regime of spinning operation is disrupted.

Vibratory cone crushers of different design concepts have been developed. They have either rectilinear displacements of the cones or cone oscillations along a helical line. The vibratory crusher with inertial drive imparting rectilinear antiphase oscillations in the vertical direction to both cones (Fig. 3.4c) consists of external *12* and internal *13* moving crushing cones. An annular crushing chamber having changing cross section along its height is formed between the cones. The internal and external cones are mounted on the frame by means of elastic elements *14* and *15*. The frame is supported on the load-carrying structure via amortizing mounts *16*. Mounted on the internal and external cones are the unbalanced masses *17* and *18* connected to each other by the Cardan shafts *19* and operating in antiphase. The crusher is equipped with loading bin and discharge hopper.

When the unbalanced vibration exciter rotates, the crushing cones oscillate in antiphase, crushing the material delivered between them from the loading bin. The crusher is adequately balanced, since the reactions of the elastic element of the moving cones on the frame are approximately equal in magnitude and opposite in direction.

The vibratory jaw crusher with eccentric drive which imparts oscillations to the internal moving cone in a helical line is shown in Fig. 3.4*d*. Such oscillations of the crushing cone increase the effectiveness of crushing as a result of formation of shear strains in the chunks of the crushed material. The crusher consists of a frame *20* on which an external cone *21* is rigidly mounted. An internal moving cone *23* is mounted on the frame by means of elastic elements *22*. The elastic elements *22* are arranged along the perimeter of the cones at an angle to their vertical axis. Oscillations are imparted to the moving cone by an eccentric vibration exciter *24* with driving elastic connection. The crusher can be mounted on the carrying structure either rigidly or on elastic elements *25*.

### 3.3  DESIGN SCHEMATICS OF VIBRATORY CONVEYING AND PROCESS-CONVEYING MACHINES

Vibratory conveying machines are widely used for the movement of bulk and lump loads in various industries. Process-conveying machines, which also execute load processing in the course of conveying, are also widely developed (drying, dust removal, classification, granulation, dewatering, and so on). Vibratory conveying machines include vibratory conveyers, feeders and screen-feeders, and also elevators and batching dosating hoppers.

Conveyers and feeders of standard designs for the movement of bulk and lump loads are the most widespread types of vibratory conveying machines. They are distinguished by the wide diversity of fundamental and structural designs and are produced for different capacities. For the movement of loads in the vertical direction, elevators are used (conveyers with helical load-carrying element).

In view of the fact that the principal design features of vibratory conveying machines are determined by the type of drive used in them, the schematic arrangements are considered with respect to the following five types: electromagnetic, pneumatic, inertial, eccentric, and hydraulic.

The schematic diagrams of such machines are arranged in a sequence determined by a number characterizing the number of actually used degrees of freedom.

Electromagnetic vibratory conveying machines are divided into single-drive and multi-drive machines. Single-drive machines can be one-, two-, and three-mass machines; multiple-drive machines can be one- or multiple-mass machines (Fig. 3.5). The structural layout of a one-mass electromagnetic machine is shown in Fig. 3.5a. It is comprised of the following elements: a load-carrying element 1 to which the active section 2 of the electromagnetic vibrator is attached and elastic connections 3 connecting the active section with the reactive section 4 which is fixed on the frame. Oscillations of the load-carrying element are excited and maintained by the pulses of the electromagnetic vibrator. The main components of the two-mass system (Fig. 3.5b) are: a load-carrying element 1, to which the electromagnetic vibrator is attached, comprising active 2 and reactive 4 sections with built-in elastic connections. The vibratory machine is isolated from the load-carrying structure by means of isolating mounts 5.

The single-drive three-mass electromagnetic device consists of the following components (Fig. 3.5c): load carrying element 1 with active section 2 of vibrator and reactive section 6 of vibrator with elastic connections, reactive mass 7, auxiliary working elastic connections 3, 4, and isolator 5.

For reversing of the motion, a special drive is used comprising two vibrators oscillating in mutually perpendicular directions. The reversible machine (Fig. 3.5d) consists of the components: load-carrying element 1 with rigidly attached active parts of the vibrators, reactive parts of two vibrators 2 with elastic connections 3 and vibration isolators 4.

Conveying machine with dynamic vibration absorber mounted on the reactive section of the vibrator belongs to the group of three-mass machines. Vibration abatement of the reactive section of the vibrator enables reduction of the clearance between the poles of the electromagnets thereby increasing the output power of the drive. The machine (Fig.

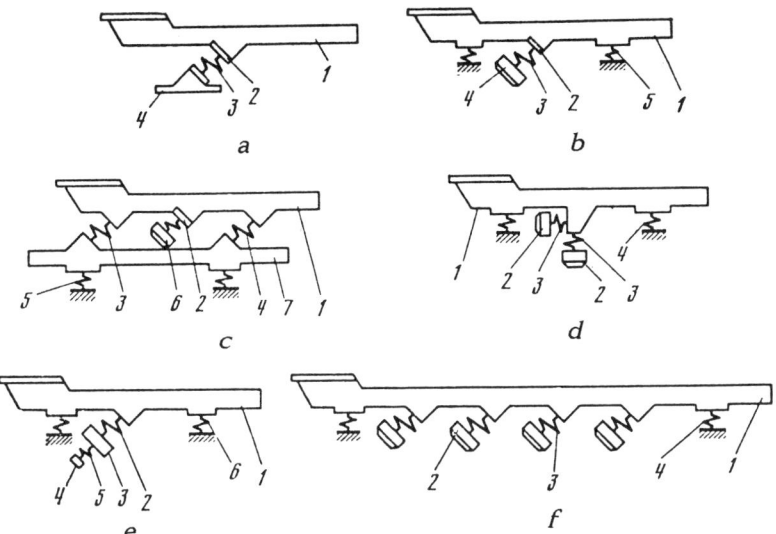

**Figure 3.5** Schematic diagrams of vibratory conveying and process–conveying machines with electromagnetic drives.

3.5e) consists of a load-carrying element *1* with the active section of the vibrator rigidly attached to it, main elastic system *2* linking the active section of the vibrator with the reactive section *3*, and dynamic vibration absorber comprising the reactive mass *4* which is linked with the reactive section of the vibrator by the elastic connections *5*. The load-carrying element is mounted on the supporting structures by means of isolators *6*.

The multiple-drive multiple-mass machine (Fig. 3.5*f*) has a load-carrying element *1* on which a number of electromagnetic vibrators *2* with elastic connections *3* are mounted; the machine is isolated from the support structures by mounts *4*.

Pneumatic vibratory conveying machines are characterized by the small number of types available and are usually made as single-drive, since the existing designs of pneumatic vibrators do not allow synchronization of their operation.

The single mass machine with an active pneumatic vibrator contains a load-carrying element, to which force impulses are imparted by a pneumatic vibrator, mounted on the foundation and elastic connections. The two-mass machine with piston reactive pneumatic vibrator includes a load-carrying element, to which a piston pneumatic vibrator consisting of active section (cylinder) and reactive section (piston) is attached, and elastic connections. The piston under the action of compressed air performs reciprocating motions. Due to reactive forces generated in this process periodic pulses are imparted to the load-carrying element which induce and maintain its oscillations.

Eccentric conveying machines have an elastic element in the drive for developing the necessary mobility in the dynamic system.

The fundamental arrangements of machines with hydraulic drives are analogous to machines with eccentric drive. The single mass machine (Fig. 3.6a) consists of a load-carrying element 1 mounted on foundation 2 by means of elastic connections 3; oscillations are imparted to the load-carrying element by an eccentric drive 4 with elastic connecting rod.

Machines of this type are distinguished by their simple design; however, due to the fact that the inertial forces of the oscillating masses are not balanced, their drive is subjected to the action of large dynamic loads. The high energy consumed in overcoming the harmful resistances acting in the system is another disadvantage.

In order to unload the drive and eliminate transmission of dynamic loads to the supporting structures, resonance balanced vibratory machines are used. The machine shown in Fig. 3.6b is supported in a stationary point of the system. Eccentric drive is mounted on an oscillating mass. The machine consists of two load-carrying elements 1 or of one load-carrying element and a reactive mass, eccentric drive 2, working elastic connections 3 and support columns 4, which are affixed to the stationary point of the entire system.

The most widely used presently are resonant balanced two-mass system installations. The vibration-isolated machine with parallel arrangement of the oscillating masses (Fig. 3.6c) consists of a load-carrying element 1 and reactive mass 2 mounted opposite each other, eccentric drive 4 with elastic connecting rod, working elastic connectors 3 and vibration isolators 5, with which the reactive mass is isolated from the supporting structures.

In a two-mass machine with a series arrangement of the oscillating masses (Fig. 3.6d) the load-carrying elements 1 are arranged one after the other. The eccentric vibrator 2 imparts force pulses to them in opposite directions as a result of which the load carrying elements supported on the foundation via elastic mounts 3 oscillate in antiphase. Such a design

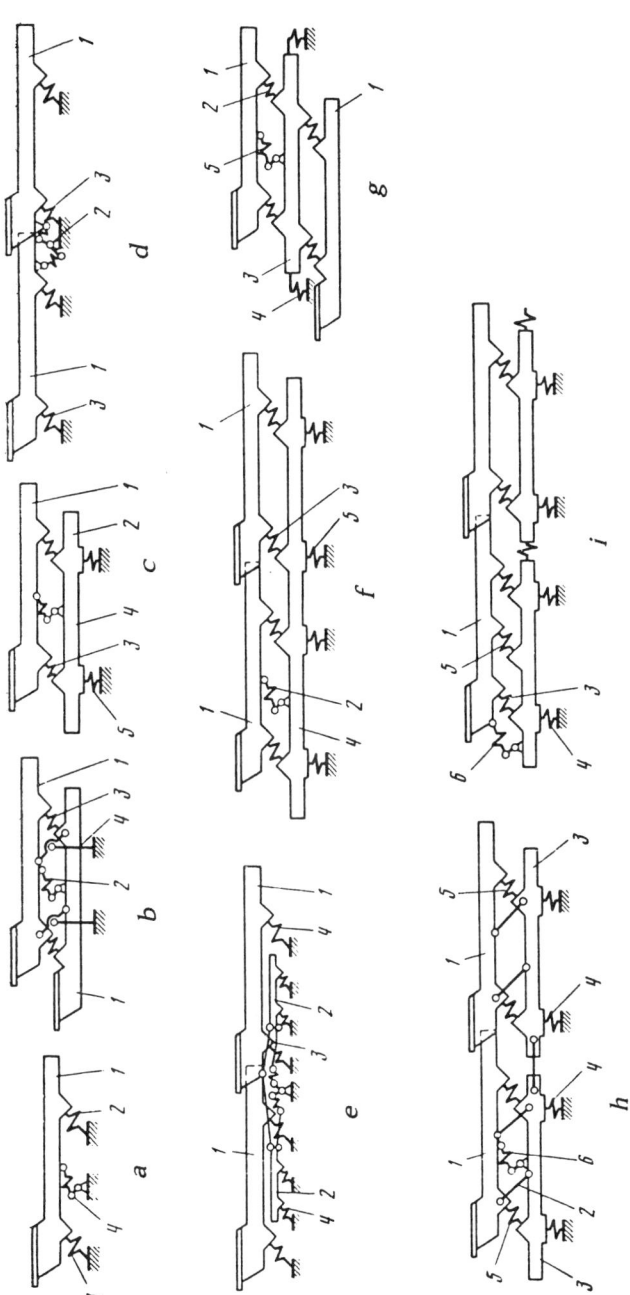

**Figure 3.6** Schematic diagrams of vibratory conveying and process-conveying machines with eccentric drive, d – with inertial drive.

ensures balancing of the horizontal components of the inertial forces.

A more complete balancing is attained in machines with a series-parallel arrangement of the oscillating masses (Fig. 3.6e). Each is comprised from load-carrying elements 1 and auxiliary beams 2 which are rigidly connected with each other by tie-rods 3. The oscillating mass is so arranged that the auxiliary beam of the second load-carrying element would be under the first load-carrying element. The load-carrying elements and auxiliary beams are supported on the bedframe via the working elastic connections 4. The unit is driven by an eccentric vibrator 5 whose connecting rods are linked with reactive masses.

In the three-mass machine (Fig. 3.6f) with a series arrangement of the load-carrying elements 1 and with eccentric drive 2, the load-carrying elements are supported with the aid of elastic connections 3 on the counter balancing frame 4 which is isolated from the supporting structures by isolators 5. Drive is effected by an eccentric vibrator connected with one of the load-carrying elements.

In a three-mass machine with a parallel arrangement of the load-carying elements and eccentric drive (Fig. 3.6g) the load-carrying elements 1 are supported by elastic connections 2 on the counter balancing frame 3 which is isolated from the supporting structures by vibration isolators 4. One of the load-carrying elements is driven by an eccentric vibrator 5. Oscillations are reactively imparted to the second load-carrying element. Large vibratory conveying machines are often designed as four-mass systems comprising two two-mass sections (Fig. 3.6h and i). The drive is installed on only one section, while the second section is driven reactively. The four-mass machine with eccentric drive consist of two sections. Each section consists of a load-carrying element 1, support columns 2, a reactive mass 3 (frame) mounted on isolators 4, and working elastic connections 5; the frames of both sections are joined together by a tie-rod; drive is effected by means of eccentric vibrator 6 mounted on one of the sections.

An advantage of machines of this structural schematic is the feasibility of making installations of a considerable length. The system allows a simple transmission of the exciting force via a connecting tie-rod.

In vibratory machines with inertia drives with directional rectilinear and rotational exciting forces are used. In the single-drive single mass machine with inertial drive the load-carrying element is mounted on the foundation with the aid of elastic connections. Oscillations are imparted to it by an inertial vibrator.

The single-mass vibratory conveying machine with inertial drive which is supported at a stationary point of the system consists of two load-carrying elements connected with each other by elastic links and by the housing of the inertial vibrator with displaced unbalanced masses. Support of the system at the stationary point on the load-carrying frame can be either direct or via isolators.

The two-mass inertial vibratory conveying installation consists of a load-carrying element, vibrator, working elastic connections, and vibration isolators; a special heavy frame or the vibrator itself can serve as the reactive mass. The design of vibratory conveying machines in accordance with the two-mass structural arrangement afford great opportunities for creating resonant installations with vibration isolation and dynamic balancing of the oscillating masses.

In the three-mass vibratory conveying installation with inertial drive, the load-carrying elements are supported by means of elastic links on a carrying frame which is isolated from the supporting structures by vibration isolators. The machine is driven by a vibrator which is rigidly attached to the carrying frame.

The multi-mass machine has a carrying frame on which a load-carrying element is mounted with the elastic links. The frame is vibration-isolated by shock absorbers, and the machine is driven by self-synchronizing vibrators.

The multi-mass machine consists of a load-carrying element to which longitudinal oscillations are imparted by an inertial vibrator with rectilinearly directed exciting force. The machine is supported on the foundation by means of shock-absorbing elastic links. In order to impart oscillations to the load-carrying element, reactive masses with elastic links are mounted along the element and at an angle to its longitudinal axis.

The working elements of the vibratory process-conveying machines have different designs according to their destination, namely for conveying, batching, and processing operations. Conveying working elements are made as tubes and troughs (chutes) which can be either open or closed. The working elements intended for batching are very diverse: with a wide chute, diagonal slot which is cut at an angle by the discharge rim, multiple-trough, and so on.

The most diverse are the working elements for process operations. Vibratory dryers use conductive and convective heat transfer, drying in infrared rays, in electrical field of industrial or high frequency; a combined drying method is also used incorporating several of these methods of heat transfer.

Drying in the vibratory boiling layer is an extremely effective process; the process of cooling is also equally effective. In vibratory units with conductive action, in which heat input into the product or heat removal is effected through the surface of the heat exchanger, the working element is made in the form of sealed unit having double walls between which flows the cooling or drying medium. In this respect, their design is similar to the structure of the load-carrying elements of vibratory conveyors, which are intended for the movement of hot products. The conductive method of heat transfer has a number of advantages, since the working chamber in this case can be completely isolated from the heat carrier. Drying can be carried out at low velocities of the gas flowing over the layer at low pressures and in vacuum. The vaporizing liquid can be recycled into the food process line.

Comparative experimental investigations on drying granulated sugar in plate, drum, and vibratory dryers with conductive heating showed that drying in a vibratory boiling layer is 2 - 3 times faster and, furthermore, lower final humidity of the product is attained.

In convective action units the processed product comes into contact with the cooling or heating stream of air or gas which simultaneously play roles of both a heat carrier and a desiccant. Accordingly the working elements of vibratory dryers and coolers of this type can be arranged without special jackets. Since the materials must be spread in a thin uniform layer in order to provide high-quality drying, working elements with very low side walls are used.

In vibratory dryers the drying agent is blown over the layer or through the layer of the processed product. In the first case the agent is moved in a counter current to, with the current, or perpendicular to the direction of the flow of the product. In units of the second type the flow of the agent can be directed from bottom to top or from top to bottom.

Gas flow over the layer is mainly used for rapidly drying products which are moved in vibratory dryers in a thin layer. Drying of the products that give away moisture slowly is carried out in a thick layer through which the agent is blown.

In this case, heat-transfer processes and the structure of the vibratory boiling layer are improved (the boiling process is intensified).

In convective action vibratory dryers, in which the drying medium is blown through the layer of the processed products in order to increase the process effectiveness, double-bottom working elements are used. The second bottom is made with transverse louver-like slots through which is passing air or gas moving in the channel between the bottoms. The whole working element is hermetically sealed, which eliminates losses of the drying medium.

The structural designs of the working element of the vibratory dryers are shown in Fig. 3.7. The dryer (Fig. 3.7a) consists of a flat bottom 1 under which there is a channel 2 along which the hot medium (water or air) moves. The dryer is covered by enclosure 3 with loading 4 and discharge 5 openings. The vaporized moisture together with the hot air is removed through opening 6. The direction of the motion of water vapors is indicated by broken arrows, and the rectilinear harmonic oscillations of the working element with amplitude $A$ and frequency $\omega$ are shown by a continuous arrow. It is pertinent to note that along with the rectilinear harmonic oscillations, the working element can perform more complex oscillatory motions: biharmonic, polyharmonic, and two- and three-component oscillations. Such oscillations usually ensure high efficiency of the drying process.

Vibratory dryers with the conductive method of heat-transfer to the processed product are widely used in food production, e.g., for drying granulated sugar. Heaters can also be arranged as a system of tubes imbedded in the layer of the processed product transversely or longitudinally to its displacement. Low-pressure vapor, products of fuel combustion, heated air, or electric heaters are usually used to heat the product. Drying granulated sugar and flour in the vibration boiling layer under conductive heat input into the product allows sharply reducing its losses from the working chamber and drying to a low final humidity under high intensity of the moisture evaporation process. It also ensures uniform drying of the flour and excludes formation of particle aggregates. When the duration of travel of the product over the installation is less than the time of the process cycle, several units installed in series or cascade arrangements are used.

The utilization of vibration boiling layer enables substantially reducing the total consumption of air needed to

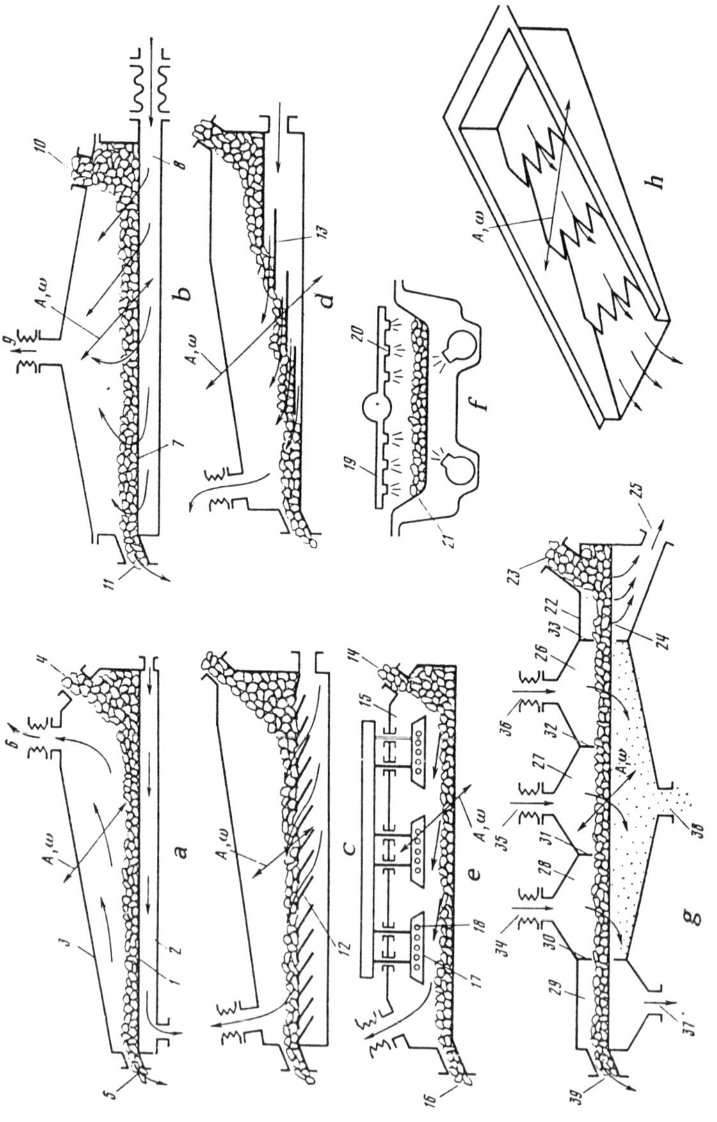

**Figure 3.7** Working elements of vibratory dryers.

create the suspended (boiling) layer; as well, the heat expenditures on the drying process are reduced.

The working element of the convective dryer whose bottom is made as a gauze 7, through which the heated air is supplied by channel 8, is shown in Fig. 3.7b. After passing through the layer of the processed product which is in a vibratory boiling state, the hot air with moisture vapors are exhausted through opening 9. The product is supplied through door 10 and the dried mass is removed through the discharge opening 11.

The bottom of the working element of the dryer can be designed as louvers 12 (Fig. 3.7c) or as a stepped surface 13 (Fig. 3.7d).

Vibratory dryers and coolers with external sources of heat or of cooling medium (Fig. 3.7e) are also used. The product to be dried is supplied via opening 14 to the working element 15 and is intensely suspended under the action of vibration while moving gradually to the discharge opening 16. Drying of the product is carried out by infrared heaters 17 with quartz tubes 18 or by other heaters mounted rigidly on the stationary non-vibrating parts of the dryer.

Vibratory conveyance of the processed product in coolers is effected over the surface of the trough. The duct formed by the double bottom is used to supply the cooling agent to the trough surface, which is in contact with the processed product. Granulated products are effectively cooled upon removal of heat from the trough by sprinkling cold water on its lower surface. The sprinkler system comprises a collector 19 and nozzles 20 from which water is sprayed onto the lower surface of the trough 21 (Fig. 3.7f). The principal benefit of such a system compared with water jackets is the following: the cooling water after absorbing the heat from the trough flows down to the bottom of the duct and fresh cold water is supplied continuously. If the processed product allows direct contact with the water, the sprinkler system comprising the collector and nozzles is mounted above the conveying trough.

In similar vibratory installations the product can be washed, dewatered, granulated, and subjected to other types of process operations. The working element of the installation executing a whole complex of process operations (separation, drying, and so on) consists of a number of chambers whose bottoms are made from mesh with various openings. Thus, the humid product delivered through door 23 is dewatered in chamber 22 on mesh screen 24 and the separated moisture is carried away through the discharge channel *25*. The dewatered product is fed successively into chambers *26, 27, 28,* and *29* which are separated from each other by partitions *30, 31, 32,* and *33* (Fig. 3.7*g*). The predrying medium (hot air) with specified various temperatures moves from the top downwards through reducers (confusors) *34, 35, 36* and is sucked out through the oulet tube *37*. Along with the process of drying, dust removal from the product is also carried out by screening it through a mesh bottom. Dust-like fractions of various classes are removed through openings *38* and *39*.

Vibratory granulators and mixers (Fig. 3.7*h*) are widely used. The trough-shaped working element has a stepped bottom and each step has sharp teeth on the discharge end. Such a design ensures intensive mixing of the processed product and helps to obtain a homogeneous mixture.

## 3.4  VIBRATORY ELEVATORS

The movement of loads vertically upwards which can be simultaneous with their processing is effected by vibratory processing conveying machines with a screw-shaped working element. These are divided with respect to their intended purpose, into three main types: vibratory elevators; designs in which the transported product is subjected to processing; feeding, accumulating, and batching units (vibratory hoppers, feeders, batchers).

Fig. 3.8*a* shows a schematic of a one-mass vibratory elevator with electromagnetic drive. It comprises a working element *1* mounted on the supporting structure by means of elastic system *2*. An electromagnetic vibrator, whose active section *3* is rigidly fixed to the working element and the reactive section *5* Is freely suspended on an elastic system *4*, imparts oscillations to the working element in the vertical direction. The vertical oscillations of the vibrator are converted into torsional-translational oscillations of the working element as a result of inclination of the principal extremal stiffness axis of the elastic system at an angle to the vertical axis of the vibratory elevator.

In some designs of vibratory elevators the reactive mass of the resonant vibrator is rigidly mounted on the machine foundation. The elastic system is attached at the same location and is made in the form of springs arranged around the working element at an angle to its longitudinal axis. Figure

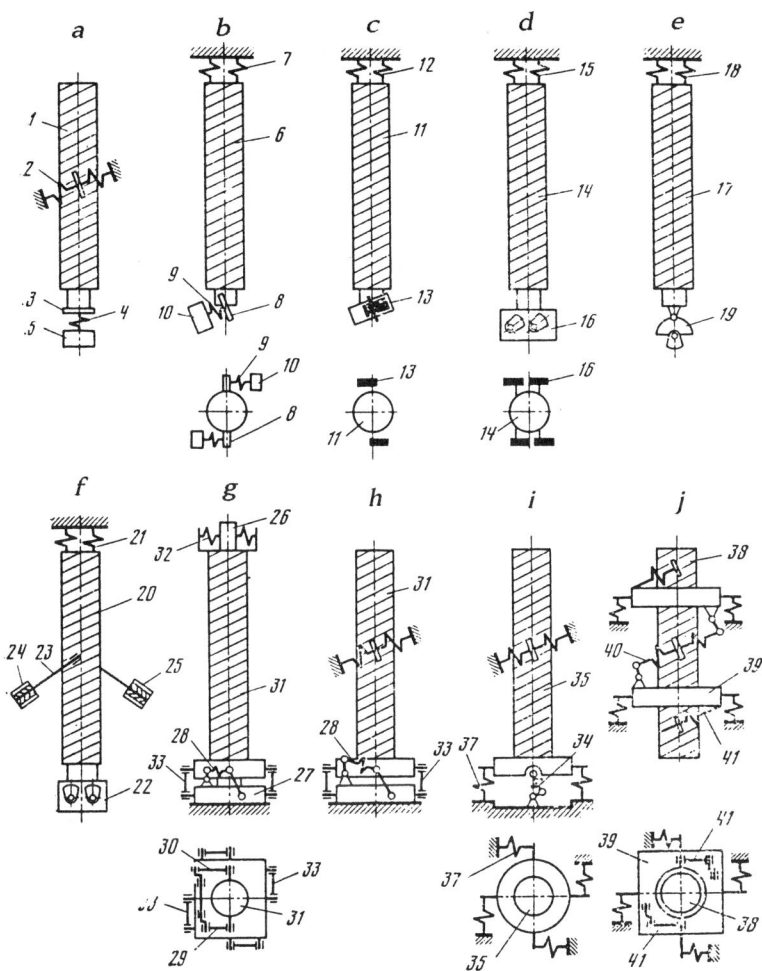

**Figure 3.8**  Schematic diagrams of vibratory elevators.

3.8*b* shows the layout of the elevator in which the electromagnetic vibrator imparts oscillations to the working element at an angle to its longitudinal axis. The elevator consists of a working element *6* suspended by vibration isolating system *7* from the supporting structure. In order to provide symmetry of the exciting force, two or three vibrators are installed on the working element. The active sections *8* of the vibrators are attached to the working element and the reactive sections *9* are connected with them through elastic elements *10*.

Resonance vibratory elevators with electromagnetic drive are usually small in size and limited in power: in this case the elastic oscillations of the working element are insignificant and pose no danger to its reliability. It must be born in mind that in vibratory elevators with electromagnetic drives, elastic oscillations are particularly intensive, since the natural frequencies of the structure are close to the forcing frequencies.

Generating drives of various basic designs (see chapter 4) are more widely used in heavy installations. Figure 3.8c depicts the layout of an elevator with motor-vibrator drive. The unit consists of a working element 11 suspended on shock-absorbing working links 12, to which oscillations are imparted by motor-vibrators 13. They are used in clusters of two or three in order to generate symmetric exciting force and exciting torque. Figure 3.8d shows the layout of a vibratory elevator with a special vibrator with four or six unbalanced masses which generate an exciting force directed along the axis of the working element and an exciting torque around this axis. The elevator consists of a working element 14 suspended from the carrying structure by means of shock-absorbing elastic links 15, with the frame of the inertial vibrator being rigidly fixed to the lower part of the load-carrying element. The exciting force and torque that are required to drive the elevator can be generated by a two-joint (double) pendulum vibrator. The schematic diagram of a vibratory elevator with double pendulum vibrator is shown in Fig. 3.8e, where 17 is the working element, 18 are the shock-absorbing links, and 19 is the pendulum vibrator.

Elevators with elongated working elements can be designed with resonators which reduce stresses in the structural components. The schematic diagram of a vibratory elevator with inertial drive and resonators is shown in Fig. 3.8f. It consists of a working element 20, which is suspended from or resting on the shock absorbing elastic link 21. At the lower cone an inertial vibrator 22 is fixed to the working element. The vibrator generates a rectilinear exciting force acting along the vertical axis of the elevator. Along the working element resonators are mounted with a specified pitch. They include springs 23 which are inclined to the working element and moving masses 24 which are connected with the springs by the elastic links 25. The vibrator imparts to the working element oscillations which excite vibrations of masses 24 of the resonators along the axis of spring 23. The reactive forces from deformations of the elastic links are transmitted to the working element via springs. The horizontal components of these forces create exciting torques around the axis of the load-carrying element. The oscillatory system of the elevator is tuned to the resonance regime, at which the motions of the reactive masses of the resonators and the load-carying element are in antiphase. The resonant mode of operation enables one to design elevators with substantially long working elements. This is explained by the fact that dangerous elastic torsional oscillations of the working element are reduced when exciting torques are uniformly distributed along the height of the installation. The sufficiently high longitudinal rigidity of the working element is rationally utilized as a force-transmitting element for transmitting longitudinal oscillations with their subsequent transformation into torsional oscillations.

Vibratory elevators with eccentric drives are also used. The latter enables generation of large amplitudes of the working element and reduction of the excitation frequency, thus achieving a better detuning from the structural natural frequencies. The smaller acting accelerations enable bringing into motion a unit of larger mass by the eccentric drive. The layout of such an elevator is shown in Fig. 3.8g. It comprises a central stationary column 26 mounted on bed 27 on which a double connecting-rod eccentric drive 28 is mounted with connecting rods 29 and 30 attached on the sides of the working element 31. The connecting rods are mounted on an eccentric shaft whose eccentrics are displaced by $180^\circ$ relative to each other. The load-carrying element of the elastic system 32 is fixed to the central column and to the bed by the inclined connecting rods 33, which determine the direction of oscillations. The connecting rods must not necessarily be mounted in the direction of oscillations of the working element and can even be arranged horizontally (Fig. 3.8h). Figure 3.8i shows the layout of a vibratory elevator whose eccentric drive 34 imparts vertical oscillations to the working element 35 mounted on bed 36 with the aid of the directional elastic system 37. The necessary translational and angular oscillations of the working element are generated by using an elastic system whose maximum stiffness axis is directed at an angle to longitudinal axis of the working element. The vibratory elevators have a rigid elastic system and eccentric drive, which together transfer the total inertial load to the foundation. To prevent this from happening, designs of two-mass resonance vibrator elevators with eccentric drives have been developed. The layout of such an elevator is illustrated in Fig. 3.8*j*. It consists of the working element *38* and the balancing frame *39* on which eccentric drives *40* are mounted. The working element and the frame are connected by the elastic system *41* which provides directional oscillations of the working element. The balancing frame and working element are mounted on the foundation with the aid of vibration isolating links.

Vertical vibratory process-conveying machines are very effective in the realization of many process operations, they occupy small working areas, and have a working element of a considerable length.

The basic designs of working elements of vibratory elevators are presented in Fig. 3.9. In order to carry out the operations of separation combined with the elevation of granulated products, the working element (Fig. 3.9*a*) is made as a two-entry helical trough, with a screening working surface *2* placed on the upper helical surface *1*, and the trough has a continuous surface *4* on the lower helical surface *3*, where the screened product is collected and transported. Should there be a need, separation into three classes can be achieved (as in a two-screen grizzly). In this case the working element is made as a three-entry helix: two helical troughs with a screening surface and one with solid surface.

The vertical vibratory grizzly differs favorably from the ordinary grizzly: firstly, it enables carrying out separation on very extended surfaces on the limited areas of industrial buildings, which ensures high separation quality and enables realizing high output per unit area of the screening surface; secondly, the design of the working element allows discharging the product below or above the screen at any point within the conveying height which facilitates the use of the machine in any processing layout of food production.

When carrying out processing of granulated products in liquids, the lower part of the working element is placed in a receptacle filled with the processing medium (Fig. 3.9*b*). As the granulated product climbs up the helical trough, it comes out of the liquid and in moving upwards along the trough it simultaneously becomes free of the liquid. For more effective separation of the processing liquid and for the dewatering of the products a slotted screen is placed on the trough or the working element is made with perforations (Fig. 3.9*c*). In this case the machine plays the role of dewatering grizzly or elevator. The dewatering process can also be carried out

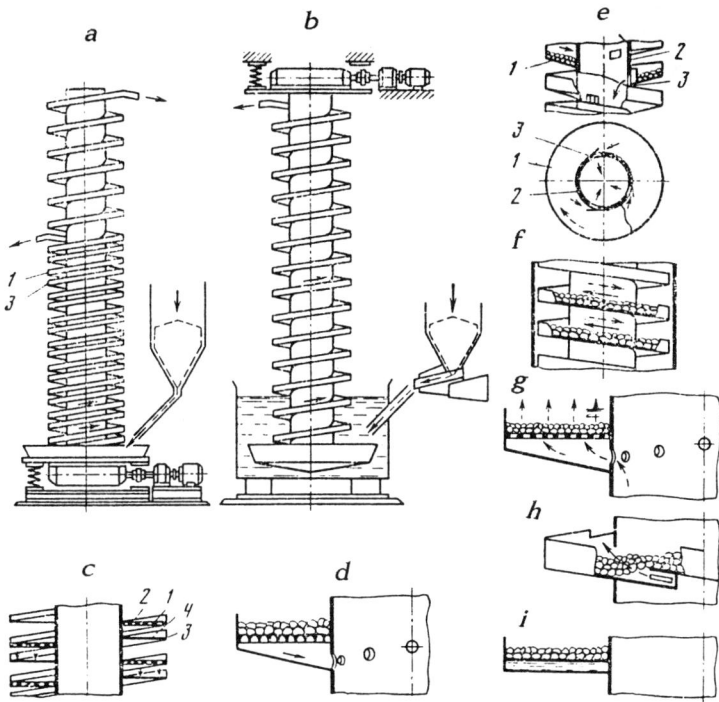

**Figure 3.9**  Working elements of vibratory elevators.

without special screens or perforations. In this case the surface of the helical trough *1* is inclined toward the tubular carrying column *2* in which holes *3* are made (Fig. 3.9*d*). The separated liquid flows in the radial direction to the joining of place the trough and the tube, then moves downward and into the tubular column through the holes.

Vertical vibratory process-conveying machines can successfully combine conveying with the process operations of drying or cooling of granulated products. The layout and the methods of carrying out product processing by a heat carrier can be very diverse depending on the specific conditions and its physicomechanical properties. Heat-carrying agents can flow in a countercurrent in the space between the lower surface of the trough and the upper free surface of the product (Fig. 3.9*f*), be drained through perforations in the bottom of the trough, and then penetrate the layer of the moving product (Fig. 3.9*g*), penetrate the product where it is poured out from one segment of the stepped extended compartmental trough to another (Fig. 3.9*h*), heat or cool the surface of the trough, and thus the product as it moves in the helical gas duct adjacent to the trough (Fig. 3.9*i*). It is essential to note that in all the described design variations the efficiency of drying or cooling increases substantially due to intensification of heat transfer in the vibration boiling layer.

## 3.5  VIBRATORY MACHINES FOR COMPACTION OF GRANULAR MEDIA

The operation of compaction of various granular media is used to increase the degree of utilization of various tanks and containers and also to strengthen some products. Considered below are installations for compaction of granular media in containers or bales of transported loads in a rolling stock (cars, trailers, and so on) and compaction of core and mold sands in casting operations. Compaction of granular media and loads increases the actual capacity of containers in transport facilities by 15 – 20%. Vibratory working of core and mold sands is enhancing strength of cores and molds produced from them.

Compaction of granular media takes place as a result of fluctuation of the frictional forces between individual particles under vibration and action of the frictional and inertial forces. It continues until the action of the forces of internal resistances and gravitational forces are balanced. To ensure the compaction of the load by means of vibrations it is necessary that the acceleration of the receptacle has a specific value. For media with small internal friction this acceleration is less than for materials with large forces of internal friction.

Vibratory installations for the compaction of transported loads are usually designed as vibratory platforms on which the vehicle being loaded (rail car, trailer) is mounted (Fig. 3.10*a*). The platform consists of a working surface *1* which, when used in railway transportation, is equipped by rails, carrying elastic system *2*, and vibratory *3*, usually of inertial type. Vibratory compactors of the suspended type (Fig. 3.10*b,c*) which impart oscillations directly to the body of the vehicle are also being used. Such compactor comprises an inertial, pneumatic, or electromagnetic vibrator *4* with pads *5* by means of which it is mounted to or pressed against the body of the vehicle. The vibratory agitator is suspended permanently by means of arm *6* at the loading station, or by a chain *7*, and is handled by a crane. In some cases, small vibratory compactors are used in railway transportation. These compactors are mounted between the rails at the loading station and oscillations are imparted to rails and to the vehicle which is placed on them (Fig. 3.10d). The compactor comprises a vibrator 8 and jacks 9 by means of which it is extended between the rails.

Vibratory compactors for core sands are most often made as platforms of various designs. The platform in Fig. 3.10e consists of frame 10 mounted on the foundation with elastic mounts 11. Vertical oscillations are imparted to the frame by means of an inertial vibrator 12. The molding box is directly mounted on the frame.

In order to intensify the compaction process of core sands the platform is subjected, along with harmonic oscillations, to impact pulses that are directed upwards or downwards. Figure 3.10f and g show units equipped with buffers 13 and 14 against which the frame with the molding box impacts during the oscillation process. Asymmetric high-frequency oscillations are induced which facilitate better compaction quality of the sand.

For a better compaction, various systems of static loading of the sand combined with simultaneous vibratory processing are also used. Presented in Fig. 3.10h is a unit in which the loading of the compacted mixture is effected by an applied weight. It consists of platform 15, mounted on a foundation by means of an elastic system 16. Mounted rigidly to the platform are an inertial vibrator 17, imparting to it vertical oscillations, and a punch 18, which enters the molding box. A weight 19 is placed on the compacted mixture in the molding box. Figure 3.10i shows a layout in which the static loading of the mixture is achieved by compressed air. The molding box (Fig. 3.10j) has a cover 20 with a tube through which compressed air is supplied during the operating process of the unit. Vibratory units have been developed in which oscillation excitation and static loading of the mixture are effected by a single device, namely pneumatic cylinder system. Such a device is presented in Fig. 3.10k. It comprises a platform 21 with a punch 22, which enters into the molding box. The platform of the pneumatic cylinder 23 is connected with the upper frame 24. During operation compressed air is delivered to the cylinder in a pulsating flow as a result of which static pressure and dynamic loading of the compacted mixture is applied through the punch.

In order to increase the capacity of the vibratory compactor and facilitate auxiliary operations which are associated with the mounting and removal of the molding box,

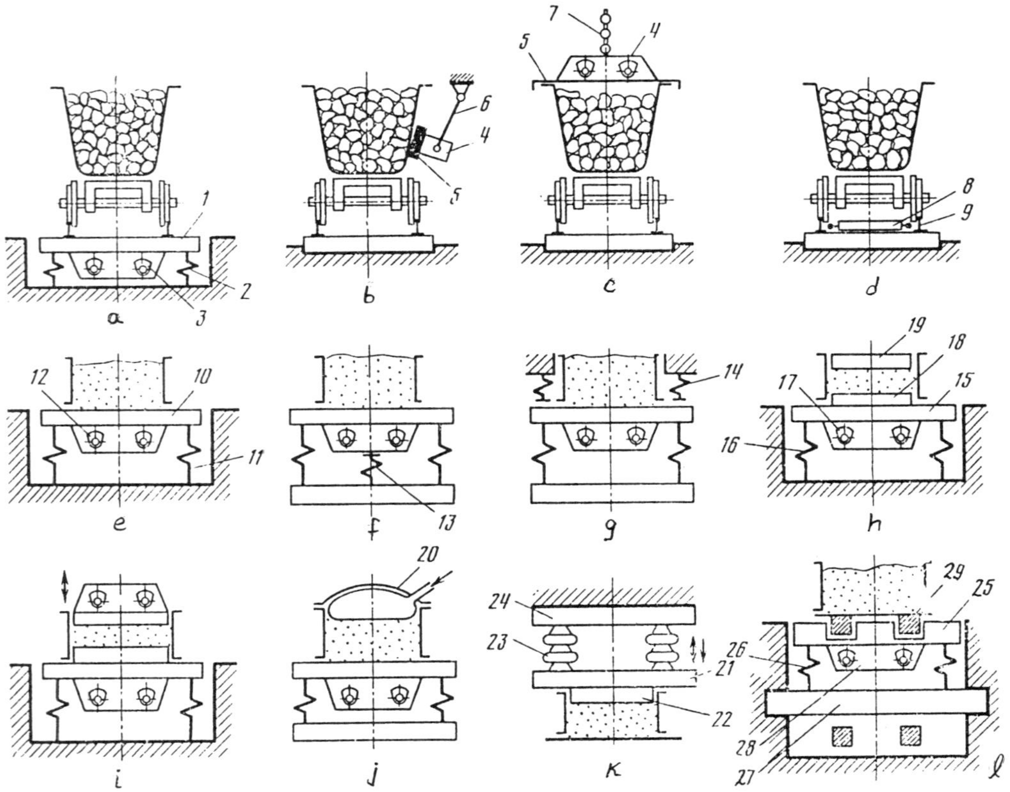

**Figure 3.10** Schematic diagrams of vibratory machines for compaction of granular media.

the vibratory unit is equipped with handling devices (3.10*l*). The unit consists of platform *25*, mounted on an elastic system *26* on bed *27*. The inertial vibrator *28* on the platform imparts to it vertical oscillations: the mounting and removal of the molding box is carried out by chain conveyer *29* whose top branch is mounted on the vibrating platform.

## 3.6  VIBRATORY MACHINES FOR PART STRENGTHENING

Vibratory processing installations are presently used widely for cleaning, finishing, grinding, and hardening (strengthening) operations. If they are to be classified by the type of vibrator, then the most widely used, both domestically and abroad, are inertial unbalanced-mass and electromagnetic systems.

The schematic diagrams of vibratory installations with unbalanced vibrators are shown in Fig. 3.11*a-g*, and with electromagnetic vibrators in Fig. 3.11*h-j*. The working chamber *1* having a U-shaped cross-sectional area is rigidly fixed on frame *2* (Fig. 3.11*a*). The housing of the unbalanced mass *3* is affixed to the frame or directly to the working chamber. The frame, together with the working chamber, is mounted on elastic elements *4* on a massive base *5*. The working chamber can also be elastically mounted to the frame. The elastic elements used are springs, cord-reinforced rubber cylinders with compressed air, and rubber vibration isolators. Unbalanced masses are driven by an electric motor, which is installed on base *5*, by means of a V-belt transmission via step pulley and flexible coupling. The exciting force, generated by the rotating unbalanced mass, imparts circular oscillations to the working chamber with the working medium and the processed components inside it.

The layout depicted in Fig. 3.11*b* of a vibratory installation with a three-mass vibrator has a working chamber *1* with an inclined side wall. Such a shape of chamber section, which is close to O-shaped section, accelerates the working process as a result of improvment of the conditions of vibratory mass flow over the chamber walls as compared with a U-shaped section. The shafts of the three unbalanced masses *3* are arranged along the bottom of the working chamber. The housings of the unbalanced masses are rigidly attached to the chamber and to frame *2* by the elastic link *4* which is mounted on base *5*. The first unbalanced-mass shaft is driven by an

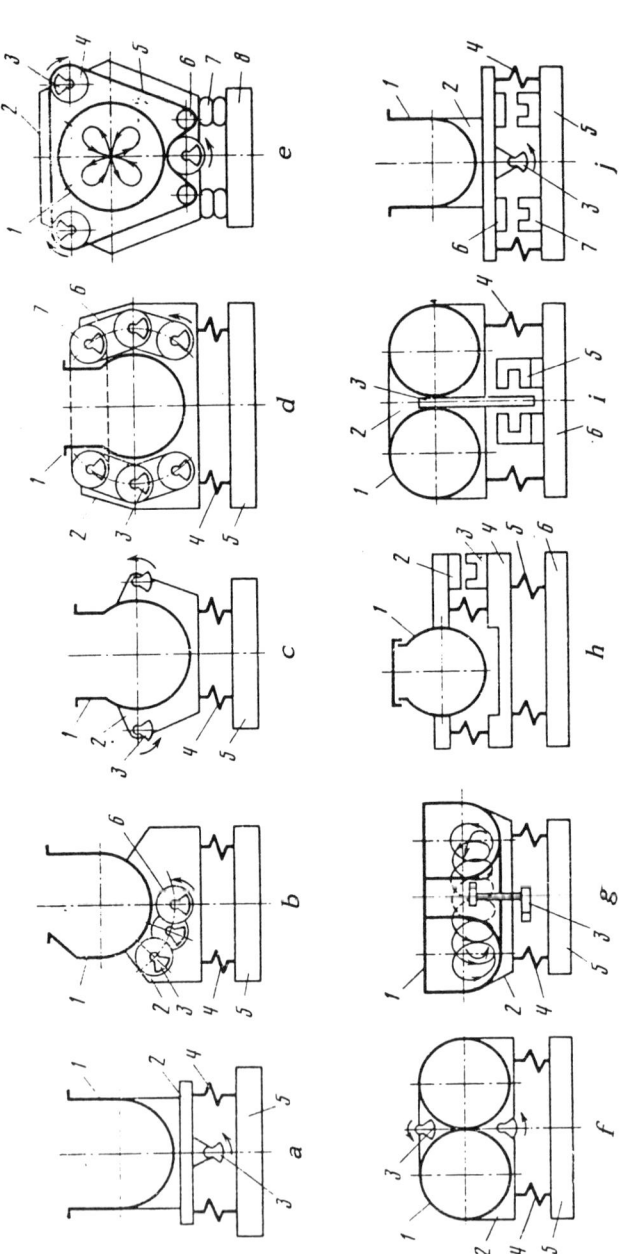

**Figure 3.11** Schematic diagrams of vibratory machines for abrasive and hardening processing.

electric motor and its motion is transmitted to the other two shafts by gears. Under the action of the generated centrifugal forces, the working chamber performs motion along a non-circular path which also contributes to the intensification of the working process.

The working chamber *1* of a unit with two unbalanced-mass vibrators (Fig. 3.11*c*) has a shape of a horizontally placed cylinder, which is open from the top. It is rigidly linked with the carrying frame *2* which is mounted on base *5* with the aid of springs *4*. The casings of the unbalanced-mass shafts *3* are attached to the carrying frame. The shaft axes are mounted on both sides of the axis of the working chamber and lie in one plane with its center of gravity. Both unbalanced-mass shafts are driven from a common driving motor. The working chamber receives vibrations from the unbalanced masses rotating in one direction.

The vibratory installation shown in Fig. 3.11*d* has the chamber *1* of the same shape as in the preceding case. The difference is to the effect that the chamber oscillations are excited by six unbalanced masses *3* whose housings are rigidly affixed to the chamber and to the frame *2*. The shafts of the unbalanced masses are linked with each other by means of time belts *6* and gears *7*. The chamber is mounted with the aid of springs *4* on base *5*. Rotation from the drive motor is transmitted to the lower right unbalanced mass. All the unbalanced masses in their motion in one direction act on the mass in the working chamber by effective vibrations which increase the output of the plant as compared with the single unbalanced-mass unit. The low degree of reliability of the system and the high noise level are the main shortcomings of this design.

The vibratory installation depicted in Fig. 3.11*e*, unlike the considered installations whose working chambers are open and shaped in the plan view as a rectangle, has a closed cylindrical horizontally placed working chamber *1*, which is rigidly linked with the carrying frame *2*. The unbalanced masses *3*, placed in tubular housings that are rigidly linked with the frame, are mounted outside on shafts parallel to the chamber axis. The axis of the shaft of the lower (larger in mass) unbalanced mass is mounted in the vertical plane passing through the axis of the working chamber. The axes of the other two (equal) unbalanced masses are located above and symmetrically to this plane. The shafts of the unbalanced masses are linked with each other by sprocket *4*, chain *5*, and auxiliary sprockets *6*. The carrying frame is fixed to base *8* by means of pneumatic shock absorbers *7*. The shaft of the lower unbalanced mass is driven by a V-belt transmission from an electric motor. The auxiliary sprockets enable one to obtain a rotation of the upper unbalanced masses in a direction opposite to the rotation of the lower unbalanced mass. The resulting exciting force, which arises with their rotation, causes oscillatory motion of the working chamber in a complicated multi-lobe trajectory, with the number of the lobes depending on the transmission ratio between the shafts of the unbalanced masses. Furthermore, the mass inside the chamber performs a circulatory motion with high speed around the longitudinal axis of the working chamber. Such a complex motion is attained by very simple means: by the rational positioning of the unbalanced-mass shafts relative to the chamber axis, by selection of the unbalanced masses, and by specific transmission ratio between the shafts.

The given unit has a number of advantages as compared with the existing machines, namely the enhanced intensfication of mixing of the feed, the possibility of highly effective working of parts with complicated external shape, and increased output.

In addition to the considered vibratory installations with one working chamber, designs with multiple chambers have been developed, i.e., with two, three, and four chambers. Their main advantage is an increase in output.

Figure 3.11*f* shows a design with two cylindrical chambers *1* driven by a vibrator with two unbalanced masses whose shafts are situated in the vertical plane of symmetry of the working chambers. There are installations with a working chamber of an annular (toroidal) type. The working chamber *1* of such a unit, which represents an open torus (Fig. 3.11*g*), is rigidly mounted on frame *2*. Under the chamber, and inside its neck a vibrator with two unbalanced masses *3* is installed. The axes of the working chamber and of the vibrator coincide. The frame of the unit rests on springs *4* located around the circumference of the base *5*. The heavier lower unbalanced mass causes vertical oscillations of the loaded mass during rotation, and the upper one causes horizontal displacement of the mass around the circumference of the chamber. As a result of the addition of these components, the mass travels around the axis of the ring of the working chamber (shown by arrows). Installations of this type are small in size, operate with less noise, and are used to process flat small components. The working chambers of similar vibratory installations are also made in the shape of a closed torus (ring), helix, or polyhedron.

The schematic diagram of a unit with electromagnetic vibrator is shown in Fig. 3.11h. A lever with armature 2 is rigidly attached to the working chamber 1. Stator 3 is affixed to frame 4 which is supported on base 6 by means of the elastic elements 5. When current flows through the windings of the stator the armature is attracted to it. Due to the elastic suspension, the chamber with the lever and armature performs oscillatory motion which is imparted to the loaded mass.

A two-chamber unit of electromagnetic action is depicted in Fig. 3.11i. Two horizontally arranged chambers 1, which are connected with frame 2, form together with the armature 3 fixed between the chambers a single rigid system mounted on elastic elements 4 on base 6. The stators 5 of the electromagnets which are placed under the chambers alternately attract the armature, forcing the combined system of the armature and the chamber frame to perform a swaying motion which is transmitted to the load. Installations with an electromagnetic vibrator have, as a rule, lower power and output compared with units of the unbalanced-mass type.

Figure 3.11j shows the layout of a single unbalanced-mass vibrator having the same constituent elements as the unit of Fig. 3.11a with the exception of the electromagnetic device. The latter serves as automatic regulator of the amplitude of oscillations and consists of an armature 6 and stator 7, which are rigidly mounted on the frame 2 and the base 5.

## 3.7  VIBRATORY SEPARATORS AND MIXERS

### 3.7.1  Processing Functions of Vibratory Separators and Mixers

Vibratory separators and mixers perform opposite processing functions.

Vibratory separators separate dispersed media according to size, shape, density, coefficients of friction, and other characteristic properties of the particles forming the dispersed system. Thus, the process of separation puts an order in the dispersed system with respect to one or several characteristic indicators.

Vibratory mixers, on the other hand, must provide uniform distribution of the mixed components over the entire mass, i.e., bring the dispersed system into a state of the highest possible disorder.

The most widely used separators are the types with perforated screening surfaces, which are usually referred to as grizzlies. By imparting vibratory motions to the screening surfaces the processed product is displaced and uniformly distributed over the working surface. Vibratory motions make it easier for the fine grains of the product to pass through the openings of the screen and ensure the removal of the retained classes (fractions) from the separator.

The working surface of the grizzly can be imparted harmonic rectilinear oscillations in the plane of the screen or at a certain angle to it. Oscillations in the plane of the screen are used for the separation of mixtures with respect to the width or thickness of the particles with the aid of perforated working surfaces with round or rectangular openings. Wattled screens can also be used. Constant contact with the screening surface increases the probability of screening the particles from the lower layer.

Rectilinear oscillations at an angle to the working surface or circular and elliptical oscillations in the vertical plane are imparted to the grizzly for the separation of particles which are characterized by disordered shapes. In such regimes the product layer is periodically tossed up by the working surface. Thus, the layer is loosened and impacts of the particles on the screen upon falling leads to an increase in the effectiveness of the process of separation.

The circular oscillations in the plane of the screening surface are imparted to the separator for the separation of granular mixtures into a larger number of fractions with respect to the thickness or width of the particles. In such systems many screens with different openings sizes are mounted.

Vibratory grizzlies in which elastic oscillations of the screening surfaces are realized are currently being actively developed. This is a grizzly with resonating grates and elastically deformed screens. In these designs predominantly transverse wave-like and wave oscillations of the screening surfaces are realized. The effectiveness of such grizzlies is extremely high, since the elastic strains of the screening surfaces prevent clogging of the openings.

When mixing various granulated media, the particles are subjected to the action of differently oriented forces which shape their motion in the mass of the medium. The mixing mechanism is determined by the construction of the mixer and the regime of its operation. In mixers with a purely vibratory principle of action, the processed medium is acted upon by forces of inertia, which are generated by the periodic motion of the working chamber, by pressure and friction forces on the walls, internal resistances and breakage of the medium under gravity forces. Under the influence of these forces, a complicated vibrocirculatory motion of the medium is obtained whose laws are mainly defined by the regime of oscillations and design features of the chamber.

Generally, a number of characteristic processes take place in the mixer. Moreover, their role in the mixing process can be very different depending on the design and operating regime of the mixer. Formation of surfaces slipping over each other in the mass being mixed is characteristic for the so-called shear mixing. During convective mixing, groups of particles of the mixed mass are displaced from one location to another. The change in locations of single particles is characteristic for diffusion mixing. Impact mixing takes place as a result of scattering of single particles under the action of collisions against the walls of the mixing chamber. Mixing can also take place in the process of reduction of the chunk size of the aggregates constituting the mixed product.

### 3.7.2 Vibratory Separators (Grizzlies) with a Rigid Working Element

The fundamental features of the design of vibratory separators are determined mainly by the type of the drive. Installations with inertial drive are the most widespread, followed by installations with electromagnetic and, finally, with eccentric drives. Installations with an eccentric drive are usually of the resonant type.

At the present time, inertial vibratory grizzlies units with unbalanced-mass and self-balanced vibratory drives as well as with vibratory motors are used. When vibratory motors are used, both rectilinear and circular oscillations can be imparted to the grizzly with an appropriate installation. Vibratory grizzlies driven by vibratory motors are the most widely used, particularly in small and medium units. They are notable for extreme structural simplicity as a result of using off-the-shelf vibratory drives, and of the absence of supplementary electric motors and intermediate transmission.

Currently produced inertial grizzlies are divided into two large groups: inclined and horizontal. In inclined grizzlies vibrators of the unbalanced-mass type imparting circular motion to the duct are used, or the self-balanced vibrators generating directional exciting force and providing rectilinear motion of the duct.

One of the most important requirements to modern vibratory machines in general and to vibratory grizzlies in particular, in addition to high performance, is simplicity of maintenance. This requirement is satisfied by vibratory grizzlies in which vibratory motors are used. This drive eliminates the need for any intermediate drive elements from the motor to the vibrator such as couplings, chains, etc. As a result, the size of the machine is reduced, assembly at the site becomes easier, and maintenance is simplified significantly.

In grizzlies with circular oscillations of the duct, one vibratory motor is usually installed. Several vibrators can also be used, but in this case their cophased operation must be provided. To obtain rectilinear oscillations of the duct, one can use kinematically or dynamically connected vibratory motors. Use of vibratory motors with adjustable magnitude of the exciting force and their mounting at different angles with respect to the plane of the screen enables one to correctly select the operating regime for the specified worked product and its chunk size.

In vibratory grizzlies having a screen with large openings, the vibratory motor is mounted in such a way that a large vibration angle would be obtained. The material in this case is intensively tossed up and performs small jumps, which enhances its effective separation. For the average value of screen openings the vibration angle is approximately $45^{\circ}$. When fine products are screened, the vibration is directed at a more gently sloping angle with respect to the screen plane. The double amplitude of oscillations is established in accordance with the properties of the worked product by changing the magnitude of the exciting force of the vibratory motor by means of rotation of the adjustable unbalanced mass.

Schematic diagrams of vibratory grizzlies with motor-vibrators are given in Fig. 3.12. Figure 3.12a shows an inclined vibratory grizzly with circular oscillations in the vertical plane driven by a motor-vibrator whose axis is

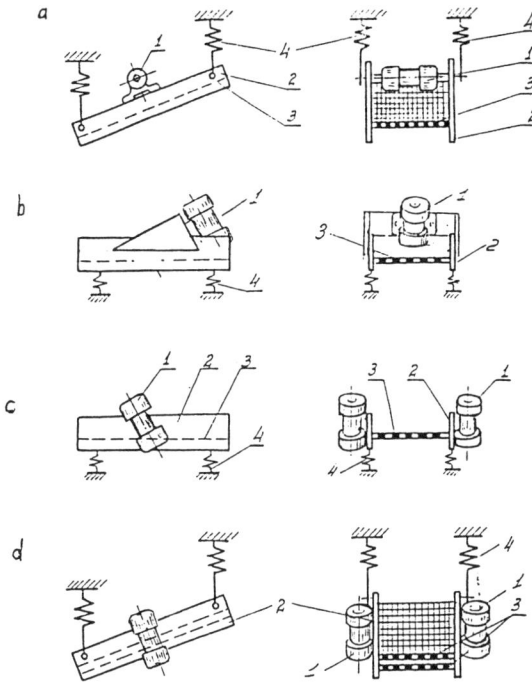

**Figure 3.12** Schematic diagrams of vibratory grizzlies with motor-vibrator drives. a) inclined: *1* – motor-vibrator, *2* – duct, *3* – screen, *4* – elastic vibroisolating connections; b) horizontal with one motor-vibrator; c) horizontal, driven from two dynamically synchronized vibrators operating in antiphase; d) inclined, driven from a dynamically synchronized motor-vibrators operating in phase.

horizontal and mounted perpendicular to the longitudinal axis of the box. Figure 3.12b shows a horizontal vibratory grizzly with two dynamically synchronized motor-vibrators whose boxes perform linear oscillations.

The vibrators are adjusted for antiphase oscillations; their axes are arranged at an angle to the longitudinal grizzly axis. In the first case the motor-vibrator is mounted in the middle part of the box on a special transverse beam, in the second case it is mounted on the box sides.

Figure 3.12c depicts a vibratory grizzly with two dynamically synchronized vibrators operating in phase and whose axes are perpendicular to the screen surface. The grizzly is mounted with a small inclination. The screening surface performs circular oscillations in its plane. Vibratory grizzlies of all types are suspended or mounted on load-carrying structures by means of soft helical springs. As a result, they do not practically transfer dynamic loads to the carrying structures.

Inertial grizzlies are manufactured with one screen and two screens. In both cases, they are characterized by exceptional simplicity and small height; they do not require large floor areas. Horizontal grizzlies require the least floor area and vertical space. Vibratory grizzles of the types under consideration can be used for the separation of moisture and for dewatering various products by installing special screens.

The fundamental design layouts of vibratory grizzlies with unbalanced masses and self-balanced vibrators of special design are shown in Fig. 3.13. As a rule, similar drives are used in large grizzlies.

Figure 3.13a shows the fundamental design layout of unbalanced horizontal and inclined vibratory grizzlies with vibrators of the self-balanced type which provide directional oscillations of the box. In the horizontal vibratory grizzly, the vibrator is installed in the upper part of the box; in the inclined grizzly it is installed below. Vibratory grizzlies of the unbalanced type operate far beyond the resonance and, since they are mounted on support structures on soft elastic elements, transfer to them insignificant dynamic loads.

Vibratory grizzlies with inertial drives can be designed in vibration-isolated embodiments (Fig. 3.13b). For this purpose they are mounted on a reactive mass, which is isolated from the load-carrying structure by soft elastic elements.

Vibratory grizzlies with unbalanced-mass drives (Fig. 3.13c) whose boxes perform circular oscillations are always mounted at an inclination in order to provide translational motion of the processed product.

All types of vibratory grizzlies under consideration can have one, two, and more screens.

Grizzlies with electric vibrators are used for screening of mainly fine materials and also for dewatering and for dust removal. They are manufactured also in two versions, namely horizontal and inclined (Fig. 3.14). In the grizzlies of the first type, oscillations are usually directed perpendicular to the screen plane, and displacement of the material over the screen is caused by gravitational forces due to the inclination of the grizzly (Fig. 3.14a). In vibratory installations of this type

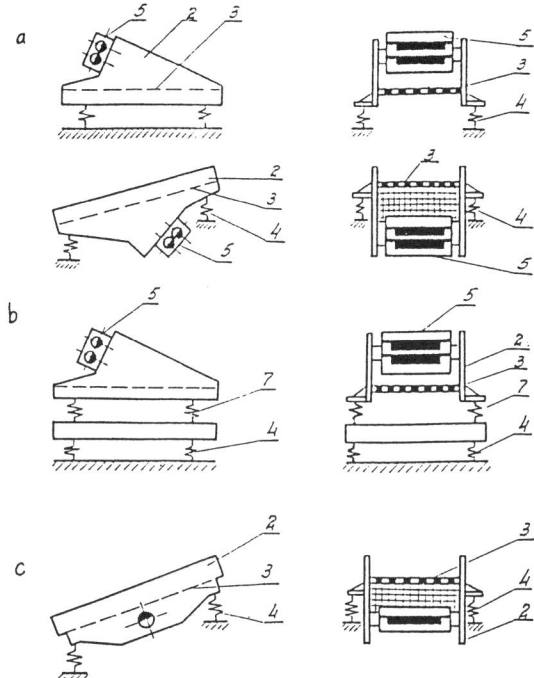

**Figure 3.13** Schematic diagrams of vibratory grizzlies driven by unbalanced and self-balanced vibrators, a) one-mass with self-balanced drive: 1 - vibrator, 2 - box, 3 - screen, 4 - isolators; b) two-mass with self-balanced drive: 1 - vibrator, 2 - box, 3 - screen, 4 - isolators, 5 - structural elastic connections, 6 - balancing frame; c) one-mass with unbalanced-mass drive.

oscillations can be imparted either to the entire box of the grizzly, or to the screen directly.

In grizzlies with electric vibrators and with horizontal positioning of the box, displacement of the separated product is induced by oscillations acting at an angle to the direction of conveying (Fig. 3.14b). The vibrator in this case is mounted at an angle to the screen.

Resonant vibratory grizzlies find wide application in industry. They are more complex in design than inertial and electromagnetic grizzlies. However, they are characterized by a number of operational merits, namely high screening quality, low energy consumption, operational reliability, and low

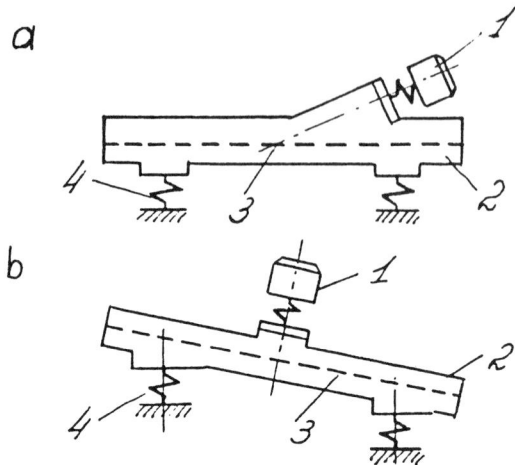

**Figure 3.14** Schematic diagrams of a vibratory grizzly with electromagnetic drive. *a)* horizontal: *1* – vibrator, *2* – box, *3* – screen, *4* – isolators, *b)* inclined.

dynamic loads on the foundation. These advantages led to wide applications of this system.

Resonant vibratory grizzlies can be divided, with respect to the fundamental design of the dynamic system, into two-, three-, and four-mass machines. Depending on the structural layout they can have series and parallel arrangement of the boxes. To balance the moving parts, both the boxes themselves and special balancing frames can be used. Resonant grizzlies can have one or two boxes.

The structural arrangements of two- and three-mass resonant vibratory grizzlies are given in Fig. 3.6. The structural arrangement of a two-mass resonant vibratory grizzly with reactive frame is depicted in Fig. 3.6c. The two-mass resonant grizzly consists of a box and reactive frame, and eccentric drive with elastic connecting rod. The box is connected with the reactive frame by leaf springs and rubber-metal elastic elements. The reactive frame is isolated from the supporting structure with spring or rubber isolators.

The structural layouts of resonant vibratory grizzlies of a three-mass system with parallel and series arrangement of the boxes are depicted in Fig. 3.6f, g. The three-mass resonant vibratory grizzly consists of two boxes supported by means of leaf springs and rubber-metal elastic elements on the balancing frame. The frame is isolated from the supporting structure by means of rollers or rubber isolators. The first box is driven by an eccentric vibrator with elastic connecting rod. Oscillations are imparted to the second box reactively, and it oscillates approximately in antiphase to the first box. When started, the drive does not immediately impart oscillations to the box, due to the presence of an elastic connecting rod, but it does so gradually by rocking it for approximately 3 - 10 seconds. As vibrations of the first box develop, the balancing frame starts vibrating and its oscillations are transmitted to the second box, which also starts oscillating. Since the second box oscillates in an opposite direction to the first, their inertial forces cancel each other on the balancing frame and are not transmitted to the load-carrying structure. However, the frame nonetheless performs small oscillations, since the suspensions of the box do not oscillate exactly in antiphase as a result of resistances and energy dissipation in the elastic elements. Oscillations of the balanced frame are transmitted to the free box and are used to compensate for the losses of energy in the elastic elements of the suspension. Thus, the power of the drive is expended only to make up for the energy losses in the elastic suspension and on screening the material.

Large resonant grizzlies are four-mass designs comprising two two-mass sections connected by a tie rod. The drive in this case is mounted only on one section, and the second is brought into action by reactive forces. The four-mass grizzly has an eccentric drive which is located on the first section and connects the reactive mass and the box (Fig. 3.6*h*). Each section consists of a box, supporting connecting rods, a frame mounted on isolators, and structural elastic connections. The supporting frames of both sections are linked to each other by means of a tie rod.

### 3.7.3  Vibratory Separators (Grizzlies) with Elastically Deformable Screening Surfaces

The schematic diagrams of inertial grizzlies with elastically deformable screening surfaces are presented in Fig. 3.15.

One of the first to be designed was an electromagnetic grizzly with elastically deformable surface made from steel bars, whose ends were rigidly attached to the frame of the grizzly (Fig. 3.15*a*). The natural frequencies of the bar screening surface were adjusted for resonance with the forced frequency of the vibratory drive. When the grizzly is operating the bars

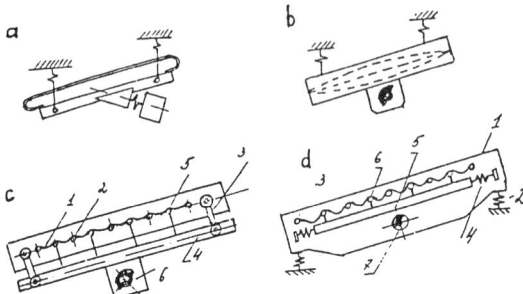

**Figure 3.15** Schematic diagrams of grizzlies with elastically deformable surfaces.

oscillate with larger amplitude than the frame due to resonance tuning. This intensifies the classification process without an increase of the loads transmitted on the supporting structure of the buildings.

Much later, vibratory grizzlies with rubber resonating surfaces (Fig. 3.15*b*) appeared which operated on the same principles as the design considered above. These grizzlies have an eccentric or inertial drive.

The design of machines with forcibly deformed surfaces was a fundamentally new direction in the development of vibratory grizzlies with elastically deformed surfaces. What is new in their design is the method of oscillation excitation of the screening surface. The screen is fixed to a system of transverse beams that are mounted on the parts of the vibratory grizzly vibrating in antiphase. Therefore, when the machine is operating, the screen either stretches or sags as a result of change of the distance between adjacent beams. Furthermore, the screen performs wave-like motion with a significant transverse amplitude. The process of separation in these units takes place with high intensity and efficiency.

The grizzly of the GDEP type belongs to this class of vibratory machines. The grizzly includes two oscillating masses: the frame of the screening surface *1* with stationary transverse bars *2* (Fig. 3.15*c*) and the reactive mass *4*, having moving transverse bars *5* and connected with the frame by connecting rods *3* with rubber hinges. The reactive mass is driven into oscillatory motion by inertial vibrator *6*. An elastic screen is mounted on the transverse bars *2* and *5*. In operation, as a result of the phase shift between oscillations of the working masses of the grizzly, the distance between the beams periodically changes. This is causing elastic straining of the screen and leads to wave-like oscillations of the surface.

A similar principle of action is successfully implemented in a vibratory grizzly by the firm Binder Co., (Austria). The vibratory grizzly known under the name B-V-Tex consists of a casing *1* with a continuous bottom mounted on elastic elements *2* on the foundation (Fig. 3.15*d*). Transverse bars *3* are installed in the frame. A moving frame *5*, also carrying transverse bars *6*, is connected with the casing by means of springs *4*. Oscillations are imparted to the casing by an unbalanced-mass inertial vibrator *7*. Elastic screening surfaces are mounted on the transverse bars. The principle of operation of the grizzly is the same as in the previous design.

### 3.7.4 Vibratory Mixers for Granular Media

The schematic diagrams of vibratory mixers for granulated products are shown in Fig. 3.16.

The simplest vibratory batch-type mixer is a cylindrical chamber, performing rectilinear oscillations in the vertical direction, into which the components to be mixed are loaded (Fig. 3.16*a*). Under the actions of the vertical oscillations rising circulation in the mixed medium are created. These circulations contribute to the mixing of the components of the mixture.

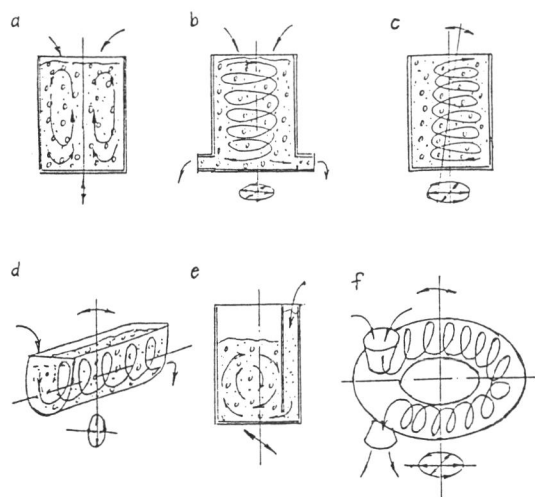

**Figure    3.16** Schematic    diagrams    of    vibratory    mixers    for granulated media.

The character of the circulations of the mixed medium can be changed and their intensity increased simultaneously by imparting to the working chamber of oscillations along the horizontal circular path (Fig. 3.16b). In this case, the mixer can operate in a continuous regime. When operating in the regime of continuous mixing, the components of the mixture are fed from above and the mixed product is discharged from below. Furthermore, the duration of mixing can be regulated by the intensity of oscillations and by the height of the chamber.

In the cylindrical working chamber motion can be imparted to the mixed products. For this purpose, circular oscillations and rocking of the central axis are simultaneously imparted to the working chamber (Fig. 3.16c). Such circulations in the mixed medium leads to an increase in mixing quality and shortens the duration of the process.

The so-called through-type mixers are also used. The working chamber of such a mixer is a sufficiently long trough, usually of V-shaped cross section (Fig. 3.16d). Circular oscillations in the vertical plane are applied to the trough perpendicular to its longitudinal axis. The components of the mixture are loaded from one end of the trough and the final product is discharged from the other. To ensure a slow motion of the processed product along the trough, fins are installed at the bottom of the trough at a small angle to its longitudinal axis. Under the influence of the circular oscillations, the mixture in the working chamber performs elliptical circulatory motion in a plane perpendicular to the longitudinal axis of the trough.

With the installation of inclined fins at the bottom of the trough the mixture performs a spiral motion slowly moving along the working chamber. The longitudinal motion of the mixture is also achieved in a smooth trough if it is mounted at a slight inclination.

A good mixing process of multi-component mixtures is also achieved in rectangular multi-section working chambers to which rectilinear oscillations are applied at an angle to the bottom (Fig. 3.16e). In this design the components of the mixture are loaded into narrow sections of the chamber which are separated from the main compartment by walls that do not reach the bottom, and the mixture is formed in the larger compartment.

Mixers with toroidal working chambers (Fig. 3.16f) are widely used. The mixing chamber is positioned horizontally and complex oscillatory motion is imparted to it. This motion comprises circular oscillations in the plane of the torus and swaying of the central vertical axis of the chamber. The combination of such oscillatory motions of the working chamber creates spiral circulation of the mixed medium. The spiral circulation consists of elliptical circulations in the cross sections of the chamber and slow motion along the annular axis of the torus. Such circulatory motions provide high degree of mixing of the final product components. Mixing of moving granular products can also be effected in vibratory conveying machines. For this purpose, load-carrying elements of special design are used. Mixers of this type are considered in section 3.3.

# FOUR

# VIBRATION EXCITERS (VIBRATORS): PRINCIPLE OF OPERATION, GENERATION OF EXCITING FORCE

## 4.1 TYPES OF VIBRATION EXCITERS (VIBRATORS)

Vibration exciters impart motion to the oscillatory system and generate the exciting force which is necessary for overcoming internal and external resistances. In transresonant and subresonant regimes, the vibration exciter also overcomes the inertia forces of the rotating masses or the restoring forces of the elastic links, while in the resonant regime these forces balance each other. Various types of vibration exciters are characterized by different actions on the driven component of the vibratory system.

The following types of vibration exciters are considered: inertial, eccentric, electromagnetic, and hydraulic.

Vibration exciters can be classified with respect to the character of their action into the following main groups.

The first group includes force vibration exciters, which apply to the driven component of the oscillatory system a force which has a specific time history depending on the position of this component or its velocity. Force vibration exciters are represented by electromagnetic and pneumatic vibrators.

The distinguishing feature of the force drive is the possibility of its application in vibratory machines with one degree of mobility.

The second group, to which the present section is mainly devoted, includes kinematic vibration exciters, i.e., such vibrators whose driving component has fully defined absolute or relative motion depending only on the geometrical dimensions of the driving mechanism. The most widely used kinematic vibration exciters are inertial, eccentric, and hydraulic types.

The third group includes impact vibration exciters, i.e., vibrators which induce the oscillations of the driven component of the vibratory system by impact. Some types of vibrators such as, for example, impact electromagnetic vibrators or inertial vibratory hammers, impart to the driven component of the machine both impact and vibratory pulses. Use of such drives is presently rather limited.


## 4.2  BASIC DESIGN FEATURES; FORMATION OF THE EXCITING FORCE

### 4.2.1 Inertial Vibrators

#### 4.2.1.1  Schematics and Formation of the Exciting Force

The exciting force in inertial vibrators is created as a result of rotation of one or several unbalanced masses. The exciting force thus created can have a rotating vector, i.e., continuously changing its direction, or directional. In vibrators with directional exciting force the latter always acts in the same direction and changes only in magnitude. There are also special types of inertial vibrators generating exciting torque or various combinations of exciting forces and torques.

Vibrators with rotary exciting force vectors include: vibrators of the unbalanced-mass type in which the exciting force is generated by one rotating unbalanced mass, and also vibrators for generation of elliptical and biharmonic oscillations.

To obtain a rectilinearly directional exciting force in inertial vibrators, two methods are used: the components acting in an undesirable direction are balanced by forces that are equal in magnitude but acting in opposite directions or the known property of the revolute joint is utilized, namely to transmit the force only in a direction perpendicular to its axis. In practice, rubber joints possessing some elasticity or springs with large transverse stiffness are usually employed for these purposes.

The directional action of the exciting force is provided by a vibrator of the self-balanced type which comprises two paired unbalanced vibrators rotating synchronously with identical angular velocity in opposite directions. If the synchronism of the rotation of the unbalanced masses is attained without a mechanical link, then they are referred to as self-synchronizing. In order to obtain a directional exciting force from one unbalanced mass, it is suspended from a revolute joint by means of a rocker in the form of a pendulum. To ensure a stable position of the rocker in space, it is fixed between two springs or a rubber bushing is used. The exciting force is imparted to the vibratory machine only in the direction of the line connecting the center of rotation of the unbalanced mass and center of suspension of the rocker on which the motor-vibrator is fixed. Such vibrators are called pendulum vibrators. The directional exciting force which changes according to a biharmonic law is generated by two paired self-balanced vibrators. One of these vibrators rotates twice as fast as the other one.

For the simultaneous generation of an exciting force and exciting moment pendulum vibrators with displaced unbalanced masses that are suspended by means of a 3-D joint and special two-shaft vibrators with four unbalanced masses are used.

It must be born in mind that we are talking here about the character of the exciting force relative to the vibrator base, and ignoring the forces created by its arm.

In mounting the vibrator on the machine, the character of excitation will depend not only on the type of the vibrator, but also on its position relative to the center of mass and the center of reduction of the restoring forces of the elastic connections. Thus, for example, if the axis of the simplest unbalanced mass vibrator does not coincide with the center of mass of the oscillatory system, then the latter will generate an exciting moment along with the rotating exciting force.

The inertial unbalanced-mass vibrator consists of an unbalanced mass rotating with a constant angular velocity on a shaft in the bearings of the support section which is rigidly affixed to the vibratory machine. Upon rotation of the unbalanced mass, a centrifugal (exciting) force which is constant in magnitude and continuously variable in direction is generated. In an unbalanced-mass vibrator the exciting force is generated by the centrifugal force of the unbalanced mass $m$ rotating with angular velocity $\omega$. If the distance from the center of rotation of this mass to its center of gravity is $r$, the exciting force is equal to $mr\omega^2$.

Thus, the exciting force of the unbalanced-mass vibrator is constantly changing its direction, while rotating with the unbalanced mass, and remains constant in magnitude.

Projections of the exciting force on axes $x$, $y$, are respectively equal to

$$F_x = mr\omega^2 \sin \omega t, \quad F_y = mr\omega^2 \cos \omega t$$

The pendulum inertial vibrator consists of an unbalanced mass rotating on a shaft fixed on a rocker which is mounted in the support with the aid of rubber-metal bushing. With the pivoted attachment of the vibrator the component of the centrifugal force passing through the center of rotation of the unbalanced mass and the rocker joint (pivot) is completely transmitted to the support which is in turn rigidly mounted on the vibratory machine. The component of the centrifugal force acting in the perpendicular direction causes the oscillation of the vibrator around the axis of the rocker pivot. Since stiffness of the rubber-metal pivot is small, the reaction transmitted to the vibratory machine is insignificant. It can practically be assumed that only the directional exciting force is imparted to the machine.

The inertial vibrator of the self-balanced type consists of two unbalanced masses rotating in opposite directions with equal angular velocity on shafts supported in a common housing.

The self-balanced vibrator with unbalanced masses rotating in opposite directions generates an exciting force which is constant in direction and variable in magnitude, $F = 2mr \, \omega^2 \sin \omega t$.

The directional exciting force can also be obtained with the aid of a vibrator with three unbalanced masses. The middle unbalanced mass rotates in one direction and the other two side masses in the opposite direction. The schematic diagram of a vibrator with three unbalanced masses is presented in Fig. 4.1a. The vibrator consists of three unbalanced masses (central $1$ and two side ones $2$) rotating on three parallel shafts located at equal distances from each other. The shafts are mounted in a common housing. The central shaft of the vibrator rotates in a direction opposite to the direction of rotation of the side shafts. All shafts rotate with constant angular velocity $\omega$ and the unbalanced masses generate a centrifugal force which is constant in magnitude ($mr\omega^2$ from the middle mass and $1/2mr$ $\omega^2$ from each side) and continuously variable in direction.

The resultant exciting force of the vibrator is equal to the vector sum of these forces. The pattern of formation of the directional force in a vibrator with three unbalanced masses is elucidated in Fig. 4.1b.

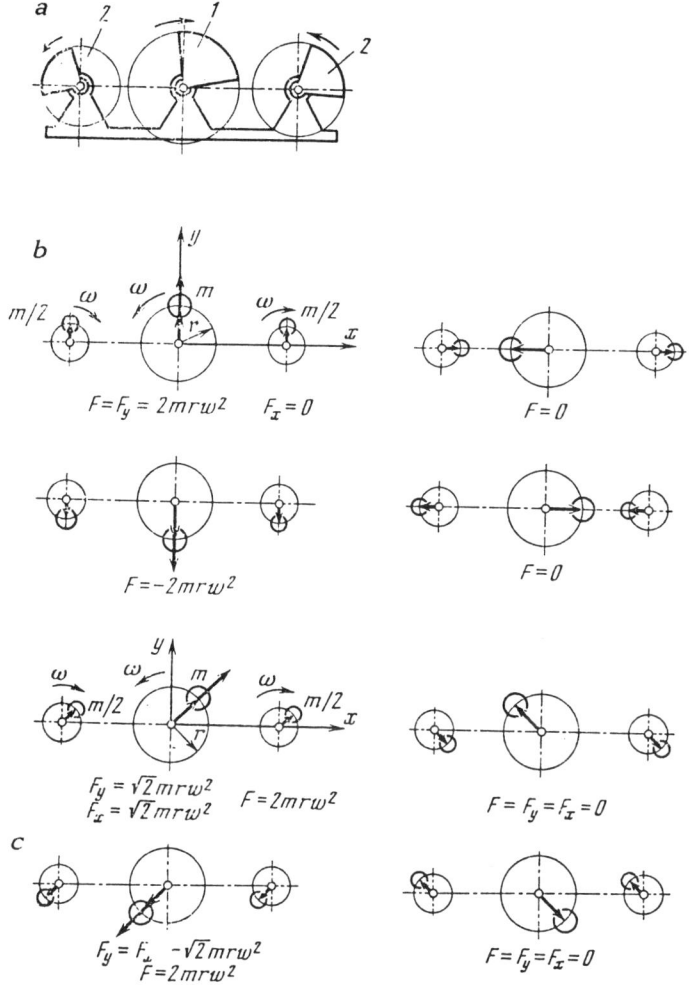

**Figure 4.1** Inertial vibrator with three unbalanced masses a) schematic arrangement; b) principle of operation of a vibrator of the self-balanced type; c) principle of operation of a vibrator with displaced unbalanced masses.

Vibrators with three unbalanced masses enable one to easily control the direction of action of the exciting force. Its angle is equal to half the angle which is formed by the directions of the side and middle unbalanced masses in the initial position. Thus, Fig. 4.1c shows the pattern of formation of the exciting force at angle $45°$ to the axis.

When it is desirable to have an exciting moment together with the exciting force, one can use a vibrator with two unbalanced masses in which the unbalanced masses are shifted relative to each other by some angle and are rotating in opposite directions.

The vibrator comprises two unbalanced masses rotating on shafts in a common supporting housing. The unbalanced masses are linked by gearing, which ensures their synchronous rotation in opposite directions. With such an arrangement of the unbalanced masses an exciting force is generated in the process of rotation directed at angle $45°$ ($90°/2$) to a straight line connecting their centers. A moment is also generated which is proportional to the magnitude of the exciting force and to the distance between the unbalanced masses. In such a vibrator, unlike the self-balanced ones, the exciting force is directed at angle $45°$ to the horizontal. The force magnitude is $F = 2mr\,\omega^2$, its projection on the vertical and horizontal axes are equal to $(\sqrt{2}/2)mr\,\omega^2$. The torque created by the vibrator is equal to $M = mr\,\omega^2\alpha$. The pattern of formation of the exciting force and the moment at other positions of the unbalanced masses are illustrated by the presented schematic diagrams.

Special designs of vibratory drives are also used in vibratory machines. These include vibrators for generation of elliptical and biharmonic oscillations and also three- and four-shaft vibrators which generate an exciting force that is changing in a complex way.

For generation of biharmonic oscillations one can use inertial vibrators with four unbalanced masses whose sychronized pairs of unbalanced masses rotate with angular velocities having a 1 to 2 ratio. The schematic diagram of a biharmonic inertial vibrator with four unbalanced masses is shown in Fig. 4.2a. It consists of two pairs of unbalanced masses 1, 2 and 3, 4 rotating synchronously in opposite directions and which are mounted in a common frame 5. The first pair of the unbalanced masses rotates with a two times higher speed.

The principle of operation of a biharmonic inertial vibrator is illustrated in Fig. 4.2b (the diagram shows the case when the phase shift between the exciting forces generated by the first and second pairs of the unbalanced masses is $90°$). The following notations are adopted: $m'$ and $r'$ are mass and eccentricity of the unbalanced masses of the first stage rotating with angular velocity $\omega$; $m''$ and $r''$ are mass and eccentricity of the unbalanced masses of the second stage.

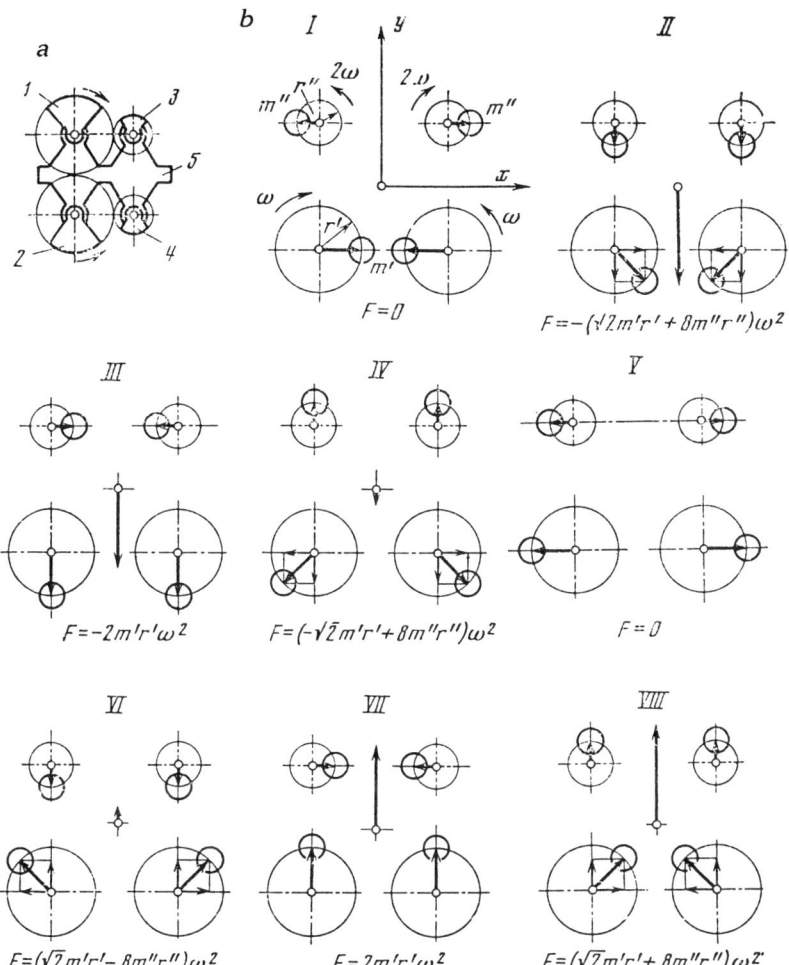

**Figure 4.2** Inertial biharmonic vibrator a) schematic arrangement; b) principle of operation.

When the unbalanced masses are in position I, the centrifugal forces of all four unbalanced masses act in the horizontal direction and in opposite directions in each pair. As a consequence, the resultant exciting force of each pair of the unbalanced masses and the whole vibrator is equal to zero. Position II corresponds to the rotation of the first pair of the unbalanced masses by 45° and the second by 90°. In this position both pairs of the unbalanced masses yield components that are directed downwards. The exciting force of the vibrator in position II will also be directed downwards and will be equal to the resultant of the components of both pairs of the unbalanced masses $F = -2m''\ r''\ 4\ \omega^2 - \sqrt{2}\ m'\ r'\ \omega^2$. In position III (the first pair of the unbalanced masses rotated by 90° and the second by 180°), the exciting force of the first pair reaches the maximum value $2m'\ r'\ \omega^2$ and is directed downwards; the resultant exciting force of the second pair is equal to zero. Therefore, the exciting force of the vibrator as a whole in position III is equal to the exciting force created by the first pair of the unbalanced masses $F = 2m'\ r'\ \omega^2$.

In position IV (the first pair of the unbalanced masses is rotated by 135° and the second by 270°), the resultant exciting forces of both pairs are in opposite directions. The exciting force of the vibrator is equal in this case to their difference, and its direction is dependent on the ratio of the exciting forces of the pairs of the unbalanced-masses: if the resultant component of the first pair is larger, then the exciting force of the vibrator is directed downwards; otherwise, it is the other way around. The formation of the exciting force of the biharmonic inertial vibrator in positions V - VIII is evident from the figure. By varying unbalanced masses of each pair and phase shifts between them, one can obtain various biharmonic excitations and select their optimum character for solution of a given process problem.

The law of variation of the exciting force of a biharmonic vibrator is

$$F = 2m'r'\omega^2 \sin \omega t + 8m''r''\omega^2 \cos (2\omega t + \gamma)$$

where $\gamma$ is the phase shift between the unbalanced masses of the second and first stages.

Special inertial vibrators with two unbalanced masses are finding application for generation of elliptical oscillations. The fundamental layout of the inertial vibrator for the generation of elliptical oscillations is analogous to the layout of a vibrator of the self-balanced type. The difference is in the fact that the kinetic moments of the unbalanced masses of the vibrator are unequal.

The vibrator with two unbalanced masses having different kinetic moments generates a rotary exciting force whose locus is an ellipse.

To obtain more complex loci of the exciting forces, a biharmonic elliptical vibrator can be used. Its fundamental design is similar to the biharmonic inertial vibrator for generation of rectilinear oscillations. However, each of the

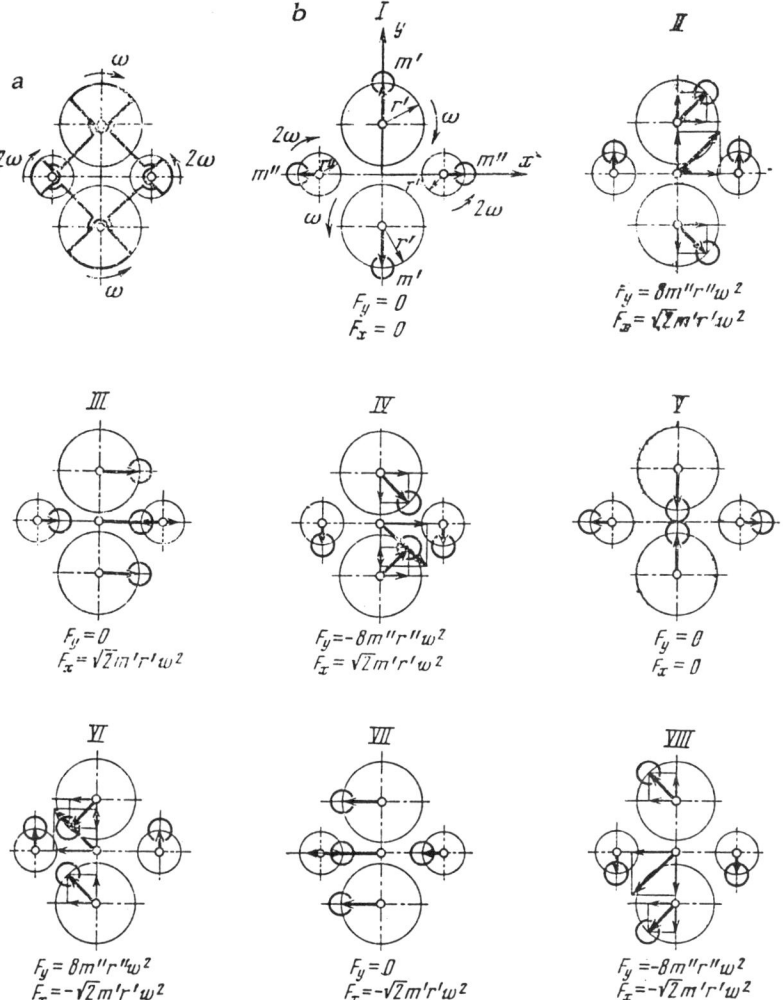

**Figure 4.3** Inertial biharmonic elliptical vibrator a) schematic arrangement; b) principle of operation.

vibrators is turned relative to the other by 90° (Fig. 4.3a). The principle of operation of the biharmonic elliptical vibrator is illustrated by schematics I – VIII in Fig. 4.3.

The schematic arrangement of a pendulum vibrator with two pivots, located in perpendicular planes, and with displaced unbalanced masses, is presented in Fig. 4.4a. The vibrator consists of two unbalanced masses *1* and *2* rotating on shaft *3* and displaced relative to each other by 90° or by a somewhat larger angle. The shaft with the unbalanced masses is mounted, by means of joints 4 whose axis is perpendicular to the shaft, in rocker 5, attached with the rubber-metal bushing 6 to support 7.

The principle of operation of a two-joint pendulum vibrator is depicted by Fig. 4.4b. In position I the horizontal components of the centrifugal forces of the unbalanced masses are directed perpendicular to the axis of pivot 6 and, as a result, are not transmitted to support 7 of the vibrator. They just cause the rocker to sway to the left. The vertical components generate a moment about the pivots axis, thus causing counterclockwise rotation of the vibrator shaft. Practically no forces are transmitted to the vibrator support and, hence, to the oscillatory system. In position II the horizontal components generate clockwise moment $M_y = \sqrt{2}\ mr\ \omega^2 a$ in the horizontal plane around axis $y$. The vertical components act in one direction and create the vertical exciting force $F = \sqrt{2}\ mr\ \omega^2$. Since neither the moment nor the exciting force is compensated for by the hinges (see Fig. 4.4a), they are fully transmitted to the support of the vibrator and then to the oscillatory system. In position III the two-joint pendulum vibrator does not impart any forces to the oscillatory system. In position IV an exciting force $F = \sqrt{2}\ mr\ \omega^2$ acting vertically downwards and a moment $M_y = -\sqrt{2}\ mr\ \omega^2 a$ in the horizontal plane directed counterclockwise are generated. Thus, the two-joint pendulum vibrator with displaced unbalanced masses creates an exciting force acting in the vertical plane and an exciting moment in the horizontal plane which vary as $F = 2mr\ \omega^2 \sin\ \omega t$ and $M = 2mr\ \omega^2 a \sin\ \omega t$.

For the simultaneous generation of an exciting force and an exciting moment, two-shaft vibrators with six unbalanced masses are also used. The fundamental arrangement of such a vibrator is shown in Fig. 4.5a. The vibrator consists of two shafts 1 and 2 sychronously rotating in opposite directions. Three unbalanced masses 3, 4 and 5, 6 and 7, 8 are attached to the shafts. Since the resultant of exciting forces from the middle unbalanced masses passes through the center of rotation of the system, they do not generate an exciting moment. In the design under consideration the opportunity is provided to separately adjust the kinetic moments of the middle and side unbalanced masses and thus to change independently magnitudes of the exciting force and moment.

The principle of operation of a two-shaft vibrator with six unbalanced masses is illustrated in Fig. 4.5b. In position I the horizontal components of the centrifugal forces of the front

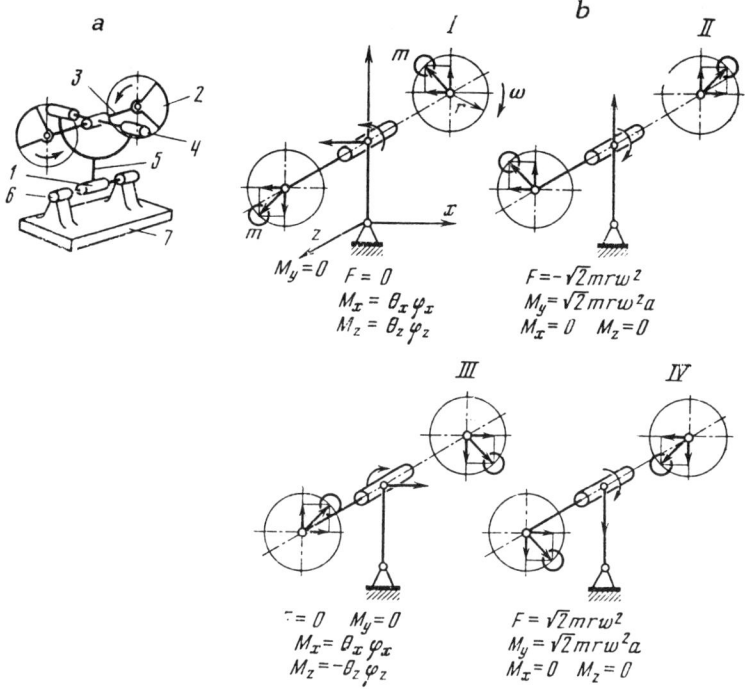

**Figure 4.4** Inertial two-joint pendulum vibrator a) basic arrangement; b) principle of operation.

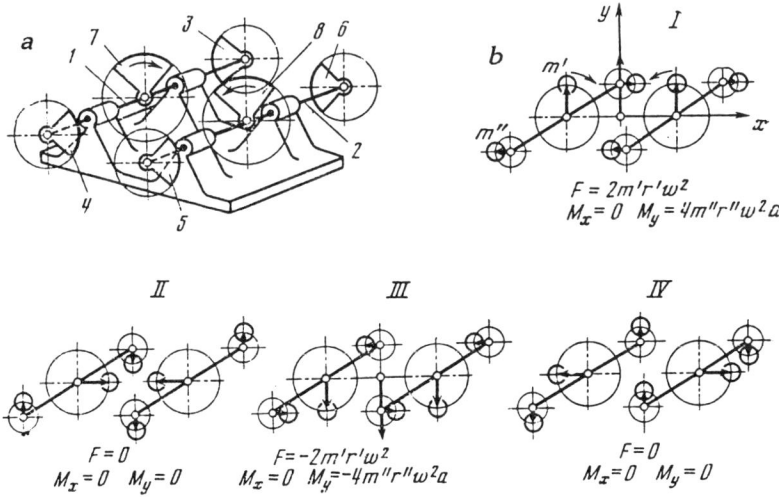

**Figure 4.5** Inertial two-shaft vibrator for generation of exciting force and moment a) basic arrangement; b) principle of operation.

unbalanced masses are directed to the left and of the rear masses — to the right. The resulting forces of the horizontal components create a moment in the horizontal plane directed clockwise and equal to $M = 4m'' r\omega^2 a$. The vertical components create a vertical exciting force $F = -2m'r\omega^2$ directed downwards. In position II the horizontal components of the cetrifugal forces of the left, front, and right rear unbalanced masses generate a moment in the horizontal plane directed clockwise. The horizontal components of the other two unbalanced masses generate moments in the horizontal plane which are equal in magnitude but opposite in direction. Therefore, the resultant moment of four unbalanced masses is equal to zero. The moments generated by the vertical components of the centrifugal forces of the unbalanced masses are balanced in exactly the same way. Thus, the resultant moment of the centrifugal forces of all the unbalanced masses in position II is equal to zero. In position III the vibrator generates a vertical exciting force directed upwards $F = 2m'r\omega^2$ and a moment in the horizontal plane $M = 4m''r\omega^2 a$ acting counterclockwise. In position IV the vibrator does not transmit any excitations. The exciting force and moment are varying as follows: $F = 2m''r\omega^2 \cos \omega t$, $M = 4m'r\omega^2 a \cos \omega t$.

Discussed above was the character of formation of the exciting force when the axis of rotation of the unbalanced mass or the center of the vibrator base coincides with the center of mass of the oscillatory system. In practice, this condition is not always satisfied either due to incorrect vibrator mounting or due to imparting complex oscillations to the oscillatory system, which are created, for example, to achieve higher processing effectiveness.

Figure 4.6a shows the schematic diagram of unbalanced-mass vibrator for which the plane of rotation of the unbalanced mass does not coincide with the plane of extremal stiffnesses of the elastic connections (plane $x$, $y$). In this case the vibrator generates excitations in the direction not only of axes $x$ and $y$, but also of axis $z$. Figure 4.6 considers the case when the plane of rotation of the unbalanced mass passes through axis $y$ and forms angle $\alpha_x$ with axis $x$. In this case projections of the radius of the unbalanced mass on axes $x$ and $z$ will be equal, respectively, to $r_x = r \cos \alpha_x$ and $r_x = r \sin \alpha_x$, and projection of the radius on axis $y$ is equal to $r_y = r$. In the general case the plane of rotation of the unbalanced mass can form with axes $x$, $y$, $z$ angles $\alpha_x$, $\alpha_y$, $\alpha_z$.

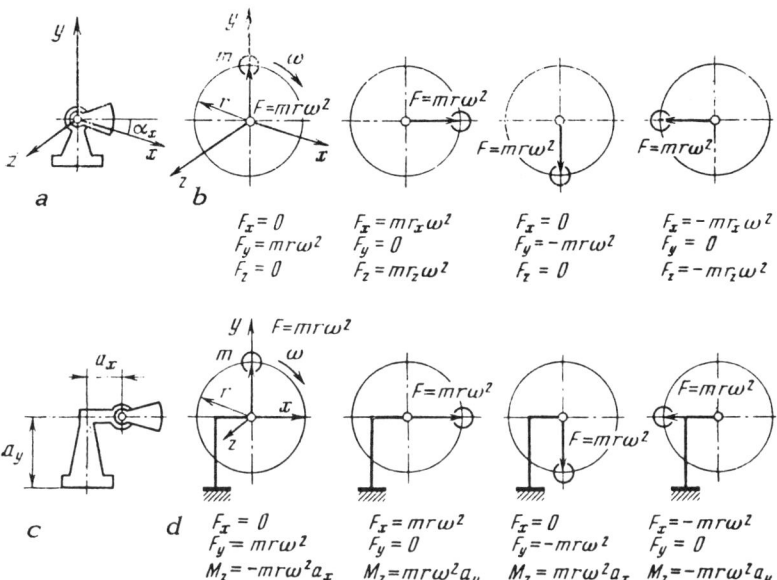

**Figure 4.6** Unbalanced-mass vibrator for generation of three-component translational and torsional oscillations *a)* fundamental arrangement of a vibrator for translational oscillations; *b)* principle of operation; *c)* basic arrangement of a vibrator for translational and torsional oscillations; *d)* principle of operation.

The principle of operation of the unbalanced-mass vibrator under consideration is illustrated by the schematic diagrams presented in Fig. 4.6$b$. The exciting force generated by the rotating unbalanced mass is equal to $F = mr\,\omega^2$. In position I the components of the exciting force are equal to $F_x = 0$, $F_y = mr\,\omega^2$; in position II the unbalanced mass is turned to the right and the components of the exciting force are respectively equal to $F_x = mr_x\,\omega^2$, $F_y = 0$, $F_z = mr_z\omega^2$; in position III the unbalanced mass is below and the components of the exciting force are equal to $F_x = 0$, $F_y = -mr\omega^2$, $F_z = 0$. In position IV the unbalanced mass is directed to the left, the components of the exciting force are equal to $F_x = mr_x\,\omega^2$, $F = 0$, $F = -mr_z\omega^2$.

Figure 4.6$c$ shows the schematic diagram of unbalanced-mass vibrator for which the axis of rotation of the unbalanced mass is at distance $a_x$, $a_y$ from the center of mass of the oscillatory system. In this case the vibrator generates not only a rotary exciting force $F = mr\omega^2$, but also an exciting moment. The principle of operation of the unbalanced vibrator, for which the axis of rotation of the unbalanced mass does not coincide with the center of mass of the oscillatory system, is elucidated by the diagrams presented in Fig. 4.6d. In position I the unbalanced mass is directed upwards, the components of the exciting force and moment are equal to $F_x = 0$, $F_y = mr\omega^2$ and $M_z = -mr\omega^2 a_x$. In position II the unbalanced mass is directed to the right, the components of the exciting force and moment are equal to $F_x = mr\omega^2$, $F_y = 0$, $M_z = mr\omega^2 a_y$. In position III the unbalanced mass is directed downwards, furthermore the components of the exciting force are equal to $F_x = 0$, $F_y = -mr\omega^2$, and the exciting moment $M_z = mr\omega^2 a_x$. In position IV the unbalanced mass is directed to the left, the components of the exciting force and moment are equal to $F_x = -mr\omega^2$, $F_y = 0$, $M_z = -mr\omega^2 a_x$.

Thus, if the axis of rotation of the unbalanced mass does not pass through the center of mass of the oscillatory system and the plane of rotation of the unbalanced mass does not coincide with the extremal values of the elastic connection, one can generally induce translational oscillations along axes $xyz$ and angular oscillations around these axes of the oscillatory system using the simplest unbalanced-mass vibrator.

In many cases the process effectiveness increases if special excitation laws are used.

Formation of an exciting force having a cardioid-shaped locus can be provided by a biharmonic elliptical vibrator. By changing kinetic moments of the unbalanced masses and phase shift between pairs of unbalanced masses, one can obtain diverse exciting forces.

For generation of an exciting force that varies according to a multi-lobe locus which can rotate around its center, V. D. Zemskov had proposed use of vibrators with three unbalanced masses.

The fundamental arrangement of such a vibrator has been considered above. Each of the unbalanced masses in rotating with its own angular velocity generates a centrifugal force which is constant in magnitude but variable in direction; the middle unbalanced mass generates force $m'r'\omega'^2$ and the side mass-force $2m''r''\omega''^2$ (the notations adopted here: $m'$, $m''$ are masses, $r'$, $r''$ are radii, $\omega'$, $\omega''$ are angular velocities of the middle and side unbalanced masses). The resulting force of the vibrator is equal to the vector sum of these forces.

The angular velocities of rotation and centrifugal forces of the central and side unbalanced forces are related as $\omega' = \omega''i$; therefore, the total force generated by the side unbalanced masses is equal to $F'' = 2m''r'' (\omega'/i)^2$, and the exciting force of the middle unbalanced mass is equal to $F' = m'r'\omega^2$.

Upon rotation of the first unbalanced mass by angle $\varphi'$, the second unbalanced mass turns by angle $\varphi'' = \varphi'/i$. By summing the vectors of forces $F''$ and $F'$, the resultant is

$$F = \sqrt{F'^2 + F''^2 + 2F'F'' \cos \alpha}$$

where $\alpha$ is the angle between vectors $F'$ and $F''$.

The resultant becomes zero when $\alpha = 0$, $F' - F'' = 0$, $m'r'\omega^2 = 2m''r''(\omega'/i)^2$. From the same condition, the relation between the unbalanced masses is determined assuming $r' = r''$

$$m'/2m'' = 1/i^2$$

For transmission ratio $i = 1:2$

$$m'/2m'' = 4, \quad m'' = 1/8m'$$

i.e., for the resultant exciting force to pass through zero, the mass of each second and third unbalanced masses must be eight times less than the first unbalanced mass, and for transmission ratio of $i = 1:3$ it should be 18 times less

$$m'' = 1/18m'$$

By graphical constructions the analytical dependence of angle $\alpha$ between the force vectors $F'$ and $F''$ on the angle of rotation of the first unbalanced mass $\varphi'$ has been derived

$$\alpha = | (2n + 1)\pi - (1 + p)\varphi' | \quad \text{when} \quad n2\pi < (1 + p)\varphi' < (n + 1)2\pi$$

where $n$ are integers ($n = 0, 1, 2, 3, ...$); $p$ is the denominator of the fractional expression for the transmission ratio ($i = 1/p$).

The presented formula is valid for any transmission ratios. Equating the obtained expression to zero we obtain a formula for determination of the position of the resultant vector with a maximum value ($F_{max}$). If this expression is equal to $\pi$, then a formula is obtained for determination of the position of the resultant vector of the exciting forces with minimum value ($F_{min}$)

$$\varphi'(F_{max}) = \frac{\pi}{1+p}(2n+1), \; \varphi'(F_{min}) = \frac{\pi}{1+p} 2n$$

The values of angles $\varphi'(F_{max})$ and $\varphi'(F_{min})$ enable one to determine the number of maximum and minimum values of the resultant $F$ when plotting the locus in polar coordinates.

For example, when $i = 1:2$ the graph of function $F = f(\alpha, \varphi')$ has three maxima and three minima

$$\varphi'(F_{max}) = \frac{\pi}{3}, \; \pi, \; \frac{5\pi}{3}$$

$$\varphi'(F_{min}) = 0, \; \frac{2\pi}{3}, \; \frac{4\pi}{3}$$

when $i = 1:3$ the graph has four maxima and four minima

$$\varphi'(F_{max}) = \frac{\pi}{4}, \; \frac{3\pi}{4}, \; \frac{5\pi}{4}, \; \frac{7\pi}{4}$$

$$\varphi'(F_{min}) = 0, \; \frac{\pi}{2}, \; \pi, \; \frac{3\pi}{2}$$

By analyzing the graphs of function $F = f(\alpha, \varphi')$ constructed in polar coordinates for $i = 1:2$ and $i = 1:3$ (Fig. 4.7), we can see that they represent closed curves with several lobes. Moreover, the number of lobes is equal respectively to three and four, i.e., to the sum of the numerator and denominator of the transmission ratio: for $i = 1:2$ $k = 1+2 = 3$; for $i = 1:3$ $k = 1+3 = 4$. This pattern occurs for any transmission ratio including the ratio which has a decimal fraction in the denominator. For example, the graph $F = f(\alpha, \varphi')$ for $i = 1:6.5$ has 7.5 lobes.

Analysis of the loci of the exciting force for the transmission ratios from $i = 1:2.1$ to $i = 1:2.9$ per revolution of the first unbalanced mass presents the opportunity to observe the initiation and development of a new fourth lobe which is completely formed when $i = 1:3$.

Simultaneously with the increase of the number of lobes, one can notice their narrowing and clockwise rotation and also the reduction of the angle between the adjacent vectors of $F_{max}$ with $2\pi/3$ to $\pi/2$. During each revolution of the first unbalanced mass, the vector $F_{max}$ rotates by an angle which is constant for the given transmission ratio, and in a direction opposite to the rotation of the unbalanced mass. In other words, vector $F_{max}$ rotates with a constant angular velocity which increases as the fraction in the denominator of the transmission ratio increases.

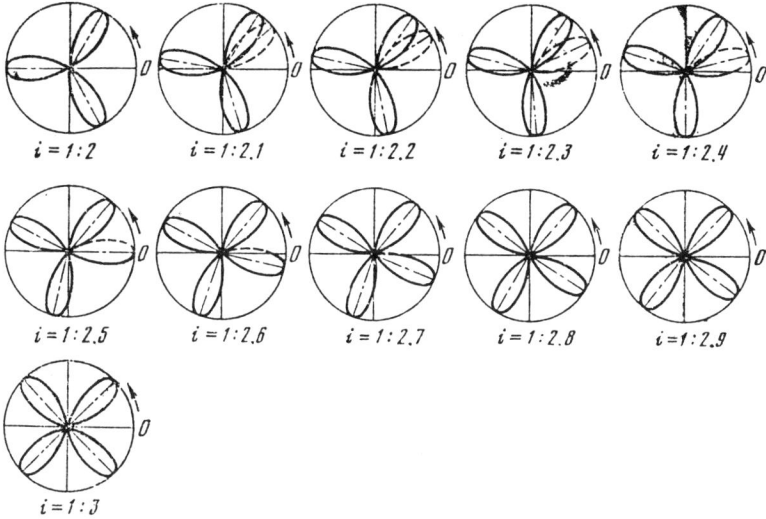

**Figure 4.7** Formation of exciting force patterns at various transmission ratios.

The angular velocities of rotation of the vector of $F_{max}$ at the values of the transmission ratio 1:2.1 $\leq i \leq$ 1:2.9, 1:3.1 $\leq i$ $\leq$ 1:3.9 are practically linearly dependent on its magnitude.

Inertial vibrators are used at medium oscillation frequencies. Use of vibrators of this type in low-frequency regimes is not rational, since in this case one must significantly increase the unbalanced masses to obtain the required magnitude of the exciting force. At high frequencies reactions in the bearings of the vibrator increase considerably which causes their rapid failure. Inertial vibrators are capable of generating significant exciting forces at moderate dimensions and weight and enable one to obtain various laws of variation of the exciting force by simple means. As a result of the absence of a rigid connection between the moving components of the vibrator and the oscillatory system drive, there is no breakdown even in the case of system jamming. The shortcoming of the majority of designs of inertial vibrators is the increased time for starting and stopping. This hampers their application in vibratory machines operating with frequent starts and stops.

Modern designs of inertial vibrators enable adjustment of the operating regimes. Variation of oscillation frequencies is effected by controlling the rotational speed of the drive motor.

Control of the magnitude of the exciting force (moment) is carried out by changing the mass or eccentricity of the unbalanced unit. Rotating unbalanced masses are also used in some designs. By changing relative positions of the unbalanced masses the resultant exciting force can be regulated. Vibrators with automatic control of the exciting force magnitude also find application. In particular, for vibratory machines operating in transresonant regimes, one can use inertial vibrators with sliding and rotating unbalanced masses. In the initial position the magnitude of the kinetic moment of the unbalanced mass of such a vibrator is small or even equal to zero, and upon rotation of the vibrator shaft with a speed corresponding to the natural frequency of oscillation of the vibratory machine (at resonant frequency), it generates insignificant exciting force or no force at all. The exciting force reaches nominal value only after crossing the resonant region with the unbalanced mass extended to the limit of the eccentricity. Use of such vibrators eliminates excessive rocking of the vibratory machine when the resonant regime is reached.

### 4.2.1.2  Mechanisms for Automatic Variation of Kinetic Moment of Unbalanced Masses

For elimination of resonant oscillations during starting and stopping of vibratory machines with inertial drives, mechanisms for automatic variation of kinetic moment of the unbalanced masses are used. When the vibratory machine is started, the mechanism automatically switches on, following the passage of the resonant region. When the machine is turned off, the mechanism is switched off until the resonant regime is reached.

     These mechanisms can be grouped with respect to the character of motion of the unbalanced masses as mechanisms with sliding, folding, and freely rotating unbalanced masses.

     The schematic diagram of the mechanism with a sliding unbalanced mass is presented in Fig. 4.8a. The moving unbalanced mass 1 of the vibration exciter is mounted in a sleeve which is fixed on shaft 2 by means of rod 3 held by spring 4. When the shaft of the vibration exciter is rotated with angular velocities below the design critical value, the unbalanced mass remains pressed to the sleeve by the spring (position I in the figure). In this position the kinetic moment of the unbalanced mass with respect to the axis of rotation is

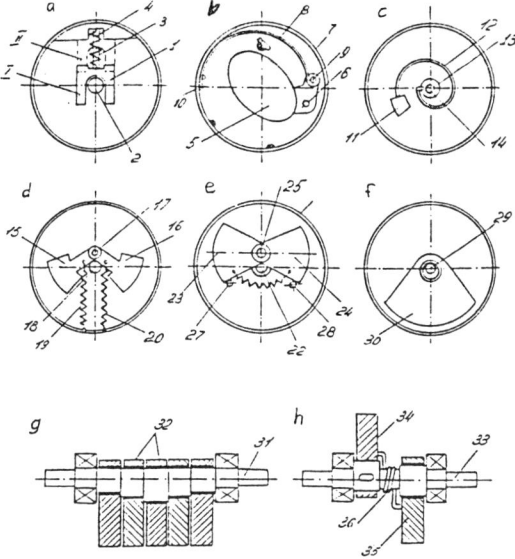

**Figure 4.8** Schematic diagram of mechanisms for automatic variation of kinetic moment of unbalanced masses.

equal to zero. At angular velocities above the design critical velocity, the centrifugal force exceeds the elastic force of the spring, thus the unbalanced mass slides and takes the operating position II shown in the figure by the dashed line. Adjustment of the exciting force magnitude is realized by a nut screwed on the rod of the unbalanced mass. It reduces the stroke of the unbalanced mass and, consequently, the magnitude of the exciting force. The sliding mechanism of the unbalanced mass consists of a cup which is attached to the shaft of the vibration exciter and, connected with it by means of a rod and a spring, a sleeve with the unbalanced mass. The spring tension is regulated by a nut screwed on the rod.

The tendency of the unbalanced mass to jam during starting and run-out periods is one of the shortcomings of this mechanism, since the vector of the relative speed of the unbalanced mass rotates and a Coriolis force is generated in a direction perpendicular to its displacement. A static balance of this mechanism can be realized only in one position of the unbalanced mass corresponding, for example, to the maximum magnitude of the exciting force. When the vibrator is adjusted for a lesser force, the general balance of the mechanism is disturbed.

The inertial vibration exciter with folding unbalanced masses, which is schematically depicted in Fig. 4.8b, is analogous in principle but different in design.

The mechanism has an unbalanced mass 5 of elliptic shape mounted on axis 6 which is shifted relative to the rotation axis of pulley 7. In the static condition the unbalanced mass is held at the center of the pulley by preloaded spring 8. One end of the spring is attached to the unbalanced mass by means of pin 9 and the other is attached to support 10 mounted on the pulley.

Unlike the mechanism with sliding unbalanced mass, this design is not adjustable with respect either to magnitude of the exciting force or the value of angular velocity corresponding to the start of turning (unfolding) of the unbalanced masses from the initial position to the operating position.

Figure 4.8c shows the schematic diagram of a vibration exciter in which unbalanced mass 11 is attached to the free end of helical spring 12, whose other end is clamped in sleeve 13, which is mounted on drive shaft 14. The entire mechanism is housed in an enclosure which is also mounted on the drive shaft. This design of the vibration exciter provides gradual increase in the value of the kinetic moment of the unbalanced mass with increasing rotational speed of the drive shaft. This occurs due to continuous unwinding of the helical spring up to the moment when the unbalanced mass leans against the enclosure of the vibration exciter.

The vibration exciter with a folding mechanism consists of two unbalanced masses 15 and 16 mounted in a bracket on axis 17, which is offset relative to the center of the drive shaft 18. Unbalanced masses are pressed to the enclosure hub by springs 19 and 20 in such a position that without rotation they completely balance each other (Fig. 4.8d). When the shaft together with the enclosure reach a rotation speed exceeding the resonant speed, centrifugal force throws the masses into the middle position, which is shown in the drawing by the dashed line, creating an unbalanced mass.

The drawbacks of the last three designs include the total absence of any system of adjustment of moment of time of actuation; and also the presence of springs which are loaded in the operating regime.

Mechanisms with swinging (rotating) unbalanced masses have been developed in two modifications with constant and adjustable eccentricity. Adjustment of the eccentricity is carried out with the aid of two eccentric sleeves fitted on one another.

The mechanism of turning of the unbalanced masses comprises disc *21* mounted on the drive shaft *22* and two unbalanced masses *23* and *24* mounted on pin *25*, which is offset by some distance from the axis of the drive shaft (Fig. 4.8e). In the initial position, the unbalanced masses are pressed to stops *27* and *28* on the disc by spring *26*. The mechanism operates as follows. When the vibration exciter is accelerated, the centrifugal forces of the unbalanced masses generate a torque which overcomes frictional forces in the joints and the force of pretwist of the spring upon reaching a specific speed of rotation and turns the unbalanced masses to the operating position. During run-out, as the angular rotational velocity of the vibration exciter is decreased, the spring returns the unbalanced masses to their initial positions.

A very simple unbalanced-mass mechanism with automatically varying kinetic moment in start and run-out periods was developed with the use of freely rotating unbalanced mass. The mechanism consists of a drive shaft *29* on whose eccentric neck unbalanced mass *30* is loosely mounted (Fig. 4.8f). The principle of operation of the freely rotating unbalanced mass is as follows. Prior to start, the unbalanced mass suspended on the eccentric neck is only under the action of the gravity force. Therefore, the radius vector of the unbalanced mass (the line connecting the center of rotation of the drive shaft with the center of mass of the unbalanced mass) is in a vertical position. When the drive shaft rotates, the eccentric neck imparts a circular motion to the point of suspension of the unbalanced mass. A centrifugal force is generated which is added to the gravitational force of the unbalanced mass. A force of normal pressure develops over the contact area between the hole of the unbalanced mass and the eccentric neck, which generates a friction force on this surface. The resultant of these forces on the arm of the "friction circle" of the trunnion of the unbalanced mass generates a frictional torque which tends to turn the mass in the direction of rotation of the drive shaft. This rotation is hindered by the moment of the gravitational forces of the unbalanced mass. Entrainment of the unbalanced mass and its transition from swinging to rotational motion occurs when the frictional moments and gravitational forces become equal. The entrainment moment is dependent on the value of eccentricity of the shaft neck, the coefficient of friction in the trunnion of the unbalanced mass, and the square of the angular velocity of the drive shaft. The first two design parameters of the unbalanced-mass mechanism are selected so that during the start of the vibratory machine, entrainment of the unbalanced mass would occur at a frequency higher than the resonant frequency. Switching off the mechanism of the freely rotating unbalanced mass takes place in a similar way.

A number of mechanisms of automatic control of the kinetic moment of the inertial vibratory drive has been developed using the freely rotating unbalanced mass principle.

Figure 4.8g shows schematic diagram of the mechanism whose drive shaft 31 has several journals with different eccentricities. Each journal has its own unbalanced mass 32. In such a mechanism the maximum moment is generated gradually, since the unbalanced masses are not gripped by the shaft simultaneously. Actuation of the unbalanced masses takes some time and commences with the mass mounted on the journal with the largest eccentricity.

Figure 4.8h shows the schematic diagram of a mechanism consisting of a drive shaft 33 with eccentric journals. On one journal, the unbalanced mass 34 is fixed, on the second one the unbalanced mass 35 rotates freely (the shaft has stops limiting the angle of rotation to 180°). The fixed and moving unbalanced masses are connected by a helical spring 36. The spring is mounted and adjusted in such a way that the moving unbalanced mass under the action of the tangential component of the inertial force rotates on the shaft in the opposite direction to the shaft rotation only when the drive shaft reaches an angular velocity exceeding the natural frequency of the vibratory machine. The spring in this case is additionally twisted. At the end of the starting period, the moving unbalanced mass rotates by a half of a revolution, and the kinetic moment of the mechanism reaches a maximum value. During run-out, the spring returns the unbalanced mass to the initial position and the machine passes the resonant region with a balanced vibration exciter.

### 4.2.2 Eccentric Vibrators

Eccentric vibrators are grouped according to their fundamental design as drives with an elastic connecting rod and with a driving damper.

From the standpoint of the possibility of adjusting the drive, one can distinguish drives with regulated and unregulated oscillation amplitudes. The regulated drive can be regulated without stopping the machine, or regulated only when inoperative. With respect to the character of regulation, one can distinguish drives with continuous and stepped regulation.

In vibratory machines the vibration exciter must not impose additional constraints on the motion of the oscillatory system; therefore, to obtain the required degree of mobility, elastic elements are introduced into the mechanism.

The basic design of the simplest eccentric drive with elastic connecting rod is presented in Fig. 4.9a. Eccentric 2, which is enveloped by yoke 3 of the connecting rod 4 with elastic element 5, is mounted on shaft 1. The free end 6 of the connecting rod is hinged to the working element of the machine.

In order to balance the moving masses (the eccentric and the connecting rod), and in some cases to partially balance the moving parts of the working machine, the unbalanced mass 7 (Fig. 4.9b) is mounted on the shaft in such a way that the generated centrifugal force would balance the inertial forces of the moving parts of the drive (the unbalanced components of the machine).

The most complete balancing of the inertial forces of the moving components of the drive and of the vibratory machine is achieved in a system with two eccentric shafts (Fig. 4.9c). Such a vibrator consists of two shafts 8 and 9 with eccentrics 10 and 11 on which the connecting rods 12 and 13 are mounted. The eccentric shafts are interconnected by the synchronizing transmission 14. Both shafts are driven with equal angular velocities by an electric motor. In order to balance the centrifugal forces that are transmitted to the connecting rods the unbalanced masses 15 and 16 are mounted on the shafts. Elastic elements 17 are installed in the connecting rods.

The horizontal components of the centrifugal forces of the unbalanced masses in such drives are completely balanced owing to the rotation of the shafts in opposite directions (the same principle of operation as in the self-balanced vibrators considered above). The components of the centrifugal forces of the unbalanced masses which act in the direction of oscillations still remain. With proper adjustment, they can help to balance reactions of the connecting rods on the vibrator bearings. The balanced eccentric drive transmits smaller

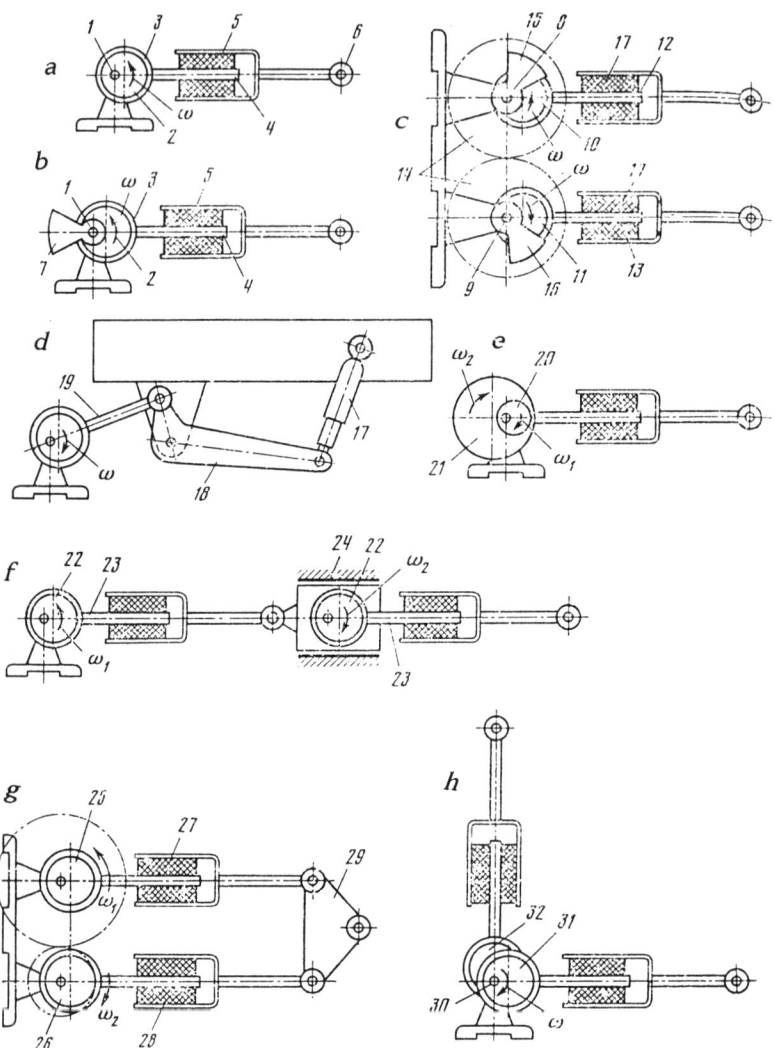

**Figure 4.9**  Schematic diagrams of eccentric vibrators.

dynamic loads to the frame and causes elimination of unfavorable parasitic oscillations.

When using an eccentric drive in vibratory machines with resonant tuning in which the inertial forces of the moving parts are totally balanced by the elastic forces of the working springs, the starting conditions become very difficult since the drive must overcome the compression force of the high stiffness springs. In a drive with rigid connecting rod, this is achieved by one of two methods.

In the first method, to facilitate startup a special starting flywheel is used. Such a drive includes an eccentric shaft consisting of two parts that are connected by a clutch on one half of the shaft. An eccentric with connecting rod is mounted and on the other half - a flywheel which is rotated by a motor through a V-belt drive. During startup both parts of the shaft are disconnected by a clutch and the motor rotates the flywheel. When the maximum rotational speed is reached the clutch is engaged and the two parts of the shaft are connected. The kinetic energy stored in the flywheel is expended on straining the working springs which makes starting easier. In the steady-state regime the flywheel facilitates the smooth operation of the machine.

In the second method starting is made easier by means of an eccentric with adjustable eccentricity during the starting process. Such a drive consists of two eccentric sleeves, one inside the other. By turning the sleeves relative to each other, the magnitude of the eccentricity can be regulated from zero to maximum. In other respects, the design of the drive is as usual. Starting of a machine having such a drive is realized as follows. One of the sleeves is turned so as to minimize eccentricity. Then the motor is started and easily compresses the springs at small stroke of the connecting rod of the drive. As the machine accelerates, the eccentricity is increased up to the operating value.

Various systems of elastic connecting rods are used in the eccentric drives. Depending on their design, two modifications of drives of this type are distinguished. In the first type, connecting rods are elastic in the entire range of operating loads. The second type includes drives with the so-called semi-elastic connecting rods. Such a rod consists of two halves pressed to each other by precompressed helical springs. The precompression force only slightly exceeds the resistance force in the steady-state regime of the vibratory machine. When a drive of this type is used in resonant vibratory machines it operates as an elastic drive during starting and as a kinematic drive during the steady-state operation. A drive with a semi-rigid connecting rod facilitates the starting of resonant vibratory machines and simultaneously provides kinematically defined motion during the steady-state regimes of operation. The shortcomings of the drive include the presence of impact forces and intensive noise during starting when the springs are strained and the halves of the split connecting rod collide.

The vibration exciter with a damper is depicted in Fig. 4.9d. It comprises a shock absorber 17, one end of which is fixed on the oscillating mass and the other which is connected with a three-joint rocker (balance beam) 18. The rocker is also fixed to the oscillating mass by means of a center joint (hinge). The connecting rod 19 of the vibrator is connected to the second (short) end of the rocker.

Recently, drives with an elastic eccentric were adopted instead of drives with elastic connecting rods or dampers. The drive comprises an eccentric with an elastic ring. The yoke of the connecting rod is connected with the eccentric through an elastic ring; the rigid connecting rod is attached to the yoke. During operation the elastic ring is twisted, which reduces the amplitude of displacement of the connecting rod and facilitates starting of the machine.

For generation of biharmonic oscillations, one can use a drive consisting of two eccentric sleeves fitted one inside the other and rotating with different speeds. The basic design of the drive is shown in Fig. 4.9e. It consists of two eccentric sleeves 20, 21 driven with angular velocities $\omega_1$, $\omega_2$ and specified phase shift angle. Polyharmonic vibrators can be generated by a sectionated eccentric vibrator (schematic in Fig. 4.9f). The number of the sections of the vibrator is equal to the desired number of harmonics. Each section consists of an eccentric 22, connecting rod 23, and guide 24. The eccentric shaft of each subsequent section is mounted on the guide of the preceding section. Each eccentric is rotated with a speed corresponding to the order of harmonic it has to generate.

Polyharmonic exciting force is obtained on the connecting rod of the last section. In order to generate biharmonic oscillations an eccentric drive with a parallel arrangement of the connecting rods (Fig. 4.9g) is used comprising two parallel synchronously rotating eccentric shafts 25 and 26 on which elastic connecting rods 27 and 28 are mounted. Displacements of the individual connecting rods are added on the pendulum balance beam 29, which is connected with the driven mass and imparts biharmonic motion to it.

Elliptical oscillations can be excited by an eccentric vibrator with two displaced eccentrics rotating at the same angular velocity (Fig. 4.9h). The vibrator consists of an eccentric shaft 30 with two eccentrics 31 and 32 shifted relative to each other by $90^\circ$. Elastic connecting rods are mounted on the eccentrics and are also shifted by a quarter of a circle and attached to the oscillating mass.

The eccentric drive is most rationally used in low-frequency oscillating systems. This type of drive is capable of generating large exciting forces at low oscillation frequencies. At high frequencies large inertial forces are generated which are transmitted to the bearings of the eccentric drive shaft. It causes considerable frictional forces in the bearings leading to their failure relatively more rapidly. At higher frequencies the eccentric drive is used only in balanced oscillatory systems operating in resonant regimes. In this case, the inertial forces of the oscillating masses are practically totally balanced and insignificant loads are transmitted to the bearings.

The eccentric drive with a rigid connecting rod (drive with kinematically determined motion) provides constant oscillation amplitude of the working element over the entire frequency range of machine operation. In the meantime, difficult starting conditions are a serious demerit of the rigid drive. During starting, the drive must overcome, over one revolution, the restoring forces of the elastic system of the vibratory machine and impart to it the required initial acceleration. In such drives, special electric motors with a high (10 - 12 times rated torque) starting torque are used, or two motors are used, one of which is disengaged after starting, or special energy accumulators are used, for example, in the form of a preaccelerated flywheel. The eccentric drive with rigid connecting rod usually operates at the natural frequencies of the oscillatory system. Minimum exciting force is developed by the drive in this case.

Decreasing or increasing the frequency in relation to the natural frequency of the system is immediately manifested as a sharp increase in the excitation forces. Energy expenditure for maintaining the oscillations of the system when using an eccentric drive with rigid connecting rod is directly proportional to the frequency.

The eccentric drive with elastic connecting rod does not have the shortcomings of the drives with rigid connecting rod. Due to presence of the elastic link, starting of the vibratory machine with such a drive becomes prolonged, amplitude of the oscillating masses increases gradually; thus the drive is subjected to insignificant loads during the transient regimes. The elastic drive is also tuned to the natural frequency of the oscillatory system in order to decrease the required exciting force. In this regime, the amplitude of oscillations of the working element is somewhat less than the eccentricity of the drive; the closer they are, the lower the resistances in the system. When the elastic drive is tuned to the resonant regime, the exciting force somewhat increases and the oscillation amplitude and energy expenditure in the oscillatory system increase substantially.

It is essential to note that when operating in the resonant regime, the eccentric drive consumes less energy than the eccentric drive with rigid and viscous connecting rods and considerably less than the inertial drive for the same amplitudes. From the standpoint of energy consumption in resonance regimes, the best is an eccentric drive with elastic conrod, then the eccentric drives with rigid and viscous conrods, and the highest energy consumption is characteristic for inertial drive.

The disadvantages of the elastic drive which are manifested when operating at the natural frequencies of the dynamic system include the dependence of the amplitude on the present resistances and degree of the machine loading.

The eccentric drive with viscous connecting rod provides greater stability of operation in the resonant regimes; however, in order to achieve equal amplitudes, such a drive must generate larger exciting forces than a drive with elastic connecting rod. In subresonant and transresonant regimes, this difference is even more significant. The same applies to energy expenditure.

The eccentric drive is good for low-frequency machines, since it can generate in this case the required large amplitudes of oscillations. This type of drive is capable of generating large exciting forces at low rotational speeds of the drive shaft. Also in low-frequency machines the dynamic loads on the drive are lower, which enables operating with some detuning from the natural frequency of the system, thus ensuring greater stability.

By using an adjustable speed motor, for example, a dc electric motor or three-phase commutator electric motor, a drive with a wide frequency range can be obtained. Using adjustable eccentrics, the oscillation amplitude can also be regulated.

### 4.2.3  Hydraulic Vibrators

Hydraulic vibration exciters impart oscillations to the working element of the machine either by using a pulsating source of the working fluid, or by interrupting the flow of the working fluid with constant flowrate with the aid of control valves. The latter can be controlled either by an external drive or by the vibrator itself, according to the position of its actuator.

According to their operation, hydraulic vibrators can be of pulsating, self-excited follower, and self-regulating types.

Hydraulic vibrators of the first type are designed on the principle of excitation of the actuator (hydraulic cylinder) by the pulsating pressure generated by the variable flow of the working fluid. There are pulsating hydraulic vibrators with a through flow of the working fluid, and with closed working volume. Single- and double-acting hydraulic vibrators are used. In the former, the working fluid does work only during the forward stroke and the reverse stroke is actuated by the machine elastic system. In double-acting vibrators, the reverse stroke also occurs under the action of the working fluid.

The hydraulic drive which has a rigid negative feedback with respect to the displacement between the hydraulic distributor and the working element of the machine is called follower drive.

In self-excited and self-controlled hydraulic vibrators, the periodic exciting force is generated under constant-pressure supply due to the presence of a special system which automatically carries out periodic supply and drain of fluid. The piston of the hydraulic cylinder itself controls the motion of the distribution valve ensuring continuous reciprocating motion.

In auto-oscillating (self-excited) vibrators oscillations are generated due to the presence of a nonlinear element in the system, namely a clearance, motion limiter, dead zones of feedback. In self-controlled vibrators, oscillations are generated due to the presence of special devices providing a flip of the control valve at the moment the piston of the hydraulic cylinder is in its extreme position. Frequency of vibrators of this type is regulated by the input pressure, and amplitude—by the magnitude of the clearance in the feedback of the auto-oscillating vibrator or by adjusting the stops of the switching devices of the self-regulating vibrators.

Pulsating vibrators are divided into two groups, according to the principle of excitation—into generators with pulsating pumps and into ones with hydraulic control valves. As hydraulic control valves, one can use rotary or translational valves which are driven externally. Frequency of the vibrator is regulated by the speed of rotation or reciprocating motion of the hydraulic control valve. Variation of the working fluid pressure is controlling amplitude.

The schematic diagram of a double-acting pulsating hydraulic vibrator for generation of harmonic oscillations is presented in Fig. 4.10a. The vibrator comprises hydraulic cylinder *1* in which piston *2* moves under the pressure of the working fluid supplied by a two-piston pulsator *3*. During the first half of the stroke, the pulsator supplies the working fluid to one side of the piston (tube *4*) and discharges it from the other side (tube *5*). During the other half of the stroke, the direction of fluid supply is changing. The oscillatory system is connected with the piston of the vibrator by rod *6* having elastic element *7* which is required in this case to impart to the system the necessary degrees of mobility. Some additional elasticity is generated as a result of the fluid compressibility and elasticity of the connecting pipes (hoses).

One of the most significant disadvantages of hydraulic vibrators is leakage of the working fluid during operation through clearances between the piston and the cylinder, the seal and the rod. Hydraulic vibrators without sliding pairs and not having this shortcoming are used. In such machines instead of a piston a rubber elastic element, deforming in shear, is used. The merit of this design is the organic union of the vibratory motor with the elastic system which enables one to design a universal modular vibratory machine.

To generate biharmonic oscillations one can use hydraulic vibrator with a two-piston pulsator in which one of the piston moves with a double frequency. The schematic layout of the biharmonic single-action hydraulic vibrator is shown in Fig. 4.10b. It consists of a hydraulic cylinder *8* containing piston *9*, whose rod *10* has an elastic element *11*. The working fluid is supplied to the hydraulic cylinder by a two-piston pulsator *12* through pipes *13* and *14*. Since the flow rates from each cylinder of the pulsator are combined in the working cylinder of the hydraulic vibrator, the piston of the latter moves according to biharmonic law.

Figure 4.10c depicts the layout of the pulsating single-action vibrator for generation of elliptical oscillations. The vibrator consists of two working hydraulic cylinders *15, 16* forming a right angle and containing pistons *17, 18* under the pressure of the working fluid which is supplied by two piston pulsators whose pistons *19, 20* are moving with a phase shift by the eccentric mechanisms *21, 22* rotating in the same phase

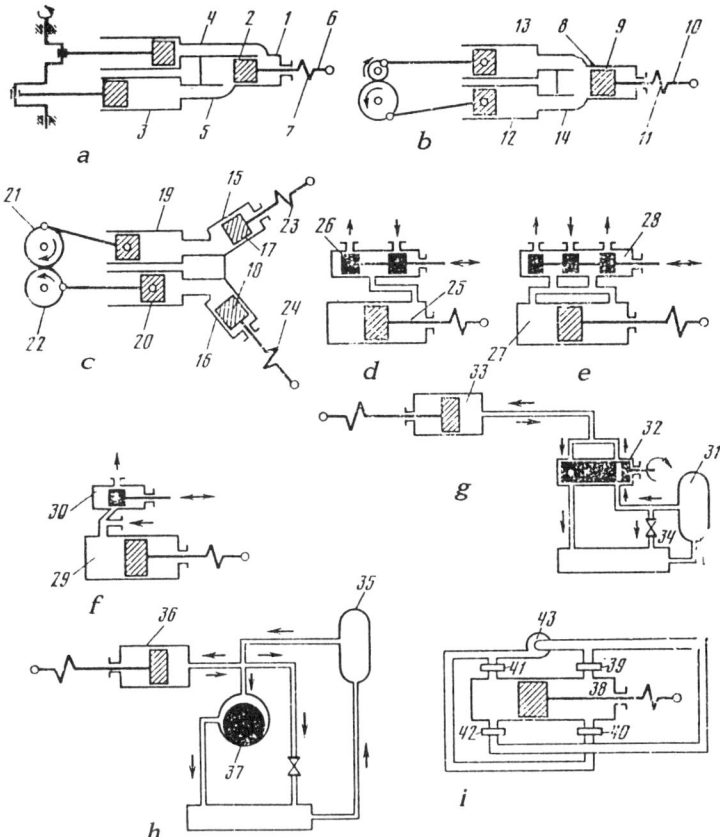

**Figure 4.10** Schematic diagrams of hydraulic vibrators of pulsating action.

and with equal angular velocities. The pistons of the hydraulic cylinders transmit displacements to the oscillatory system in mutually perpendicular directions via a rod with the elastic elements *23* and *24*.

The hydraulic vibrators shown in Fig. 4.10*b*, *c* can also be made as double-acting machines. In this case, for each working cylinder two cylinders must work in the pulsator.

The main advantage of pulsating hydraulic vibrators with pulsating pumps is flatness of the amplitude-frequency characteristic of the piston of the actuating hydraulic cylinder. It is a volumetric-action hydraulic drive. Amplitude of the rod of the vibrator hydraulic cylinder is determined by the volume displaced by the pistons of the pulsating pump and by design parameters of the machine, regardless of the working load.

Control of the operating regimes of pulsating hydraulic vibrators is effected by changing either frequency or oscillation amplitude. The frequency is regulated by changing rotational speed of the pulsator shaft, and the oscillation amplitude (output of the pulsator) is regulated using a throttle device.

One of the most serious shortcomings of hydraulic pulsating drives is heating of the working fluid since the volume in the subsystem pulsator piston-vibrator piston is closed.

There are two varieties of the sliding-valve hydraulic pulsating drive, namely hydraulic drive with a valve at the inlet to the actuating hydraulic cylinder (vibrator) and with a valve at the outlet (with adjustable discharge).

Figure 4.10d shows the hydraulic single-acting vibratory drive of the pulsator type with a translationally moving sliding valve. It consists of hydraulic cylinder 25 and sliding valve 26 which periodically connects the working space with the pressure or with the discharge manifolds. The double-acting hydraulic vibrator consisting of a hydraulic cylinder 27 and sliding valve 28 is shown in Fig. 4.10e. The sliding valve periodically connects one working space of the hydraulic cylinder with the discharge manifold and, at the same time, the other working space with the pressure manifold; then the direction of motion of the working fluid changes. Figure 4.10f shows a hydraulic vibrator comprising a hydraulic cylinder 29 and sliding pilot valve 30 which controls discharge of the working fluid.

The fundamental layout of a hydraulic vibratory drive of the pulsating type with pilot valve generator of pulsations at the inlet is shown in Fig. 4.10g. The hydraulic drive consists of a pump of constant or variable flow rate 31 which supplies the working fluid to the inlet of the valve 32. The valve can be made, for example, as a rotating plug with a number of holes arranged so that the cavity of the actuating cylinder 33 is alternately connected over one revolution with the pressure manifold and with the discharge manifold. Pressure pulsation is generated in the actuator cavity causing reciprocating motion of the piston. Control of the amplitude of oscillations is effected by a pressure regulator 34 and of the frequency by changing the valve rotation speed. The valve can be driven by hydraulic motors, mechanical vibrators, and small dc motors, since the pilot valve is only a control element.

In the pulsating hydraulic drive under consideration the pressure in front of the valve is maintained constant and is determined by the pump characteristic. The working fluid is supplied to the chamber of the actuating cylinder after being throttled in the passages of the pilot valve with a pressure which is dependent on the angular velocity of the valve. The throttling resistances of the working fluid being discharged from the cylinder are insignificant; the character of pressure change in the working chamber of the hydraulic cylinder will be determined by hydraulic resistances of the discharge manifold and by dynamic characteristics of the oscillatory system which is coupled with the hydraulic drive.

This hydraulic drive has a characteristic shortcoming: when the resistances in the oscillatory system are increased, causing an increase of pressure in the hydraulic cylinder, the pressure drop in the throttling passage of the pilot valve decreases and the flow rate of the working fluid drops. As a result, the amplitude of oscillations decreases with increasing load, i.e., stiffness of the drive is inadequate. The hydraulic vibratory drive with a pilot valve at the oulet of the actuating cylinder has a significantly stiffer load characteristic.

The fundamental layout of a hydraulic pulsating vibratory drive is shown in Fig. 4.10*h*. It consists of a pump *35* of constant or variable flow rate which feeds the working fluid to the chamber of the actuating cylinder *36*. At the outlet of the latter in the discharge manifold, a pilot valve with rotating plug *37* is installed. When the plug (having a special profile) rotates, the valve passage cross section changes and pressure pulsation is generated in the chamber of the actuating cylinder. Its frequency is regulated by the plug angular velocity, and the amplitude is regulated by a throttle or by varying the pump flow rate. In a vibratory drive of this type the oscillation amplitude of the piston of the actuating cylinder depends on the flow rate through the valve and through the throttle. The working fluid supplied by the pump is fed into the cylinder and into discharge tank via pilot valve and throttle. If the throttle is totally closed, the oscillation amplitude of the cylinder will depend only on the output of the pilot valve. The pilot valve plug can be shaped so that the valve would be closed during the greater part of the forward stroke of the piston. In this case (under sufficient pump pressure), the amplitude of the piston during forward stroke is independent of resistances in the oscillatory system. When the piston moves back under the action of restoring forces of the elastic elements of the oscillatory system, the pilot valve has sufficient opening to allow the passage of the total flow rate

of the working fluid supplied by the pump and flowing from
the cylinder. Since amplitude of the steady-state oscillations in
the hydraulic drive is determined by the magnitude of the
piston forward stroke, the drive will have a sufficiently flat
(stiff) loading characteristic.

The hydraulic vibratory drive of the pulsator type with a
generating pilot valve is simple in construction and allows
relatively simple adjustments of amplitude and frequency; the
pulsations from one pump can be transmitted to the pistons of
any desired number of actuating cylinders.

For reduction of inertia of the control system, the
hydraulic vibrator of the pulsator type in which flow direction
of the working fluid is controlled by an electro-rheological
control system is of interest. As a working fluid, a special
medium is used which has the property of instantaneous
solidification in the electrostatic field. The schematic layout of
a hydraulic vibrator with an electro-rheological control system
is shown in Fig. 4.10*i*. It comprises a double-acting hydraulic
cylinder *38*. The working fluid is supplied to its working
chamber via four electrostatic control devices *39*, *40* and *41*, *42*
from pump *43*. When the control device is energized flow of
the working fluid through it stops. The hydraulic manifold is
designed so that when one pair of hydraulic distributors (*39*,
*40* or *41*, *42*) is energized one chamber of the hydraulic
cylinder is connected with the pressure manifold and the other
with the discharge manifold. When the hydraulic distributors
are reversed the direction of motion of the working fluid is
changing.

The hydraulic vibratory drive is more suitable for
application in machines requiring significant excitation power
with limited structural dimensions, since it can generate
considerable forces at large oscillation amplitudes. The
disadvantage of such vibratory drives is their relative
complexity, heat generation, and leakage of the working fluid.

## 4.2.4 Electromagnetic Vibrators

All the main types of electromagnetic vibrators can be
considered as two-mass systems. The majority of them
generate harmonic exciting forces, while some types generate
an additional exciting moment or transmit impact impulses.

Vibrators can be of single- and double-stroke types. In
single-stroke vibrators there is an electromagnet whose

armature is attracted in one direction, while the reverse stroke
is completed by the restoring elastic forces. In two-stroke
vibrators, two electromagnets are used which alternately
attract the armature in different directions.

In single-stroke vibrators, undesired additional loading of
the elastic system occurs as a result of the unidirectional
action of the constant component of the attractive force of the
electromagnet. In two-stroke vibrators, the attractive force of
the electromagnets (the exciting force) act in both directions
and does not create additional load on the elastic system. In
special types of vibrators, for example with permanent
magnets, the electromagnet generates an attractive force that
is variable in magnitude and direction and is acting parallel to
the plane of oscillations of the vibrator mass. Under the
influence of such forces the vibrator generates not only a
directional pulsating force, but also a variable moment. This
moment arises due to the fact that the motion of the centers
of gravity of the two moving parts of the vibrator occurs not
in one plane as usual, but in two parallel planes, the distance
between which determines the arm length. In a number of
cases, such combination of action is found to be very effective,
for example, when using the vibrator as a hopper exciter.

By their action, resonant and impact-action vibrators are
distinguished. In the first group, tuning of the operating
regime, the attractive force of the electromagnets, and the
magnitude of the air gap are selected so that the armature and
stator do not come into contact during operation. In vibrators
of impact action the armature, equipped with special strikers,
collides with the buffer mounted on the stator. Consequently,
the vibrator imparts to the working element of the machine
not only harmonic oscillations, but also additional impact
pulses.

With respect to the method of electrical energy supply,
synchronous vibrators that are predominantly applied in
practice can be divided into reactive; with a rectifier; with a dc
excitation (with magnetic bias); and with permanent magnets.

The simplest vibrator is the reactive electromagnetic
vibrator (Fig. 4.11a), which consists of stator (electromagnet) *1*
connected to an ac source and armature *2* mounted on elastic
system *3*. During each half period when the maximum value of
the current is reached the armature is attracted, and at a
small value of the current it is repelled as a result of the
elasticity of the springs; therefore, frequency of the vibrator is
double the frequency of the power supply. If the current is

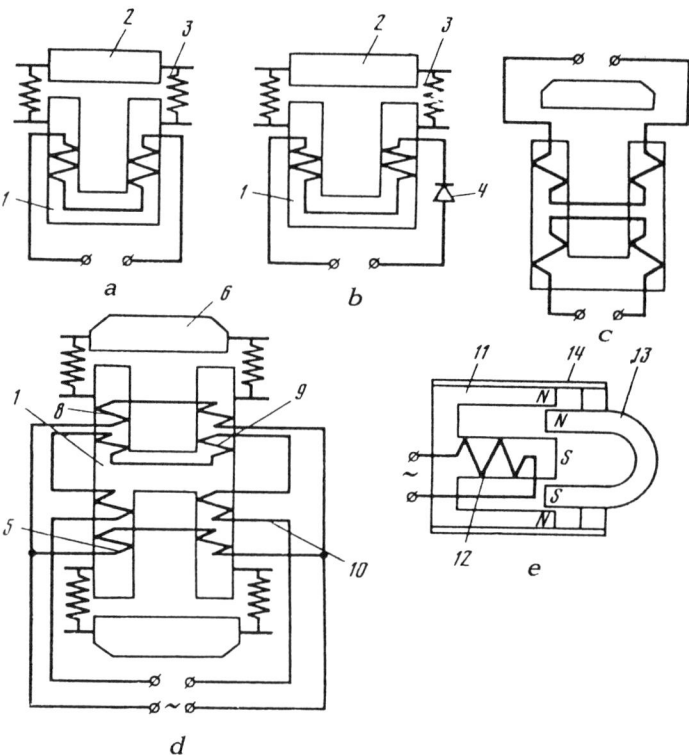

**Figure 4.11** Schematic diagrams of electromagnetic vibrators.

supplied from an ordinary source having a frequency of 50 Hz, such vibrators yield 6000 osc/min. Reactive vibrators can also operate on interrupted dc. Their frequency in this case is dependent on the pulse frequency of the dc.

A vibrator with rectifier (Fig. 4.11b) differs from a reactive vibrator only by the presence of rectifier 4 connected in series (i.e., using the single half-wave circuit) in the coil circuit of the electromagnet. The current after the rectifier reaches maximum once per period; therefore, frequency of the vibrator is equal to the frequency of the current.

The principle of operation of vibrators with excitation by dc (Fig. 4.11c) is to the effect that with two windings in the stator and their excitation by ac and dc pulsating magnetic flow develops in the stator, which acts on the armature of vibrator and brings it into oscillatory motion; frequency of the armature is equal to the ac frequency. When biased by direct current, the sinusoid of the alternating current is displaced in the direction of the bias field and the magnetic flux varies with a frequency equal to half the frequency of the feed current.

Two-stroke vibrators are the best, since they enable maintaining symmetry of current, exclude possibility of introducing distortions in the power supply, and increase the power of the vibratory machine. The layout of a vibrator with a generalized magnetic system and separate ac and dc windings is depicted in Fig. 4.11d. It comprises an H-shaped stator (core) 5, two armatures 6 which are mechanically rigidly connected with each other, and four coils with ac windings 7, 8 and dc windings 9, 10. With combined excitation by ac and dc both alternating and constant magnetic fields are generated in the stator (core). The alternating fields interact with the constant fields, which are in opposite directions in each half of the vibrator, causing in these sections a phase shift of the overall pulsating magnetic fluxes by 180° and, consequently, the same phase shift of the mechanical forces corresponding to these fluxes.

Electromagnetic vibrators with permanent magnets, just like vibrators with rectifiers or with dc excitation, generate oscillations with a frequency corresponding to the frequency of the power supply. However, compared with the latter, they possess the same merits so that their energy supply can come directly from the industrial ac system. The absence of rectifiers in the design of such vibrators significantly increases their operating reliability and eliminates an additional potential source of failure. The fundamental design of an electromagnetic vibrator with permanent magnets is presented in Fig. 4.11e. It consists of an E-shaped core 11 with dc windings 12, horseshoe-shaped permanent magnet 13, and a spring system 14. The core is affixed to the working element of the vibratory machine. The permanent magnet and the parts attached to it represent the reactive mass. The vibrator is arranged in such a manner that the poles of the permanent magnet are placed in the air gaps of the electromagnet. In this case, as a result of the polarity of the permanent magnet being always the same, and the polarity of the electromagnet being variable with the current frequency, periodic mutual attraction and repulsion of the magnets takes place.

The disadvantage of these designs of electromagnetic vibrators is their small armature stroke. At the present time, electric vibrators with a large stroke are being developed, in which the force of mutual interaction of the armature and the core is directed across the magnetic field which is formed in the gap between the poles.

Adjustment of the operating regimes of electromagnetic vibrators is realized as follows: when supplied with alternating or pulsating (alternating rectified by the single half-wave circuit) current adjustment is by a resistor connected in series, by induction regulator, by excitation current of generator, or by switching sections in the transformer windings in case of autonomous power supply. When power is supplied by interrupted dc, adjustment is performed by a resistor connected in series, by changing pulse frequency, and by duration of energized period.

### 4.2.5  High-frequency Vibrators

High-frequency vibration exciters operate in the range of ultrasound oscillations from 20 thousand Hz to 20 million Hz, hence the name ultrasound exciters. These exciters convert all forms of energy, most frequently electrical or mechanical energy, into sound field energy. Electroacoustic converters are the most widely used. In these devices electric oscillations are converted into mechanical oscillations of some solid body which emits acoustic waves to the surrounding medium. The following main types of converters (transducers) of electrical oscillations into mechanical find application: piezoelectric, magnetostriction, and electrodynamic converters.

Piezoelectric converters are based on the utilization of the piezoelectric effect.

Artificial materials have been developed lately which possess pronounced piezoelectric properties. The piezoeffect in these materials is created artificially as a result of special treatment by an electric field. The strongest piezoeffect is manifested by piezoceramics.

The piezoelectric converter, depending on its intended purpose, is made up from various groups of electrically and mechanically bonded piezoelements. For solution of specific processing problems, piezoelectric converters are structurally joined with mechanical linings, plates, membranes, concentrators, and so on.

The vibratory convertor usually consists of two piezoceramic discs or plates between which an electrode is compressed. The discs are grounded. The plates are tightened with bolts with enough force for the piezoelements to be initially precompressed. This ensures their operation without tensile stresses, in the region of pure compression for which piezoelements have high ultimate strength.

Piezoexciters are tuned to the resonance of their mechanical system. Since the mechanical part of the vibratory exciter is a system with distributed parameters, it has infinite discrete sequence of natural frequencies. Resonance in such a vibratory system develops when the exciting force frequency coincides with one of the natural frequencies. Resonant regimes enable generating maximum oscillation amplitudes. Excitation of oscillations in different modes enables synthesizing various operating regimes of vibratory machines with high-frequency exciters.

Taking into consideration that even low natural frequencies of piezoconverters are sufficiently high, special measures are taken in those cases when excitation is required at lower frequencies.

It is well known that natural frequencies decrease with increasing mass of the vibratory system. This property is used for reduction of natural frequencies of vibratory exciters. For this purpose metallic linings are attached by bolts to the emitting surfaces of the vibratory exciter (Fig. 4.12a). Thus the mass of the converter is increased and the frequency of natural oscillations reduced. Usually one plate is made from steel and the other from a strong aluminum alloy. The vibratory converter in this case emits energy of acoustic oscillations via the lighter plate.

Depending on the frequency range, operating conditions, and application, piezoelectric converters of different types are used (Fig. 4.12b-e). In the range of very low frequencies, one can use piezoconverters in the shape of bimorph plates performing transverse bending and torsional oscillation. The properties of such piezoconverters are considerably dependent on conditions of mounting of the plates. In the region of low frequencies of the ultrasound range composite piezoconverters are used, shaped as rods with passive linings. At higher frequencies rods performing longitudinal oscillations are used. For high frequencies piezoconverters shaped as plates and shells performing oscillations across their thickness are used. Converters shaped as hollow piezoceramic spheres that are polarized across thickness also find application. Radial oscillations are generated in them, usually in the subresonant region.

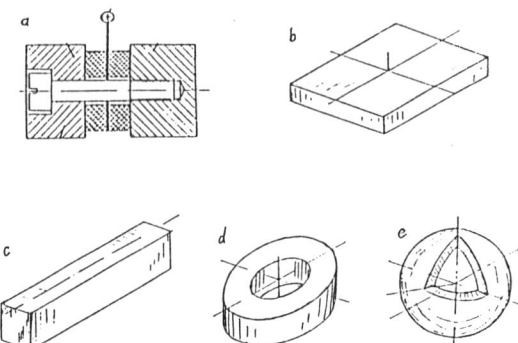

**Figure 4.12** Piezoceramic vibratory converters a) low-frequency with linings; b) plate; c) rod; d) annular; e) hollow spherical.

The effectiveness of the high-frequency vibration exciter is largely dependent on the additional devices matching impedances of the vibratory converter with the specific load. In many process plants this role is played by mechanical velocity transformers. In order to increase oscillation amplitudes of vibratory devices, concentrators shaped as variable cross-section rods are used. For the treatment of liquid media, emitting diaphragms are used.

High-frequency converters in a number of applications do not provide the required deformations of the processed medium and the required energy density in the working zone. This is due to the fact that displacement amplitudes of the converters themselves are insignificant and the power emitted per unit surface area is limited.

The intensity of the emitted power and amplitude of displacements can be substantially increased with the aid of special devices called concentrators. Two types of concentrators are used: rod-shaped and focusing concentrators. Rod concentrators are used in devices operating in the low frequency region of the ultrasound range. Focusing concentrators are effective for considerable wave lengths, since a significant part of the wave energy is lost upon reflection. They are used in high-frequency regions of the ultrasound range.

The low-frequency concentrator is a solid rod of variable cross section of variable density. The principle of operation of the rod concentrator is based on the law of conservation of momentum. Momentum is proportional to the moving mass and its velocity. In conformity to the conditions of passage of the elastic wave along the rod, the momentum for each section of the rod is determined by the area, material density in the section under consideration, and amplitude velocity values of the elastic wave.

The devices serving to transmit waves are known as waveguides. A concentrator is also a waveguide of variable section along which the wave, characterized by constant amplitude across the cross section, propagates. Such waves are referred to as zero mode oscillations. In order to ensure propagation of only the zero mode waves along the concentrator, it is essential that the dimension of its wide end was less than half the wavelength in the concentrator material. Hence, in order to ensure dimensions of the concentrator resulting in a sufficient strength, one must operate with adequately long waves. Therefore, wave concentrators can be used within a limited frequency range from 20 to 100 thousand Hz.

To provide effective energy transfer, concentrators operate at one of their resonant frequencies. Tuning of the concentrator at resonant frequency of oscillations is achieved by making its length equal to a multiple of one half of the wavelength.

Concentrators can be designed with the most diverse shapes. In the cross section they can be circular, rectangular, solid, and hollow; in the longitudinal cross section they can be stepped, conical, exponential, and ampule-shaped (similar in shape to a medical ampule) (Fig. 4.13). The main characteristic of the concentrator is the gain factor which represents the ratio of the displacements at its inlet and outlet ends. Mainly, it is defined by the ratio of the dimensions of the inlet and outlet ends. However, in concentrators of various types the change of wave velocity along their length takes place differently. The greatest gain is attained in stepped concentrators.

The focusing concentrator is based on other principles. It is used to increase the intensity of ultrasound in a limited region of the gaseous or liquid medium compared with the intensity at the surface of the emitter. In a concentrator of this type, focusing of the waves takes place, which is realized with the aid of acoustic reflectors and lenses. Focusing concentrators are used at sufficiently high frequencies, since wave absorption in this case is not intensive. Focusing is usually effected by the oscillation converter itself. Hence, its emitting surface has a spherical or cylindrical shape. Higher

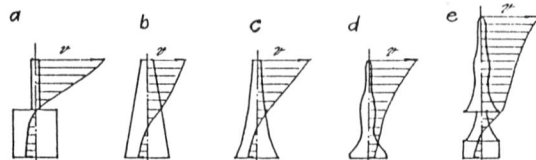

**Figure 4.13** Concentrators. a) stepped; b) conical; c) exponential; d) ampule-shaped; e) composite.

energy densities can be generated with spherical-shaped converters. Emitters with cylindrical form generate less energy concentration, but have large focal region. Energy density in the focal region can reach several kilowatts per square centimeter.

Focusing concentrators are prepared as solid elements from piezoceramics and represent a segment of a spherical or cylindrical surface. Focusing concentrators are mainly used for processing of liquid or solid-liquid dispersed systems, i.e., where there is a medium which transmits ultrasound oscillations sufficiently well.

For the processing of liquid media vibratory converters are equipped with membranes having large emitting surfaces.

High-frequency vibratory exciters can be designed on the basis of magnetostriction effect. Magnetostriction is evidenced by the fact that a rod-shaped ferromagnetic material is deformed when placed in a magnetic field which is directed along the rod. Rods from different materials can expand or contract. This property of ferromagnetic materials is utilized for the conversion of the magnetic field energy into strain energy.

The magnetostriction converter is a core from ferromagnetic material with windings. The energy of the alternating magnetic field, which is generated in the core by alternating current in the windings, is converted into energy of mechanical oscillations of the core. The ferromagnetic rod deforms at any direction of the magnetic field. Therefore, when it is excited by an alternating magnetic field, the rod will oscillate with a double frequency. In order to match the frequency of mechanical oscillations with the frequency of the power supply, a dc component of the magnetic field is additionally created in the converter. The constant component of the magnetic field is selected so that it would be somewhat greater than the amplitude of the variable component of the magnetic field. The resultant of the variable and constant components generates a pulsating magnetic field whose frequency is two times lower than the frequency of the variable component. The converter operating with such a bias is referred to as a polarized converter.

Constant bias is generated by direct current flowing through the windings or with the aid of permanent magnets that are placed in the magnetic core.

Magnetostriction converters operate in the frequency range from several hundred to tens of thousands of oscillations per second. These converters, just like piezoconverters, are usually resonance tuned. Magnetostriction converters of the rod and annular types are also manufactured. The cores of rod converters consist of several rods joined by plates in order to form a closed-loop magnetic circuit (Fig. 4.14). The windings of the rods are arranged in such a manner that the fields in neighboring rods would be opposite in direction. For polarization of the field, permanent magnets are glued between the cores.

Magnetostriction converters that are capable of generating large-amplitude oscillations are distinguished by higher operating life and greater strength than piezoelectric converters.

In addition to piezo- and magnetostriction converters, aero- and hydrodynamic emitters find application in industry.

Aerodynamic and hydrodynamic emitters generate high-power emissions and are capable of processing substantial product volumes.

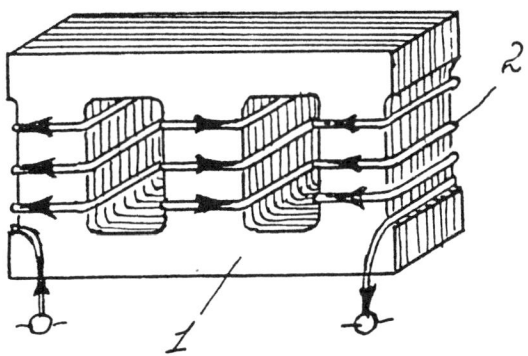

**Figure 4.14** Magnetostriction vibration converter *1)* core; *2)* windings.

Aerodynamic emitters are intended for operation in gas media in which emitters with solid oscillating surfaces cannot be used since wave impedance of the gas media does not provide transmission of oscillation energy from solid emitters. Acoustic emitters of two types find application: gas-jet and dynamic emitters based on jet interruption.

High-pressure gas-jet emitters, which allow the generation of acoustic power of several hundreds of watts, are used in industry. The operation of the gas-jet emitter is based on the generation of self-excited oscillations of the supersonic jet when decelerated by a resonator. The gas-jet emitter comprises a conical nozzle and a cylindrical resonator which is coaxially mounted in front of the former (Fig. 4.15). For supercritical pressure drop in the nozzle and in the surrounding atmosphere, the jet issuing from the nozzle moves with supersonic speed and acquires cellular structure. This means that the static pressure along the length of the jet fluctuates in the absence of a resonator. Upon emergence from the nozzle, the pressure drops and then increases again to the previous level at some distance. The distance between the nozzle exit section and the first maximum pressure occurrence determines the length of the first cell. The resonator is mounted at the end of the first cell. When the jet is slowed down by the resonator, a detached shock wave is generated ahead of it. The kinetic energy of the jet in the resonator is converted into potential energy of the compressed gas. Since the resonator is at the end of the first cell where the static gas pressure rises, the process of gas flow from the resonator becomes periodic. The interaction of the main jet and the jet flowing from the resonator generates powerful oscillations of the gas at the section between the bottom of the resonator and the shock wave.

The gas-jet emitter is usually mounted in a reflector to generate directional emission. A parabolic reflector generates a flat wave, an elliptical reflector generates a wave converging at a focal point. Gas-jet emitters are used in acoustic gas and fuel oil burners for process heat- and mass-transfer acceleration, dispersion and, in a number of cases, for other processes.

Membrane gas-jet emitters are operating on the principle of excitation of the membrane by gas flow. The emitter consists of a cylindrical nozzle with a flange to which an elastic membrane is pressed. Under the influence of the jet the membrane deforms and separates from the flange, the gas flows out and the pressure under the membrane drops. Under

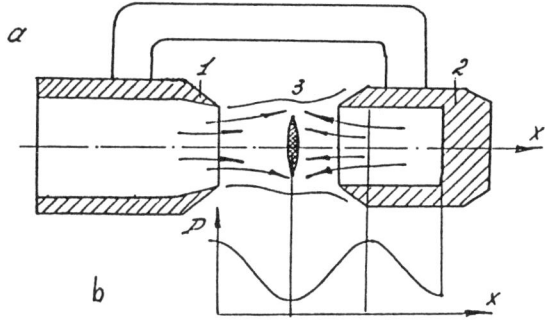

**Figure 4.15** Gas-jet emitter. a) schematic diagram: 1 - nozzle; 2 - resonator; 3 - detached shock wave; b) static pressure P distribution in the free jet.

the influence of elasticity forces, the diaphragm returns to its initial position. Membrane gas-jet emitters operate in the self-oscillating regime.

In dynamic emitters strong acoustic oscillations are generated as a result of the periodic interruption of the jets flowing with high velocity from the hole. Jet interruption is effected by a revolving rotor with holes or teeth. The rotor can be driven from the same source of energy supply from which the emitter of the compressed air works, or by an electric motor. In order to interrupt the jet, various pilot valve devices can also be used. Dynamic gas-jet assemblies are usually designed as an integral unit with the horn.

Dynamic emitters of axial and radial types are used. In axial emitters the rotor and stator are made as discs with holes. The gas flow is directed along the axis of rotor rotation (Fig. 4.16a). In emitters of the radial type, the rotor and stator are made as coaxial cylinders or cones whose lateral surfaces have holes for the gas outflow (Fig. 4.16b). The gas jet is supplied perpendicular to the axis of rotor rotation. The frequency of jet pulsations generated in dynamic emitters is determined by the speed of rotation and number of holes in the rotor. The frequency can be regulated from hundreds to tens of thousands of oscillations per second.

In dynamic emitters, oscillations are generated by the pulsations of the gas flowing from the holes. The character of the generated oscillations is determined by the shape of these pulsations. Harmonic oscillations are generated when round holes are used with the distance between them equal to their

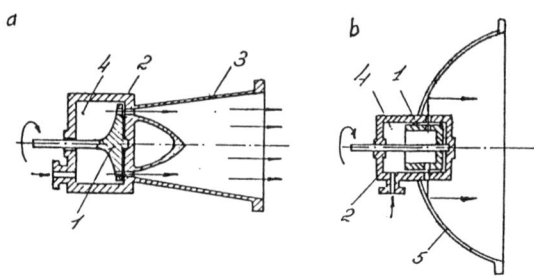

**Figure 4.16** Aerodynamic emitters. *a)* axial; *b)* radial: *1* - rotor, *2* - stator, *3* - horn, *4* - high-pressure chamber, *5* - reflector.

diameter. Holes of rectangular form generate oscillation not only of the principal frequency, but also higher harmonics. A wide spectrum can be obtained by making holes of different sizes on the rotor and positioning them unevenly.

Dynamic gas-jet emitters can emit significant acoustic power, up to tens of kilowatts.

For realization of vibratory production processes in liquid and liquid-solid dispersed systems hydrodynamic emitters are used.

In hydrodynamic emitters conversion of the energy of the turbulent submerged fluid jet into energy of acoustic waves takes place. The transformation of the turbulent flow energy into energy of oscillations occurs due to interaction of the jet flowing out of the nozzle with the elastic oscillatory system placed in the liquid. When the jet impinges onto the oscillatory system, self-excited vibrations are generated in it which are transmitted to the processed liquid medium. Under the action of these vibrations, variable fields of velocities and pressures are generated in the liquid.

The hydrodynamic emitter consists of a slit nozzle which is submerged in the liquid and a plate placed at some distance from the nozzle with the sharp edge directed towards it (Fig. 4.17a). The plate can be attached at the nodal points or supported as a cantilever. When the jet flows over the plate the latter performs bending self-excited oscillations. To provide sufficiently intensive oscillations the plate must be tuned for the resonant operating regime. Acoustic oscillations are emitted perpendicular to the plane of the plate. The nozzle can also be of annular shape (Fig. 4.17b). The resonant oscillating system in this case is a system of cantilever beams mounted opposite the annular nozzle.

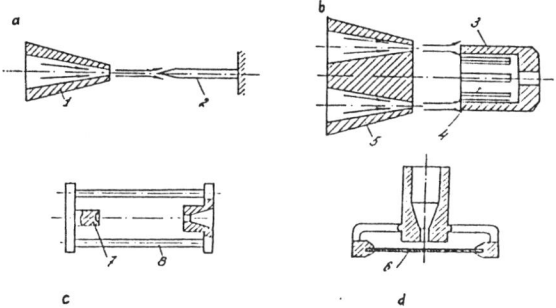

**Figure 4.17** Hydrodynamic emitters. *a*) with resonating plate; *b*) with resonating beams; *c*) rod; *d*) membrane: *1* - slit nozzle, *2* - resonating plate, *3* - cylindrical resonator, *4* - resonating beams, *5* - annular nozzle, *6* - membrane, *7* - reflector, *8* - rods (elastic elements).

Acoustic waves can also be generated by pulsations in the cavitational region which is formed between the nozzle and the special reflector mounted on the elastic resonant system (Fig. 4.17*c*). The elastic system is composed of rods connecting the nozzle with the reflector. The working surface of the reflector can be convex, flat, or concave. The cavitational region is developing better with a concave reflector surface. Oscillations in the fluid are excited under the influence of periodic ejections of the contents of the cavitational region which is formed between the nozzle and the reflector. The pulsations in the cavitational region generate periodic fields of velocities and pressures which excite oscillations of the reflector on the rods.

Oscillations of the liquid can be excited by means of a nozzle and a membrane placed at some distance from the nozzle. An annular slit forms between the nozzle and the membrane (Fig. 4.17*d*). When the liquid flows through the slit, the pressure drops as a result of transformation of the static head into dynamic head. Therefore, the membrane is pressed against the nozzle under the action of the pressure of the surrounding liquid. The flow of the liquid is interrupted, the pressure on both sides is balanced, and the membrane returns to its initial position.

Emitters with forced flow interruption are also used. Their design is analogous to gasdynamic emitters.

## 4.3  INTERACTION OF VIBRATION EXCITER WITH OSCILLATORY SYSTEM

### 4.3.1  Basic Features of Vibratory Machine Designs

Now machines of qualitatively new level are needed with low material and energy consumption, high reliability, and low-cost maintenance.

A sharp increase of the effectiveness of the machines can be attained on the basis of systems approach to their optimum design, quality fabrication, and qualified exploitation. The construction of efficient vibratory machines and their rational exploitation is inconceivable without deep understanding of their principles of design and operation.

Experience with design and operation of vibratory equipment has clearly demonstrated that inadequate effectiveness of vibratory machines is often the result of incompetent design and exploitation. The main reason for such a situation is in the fact that vibratory machines are based on different principles than traditional types of machines. Only on the basis of deep understanding of these principles and of operational principles of vibratory machines is it possible to ensure the accelerated introduction of vibration technology to industry.

The radical difference of vibratory machines from the machines of traditional technologies is dependence of formation of the law of motion of the working element on a whole set of force factors acting in the machine. In order to elucidate this situation in more detail we shall compare machines similar in their intended industrial use but different in design, and which are based on traditional principles or on concepts of vibration technology. As an example, let us compare jaw crushers of conventional and vibratory designs.

In a conventional crusher the operating regimes of the crushing jaw are completely determined by parameters of the rigid driving mechanism. At idling and under nominal load, the jaw moves according to the same law. When the crusher is overloaded the rigid kinematics of jaw motion causes an anticipated failure of the safety link.

In the vibratory crusher there is no rigid link between the rotating elements of the drive and the jaw. As a consequence, their relative motion becomes possible. Thus, for example, if the load in the vibratory crusher is so large that the jaw does not move at all under the action of the forces generated by the vibration exciter, then no breakdown takes place, although the drive continues to work. The law of jaw motion in the vibratory crusher is formed under the joint action of the exciting force of the drive and the resistances from the crushed material. In this case, the effect of the process load is not always unique. In some regimes the increase in process resistances shortens the jaw motion and in others, on the contrary, with increasing load jaw oscillations are intensifying and the crushing process progresses more effectively.

It must be born in mind that the load is not the only factor influencing the law of motion of the working element. Under the action of the actuator on the process load with various intensity, the magnitudes of the forces acting in the system also vary. Thus, there exists a close interaction and mutual influence of the working element of the vibratory machine and the process load. The vibratory machine cannot generally be considered separately from the process load and from the energy source. Only consideration of the overall system vibratory machine-load-motor enables one to obtain reliable results.

It is of importance that the closer the loading parameters of the vibratory machine to rated parameters of the energy source, the stronger their interaction. Thus, if the installed power of the motor is close to the mean power consumption by the machine, the effect of interaction between elements of the system is manifested very sharply. In vibratory machines with an energy source of limited power in which the mass of the worked medium is commensurable with the mass of the working element, the interaction of loads is so significant that, in essence, the machine in idling and the machine under load represent systems with totally different properties. As already noted, the correct selection of the load in some regimes can be a factor in increasing effectiveness of the machine operation.

The study of vibratory machines with an energy source of limited power when operating under large loads is needed not only to determine the strength parameters of the machine or more accurate calculation of power consumption, but also to enhance production efficiency of its operation.

To develop a sound and cost-effective design, one must select the installed power of the drive motor in such a manner that it would correspond to the requirements of the machine under load but not exceed them several fold. Overrating the motors of vibratory machines is allowed in some cases to cover design inaccuracies. The drive in a machine designed on the basis of maximum reserve utilization would not have excessive power. This leads to establishment of a regime of strong interaction between the oscillatory system, the load, and the energy source.

Thus, the design of the vibratory machine for real life loading conditions is reduced to consideration of the system— oscillatory unit-load-drive-motor. Character of motion of the working element of the vibratory machine determines the processing regime. Reaction of the worked medium, in its turn, affects oscillations of the working organ. The motor, on one hand, supplies energy to maintain oscillations of the working element and, on the other hand, is influenced by the vibratory action. Under these conditions, the operation of the vibratory machine is determined by the extent and properties of the process load and also by characteristics of the drive (vibration exciter) and the energy source (motor). Effective machines can only be created on the basis of consideration of the entire system.

To describe forces acting on the working element from the process load, the theoretical principles of vibratory technology are used, with application of viscoelastoplastic phenomenological models of the process medium. Application of the vibratory rheology techniques enables development of design methods for various vibratory machines. Combined mechanorheological models also enable reproduction of the structural losses in the machine.

A vibratory machine under load is acted upon by inertial forces, restoring (elastic) forces, resistance forces, and an exciting force generated by the drive. The resistance forces act over the contact areas, having relative displacement, and also in the deformable elements of the medium. These are: internal resistances in the contact zone between the working element with the process load, hysteretic resistances in deformable elastic elements, resistances in kinematic pairs, and structural hysteresis in the machine components. In the majority of vibratory machine designs structural hysteresis is small and usually neglected in engineering calculations.

In vibratory machine design practice, two approaches have been developed. In approximate engineering approaches, the process load is modeled by classical resistances, namely by viscous resistances, by velocity-independent resistances, or by their combinations. However, the weakness of this approach is that elastic and inertial properties of the load are not taken into account. As a result, effective regimes of operation of the system: load-vibratory machine cannot be formulated. In the second approach, developed in the present book, the load is described by phenomenological models reflecting all variety of its inertial and viscoelastoplastic properties. This allows, as will be shown later, establishing extremely effective operating regimes, for example at the resonances of the process load.

In the present section, the first approximate approach is used in order to solve a limited problem of interaction of the vibratory drive with the oscillatory system.

The nature of the resistance forces and the dependence of their magnitudes on motion parameters of the machine, time and behavior of the production process can be most diverse. However, two main types (groups) are identified in the practice of vibratory technology calculations. The first group can include the resistances of plastic deformation or dry friction of the bodies having relative motion at the contact surfaces. These resistances are constant; they are independent of both time and the motion parameters.

The second group includes the resistances that are proportional to the speed of the relative motion or deformation. The resistances of both types are directed opposite to the velocity vector and require constant flow of energy into the system to be overcome.

Usually both forms of resistances are present in the system. Contributions of each of the resistance forms in different specific cases vary considerably. The most realistic loading characteristic of the system can be simulated by the approximate methods by the combination of both types of principal resistances.

The inertial forces acting in the system are proportional to the accelerations of deformations of the medium or the elastic connections. The inertial forces are directed opposite to the acceleration vector.

The restoring (elastic) forces acting in the system are proportional to deformations of the medium or stiffnesses of the elastic elements. Since the elastic deformations are restored, the restoring forces exist during both loading and unloading of the elastic element. The restoring forces are directed opposite to the deformation vector. With periodic motions and deformations, the inertial and restoring forces act in opposite directions.

The characteristics of the restoring forces of both processed load and working elastic connections are usually nonlinear, although they can be assumed linear with sufficient degree of validity for some types of elastic connections. Elastic connections with substantially nonlinear characteristics also find application. This pertains, in the first place, to elastic systems with buffers.

In the majority of types of vibratory machines elastic connections with linear characteristics are used. Working elastic connections with nonlinear characteristics, generated usually with the aid of rigid buffers mounted with clearances, are used only in some heavy vibratory machines working with limited loads.

Operation of such machines in resonance regimes with conventional linear elastic connections would be extremely unstable. Small variations of the load on their working element would cause sharp changes of vibration amplitudes of the working element. When these machines are working with nonlinear elastic connections, a stable operation in conditions of variable loading can be realized due to a more shallow slope of the system amplitude-frequency characteristic.

The exciting forces are generated by drives which are called vibration exciters or vibrators in vibratory machines. As was noted vibration exciters of various fundamental designs are used. Drives of different types are characterized by significantly different parameters and characteristics. Accordingly, vibratory machines with inertial, eccentric, hydraulic, electromagnetic, and pneumatic vibration exciters require special methodical approaches during design process. Vibratory machines with different types of vibration exciters have considerably different properties. The contribution of the vibration exciter to the behavior of the system: load-machine-drive-energy source is a determining factor. Therefore, vibratory machines are usually subdivided accordingly: inertial, eccentric, electromagnetic, and so forth.

The energy to the vibratory machine is supplied by a motor, usually by an electric induction motor, and in some cases by electric motors of other types. Also used are compressed air and hydraulic drives of various types. However, induction motor drive is the most widely used. Its parameters are described either by the static characteristic or by equations connecting electrical and mechanical parameters.

Since the principal structural and operational characteristics of the vibratory machine are determined by the type of drive, the machines will henceforth be classified accordingly. The type of resistances acting in the structural elements and on the working element will be the next parameter considered. Analysis is performed for the cases of using a motor with large and limited power.

Interaction of a vibratory drive with the oscillatory system of the machine are considered for inertial and eccentric drives, since they have wide application in industry. The purpose of the present section is to consider only the fundamental features of vibratory machines of these two types. Therefore, the study is conducted on the simplest devices modelled by one-mass systems. This is necessary in order that the designer of the vibratory machine using multicriterial optimization technique would understand the fundamentals of this technology and would be able to competently interact with the computer.

The study is conducted for cases of both probable types of resistances: constant velocity-independent resistances and viscous resistances that are proportional to velocity. Considered are the effects of each type of resistance and their combinations.

### 4.3.2 One-mass Oscillatory System with an Inertial Vibration Exciter of the Unbalanced-mass Type

### 4.3.2.1 Machine Schematic and Acting Forces

The inertial one-mass vibratory machine consists of a working element (mass $M$) mounted on elastic elements that are characterized by stiffness coefficients $k_x$ and $k_y$ (Fig. 4.18). The working element of the vibration exciter of the unbalanced-mass type is an unbalanced mass $m$ with its center of mass at distance $r$ from the axis around which it rotates with angular velocity $\omega$. The shaft of the unbalanced mass is at the center of mass of the working element.

The working element of the vibratory machine is subjected to the following forces. From elastic elements there are restoring forces that are proportional to stiffnesses $k_x$ $k_y$ and to deformation magnitudes $x$ and $y$, and are directed opposite to the deformation vector, hence the negative signs $k_x x$, $-k_1 y$. The forces in the structural elements and originating from the process load are resistances $F_x$, $F_y$ which are independent of the motion parameters, and resistances proportional to velocity components $\dot{x}$ and $\dot{y}$ of the vibratory machine with the resistance coefficients $c_x$ and $c_y$. Both forms of resistances are directed opposite to velocity vectors and, therefore, have an

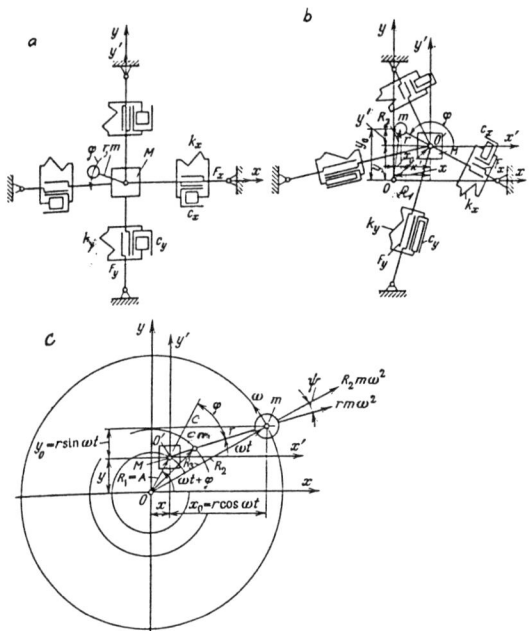

**Figure 4.18** One-mass vibratory machine with inertial drive of the unbalanced-mass type. a) schematic diagram of the machine under load; b) design layout; c) trajectory of mass motion.

opposite sign to velocity sign$(-\dot{x})$ and sign$(-\dot{y})$: sign$(-\dot{x})F_{cx}$ sign$(-\dot{y})F_{cy}$ $-c_x\dot{x}$, $-c_y\dot{y}$.

During the vibratory process, inertial forces are generated that are proportional to mass $M$ and to accelerations $\ddot{x}$ and $\ddot{y}$ of the working element. The inertial forces are directed opposite to the acceleration vectors (have opposite signs): $-M\ddot{x}$, $-M\ddot{y}$. Upon rotation of the unbalanced mass, a centrifugal force is generated which is proportional to its mass $m$, to square of angular velocity $\omega^2$, and to magnitude of position vector $r_m$ of the unbalanced mass in the stationary system of the coordinate axes $yox$, i.e., $mr_m\omega^2$.

The interaction of these forces determines kinematic, dynamic, and force characteristics of the machine. When the vibrator is not working, the vibratory machine is subjected to a gravitational force that is proportional to its total mass $M + m$, $-(M + m)g$. This force deforms the elastic elements, resulting in a restoring force equal in magnitude and directed opposite to the gravity force. Thus, static deformation of the elastic elements is $y = (M+m)/k_y \; g$.

For convenience we shall use stationary coordinate frame $xoy$, whose origin $o$ is coincident with the center of mass of the working element of the vibratory machine $M$ when it is static equilibrium under the gravity force and the restoring force of elastic elements. Axes $x$ and $y$ are directed along directions of extremal stiffnesses of the elastic elements (in our case horizontally and vertically). When the vibratory machine is working, the mass of the working element will deviate from the origin of the stationary coordinate system. Its position in space can be described knowing coordinates $x$ and $y$ of mass $M$ in the stationary coordinate frame.

Position of the center of mass of the unbalanced mass $m$ is convenient to describe in the coordinate frame $x'o'y'$, whose origin $o$ is coincident with the axis of rotation of the unbalanced mass. Axes $x'$ and $y'$ are parallel to axes $x$ and $y$. In the inoperative machine, the origins of both coordinate frames coincide. When the machine is operating, the coordinate frame $x'o'y'$ moves with mass $M$; therefore, it is called the moving coordinate frame. In the moving frame, position of the center of mass of the unbalanced mass is defined by coordinates $x$ and $y$. The position of the center of mass of the unbalanced mass can also be described by the angle of its rotation $\varphi$, which will be counted from axis $ox$ in counterclockwise direction.

Displacements of the center of mass of the unbalanced mass in the stationary frame is determined by expressions $x_0 = x + r\cos\varphi$ and $y_0 = y + r\sin\varphi$, velocities as $\dot{x}_0 = \dot{x} - z\dot{\varphi}\sin\varphi$ and $\dot{y}_0 = \dot{y} + z\dot{\varphi}\cos\varphi$ and accelerations as $\ddot{x}_0 = \ddot{x} - z\ddot{\varphi}\sin\varphi - z\dot{\varphi}^2\cos\varphi$ and $\ddot{\varphi}_0 = \ddot{y} + z\ddot{\varphi}\cos\varphi - z\dot{\varphi}^2\sin\varphi$.

In addition, in vibratory machines frictional forces also act in the vibrator bearings. These are proportional to the projection of the exciting force, which is generated upon rotation of the unbalanced masses in the stationary frame, onto the radius of the unbalanced mass $F_6 = m(-\ddot{x}\cos\varphi - \ddot{y}\sin\varphi + z\dot{\varphi}^2)$. This force creates a resistance torque on the vibrator shaft, which is proportional to the radius of friction circle $\rho$, which depends on the effective radius of the bearings $D/2$ and coefficient of friction in bearings, $fM_b = \rho m(z\dot{\varphi}^2\ddot{x}\cos\varphi - \ddot{y}\sin\varphi)$. The resistances on the vibrator shaft are overcome by the moment $M(\varphi)$ of the motor.

The steady-state motion of the vibratory machine when acceleration $\ddot{\varphi} = 0$ and the shaft rotates with constant angular velocity $\omega$ is described by the set of nonlinear differential equations

$$(M + m)\ddot{x} + k_x x = mr\omega^2 \cos\omega t + F_x$$

$$(M + m)\ddot{y} + k_y y = mr\omega^2 \sin\omega t + F_y$$

In these equations the first terms represent the projections on the corresponding axes of inertial forces of the total oscillating mass of the machine. The second terms represent restoring forces of elastic connections. The third and fourth terms are projections of the exciting force, generated by the rotating unbalanced mass in a stationary vibratory machine, and the resistance forces acting in the elastic connections and on the working element of the vibratory machine. Let us determine respectively motion parameters of the vibratory machine when viscous resistances, constant resistances, or both types of resistances simultaneously are present.

### 4.3.2.2 Parameters of Vibratory Machines with Resistances Proportional to Velocity

With resistances that are proportional to velocity, the time history of the working element motion will be $x = A_x \cos(\omega t + \varphi_x)$ and $y = A_y \sin(\omega t + \varphi_y)$, where the components of the amplitude of the forced oscillations are equal to

$$A_x = \frac{qr\omega^2}{\sqrt{4n_x^2 \omega^2 + (p_x^2 - \omega^2)^2}}$$

$$A_y = \frac{qr\omega^2}{\sqrt{4n_y^2 \omega^2 + (p_y^2 - \omega^2)^2}}$$

and the phase shifts between the displacements of the working element and the displacements of the unbalanced masses of the vibrator are

$$\varphi_x = \text{arctg} \frac{2n_x \omega}{p_x^2 - \omega^2}$$

$$\varphi_y = \text{arctg} \frac{2n_y \omega}{p_y^2 - \omega^2}$$

The following notations are adopted in these formulas: $n_x$, $n_y$ are reduced coefficients of the viscous resistances acting on the vibratory machine along the corresponding axis

$$n_x = \frac{c_x}{2(M + m)}, \quad n_y = \frac{c_y}{2(M + m)}$$

$p_x$, $p_y$ are the natural frequency of the vibratory machine on elastic elements in the corresponding direction

$$p_x = \frac{k_x}{M + m}, \quad p_y = \frac{k_y}{M + m}$$

$q$ is the ratio of rotating and total masses of the vibratory machine

$$q = \frac{m}{M + m}$$

It follows from the presented expresions that the displacement amplitude of the working element of the vibratory machine is proportional to the magnitude of exciting force of the stationary vibrator, inversely proportional to the magnitude of the total mass of the vibratory machine, and also considerably dependent on detuning of the oscillatory system, i.e., ratio of natural and forcing frequencies $\omega/p_x$, $\omega/p_y$. If $p_x > \omega$ and $p_y > \omega$, the regime is called subresonant; if $p_x < \omega$ and $p_y > \omega$, the regime is interresonant; when $p_x = \omega$ and $p_y = \omega$, resonance occurs in the direction of the corresponding axis; if $p_x < \omega$ and $p_y < \omega$, the regime is called transresonant. From the expressions for the components of the amplitude it is evident that the second term vanishes in the case of resonance and the displacement amplitude becomes maximum. In the resonant regime, the amplitude is strongly dependent on the magnitude of the resistances acting in the system. Increasingly the resistances leads to a decrease of the amplitude of oscillations. These properties of the considered system have great influence on the operational regimes of the vibratory system. Resonant vibratory machines under variable loads are extremely unstable in operation unless special measures are taken. Transresonant machines are characterized by high operating stability at high level variable loads which mainly determine magnitudes of resistances acting in the system.

Resonant vibratory machines can also work under conditions of large variable loads if they are equipped with an automatic control system for kinetic moment of the unbalanced mass. When the load is increased, the kinetic moment of the unbalanced mass must automatically be increased so that the magnitude of the oscillation amplitude remains unchanged.

From correlation between displacement of the main mass of the vibratory machine and rotational velocity of the unbalanced mass, it follows that large displacements take place not only when the natural and forced oscillation frequencies coincide, but in a frequency range. The boundaries of this range are rotational velocities of the unbalanced masses $\omega_1 = \sqrt{k/M+2m}$ and $\omega_2 = \sqrt{k/m}$ at which the displacements of the main mass are equal to the eccentricity of the unbalanced mass, $A = r$. It is interesting to note that the resonant range is the wider, the higher the natural frequencies of the system.

These boundaries are established for operation of the vibratory machine without load. Under load the relationships are somewhat different. However, the substance remains unchanged. With decreasing rotational speed of the unbalanced mass, the displacements of the working mass of the vibratory machine are decreasing, and at very high speeds they are proportional to the eccentricity of the unbalanced mass and to the ratio between rotating and total masses of the vibratory machine, and depend little on the magnitude of the acting load. This allows an important practical conclusion to be drawn. To ensure high stability of operation of resonant vibratory machines, they must be designed for high-frequency regimes.

The operating regimes of a vibratory machine without load at the borderline angular velocities have some characteristic properties and, therefore, deserve special attention. At the lower frequency boundary, the unbalanced mass moves in a circle whose radius is equal to the double eccentricity of the unbalanced mass (Fig. 4.19). In this case the centrifugal force generated by the unbalanced mass is twice as large as the centrifugal force of the unbalanced mass whose axis is stationary.

When the vibratory machine operates at the upper boundary, the unbalanced mass moves around a circle of zero radius. In this case the unbalanced mass rotates in absolute motion relative to its own center of mass, and the working mass of the machine moves along circle whose radius is equal to the eccentricity of the unbalanced mass, and whose center coincides with the center of mass of the unbalanced mass. It

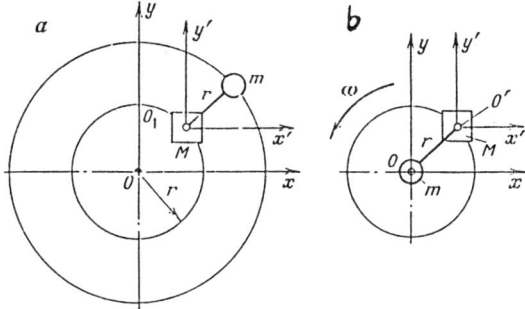

**Figure 4.19** Trajectories of motion at idling of a one-mass vibratory machine with unbalanced mass operating at the borderline angular velocities. a) at angular velocity $\omega_1$ = $k/M+2m$; b) at angular velocity $\omega_2$ = $k/M$.

is interesting to note that the centrifugal force developed by the unbalanced mass in this regime is equal to zero. Thus, in this idealized case, the centrifugal force does not affect the motion of the main mass of the vibratory machine. Here synchronous-like motion of two independent masses appears to take place: rotary motion of the unbalanced mass relative to its own center of mass and translational circular motion of the main mass of the vibratory machine under the influence of a central attractive force. Connection of the main mass with the unbalanced mass plays only a synchronizing role.

Another characteristic property of this regime is change of the natural frequency of the system. The unbalanced mass does not participate in the oscillations; therefore, it cannot affect the natural frequency of the system. Thus, this regime is also a resonant regime, since the natural and forcing frequencies coincide. These regimes can be realized only when the vibratory machine is idling. When operating with load, resistance forces cause some deviation from the considered pattern of behavior.

The expressions for the phase shifts show a possibility of displacements of the unbalanced mass of the vibrator and the working mass of the vibratory machine to be out of phase. In actual vibratory machines, regimes in which there is no phase shift between the displacements of the vibratory machine and the vibrator are practically not realized. However, with subresonant tuning of the vibratory machine, the phase shift can be small. At resonance, irrespective of magnitudes of the acting resistances, the phase shift is $90^\circ$, and at extreme transresonant regimes it reaches $180^\circ$, i.e., the unbalanced mass and the working mass of the vibratory machine in this case move in opposite directions.

Let us consider in more detail interaction of the main elements of the vibratory machine: the working element, elastic connections and the unbalanced mass. This is extremely important in order to ensure the correct formulation of the problem of the subsequent optimal design of the machine. It is established that when the vibratory machine operates in a subresonant regime far removed from the resonant regime, the working mass and the unbalanced mass of the vibrator are moving along concentric circles. The working mass moves along the smaller circle and the unbalanced mass along the larger one. They are practically located on a straight line, originating from the center of the concentric circles, and are shifted to one side. If the centers of the working mass and of the unbalanced mass are connected by lines originating from the center of the concentric circles, a certain angle exists between them. This angle characterizes a displacement or a phase shift between the displacement of the main mass of the machine and direction of the driving centrifugal force which is usually referred to as the exciting force. Turning to the graphs of Fig. 4.20a we note that in the subresonant regime the angle of phase shift between the unbalanced mass and the working mass of the machine is equal to zero.

The second important feature of operation in the subresonant regime is the fact that the actual radius vector of rotation of the unbalanced mass in the stationary coordinate system is larger than its eccentricity. It is larger by the amount of amplitude of the working element.

If the rotational speed of the vibrator is increased gradually, the unbalanced mass rotates with respect to the main mass. With further increase of the rotational speed of the unbalanced mass, the phase shift is increasing. When the rotational speed of the unbalanced mass becomes exactly correspondent with the natural frequency of the vibratory system, the phase shift becomes $90°$ (Fig. 4.20b). Thus, in the resonant regime the exciting force is directed perpendicular to the displacement of its working element.

Further increase of the rotational speed of the unbalanced mass and the transition of the machine to the transresonant regime of operation is accompanied by the increase of angle of phase shift between the displacement and direction of the exciting force (Fig. 4.20c). When the rotational speed of the unbalanced mass considerably exceeds the frequency of natural

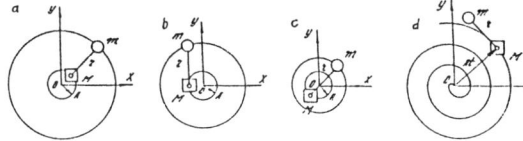

**Figure 4.20** Motion trajectory of one-mass vibratory machine with isotropic elastic connections and unbalanced mass of the vibration exciter. a) in the subresonant regime; b) in the steady-state resonant regime; c) in the transresonant regime; d) under acceleration in the resonant regime.

oscillations of the machine, such a regime of operation is referred to as deep transresonant regime; the displacement of the main mass of the machine occurs in a direction almost opposite to the direction of the exciting force. Furthermore, the less resistances act in the machine, the closer is the phase shift to 180° in the transresonant regime.

It is interesting to follow the effect of resistances on the phase shift in various regimes. In the subresonant regime increasing the resistance leads to an increase in phase shift. At resonance the resistances do not affect the phase shift, the phase angle always remaining equal to 90°. In the transresonant regime increasing resistances lead to decreasing phase shift.

Since the phase shift betwen the position of the unbalanced mass and the working mass of the machine characterizes the machine ability to do useful work, one can come to the conclusion that specifics of the considered regimes play a decisive role in selection of the type of machine for various application conditions.

Accordingly, particular attention should be given to the elucidation of the reasons causing change of the mutual positions of the main mass of the vibratory machine and the unbalanced mass and of the phase shift between the displacement of the working element and of the unbalanced mass. The main oscillating mass of the vibratory machine is acted upon by the elastic forces of springs, inertial forces, and resistance forces. The centrifugal force generated when the unbalanced mass rotates is used to overcome these forces and to ensure motion of the vibratory machine. Magnitudes and directions of the resultants of the first three forces vary depending on operating regimes. Consequently, the exciting force (in our case the centrifugal force) overcoming them must also vary in magnitude and direction. These variations are expressed in the variation of the actual eccentricity of the unbalanced mass and the phase shift with respect to the displacement of the main mass.

The restoring force is proportional to displacement of the main mass relative to its static position and is directed in the opposite direction; the inertial force is proportional to the acceleration and directed in the displacement direction. Thus, these two forces are mutually opposed and compensate each other. The resistance force acts perpendicular to them.

Let us now consider interaction of these forces with the exciting force. The vibratory machine of the subresonant type is characterized by high stiffness of the springs. Therefore, when large restoring forces are developing during its operation, they significantly exceed the inertia forces in the system. These two forces are directed opposite each other; therefore the inertial force partially compensates for the restoring forces. The difference of these forces, whose magnitude can be very significant, acts in the system. Consequently, the exciting force must have a component overcoming the excess restoring force. In addition, the exciting force has a component overcoming the resistance forces. These two components determine the magnitude and direction of action of the exciting force.

The vibratory machine of the transresonant type has soft springs. Therefore, the restoring forces acting in it are insignificant and cannot totally compensate for the inertia forces. Therefore, the exciting force must have, in addition to the component overcoming the resistance forces, a component acting against the inertial forces.

In the resonant vibratory machine working at the natural frequency, inertial forces are totally compensated by the restoring forces. Hence, the exciting force must overcome only the resistance forces. Consequently, the force is smallest in magnitude in the resonant regime and is directed opposite to the resistance forces, i.e., under viscous resistances it coincides with the direction of velocity.

If a vibratory machine which was previously operating in the subresonant or transresonant regimes needs to be changed over to the resonant operating regime by means of appropriate adjustment, then there is no need any longer to overcome restoring or inertial forces. The exciting force turns out to be larger than required for overcoming the resistance forces acting in the system. The excess exciting force would sway the vibratory machine increasing amplitude of the main mass until the growing resistance forces would balance the exciting force. The vibratory machine changes over again to the stationary operating regime which will, however, be characterized by the growing displacements of the working mass. This is a manifestation of the specific features of the vibratory machine in the resonant regime.

It must again be emphasized that in the resonant regime the amplitude is established and limited only by the acting resistances. The higher the resistances, the smaller the amplitudes. Therefore, if in the process of operation of the resonant vibratory machine, the oscillations of the working element begin to decrease due to rising resistances, one must increase the exciting force magnitude up to the values necessary to maintain the specified amplitude.

We have considered the vibratory machine whose working element is suspended on elastic connections of equal stiffness. However, a machine can have, for example, vertical elastic connections that are considerably stiffer than the horizontal ones. Then displacements of the working element of the vibratory machine in the directions of these connections will be different. It was shown that the stiffnesses of the elastic connections exert different effect on the displacement of the working organ in different operating regimes. Thus, if the machine is tuned at subresonant operating regime, then the working organ would have large displacements in the direction of the elastic connection having lower stiffness. With transresonant tuning everything will be reversed – large displacements will be in the direction of the stiffer elastic link. This is due to the fact that the restoring forces of the elastic links in subresonant regimes are large and are not balanced by the inertial forces; therefore, the vibration exciter is forced to overcome these excess forces. It is obvious that the softer connection hinders the motion less; therefore, the displacement is greater in its direction.

In the transresonant regime the inertial forces predominate, and the exciting force acts against them. The elastic connections assist in overcoming these forces; naturally, the stiffer they are, the less resistance the inertial forces exert against the displacement of the vibratory machine.

Different displacements of the working organ in different directions result in a variation of the shape of motion trajectory. Instead of a circle an ellipse is developing.

Moreover, the configuration of the ellipse varies with changing resistance and stiffness of the elastic connections.

Thus, by placing elastic links of different stiffness in different directions, one can change the shape of motion trajectory of the working part of the vibratory machine. This is important from the practical viewpoint, since it enables generation of various operating regimes of the vibratory machine with simple unbalanced-mass drive.

In the subresonant regime, the working mass of the vibratory machine and unbalanced mass move along ellipses whose axes coincide. The working part travels along the inner ellipse and the unbalanced mass along the outer ellipse (Fig. 4.21a). The characteristic feature of the relative motion of the working mass of the vibratory machine and the unbalanced mass in the subresonant regime is the fact that the direction of the radius of the unbalanced mass does not pass through the origin of the coordinates except for the cases when both masses are located on the major axes of the ellipses. The working mass of the vibratory machine and the unbalanced mass during the process of the entire motion are in the same quadrant with respect to the coordinate axes. Despite the fact that position vectors of the working and unbalanced masses traverse different angles over equal time intervals, the radius of the unbalanced mass turns by equal angles over the same time intervals. Since distance along the path of motion between successive positions of the working mass is different, its absolute velocity varies not only in direction but also in magnitude. When the working mass or the unbalanced mass are located on the principal axes of the ellipse, magnitudes of these velocities acquire extreme values. When they are located on the minor axis of the ellipse the velocity is maximum, and when located on the major axis the velocity is minimum.

In the transresonant regime the picture changes sharply. The motion also takes place along ellipses; however, their axes are mutually perpendicular and the masses are on different sides of the origin of the coordinate system (fig. 4.21b). In the transresonant operating regime, the major axes of the elliptical paths of the working and unbalanced masses are positioned perpendicular with respect to each other. The working and unbalanced masses in steady-state motion are positioned in opposite quadrants. The working and unbalanced masses are located simultaneously on the principal axes of the elliptical trajectory but on opposite sides from the coordinate origin.

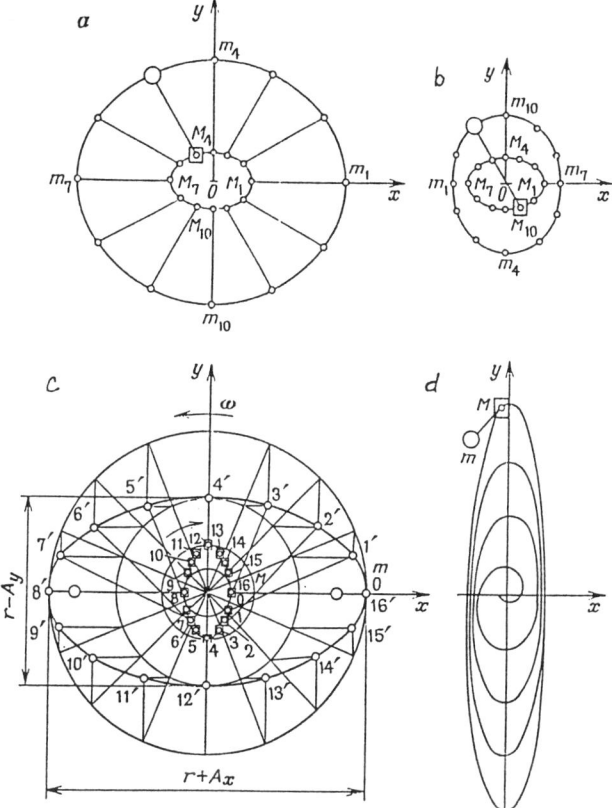

**Figure 4.21** Trajectories of motion of a vibratory machine with anisotropic elastic links and of the unbalanced mass of the vibration exciter in the regime of a) subresonance; b) transresonance; c) interresonance; d) resonance with acceleration.

The most diverse trajectories can be obtained in this case when stiffnesses of the elastic connections are selected in such a way that the natural frequency in one direction is lower and in the perpendicular direction higher than the forced frequency, i.e., there is transresonant tuning in one direction, and subresonant tuning in the other direction. This regime received the name intermediate or interresonant regime.

In the intermediate regime, when $k_x = k_y$ and $p_x > \omega > p_y$, the trajectories of the working mass and of the unbalanced mass will be ellipses, and the half-axes of the unbalanced mass trajectory are equal to $a_x = A_x + r$ and $a_y = a_y - r$ (Fig.

4.21c). In this case the trajectory is elongated along axis $X$. When plotting trajectories of both masses and determining their mutual location, the parametric angle of trajectory of the working mass must be laid off in the direction opposite to the rotation of the rotor. This is due to the fact that in the intermediate regime the working mass of the vibratory machine and the unbalanced mass move in opposite directions.

The intermediate regime of operation of the vibratory machine is characterized by the greatest diversity of trajectories of motion of the working and unbalanced masses and of their mutual position during motion. The regime presents big opportunities for synthesis of the effective processing regimes; however, it requires that special measures be taken to ensure the effective operation of the vibratory machine.

When resistance forces proportional to velocity act on the vibratory machine, the working and unbalanced masses will move along elliptical trajectories whose principal axes do not coincide with the coordinate axes. The acting resistances in the system seem to rotate the trajectory of the working mass relative to the coordinate axes. Since the production process effectiveness depends largely on the trajectory of the working mass, this should be considered for the optimum design of the machine.

Owing to the fact that, as we already know, the parameters of motion of the working mass of the vibratory machine affect the laws of motion of the unbalanced mass, its trajectory also becomes elliptical. In this case, the mutual position of the working and unbalanced masses is determined by the phase shift between them and, hence, by the operating regime of the vibratory machine and the magnitude of the resistance forces. In this case the machine is tuned for the intermediate (interresonant) regime when the detuning factors in the direction of extremal stiffnesses of the elastic connections are different, the configurations of the elliptical trajectories and their mutual position can be very diverse in each specific case. The trajectories of motion of the working and unbalanced masses for the interresonant regime $p_x > \omega > p_y$ are given in Fig. 4.21c.

If the natural and forced frequencies coincide, $p_y = \omega$, in the direction of one of the axes, resonant amplification of the oscillation takes place. The transient regime of establishing the resonant oscillations in the direction of one of the elastic connections of the vibratory machine is presented in Fig. 4.21d.

Trajectory of the working mass during acceleration of the vibratory machine with isotropic links in the resonant regime is shown in Fig. 4.21d.

From the standpoint of providing effective execution of the specified production process, the form of the trajectory of motion of the working element of the vibratory machine is of considerable interest. Analysis shows that, generally, trajectory of motion of the working element is an ellipse which can, depending on such parameters as the stiffnesses of the elastic connections and frequency, vary from circular to linear. The production process with elliptical and circular trajectories is significantly affected by the direction of trajectory following which is determined by direction of rotation of the unbalanced mass. The angle of inclination of the principal axis of the elliptical trajectory to axis $x$ and the ratio of the major and minor axes depend on angle $\gamma = (90^{\circ} + \varphi_x - \varphi_y)$ and are determined by expressions

$$\text{tg} 2\beta = \frac{A_x A_y}{A_x^2 + A_y^2} \cdot \cos \gamma$$

$$\frac{a}{b} = \sqrt{\frac{A_x^2 + A_y^2 + \sqrt{(A_x^2 + A_y^2) - 4A_x A_y \cdot \sin^2 \gamma}}{A_x^2 + A_y^2 + \sqrt{(A_x^2 + A_y^2)^2 - 4A_x A_y \cdot \cos^2 \gamma}}}$$

The direction of traversing the elliptical trajectory, which is extremely important in many practical cases, is obtained without changing its configuration by changing the phase angle from $+\gamma$ to $-\gamma$.

Another very important problem for optimum design of the vibratory machine is study of the factors influencing magnitude of the exciting force on the vibratory machine. It might appear that this is simple, and if the vibrator parameters are known, then it would not be difficult to compute the generated centrifugal force. However, this is not the case at all. It was established earlier that radius vector of rotation of the unbalanced mass is dependent not only on the structural parameters of the vibrator, but also on the regime and parameters of oscillations of the part of the vibratory machine on which the vibrator is mounted. Thus, we have established that the oscillation regime of the vibratory machine directly affects the magnitude of the exciting force generated by the vibrator. Let us determine the way the magnitude of the actual radius vector of the unbalanced mass depends on the parameters of oscillations and tuning of the vibratory machine.

Since during operation of the vibratory machine the axis of shaft rotation of the unbalanced mass is displaced in the stationary coordinate system by $x$, $y$ with acceleration $\ddot{x}$, $\ddot{y}$, and the center of the unbalanced mass performs a complicated motion (relative rotation with an angular velocity around its own axis and transfer motion together with the oscillating mass), the exciting force acting on the oscillating mass is dependent on the acceleration of the transfer motion.

The working vibratory machine is acted upon by the component of the exciting force

$$F_x = m(r\omega^2 \cos \omega t - \ddot{x}) = P_x \sin(\omega t - \psi_x)$$

$$F_y = m(r\omega^2 \sin \omega t - \ddot{y}) = P_y \cos(\omega t - \psi_y)$$

where the amplitude values of the components of the exciting force are equal to

$$P_x = mr\omega^2 \sqrt{\frac{4n_x^2\omega^2 + \left[p_x^2 - (1-q)\omega^2\right]^2}{4n_x^2\omega^2 + (p_x^2 - \omega^2)^2}}$$

$$P_y = mr\omega^2 \sqrt{\frac{4n_y^2\omega^2 + \left[p_y^2 - (1-q)\omega^2\right]^2}{4n_x^2\omega^2 + (p_y^2 - \omega^2)^2}}$$

and the phase angles between displacements of the unbalanced mass of the vibrator $x_0$ and $y_0$ and the components of the exciting force $F_x$ and $F_y$ are

$$\psi_x = \text{arctg} \frac{2qn_x \, \omega^3}{4n_x^2\omega^2 + \left[p_x^2 - (1-q)\omega^2\right](p_x^2 - \omega^2)}$$

$$\psi_y = \text{arctg} \frac{2qn_y \, \omega^3}{4n_y^2\omega^2 + \left[p_x^2 - (1-q)\omega^2\right](p_x^2 - \omega^2)}$$

From these expressions it follows that the amplitude values of the components of the dimensionless force in the regime of steady-state oscillations of the vibratory machine, just as in a stationary vibrator, are proportional to the centrifugal force of the unbalanced mass $mr\omega^2$, but also dependent on the tuning (coefficients of detuning $z_x = \omega/p_x$, $z_y = \omega/p_y$) and on the ratio $q$ of the rotating and total masses of the vibratory machine and also on the viscous resistances acting in the system which are characterized by the damping factors $v_x = n_x/p_x$, $v_y = n_y/p_y$.

The direction of the exciting force does not coincide with the position of the unbalanced masses, and is determined by the parameters and tuning of the vibratory machine.

The exciting force reaches maximum when the vibratory machine operates in the resonant regime. This is explained by the fact that in this regime amplitude of the working element is maximal and, consequently, the radius vector of rotation of the unbalanced mass is also maximum. In the subresonant regime the displacements of the working element are smaller than in the resonant regime, but since they are summed up (due to acute phase angle between displacements of the working element and the unbalanced mass), the exciting force acting in the vibratory system is greater than the exciting force generated by the stationary vibrator. In the transresonant regime the working mass of the vibratory machine and the unbalanced mass are turned to opposite sides (the phase angle is obtuse); therefore, the radius vector of the unbalanced mass decreases and the exciting force acting on the vibratory machine is always less than the exciting force generated by the stationary vibrator.

Attention should be paid to important features of correlation between magnitude of the exciting force and tuning of the vibratory machine and the acting resistance forces. In the resonant regime the exciting force increases at a faster rate if the resistance forces acting on the vibratory machine are smaller. Moreover, decreasing of the resistances leads to a sharp rise of the exciting force. The same behavior is pronounced in the subresonant and near-resonant regimes. In the transresonant regime an opposite dependence takes place – the exciting force increases with increasing resistances.

On the basis of the above stated, an important practical conclusion can be drawn. The subresonant, near-resonant, and, in particular, transresonant regimes of the vibratory machines operation are characterized by insufficient stability. Increasing load leads to decreasing exciting force and, consequently, to disruption of the oscillation regime. In the transresonant regime, the increase of load on the vibratory machine is accompanied by a rise in the exciting force. Such a characteristic of the vibratory machine in the transresonant regime provides stable operation under large variable loads.

For optimum design it is important to know not only the considered features of formation of the exciting force, but also information on the required specific value of the exciting force, which has to be applied to the vibratory machine in various tuning regimes and loads. Analysis shows that the unit exciting force changes sharply in different operating regimes of the vibratory machine. In the subresonant regime, to ensure the same parameters of oscillations much larger forces than in the transresonant regime are needed. Minimum forces for imparting oscillations to the system are required in resonance. Increasing the resistances in the system in any operating regime of the vibratory machine causes an increase of the unit exciting force. However, in the subresonant and near-resonant regimes, the exciting force increases with increasing resistances more rapidly than in the transresonant regime.

Hence the conclusion follows that under large loads the transresonant operating regime of the vibratory machine with an inertial drive is very effective. From the standpoint of reduction of the exciting force, the resonant regime has an advantage only at small and limited loads.

The exciting forces generated with the rotation of the unbalanced mass are transmitted to the vibratory machine via the vibrator bearings. Since these forces are very large, the reliability of the machine with an inertial drive is largely determined by information on the operating conditions of the bearings and by correct selection of the type and size of bearings.

Let us consider the conditions of operation of the bearings units in the unbalanced-mass vibrator. First of all, let us determine which forces act on the bearings. Having some experience of determining the magnitude of the exciting force, we shall not assert that the bearings are loaded by the centrifugal force generated by the rotating unbalanced mass of the stationary vibrator. But for conducting bearings calculations, the exciting force determined above also cannot be taken as the design load.

Let us determine the load acting on the bearings by analyzing how it is formed in the operating vibratory machine.

The pressure on the bearings is equal to the unbalanced mass multiplied by the acceleration of its center of mass in the stationary coordinate frame. The projections of the acceleration of the center of the unbalanced mass on the stationary coordinate axes are determined from double differentiation of projections of the displacement of the center of the unbalanced mass in the stationary coordinate frame. As follows from the layout of the vibratory machine presented in Fig. 4.19, the projections of the displacements of the center of the unbalanced mass in the stationary coordinate system of the axes are equal to

$$x_0 = x + z \cos \varphi$$

$$y_0 = y + z \sin \varphi$$

Therefore, the projections of accelerations of the center of the unbalanced mass on the axes of the stationary coordinate frame are

$$\ddot{x}_0 = \ddot{x} - z\ddot{\varphi} \sin \varphi - z\varphi^2 \cos \varphi$$

$$\ddot{y}_0 = \ddot{y} - z\ddot{\varphi} \cos \varphi - z\varphi^2 \sin \varphi$$

By projecting acceleration of the center of the unbalanced mass on its radius and multiplying the obtained expression by the unbalanced mass, we determine the load on the bearings

$$F_b = -m \ (\ddot{x} \cos \varphi + \ddot{y} \sin \varphi - z\dot{\varphi}^2)$$

Substituting into the obtained expression the values of the corresponding accelerations, we determine pressures in the bearings of the vibrator. When the vibratory machine oscillates along a circular trajectory, i.e., when the major and minor half-axes of the ellipse are identical $A_x = A_y = A$, pressure on the bearing units can be determined from expression

$$F_b = mr\omega^2 \left[ 1 + \frac{q\omega^2(p^2 - \omega^2)}{4n^2\omega^2 + (p^2 - \omega^2)^2} \right]$$

or from the graphs of Fig. 4.22. By analyzing the presented dependence it becomes evident that the loads in the bearing units of the vibrator are also substantially dependent on the

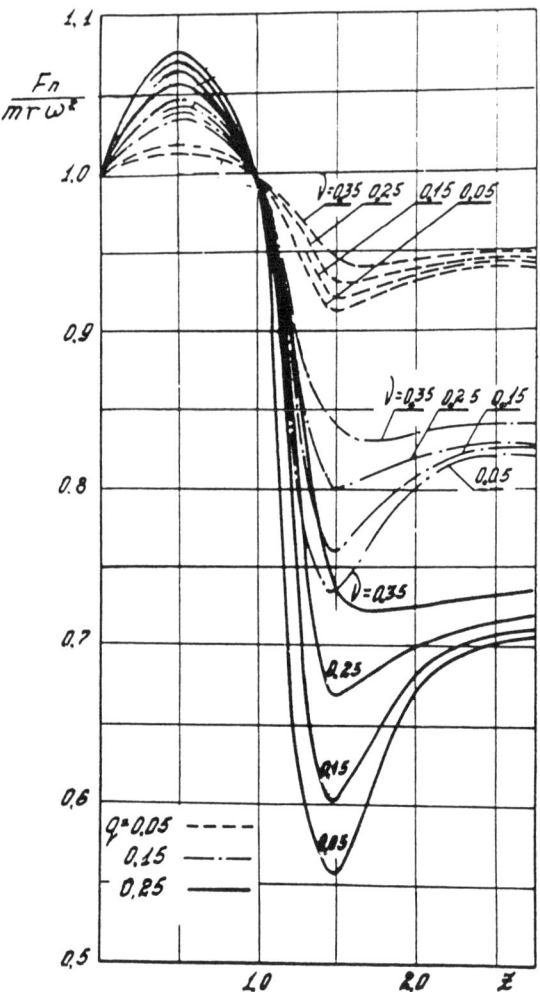

**Figure 4.22** Dimensionless loads on bearing units of the unbalanced mass of a vibration exciter of one-mass vibratory machine for different detuning $z$ and damping $\nu$ factors.

operating regime and resistances acting in the vibratory machine. When the working element of the vibratory machine is in motion along a circular trajectory, equal loads that are constant in time act on the bearings in all directions. The magnitude of the load on the bearings is determined by the centrifugal force of the unbalanced mass relative to the vibrator axis and depends on the tuning of the oscillatory system of the vibratory machine. In the resonant regime the value of the resistances does not influence the load magnitude. The forces acting on the bearings for the resonant tuning are the same as in a stationary vibrator.

In the subresonant regime, the loads on the bearings of the vibrator are increasing, reaching a maximum value at detuning factors of 0.7 - 0.75. In the transresonant regime, loads on the bearings decrease, especially at detuning factors of 1.35 - 1.50. The loads on the bearings decrease most substantially in transresonant regimes at low resistances and with the increasing ratio between the value of the unbalanced mass and the total mass of the vibratory machine.

In the subresonant regime an opposite picture is observed. Increasing the resistances and the ratios of the rotating and total masses of the vibratory machine leads to increasing loads on the bearing units. This characteristic property of vibratory machines with inertial drive must be taken into account by the designer for multicriterial design optimization.

For the resonant tuning of the machine, loads in the bearings are determined only by the magnitude of the centrifugal force generated during the rotation of the unbalanced mass. The magnitude of the loads of the bearings of the vibrator in the resonant regime is not influenced by either the resistances acting in the system or by the ratio of the rotating and total masses of the machine. In the resonant regime, the loads in the vibrator bearings and, hence, the effort involved in overcoming them, are lower than in the subresonant regime but higher than in the transresonant regime. This is an indication that the opinion held among designers that resonant machines are more energy efficiency is not substantiated with respect to machines with inertial drive.

In view of the above, the expression for determining the loads in the bearings units of vibratory machines is convenient to represent as a coefficient of variation of bearings loading in the subresonant and transresonant regimes compared with the resonant regime

$$\frac{1}{q}\left(\frac{F_b}{mr\omega^2} - 1\right) = \frac{\omega^2(p^2 - \omega^2)}{4n^2\omega^2 - (p^2 - \omega^2)^2}$$

This expression and the graph in Fig. 4.23a can be used for vibratory machines whose working element travels along a circular path. In this case $n_x = n_y = n$, $p_x = p_y = p$.

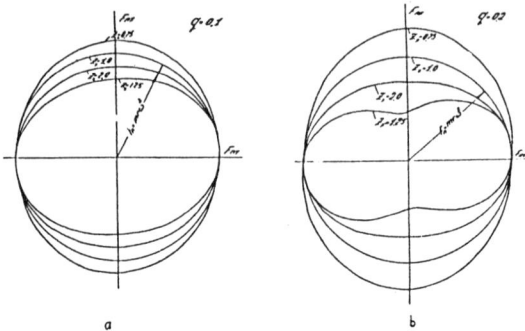

**Figure 4.23** Loci of the loads on the bearing unit of the vibration exciter of one-mass vibratory machine at different detuning factors z and mass ratios q = m/M+m in a regime of rectilinear oscillations in the direction away from the axis; a) $q\sqrt{1 - f^*}$ = 0.1; b) q $\sqrt{1 - f^*}$ = 0.2.

If the trajectory of the working element differs from the circular path, computations must be conducted by the formula

$$\frac{1}{q}\left(\frac{F_b}{mr\omega^2} - 1\right) = \frac{\omega^2(p_x^2 - \omega^2)}{4n_x^2\omega^2 + (p_x^2 - \omega^2)^2}\cos\omega t$$

$$+ \frac{\omega^2(p_y^2 - \omega^2)}{4n_y^2\omega^2 + (p^2 - \omega^2)^2}\sin^2\omega t$$

$$+ \frac{\omega^2}{2}\left[\frac{2n_x\omega}{4n_x^2\omega^2 + (p_x^2 - \omega^2)^2} - \frac{2n_y\omega}{4n_y^2\omega^2 + (p_y^2 - \omega^2)^2}\right]\sin 2\omega t$$

It follows from the presented expression that with elliptical trajectories of the working element of the vibratory machine, the loads on the bearing units are changing in different directions. Furthermore, in addition to the constant load from the centrifugal force of the rotating unbalanced mass, also periodic loads act on the bearing units: pulsating (not changing sign) loads $\sin^2\omega t$ and $\cos^2\omega t$ with frequency $\omega$, and alternating loads with frequency $2\omega$. Figure 4.32b shows the loads on the bearing unit of the vibratory machine performing rectilinear and circular oscillations in the $x$ direction at different detuning factors and values of the rotating and total masses of the vibratory machine. It is evident from the graphs that in the resonant regime in both cases the bearing units are subjected to identical forces in all directions. In the subresonant regime under circular oscillations the loads on the berings in all directions are increasing in an identical way; for rectilinear oscillations the loads are increasing only in the direction of oscillations while remaining unchanged in the perpendicular direction. In resonant regimes with circular oscillations the loads on the bearing units are decreasing in all directions; while with rectilinear oscillations they are decreasing only in the direction of oscillations, while remaining unchanged in the perpendicular direction.

Conditions for bearing units of vibratory machines in the regime of periodic loading and different loads in different directions are more demanding and lead to a faster failure. Therefore, from the standpoint of creating favorble conditions of operation of the vibratory machine, regime of circular oscillations is more desirable.

From the standpoint of ensuring favorable operation of the vibrator bearing units, it is also extremely important to know the loads on the bearing units in various regimes and under the action of diverse loads, but for the same amplitude of oscillations of the vibratory machine, i.e., changes of the unit load on the bearings.

From the above presented dependences, it follows that the load in resonance is equal to the nominal load, while in subresonant regimes the load is increasing and in the transresonant regimes – decreasing. The resistances acting in the system in all regimes lead to a decrease of the loads on the bearing units. The effect of the resistances is particularly evident in the resonance regime.

From the viewpoint of operational durability, it is very important to know which loads act on the elastic system of the vibratory machine and are transmitted through it to the supporting structure. The loads on the elastic system are generated by the inertial forces of the working element of the vibratory machine and by the inertial forces of the unbalanced mass which act via the bearings supports on the working element and then transmitted to the elastic connections. These forces are deforming the elastic connections and internal stresses generated in them are determining their service life. Two types of stresses are created in the elastic connections: elastic stresses whose values are proportional to deformations and viscous stresses which are proportional to the rate of deformation.

In the absence of external resistances the inertial forces acting in the vibratory machine are completely balanced by reactions in the elastic connections and by the exciting force acting opposite to the inertial force vector. The more the inertial forces are compensated by the component of the exciting force, the less the loading of the elastic system. The reactions of the elastic connections are in their turn transmitted to the supporting elements of the structure on which the vibratory machine is mounted. Thus, for durability of the elastic system and for reduction of loads on the supporting structure (reduction of transmission of harmful vibrations to the surrounding medium), it is desirable to reduce loading of the elastic connections as much as possible. Since the elastic elements have hysteretic losses, (usually they are modeled by the forces of viscous resistances), not only the inertial forces of the vibratory machine, which are not compensated by the component of the exciting force, are transmitted to the elastic system, but also the viscous resistance forces. As a result, the total stresses in the elastic system, made up from elastic and viscous components, are increasing and larger dynamic loads are transmitted to the base. It should be noted that in some types of elastic elements, for example, in rubber elements, hysteresis losses are large. Therefore, their contribution to the inertial stresses can be significant and it would be incorrect to ignore it.

Thus, in the general case, the elastic system is loaded by the inertial forces of the oscillating masses of the vibratory machine, which are not compensated by the drive, and by the hysteresis resistances in the material of the elastic elements. This circumstance ought to be taken into account for the optimum multicriterial design of the vibratory machine.

Internal stresses in the elastic connections and reactions on the load-carrying structure are determined by expressions

$$R_{cx}\left[k_x \cos(\omega t - \varphi_x) + c_x \omega \sin(\omega t - \varphi_x)\right] A_x = R_x^* \cdot \cos(\omega t - \xi_x)$$

$$R_{cy} = \left[k_y \sin(\omega t - \varphi_y) + c_y \omega \cos(\omega t - \varphi_y)\right] A_y = R_y^* \cdot \sin(\omega t - \xi_y)$$

where the unit (reduced to unified inertial force of the total oscillatory mass of the vibratory machine) amplitudes of the values of the pressure components are equal to

$$R_x^* = mr\omega^2 \sqrt{\frac{p_x^4 + 4n_x^2 \omega^2}{4n_x^2 \omega^2 + (P_x^2 - \omega^2)^2}} \quad \text{and} \quad R_y^* = mr\omega^2 \sqrt{\frac{p_y^2 + 4n_y^2 \omega^2}{4n_y^2 \omega^2 + (p_y^2 - \omega^2)^2}}$$

and the angles of phase shift between the displacements of the vibratory machine and the pressure components of the elastic elements are

$$\vartheta_x = \text{arctg} \; \frac{2n_x \omega}{p_x^2}, \quad \vartheta_y = \text{arctg} \; \frac{2n_y \omega}{p_y^2}$$

Here, $n_x$ and $n_y$ are the coefficients of resistance of the elastic links referred to a unit mass of the vibratory machine in the directions of axes $x$, $y$.

From the presented expressions it follows that the pressure on the elastic links is proportional to the inertial force of the total mass of the vibratory machine and is substantially dependent on the tuning of the oscillatory system. The pressures are particularly high during or near resonance. During operation in transresonant regimes the pressure drops sharply.

A double effect of the resistances on the value of loading of the elastic connections would be noted. In subresonant and near-resonant regimes the loading is incresing with decreasing resistances. In the transresonant regime this is reversed and decreasing resistances results in reduced loading of the elastic connections.

The phase shift between displacements of the vibratory machine and the force components in the elastic connections varies from zero degrees for very stiff elastic connections to $90^\circ$ for a case when they have practically no stiffness; for example, when the vibratory machine is mounted on air cushion. The phase shift in the resonant regime is determined by the resistances in the elastic system. With increasing resistances, the phase shift is increasing.

Phase shifts between the displacements of the unbalanced masses of the vibration exciter and components of forces in the elastic connections are determined by the following expressions

$$\xi_x = \varphi_x + \vartheta_x$$

$$\xi_y = \varphi_y + \vartheta_y$$

Thus, in order to reduce loads in the elastic connections and on the carrying structure, and to limit propagation of vibrations into the surrounding medium, it is expedient to tune the vibratory machine for transresonant operation.

### 4.3.2.3 Parameters of the Vibratory Machine with Constant Resistances Independent of the Machine Motion

Let us consider the pattern of operation of the vibratory machine with constant resistance forces in the system.

Constant resistances are applied to the working element of the vibratory machine in a direction opposite to the velocity of its displacement and remain constant in time. The resistance forces are expressed in terms of their projections on the coordinate axes $F_x$, $F_y$. Although the resistance forces are constant, their projections on the coordinate axes will be variable quantities, since in the motion process of the vibratory machine direction cosines of velocity vectors vary. Therefore, the components of the constant resistance forces can be written as

$$F_{cx} = F^* \sin (\omega t + \partial_x), \quad F_{cy} = F^* \cos (\omega t + \partial_y)$$

Here $\partial_x$ and $\partial_y$ are the phase shifts between the position of the unbalanced mass and direction of the components of the constant resistances forces. Motion of the vibratory machine under the action of constant resistance forces is described by the following differential equations

$$(M+m)\ddot{x} + k_x x = mr\omega^2 \cos \omega t + F^* \sin (\omega t + \gamma_x)$$

$$(M+m)\ddot{y} - k_y y = mr\omega^2 \sin \omega t + F^* \cos (\omega t + \gamma_y)$$

Solution of these equations enables one to determine displacements parameters of the working element

$$x = A_x \cos (\omega t + \gamma_x)$$

$$y = A_y \sin (\omega t + \gamma_y)$$

where the components of the amplitude of the forced oscillations of the working element are

$$A_x = \frac{qr\omega^2 \sqrt{1-f^{*2}}}{p_x^2 - \omega^2}$$

$$A_y = \frac{qr\omega^2 \sqrt{1-f^{*2}}}{p_y^2 - \omega^2}$$

Phase shifts between the displacements of the working element and the travel of the unbalanced mass of the vibrator are

$$- \gamma_x = - \gamma_y = -\gamma = \arcsin f^*$$

The following notation is assumed here

$$f^* = \frac{F^*}{mr\omega^2}$$

Quantity $f^*$ characterizes the ratio of the constant resistance force and the exciting force generated by the vibration exciter.

The presented expressions indicate that constant resistances lead essentially to the reduction of the effective force generated with the rotation of the vibrator unbalanced mass. Therefore, the amplitude of displacement of the working element of the vibratory machine is proportional to the difference between exciting and resistance forces.

Thus, the motion of the vibratory machine is possible only if the forces of constant resistances do not exceed the exciting force generated with the rotation of the unbalanced mass of the vibrator. The quantity $f^*$ in the cases that are of practical interest can change from zero (constant resistance forces are absent), in which case the machine is idling, to unity (constant resistance forces are equal in magnitude to the amplitude value of the exciting force generated by the vibrator), in which case the machine stops.

The tuning of the vibratory machine considerably affects the magnitude of the oscillation amplitude. The oscillation amplitude increases sharply in the near-resonant regimes, reaching infinity at resonance. It must be mentioned that the constant resistances, if they are less than the exciting force in magnitude, do not restrict the oscillation amplitude in the resonant regime as in the case of viscous resistances in the system. In the regimes other than the resonant regime the oscillation amplitude decreases with increasing constant resistances.

The phase shift between the unbalanced mass and displacement of the working element in the case of constant resistances is independent of the tuning of the oscillatory system and is determined only by the ratio of the resistance force and the exciting force of the vibrator. As investigations show, the phase shift cannot exceed $45°$, since further increase of the resistance causes a decrease in the power supplied by the vibrator to overcome the resistances. The motion trajectories of the working mass and unbalanced masses are ellipses. The motion trajectory of the working mass is an ellipse whose axes of symmetry coincide with the coordinate axes. Thus, the constant resistances uniformly change the dimension of the ellipse, but do not turn it relative to the axis of symmetry.

Figure 4.24 shows the motion trajectory and the mutual positions of the vibratory machine mass and the unbalanced mass when the unit is operating in the interresonant regime under the action of significant constant forces that are equal to $f^* = 0.5$. In this case, the constant forces are equal to half the amplitude value of the exciting force of the unbalanced-mass vibration exciter.

In the presence of constant resistances, the following components of the exciting force act on the vibratory machine with anisotropic elastic elements

$$F_x = P_x \sin (\omega t + \psi_x), \quad F_y = P_y \cos (\omega t + \psi_y)$$

where the amplitude values of the components of the exciting force are equal to

$$P_x = mr\omega^2 \frac{\sqrt{-f^* \omega^4 + \left[ p_x^2 - (1-q)\omega^2 \right]^2}}{p_x^2 - \omega^2}$$

$$P_y = mr\omega^2 \frac{\sqrt{-f^* \omega^4 + \left[ p_y^2 - (1-q)\omega^2 \right]^2}}{p_y^2 - \omega^2}, \quad \psi_y = \psi_x = -\psi$$

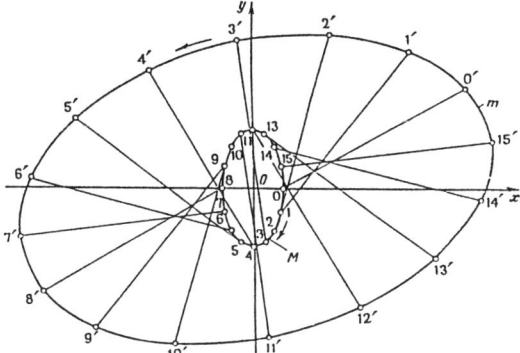

**Figure 4.24** Trajectory of motion and mutual positions of the mass of the vibratory machine $M_{(0,1,...15)}$ and the unbalanced mass $m_{(0',1',...15')}$ in the interresonant regime under the action of constant resistances $f^* = 0.5$.

It follows from the derived expressions that the exciting force generated by the vibrator, in case of constant resistance forces, is proportional to the centrifugal force of the unbalanced mass and depends on the magnitude of resistances and tuning of the oscillatory system of the vibratory machine. In the subresonant regimes, the amplitude of the exciting force increases proportionately to the magnitude of the centrifugal force of the unbalanced mass. The exciting force increases particularly sharply in the near-resonant regimes. In the transresonant regimes a reduction of the amplitude value of the exciting force takes place. The variation of intensity of the exciting force, both rise and fall, is determined by the magnitude of the resistances in the system. The larger the resistances, the less significant the variation of the exciting force in different regimes.

The phase shift between position of the unbalanced mass and direction of the exciting force is zero if there are no resistances in the system. With increasing resistances the phase shift increases.

The loads on the bearing units of the vibratory machine with elastic links of equal stiffness and with constant resistances are directed along the radius-vector of the unbalanced mass and are proportional to the centrifugal force. The dimensionless excess load on a bearing unit, i.e., the load ascribed to the inertial force of the oscillating mass of the vibratory machine and which exceeds the load generated by the unbalanced-mass centrifugal force, is determined from expression

$$\left(\frac{F_b}{mr\omega^2} - 1\right)\frac{1}{q} = \frac{(1 - f^*)\,\omega^2}{p^2 - \omega^2}$$

The excess load on the bearings unit decreases with increasing constant resistances in the oscillating system. In the subresonant operating regimes of the vibratory machine the loads increase proportionately to the loads due to centrifugal force. In the transresonant regimes loads on the bearing units are reduced. The bearing units are loaded particularly heavily in the near-resonant regimes. In the far transresonant regimes, the loads on the bearing units decrease significantly. The formula for determination of the excess load acting on the bearings cannot be used for calculations in resonance when $p = \omega$, since the phase shift in this case is $90^\circ$ irrespective of the acting resistances. In this case, the direction of the eccentricity is perpendicular to the radius vector of displacement of the vibratory machine working mass and transition takes place from positive to negative excess loads, i.e., transition through zero.

The loads transmitted by the elastic connections to the supporting structures are proportional to their stiffness and magnitude of deformation. These loads coincide in phase with the displacements and are, therefore, phase shifted relative to location of the unbalanced mass

$$R_x = R^*\,(\cos\omega t + \gamma), \quad R_y = R^*\,\sin(\omega t + \gamma)$$

where the amplitude values of the load components are

$$R_x = \frac{p_x^2\,\sqrt{1 - f^{*2}}}{p_x^2 - \omega^2}$$

$$R_y = \frac{p_y^2\,\sqrt{1 - f^{*2}}}{p_y^2 - \omega^2}$$

Here the resistances in the elastic connections are assumed absent; therefore, the stresses in them are determined only by elastic deformations. Loads sign $(-\dot{x})$ $F_{cx}$ and sign $(-\dot{y})$

$F_{cy}$, which are in antiphase with the velocity, are transmitted through the pairs with constant resistances. If the constant resistances act in the elements of the suspension assembly, then they are transmitted to the supporting structure. When the constant resistances act along the contact of the working element with the process load, they are transmitted to the processed medium. Strictly speaking, it is necessary in the latter case to take into account the laws of motion of the processed medium. In the calculations, one must take into consideration the relative velocity between the working element and the process load instead of velocity of the working element on which constant resistances act. Exact methods of calculation of vibratory machines under load, which are developed using rheological models of the process load, are presented in section 5.

### 4.3.2.4  Parameters of a Vibratory Machine with Combination of Constant and Velocity Dependent Resistances

The motion of a vibratory machine when velocity-proportionate and constant resistances act simultaneously is described by the following system of differential equations

$$(M+m)\ddot{x} + c_x\dot{x} + k_x x = mr\omega^2\cos\omega t + F^*\sin(\omega t + \gamma_x)$$

$$(M+m)\ddot{y} + c_y\dot{y} + k_y y = mr\omega^2\sin\omega t + F^*\cos(\omega t + \gamma_y)$$

The right-hand sides of these equations include the projections of the centrifugal force of the unbalanced mass and the constant resistances. By analyzing the right hand sides of these equations, it can be seen that they contain expressions that are independent of the motion parameters of the vibratory machine. All these quantities are harmonic functions of the angle of rotation of the unbalanced mass and are shifted with respect to each other by angle $\gamma$. These harmonic functions can be reduced to one function. The component force of the constant resistances reduces the component from the exciting forces.

When constant resistances and velocity proportionate resistances act in the oscillation system each one influences motion of the vibratory machine. Therefore, the general solution of the equation comprises two parts. The working element performs displacements

$$x = A_{x1} \cos (\omega t - \varphi_x) + A_{x2} \sin \left[ \omega t - (\varphi_x + \gamma) \right]$$

$$y = A_{y1} \sin (\omega t + \varphi_y) + A_{y2} \cos \left[ \omega t - (\varphi_y + \gamma) \right]$$

where amplitudes of components of the forced oscillations are respectively equal to

$$A_{x1} = \frac{qr\omega^2}{\sqrt{4n_x^2 \omega^2 + (p_x^2 - \omega^2)^2}} \ ; \quad A_{x2} = \frac{qr\omega^2 \cdot f^*}{\sqrt{4n_x^2 \omega^2 + (p_x^2 - \omega^2)^2}}$$

$$A_{y1} = \frac{qr\omega^2}{\sqrt{4n_y^2 \omega^2 + (p_y^2 - \omega^2)^2}} \ ; \quad A_{y2} = \frac{qr\omega^2 \cdot f^*}{\sqrt{4n_y^2 \omega^2 + (p_y^2 - \omega^2)^2}}$$

and the phase angles between the working element and the unbalanced mass of the vibrator are

$$\varphi_x = \text{arctg} \ \frac{2n_x \omega}{p_x^2 - \omega^2} \ ; \quad \varphi_y = \text{arctg} \ \frac{2n_y \omega}{p_y^2 - \omega^2} \ ; \quad - \gamma = \arcsin f^*$$

The presented expressions indicate that the first components of the oscillation amplitude of the working element are determined only by the velocity-proportionate resistances and are exactly equal to the components of the displacements which are determined in the oscillation system under the action of only this type of resistances. Furthermore, they have the same phase shift with respect to the position of the unbalanced mass. The second components of the oscillations of the vibratory machine are dependent on both the viscous and constant resistances and their phase shift is determined only by the constant resistances.

By analyzing the effect of each type of the resistances on oscillations of the vibratory machine, one can determine the components of the complex oscillation amplitude of the working element which are formed due to both types of resistances

$$x = A_x \cos \left[ (\omega t - \varphi_x) + \gamma^* \right] \text{ and } y = A_y \sin \left[ (\omega t - \varphi_y) + \gamma^* \right]$$

where the components of the complex amplitude of the forced oscillations are respectively equal to

$$A_x = \frac{qr\omega^2\sqrt{1 - f^{*2}}}{\sqrt{4n_x^2\,\omega^2 + (p_x^2 - \omega^2)^2}}; \quad A_y = \frac{qr\omega^2\sqrt{1 - f^{*2}}}{\sqrt{4n_y^2\,\omega^2 + (p_y^2 - \omega^2)^2}}$$

and the phase shifts between the working element and the unbalanced mass of the vibrator are $(\varphi_x - \gamma^*)$ and $(\varphi_y - \gamma^*)$, where

$$\gamma^* = \text{arctg}\,\frac{f^{*2}}{1 + f^*\sqrt{1 - f^{*2}}}$$

Analysis of the presented expressions shows that the components of the motion amplitude of the working element are proportional to the effective magnitude of the exciting force which takes into account values of the constant resistances. The resistances that are dependent on the vibratory velocity and tuning of the oscillation system significantly affect the amplitude. In the subresonant and transresonant regimes, the amplitude decreases with the increase of both viscous and constant resistances. In the resonant regime the amplitude of oscillations of the working element is restricted only by the velocity-depending resistances. The constant resistances only increase the duration of the period of the transient process for the establishment of resonant oscillations, since they reduce the magnitude of the effective exciting force.

The one-mass vibratory machine with anisotropic elastic connections with constant resistances and velocity-dependent resistances is acted upon by the components of the exciting force

$$F_x = P_x\,\sin(\omega t + \psi_x) \text{ and } F_y = P_y\,\cos(\omega t + \psi_y)$$

where the amplitude values of the exciting force are equal to

$$P_x = \sqrt{\frac{4n_x^2\,\omega^2 - f^{*2}\omega^4 + \left[p_x^2 - (1-q)\omega^2\right]^2}{4n_x^2\,\omega^2 + (p_x^2 - \omega^2)^2}}$$

$$P_y = \sqrt{\frac{4n_y^2\,\omega^2 - f^{*2}\omega^4 + \left[p_y^2 - (1-q)\omega^2\right]^2}{4n_y^2\,\omega^2 + (p_y^2 - \omega^2)^2}}$$

and the phase angles with respect to the position of the unbalanced mass are

$$\psi_X = -90^\circ + \varphi_X - \psi_X^*, \qquad \psi_y = 90^\circ + \varphi_y - \psi_y^*$$

Here the phase angles between the exciting force and the machine velocity are equal to

$$\psi_X^* = \text{arctg} \ \frac{p_X^2 - (1-q)\,\omega^2}{2n_X\omega + f^*\omega^2}$$

$$\psi_y^* = \text{arctg} \ \frac{p_y^2 - (1-q)\,\omega^2}{2n_y\cdot\omega + f^*\omega^2}$$

From the expressions given above it follows that the exciting force generated by the vibrator, when both constant and velocity-dependent resistances are acting, is proportional to the centrifugal force developed by the unbalanced mass and dependent on the total value of the acting resistances. In subresonant regimes the amplitude of the exciting force rises as compared with the value of the centrifugal force of the unbalanced mass. The exciting force increases most significantly in near-resonant regimes. In transresonant regimes the exciting force amplitude decreases. The degree of variation of the exciting force in the subresonant and transresonant regimes is dependent on both types of resistances. Increasing both types of resistances reduces the magnitude of the exciting force in the subresonant and near-resonant regimes. In the transresonant regime the resistances, which are velocity-dependent, increase the magnitude of the exciting force, while the constant resistances facilitate its reduction. In the resonant regime the magnitude of the exciting force is restricted only by the components that are velocity dependent.

The loads on the bearing units of the vibrator under the simultaneous action of viscous and constant resistances in the vibratory machine, for which the stiffness of the elastic connections is the same in both axes directions, coincide in direction with the radius-vector of the unbalanced mass. The loads on the bearings are proportional to the centrifugal force of the unbalanced mass developed by the moving vibrator. The dimensionless excess load on the bearing units of the vibrator is determined from expression

$$\frac{1}{q}\left(\frac{F_b}{mr\omega^2} - 1\right) = \frac{(1 - f^{*2})(p^2 - \omega^2)\omega^2}{4n^2\omega^2 + (p^2 - \omega^2)^2}$$

The ratio of the rotating and total masses of the vibratory machine significantly affects the excess load magnitude. Increasing this ratio when the machine is operating in the subresonant region leads to an increase of load on the bearing units. But on the other hand, the bearing units can be significantly unloaded in the transresonant regimes by increasing the unbalanced mass while keeping the mass of the vibratory machine constant.

On the basis of the above stated, in order to unload the bearing units, one can recommend for transresonant machines a reduction of the oscillation frequency and an increase of the unbalanced mass without changing the exciting force magnitude. Increasing constant resistances in the vibratory machine leads to a decrease of the load on the bearing units both in the subresonant and transresonant regions. In the regimes near resonance the loads are mainly affected by the constant resistances. The velocity-dependent resistances reduce loads on the bearing units in the subresonant regimes and increase in the transresonant regimes.

### 4.3.2.5 Operation of Vibratory Machines under Translational and Torsional Oscillation

In a number of cases, the axis of rotation of the unbalanced mass does not coincide with the center of mass of the working element of the vibratory machine. Then the vibration exciter generates an exciting torque, which leads to the creation of torsional oscillations of the working element. The torque magnitude depends on coordinates $a$, $b$ of the axis of rotation and is determined by the following expression

$$M = mr\omega^2 (a \cos \omega t + b \sin \omega t)$$

The torsional oscillations of the working element of the vibratory machine are described by the equation

$$I\ddot{\theta} = c_\theta \dot{\theta} + k_{\hat{\theta}}\theta = mr\omega^2 (a \cos \omega t + b \sin \omega t)$$

where $c_{\hat{\theta}}$, $k_{\hat{\theta}}$ are viscous resistances and stiffness coefficients of the elastic system for torsional oscillations; $I$ is the moment of inertia of the working element of the vibratory machine.

The generation of torsional oscillations causes various points of the working element to oscillate along different trajectories. The laws of formation of the trajectory fields of the working element when it is excited by two synchronously rotating unbalanced vibration exciters are considered in work [45].

Let us consider the formation of the trajectory field of the working element of a vibratory machine that is excited by an unbalanced-mass vibrator whose axis is located at different points with respect to the center of mass of the working element.

If the axis of rotation $O_1$ does not coincide with the center of mass $O$ of the working element of the vibratory machine, then it will perform, in the general case, oscillations along elliptical trajectories which at some points degenerate into circles or straight lines. The distribution of the oscillation trajectories of various points of the working element of the vibratory machine which is excited by an unbalanced-mass vibration exciter, whose axis of rotation is located at distance $h$ from the center of mass, is presented in the schematic diagram on Fig. 4.25. From this diagram, it follows that there are some characteristic points of the working element which have circular or rectilinear oscillation trajectories.

Thus, the center of mass of the working element oscillates along a circular trajectory whose amplitude is $A = mr/M$, where $m$ and $r$ are mass and eccentricity of the unbalanced mass; $M$ is mass of the working element (this dependence is true for an operating regime of the vibratory machine far into the transresonant regime).

In addition, point $O_2$ also oscillates along a circular trajectory. This point is removed from the center of oscillations of the working element $K$ by a distance $H = I/Mh$, where $I = M\rho_0^2$ is the moment of inertia of the working element with respect to axis $z$ perpendicular to the plane of oscillations; $\rho_0$ is the radius of inertia of the working element.

The rocking center, located at distance $H$ from the center of mass of the working element, oscillates along a rectilinear trajectory in the direction of line $KOO_1$ with an amplitude of $A = mr/M$. All the points of the working element located on the straight line passing through the rocking center perpendicular

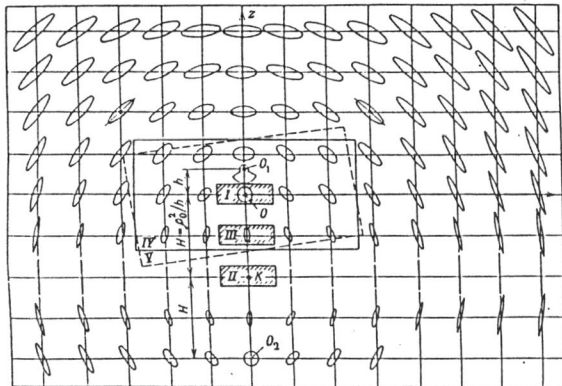

**Figure 4.25** Field of vibratory trajectories of a point on the working element of a vibratory machine excited by an unbalanced mass whose rotational axis does not coincide with the center of mass of the working element.

to line $OO_1K$ also perform rectilinear oscillations parellel to line $OO_1K$, but with larger amplitudes. The straight line $OO_1K$ is the axis of symmetr of the vibratory field. One of the axes of the elliptical trajectories of the points of the working element lying on this axis of symmetry is always directed along it. The lengths of the corresponding half-axes are the same for all these points and are equal to $A$.

In practice, vibratory machines with a uniform field of vibratory trajectories (rectilinear, circular, and elliptical) of the working elements are often used.

How should the unbalanced-mass vibration exciter be mounted on the working element of the vibratory machine in order to obtain uniform oscillations along trajectories of the same kind?

In order to obtain circular uniform trajectories of the working element, the axis of rotation of the unbalanced mass of the vibration exciter ought to coincide with the center of mass of the working element. Furthermore, the plane of rotation of the unbalanced mass must coincide with the plane of longitudinal symmetry of the working element. If excitation is realized by two unbalanced masses synchronously rotating in one direction, then the axes of their rotation must be located at the same distance from the center of mass of the working element (Fig. 4.26a).

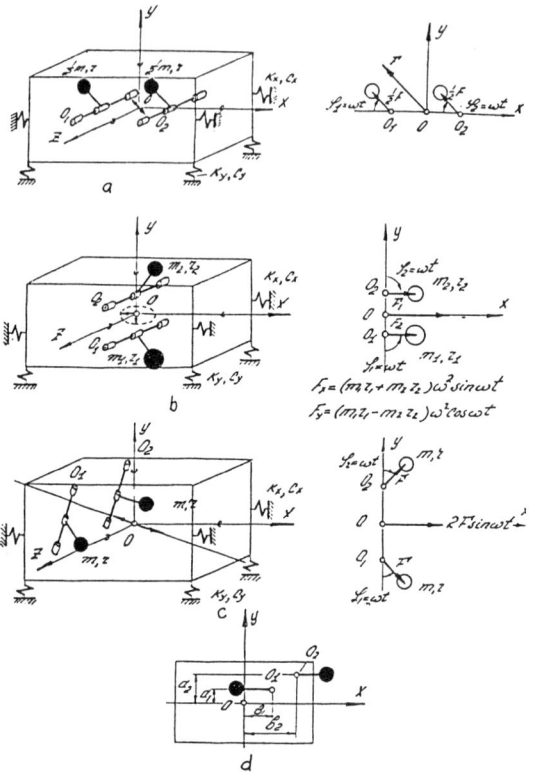

**Figure 4.26** Schematic diagrams of vibratory machines with inertial vibrators.

For the generation of rectilinear uniform oscillations trajectories of the working element, vibrators with two unbalanced masses, whose shafts rotate in opposite directions (Fig. 4.26b), are used. The vibrator must be mounted in such a way that the resultant of the centrifugal forces of the unbalanced masses passes through the center of mass of the working element of the vibratory machine.

A uniform field of elliptical trajectories can be created by a two-shaft vibrator with kinetic moments of the unbalanced masses rotating in opposite directions being not equal to each other (Fig. 4.26c). If the axes of rotation of the unbalanced masses coincide, then they must also coincide with the center of mass of the working element of the vibratory machine. If the axes of rotation are different, then the relation $a_1/a_2 = b_1/b_2 = m_1 r_1 / m_2 r_2$ must be observed (Fig. 4.26d).

### 4.3.2.6 Energy Parameters of Vibratory Machines

The power which is required to operate the vibratory machine in the given regime and the power which can be transmitted by a vibrator of a specific type are determined by a whole complex of factors: vibrator parameters, machine characteristics, and the acting loads in the machine. It is not possible to impart additional power to the vibratory machine by simply increasing motor output. Each vibratory machine consumes strictly determined power, whose value is dependent on a whole set of factors acting in the vibratory machine-vibration exciter-load system.

Let us examine the work of one-mass vibratory machine driven by an unbalanced-mass vibration exciter with a motor having sufficient power reserve. To eliminate torsional oscillations of the vibratory machine, the axis of rotation of the vibration exciter is located in the center of mass of the oscillation system. The elastic system is designed in such a way that the movement in the direction of one of the extremal stiffnesses would not strain the elastic system in the direction of the second extremal stiffness. In this case, the oscillations of the vibratory machine in the direction of the extremal stiffnesses of the elastic system will be independent. For exclusion of torsional oscillations, the resultant of the restoring forces of the elastic system must pass through the center of mass of the vibratory machine. Let us make use of the computational schematic of the vibratory machine with inertial vibration exciter of the unbalanced-mass type shown in Fig. 4.18.

The power of the vibratory machine drive is easier to determine by considering the equation of rotation of one drive shaft of the vibration exciter.

As noted earlier, due to the fact that the vibration exciter shaft performs oscillatory motion along with the vibratory machine, the amplitude of the exciting force and the pressure on the bearings of the vibration exciter are not equal in amplitude to the centrifugal force generated by the unbalanced mass when it rotates around an axis not performing oscillations. The dependences enabling one to estimate the character of change of the value of these forces during machine operation as a function of its parameters, load, and tuning in a specific operating regime, have been considered in the previous sections.

Let us analyze in more detail one more feature of the formation of the exciting force in a working vibratory machine. Analysis of the graphs representing the trajectory of motion of the centers of mass of the vibratory machine and the unbalanced mass indicate that the exciting force does not always pass through the center of mass of the oscillatory systems.

The forces of friction and the oscillations of the center of the unbalanced mass also affect the direction of the resultant force on the bearing unit. These conditions must be remembered when designing vibratory machines with inertial vibration exciters.

If effective coefficient of friction in the bearings of the vibration exciter is $\mu$, then the resultant of the forces loading the bearing unit is equal to $F_n^* = \sqrt{1 + \rho^2}\, F_n$ and is directed at angle $\theta = $ arctg $\mu$ to the pressure acting on the bearing unit $F_n$. Here, $\rho$ is the radius of the friction circle, $\rho = \mu = D/2$; $D$ is the diameter of the rotating race of the vibration exciter bearing. The frictional moment in the exciter is equal to

$$M_b = \rho m\, (z\dot\varphi^2 - \ddot{x}\, \cos\varphi - \ddot{y}\sin\varphi)$$

Taking this into consideration, the equation of motion of the rotating parts of the vibratory machine in unsteady motion is

$$(Y + mr^2)\, \ddot\varphi + \rho mg + m\ddot{x}\, (\rho\cos\varphi - z\sin\varphi) + m\ddot{y}\, (\rho\sin\varphi + r\cos\varphi)$$
$$+ mgr \cos\varphi = L(\dot\varphi)$$

where $L(\dot\varphi)$ is torque on the shaft of the driving electric motor.

This equation expresses the dependence between the oscillatory motion of the vibratory machine and the time history of rotation of the exciter shaft. The first term in the left-hand side of the equation is the moment of inertia of all the rotating parts of the vibration exciter. In steady-state motion at constant angular velocity of the unbalanced mass, this moment of inertia is zero. The second term of the equation characterizes the constant component of the moment of the resistance forces acting in the bearing units of the vibration exciter. The third and fourth terms of the equation are expressions of the moments of resistance from the periodic components of inertial and resistance forces acting on the rotor and bearing unit of the vibration exciter. Sometimes, these moments are referred to as vibratory moments. They include two components: a moving component which is proportional to the eccentricity of the unbalanced mass, and a component of the resistance forces which is proportional to the radius of the friction circle. The fifth term of the left-hand side of the equation is the moment from the gravitational force acting on the unbalanced mass of the vibration exciter.

The presented equation enables one to determine the moment on the shaft of the vibration exciter in the steady-state operating regime of the vibratory machine by substituting into it the corresponding motion parameters

$$L(\dot{\varphi}) = mr\omega^2 \left[ \left( \frac{A_x^* \cos \varphi_x - A_y^* \cos \varphi_y}{2} \right. \right.$$

$$\left. - S \frac{A_y^* \sin \varphi_y - A_x^* \sin \varphi_x}{2} \right) \sin 2\omega t + \frac{g}{\omega^2} \cos \omega t + (A_x^* \sin \varphi_x$$

$$\left. - sA_x^* \cos \varphi_x) \sin^2 \omega t + \left( A_y^* \sin \varphi_y - sA_x^* \cos \varphi_x \right) \cos^2 \omega t + \frac{sg}{2} \right]$$

Here, $A_x^*$, $A_y^*$ are dimensionless amplitudes of oscillations of the vibratory machine along axes $x$ and $y$, $s$ is the ratio of the radius of the friction circle and the radius of the center of mass (eccentricity) of the unbalanced mass, $s = \rho/r$.

The moment (torque) on the unbalanced-mass shaft represents a complex periodic function containing components that are proportional to frequencies $2\omega$, $\omega$, $\omega/2$. The first two terms represent the reactive moment characterizing circulation of energy in the system. The last terms are the active component of the moment and determine the mean energy expenditures that are caused by the actions of the resistances and friction in the bearings of the vibration exciter; the very last term is independent of the vibration regime of the vibratory machine.

By integrating the torque on the shaft of the vibration exciter, we determine the energy expenditure of the vibratory machine per one revolution of the exciter shaft

$$W = \pi mr\omega^2 \left[ A_x^* \sin \varphi_x + A_y^* \sin \varphi_y) + S(A_x^* \cos \varphi_x + A_y^* \cos \varphi_y) \right.$$

$$\left. + \frac{2sg}{\omega^2} \right]$$

The first term in the square brackets characterizes the dimensionless work spent on overcoming the resistances in the vibratory machine. Included here are resistances generated by the process load and hysteresis losses in the elastic system. Attention should be paid to the universality of the presented expression for the determination of energy consumption of the vibratory machine. It is valid for any types of process resistances and any resistance patterns in the elastic system. All the possible features of vibratory machine operation are taken into consideration by the components of the oscillation amplitude and by the phase shifts between the displacements of the vibratory machine and the position vector of the unbalanced mass.

By substituting into this expression, parameters characterizing motion of the vibratory machine under the action of viscous resistance forces and after some transformations, we determine components and total dimensionless power consumption on overcoming viscous resistances acting in the machine

$$N^* = \frac{N_x + N_y}{mr^2\omega^3} = \frac{n_x\,\omega^3}{4n_x^2\,\omega^2 + (p_x^2 - \omega^2)^2} + \frac{n_y\,\omega^3}{4n_y^2\,\omega^2 + (p_y^2 - \omega^2)^2}$$

and the dimensionless components and total power consumption on overcoming the resistances to rotation of the vibration exciter

$$N_b^* = \frac{N_b}{m\rho r\omega^3} = 1 + q\left[\frac{\omega^2(p_x^2 - \omega^2)}{4n_x^2\,\omega^2 + (p_x^2 - \omega^2)^2} + \frac{\omega^2(p_y^2 - \omega^2)}{4n_y^2\,\omega^2 + (p_y^2 - \omega^2)^2}\right]$$

If the working element is stationary, the dimensionless power expended by the vibration exciter to overcome the resistances in the bearings is equal to unity. Power expenditures during machine operation vary compared with the idling regime. In order to assess the effect of the operating regime and parameters of the vibratory machine on power expenditures as compared with idling, it is convenient to rewrite the formula as follows

$$\frac{N_b^* - 1}{q} = \frac{\omega^2(p_x^2 - \omega^2)}{4n_x^2\,\omega^2 + (p_x^2 - \omega^2)^2} + \frac{\omega^2(p_y^2 - \omega^2)}{4n_y^2\,\omega^2 + (p_y^2 - \omega^2)^2}$$

It is interesting to note that energy expenditures per unit displacement of the vibratory machine under the action of viscous resistances are proportional only to the coefficient of viscous resistance and oscillation frequency $n\omega$.

Energy expenditures in the bearing units are proportional to the radius of the friction circle, total mass of the vibratory machine, eccentricity of the unbalanced mass, and are dependent on the regime of operation. In the subresonant regime energy expenditures increase, and in the transresonant regime they decrease.

When operating in resonance, energy expenditures do not change in relation to the energy expenditures of the stationary vibration exciter. The variations of the given unit energy losses due to oscillations of the vibration exciter shaft are proportional to the ratio of the rotating and total masses of the vibratory machine, inversely proportional to the eccentricity of the unbalanced mass, and dependent on the tuning of the vibratory system

$$N_b - 1 = \frac{q(p^2 - \omega^2)}{\sqrt{4n^2 \omega^2 + (p^2 - \omega^2)^2}}$$

In the resonant regime the additional energy expenditures are equal to zero, in the subresonant regime they tend to approach zero, and in the far transresonant regime, particularly under the action of insignificant resistances that are proportional to velocity, they tend to approach $q$.

Analysis of the presented dependences indicates that the largest energy expenditures are in the resonant and near-resonant regimes of machine operation. Moreover, energy expenditures in the subresonant and near-resonant regimes increase with decreasing damping factor. Energy expenditures in far transresonant regimes drop and vary insignificantly with changing detuning factor. Increasing the load in the transresonant regime leads to increase of the required vibratory power.

This analysis enables one to conclude that the subresonant and near-resonant regimes of vibratory machine operation are unstable, since increasing the resistances reduces the capabilities of the machine with respect to power requirements. This property of a machine with inertial drive is particularly manifested in the resonant and near-resonant regimes. In the transresonant regimes the operation is stable, since when the resistances are increasing more energy is imparted to the machine from the motor to overcome them. If, by analogy with the induction motor, one analyzes characteristic of the vibratory machine with an inertial drive, then two modes can be identified, namely unstable (region of subresonant and near-resonant regimes) and stable (region transresonant regimes).

By analyzing these correlations it can be seen that in the subresonant regimes, the total energy expenditures increase, and in the transresonant regime, they decrease. The increase and decrease of energy expenditures are particularly sharp in the near-resonant regimes. Decreasing damping coefficients results in broadening the range of variation in energy consumption. Energy expenditures are also significantly affected by the ratio of the rotating and total mass of the vibratory machine. The larger this ratio, the wider the range in which energy expenditures in the bearing units vary.

As already noted above, both active and reactive power act in the vibratory machine. The ratio of these powers can be assessed by the angle between the displacements and the exciting force ($\varphi - \psi$). The active work done by the exciting force is determine by the expression $\pi$ $PA$ sin ($\varphi - \psi$); therefore, the closer angle ($\varphi - \psi$) gets to $90^{\circ}$, the higher the ratio of the active and reactive components of energy expenditures. The phase shift between the exciting force and the displacements is determined by the expression

$$\varphi - \psi = \text{arctg} \frac{g\omega^2}{\sqrt{4n^2\omega^2 + (p^2 - \omega^2)^2}}$$

By analyzing correlation of the phase shift between the exciting force and the displacement and the detuning factor $z = \omega/p$ at different damping factors $v = n/p$, we can see that the angle between the exciting force and the displacement in resonance is small. The angle increases with decreasing resistances in the system and increasing ratio of the rotating and total mass of the vibratory machine. It is pertinent to note, however, that in practically realized machines, i.e., with sufficiently large resistances and limited values of the mass ratio, the phase shift between the exciting force and displacement does not reach $90^{\circ}$.

Thus, in vibratory machines with inertial drive the reactive power is practically always present. Moreover, in heavily loaded machines it might be very substantial. It must also be noted that energy circulation in a real system is always accompanied by losses.

These formulas are universal and can be used for determination of the resistance torque on the shaft of the vibration exciter and energy expenditures of the vibratory machine with any types of resistances.

When constant resistances, which are velocity independent, act in the vibratory machine, the consumed power will be determined by expression

$$N = \frac{qr^2 \, \omega^5 m \, (p_x^2 + p_y^2 - 2\omega^2)}{204 \, (p_x^2 - \omega^2)(p_y^2 - \omega^2)} \, \sin 2\gamma$$

If the stiffness of the elastic suspension in the vibratory machine is the same in all directions $p_x = p_y = p$, then the formula for the consumed power becomes

$$N = \frac{2gz^2 \omega^5 m}{204 \, (p^2 - \omega^2)} \, \sin 2\gamma$$

The power required to bring the vibratory machine with constant resistances into oscillatory motion is determined by the kinematic and dynamic parameters of the system and by magnitudes of the resistances. From the presented expression it is evident that the power which the vibratory machine can develop under the action of constant resistances will be a maximum at phase shift $45^\circ$ between the displacements of the unbalanced mass and the working element.

At maximum power consumption components of the amplitude of oscillations of the working element will be equal to

$$A_x = \frac{0.707 \, qr \, \omega^2}{p_x^2 - \omega^2}$$

$$A_y = \frac{0.707 \, qr \, \omega^2}{p_y^2 - \omega^2}$$

Thus, if constant resistances act in the vibratory machine, then at the maximum power developed by the vibration exciter, the oscillation amplitude is reduced under the influence of the resistances by approximately 30% both in the subresonant and transresonant regimes. In the resonant regime, as already mentioned earlier, the resistances do not limit the oscillation amplitude. Only the rate of rise of oscillation amplitude decreases.

Within some range of loading, the oscillation amplitude decreases when the constant resistances increase, but in the same time required power is increasing. Such an operating regime occurs at loads below the critical, i.e., until the oscillation amplitude is reduced by about 30%. With further loading, the reduction of the oscillation amplitude will be accompanied by a decrease of the power used to overcome the acting resistances. This operating regime is unstable and eventually the machine stalls despite the fact that the unbalanced mass continues to rotate.

Let us compare the specific features of the operation of a vibratory machine under the action of constant and velocity-dependent resistances. In both cases the oscillation amplitude decreases with increasing resistances. Under the action of loads that are velocity-dependent, the consumed power decreases in all the cases with increasing load in the subresonant and near-resonant regimes, i.e., these regimes are unstable within any load range. Under the action of constant resistances in the region of subcritical loads, all regimes are unstable. Operation with velocity-dependent loads is stable only at transresonant regimes with above critical constant loads, all operating regimes of the vibratory machine are unstable.

Analysis of operation of the vibratory machine under constant velocity-independent loads, shows practically complete analogy with operation of an induction motor. At subcritical loads operation takes place on the stable branch of the working characteristic, at transcritical loads, operation is on the unstable branch. By analogy with the induction motor, the power corresponding to the critical load is referred to as breakdown power.

However, there are differences in the operation of an induction motor and the inertial vibrator, since in the motor the rotation frequency is changing, and in the vibrator the amplitude is changing at constant oscillation frequency. In order to maintain constant vibrator frequency the breakdown power of the induction motor must exceed the breakdown power of the vibration exciter.

The breakdown power of the inertial vibrator in the vibratory machine with isotropic (equal stiffness) elastic links is determined from expression

$$N_d = \frac{qr^2 \omega^5 m}{204 \left(p^2 - \omega^2\right)} \quad kw$$

It follows from the presented expression that the magnitude of the breakdown power is dependent on parameters of the vibration exciter, the ratio of the rotating and total masses of the vibratory machine, the working frequency, and the tuning of the oscillation system. The breakdown power increases in the near-resonant regimes and decreases in the subresonant and transresonant regimes. In the resonant regime the oscillation amplitude and the breakdown power are increasing simultaneously (theoretically, infinitely). However, in practice limited oscillation amplitudes are characteristic for the resonant regimes. Consequently, the breakdown power is limited in the resonant regime.

Since the oscillation amplitude is determined by the process requirements irrespective of the operating regime of the vibratory machine, in any operating regime of the vibratory machine - resonant, subresonant, or transresonant - it has to be of the same value. Accordingly, comparison of the breakdown power in the vibratory machine at various regimes of its tuning is convenient to conduct at unitary oscillation amplitude.

The unitary breakdown power, i.e., the power required to excite oscillations of the vibratory machine with a unit amplitude is determined by the expression

$$N_d^* = \frac{mr\omega^3}{204}$$

Thus, the unit breakdown power is independent of the operating regime of the vibratory machine and is determined only by the parameters of the vibration exciter and the operating frequency of the machine.

The unit breakdown power can also be represented as a function of parameters of the vibratory machine and its tuning in the following form

$$N_d^* = \frac{(M+m)\left(z^2-1\right)\omega^2}{204}$$

This expression characterizes stability of operation of the vibratory machine with a specified amplitude in various tuning regimes that are characterized by the detuning factor $z$. This breakdown power is recommended for use as a characteristic of stability of the operating regime.

By analyzing this expression, we can see that the stability margin of vibratory machine operation in the near-resonant regimes is small. This is explained by the fact that in the near-resonant regimes the vibratory machine with an unbalanced-mass drive with small oscillation amplitudes cannot do much work due to low value of the kinetic moment of the unbalanced masses of the vibration exciter. Therefore, the operation of a vibratory machine with inertial drive and linear elastic system in the near-resonant regimes under conditions of variable loads is characterized by instability.

Experience with the use of vibratory machines with inertial drives having elastic links of different stiffness indicate that in addition to the resonant regimes, interresonant operting regimes can also be unstable. For analysis of interresonant operating regimes of the vibratory machine, the expression for the breakdown power of a vibratory machine with elastic links is used

$$N_d = \frac{qrm\omega^5(p_x^2 + p_y^2 - 2\omega^2)}{204\ (p_x^2 - \omega^2)(p_y^2 - \omega^2)}$$

From this expression, it is evident that at some ratios of the natural frequency of the vibratory machine and angular velocity of the unbalanced mass, the breakdown power can be equal to zero. This happens when the condition $p_x^2 + p_y^2 - 2\omega = 0$ is satisfied. This expression can be rewritten as $(p_x^2 - \omega^2) + (p_y^2 - \omega^2) = 0$, i.e., the condition is satisfied when the differences $(p_x^2 - \omega^2)$ and $(p_y^2 - \omega^2)$ are equal in magnitude but opposite in sign. By turning to the dependences presented earlier, it is evident that the oscillation amplitudes $A_x$ and $A_y$ in this case are equal despite the inequality $p_x \neq p_y$ but opposite in sign.

The oscillation amplitudes have different signs in the interresonant regime. Moreover, the vibratory machine travels along a circular tarjectory despite the different principal (extremal) stiffnesses of the elastic links. This regime corresponds to the points of intersection of two amplitude-frequency characteristics; with one branch of the characteristic being transresonant and another subresonant.

Such an operating regime of the vibratory machine cannot be realized in practice, since it has zero breakdown power. The operating regimes of the vibratory machine in the vicinity of this point can be practically realized; however, they have low stability due to inadequacy of the power imparted to the vibration exciter of the vibratory machine in order to maintain its operation.

By analyzing the expressions for the breakdown power, it is noticed that the closer the trajectory of motion of the vibratory machine approaches a circle, the lower is this power. When the vibratory machine operates on elongated elliptical trajectories, breakdown power of the vibration exciter increases.

On the whole, the interresonant operating regimes of the vibratory machine are characterized by lower stability than subresonant or transresonant regimes.

Comparing the magnitudes and the character of change of the energy consumption of a vibratory machine with inertial drive under different forms of acting resistances, it can be noted that the character of the load influences not only the energy consumption, but also stability of operation of the vibratory machine.

### 4.3.3 Transient Regimes of Transresonant Vibratory Machines with Inertial Drives

### 4.3.3.1 Methods of Elimination of Resonant Oscillations in Transient Regimes

The inertial vibratory drive is mainly used in vibratory machines of the transresonant type. Such an operating regime ensures stable operation of the vibratory machine under large variable loads. However, during starting and acceleration, the vibratory machine crosses the resonant region which causes a sharp increase in oscillation amplitude of the working element. Furthermore, crossing the resonant regime is accompanied by a sharp increase in energy consumption. In the resonant regime the drive consumes several times more energy than in the working transresonant regime.

These features of vibratory machines with inertial drive create a number of serious problems including limiting amplitudes of the working element and energy consumption during transition through the resonant region.

A number of methods for limiting the amplitudes during transition through the region of resonance is known: these include methods associated with the introduction of damping elements into the oscillation system; methods based on impact damping, and on the use of an elastic system with nonlinear characteristic; methods that are based on increasing the speed of transition through resonance and on use of self-regulating vibrators.

The first method is based on the dissipation of part of the energy of mechanical oscillations of the working element during resonance transition with the aid of various damping or braking devices connected in parallel to the elastic elements of the vibratory machine, or using the working load. The shortcoming of the method of damping is the reduction of the coefficient of vibration isolation, since the dynamic loads are transmitted to the supporting structures via the elastic elements and damping devices simultaneously. Furthermore, in a number of cases, the unproductive losses increase not only during the period of resonance, but also during the working regime.

Starting the vibratory machine under a large load is found to be effective. In this case, the machine easily reaches the working regime without notable rise in oscillations and energy consumption during the transition through the resonance region. However, implementation of this method is not always feasible in production conditions.

The use of impact vibratory dampers (bumpers) and a nonlinear elastic system is reasonably effective. However, this method is limited mainly by the high-frequency operating regimes and is associated with additional dynamic loading of the elements of the structure and with energy dissipation. Another disadvantage of impact vibratory dampers is noisy operation.

It is known that the motion amplitude of the oscillatory system during transient regime is correlated with the angular acceleration by an inverse relationship: the lower the angular acceleration with which the system passes the resonant region, the higher the resonant amplitude. By changing the value of the angular acceleration, it is possible to effectively influence the resonant oscillation amplitudes. This is a basis for the method of elimination of the resonant increase of oscillations by increasing the speed with which the system passes through resonance. However, to realize this method the installed power of the drive must be increased with respect to the power required for normal operation in the working regime.

For elimination of resonance oscillations, a method based on the use of special vibrators with automatically variable kinetic moment is becoming widespread. These vibrators combine the functions of the oscillation source and the device for suppressing the resonant oscillation amplitudes during the transient regimes of the vibratory machine.

This principle of elimination of resonant oscillations is based on the fact that during starting, the automatic actuation of the vibrator mechanism takes place following transition of the resonant region, and disengagement (balancing) during run-out is effected prior to reaching the resonant region at the moment when the kinetic energy of oscillations of the working element is minimal. Thus, forward and backward transition through resonance is effected in a totally balanced vibrator (the kinetic moment of the unbalanced mass is equal to zero).

### 4.3.3.2 Determination of Working Parameters of Mechanisms for Changing the Unbalanced-Mass Kinetic Moment

**Mechanism with sliding unbalanced mass**. The design schematic of an inertial vibration exciter with sliding unbalanced masses is shown in Fig. 4.27. The vibration exciter consists of a drive shaft *1*, and disc *2* with rim *3*. The shaft of the vibration exciter is mounted in bearings *4*, which are assembled in the frame of the vibrator. An unbalanced mass *5* is mounted in guides on the disc and is pressed by spring *6* to the hub by force $P_0$. In this position of the unbalanced mass the vibration exciter rotor is completely balanced. Upon starting of the vibration exciter, the rotor, as a result of the initial tension of the spring, remains balanced for some time without generating an exciting force. The force of spring pretension holds the unbalanced mass at the internal stop until the angular velocity of the vibration exciter exceeds the natural frequency of the vibratory machine oscillatory system. The angular velocity at which the unbalanced mass begins to move from the internal to the external stop is determined by the condition of equality of the centrifugal force of the unbalanced mass to the spring pretension

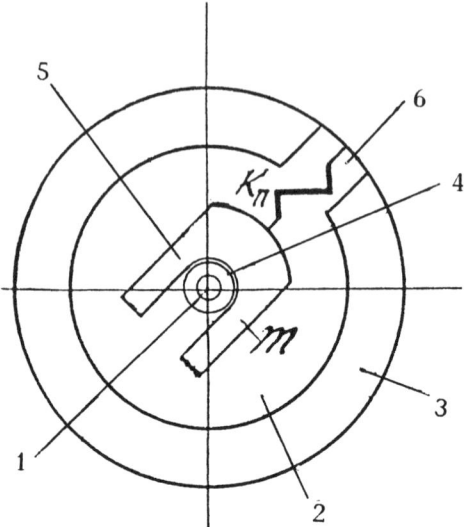

**Figure 4.27** Design schematic of a vibration exciter with sliding unbalanced mass.

$$\omega_{start} = \sqrt{\frac{k_{sp} f_0}{r_0 m}}$$

where $k_{sp}$ is spring stiffness; $f_0$ is initial spring tension; $m$, $r_0$ are the mass and eccentricity of the center of gravity of the unbalanced mass in the initial position at the inner stop.

    During run-out of the oscillatory system, the process takes place in the reversed order. When the angular velocity of the vibration exciter is reduced to the specified value, the unbalanced mass is returned from the outer to the inner stop by the force of spring tension. The vibration exciter is then balanced and the vibratory system passes the resonant region without the exciting force. In order to obtain minimum resonant amplitudes, the angular velocity of the vibration exciter at which the unbalanced masses must be disengaged at run-out is determined from the relationship $\omega_{stop} = \sqrt{3}\ p$ (where $p$ is the natural frequency of the vibratory machine).

    The angular velocity of disengagement of the unbalanced masses is determined from the condition of equality of the forces of spring compression and the centrifugal forces of the unbalanced mass on the outer stop

$$\omega_{stop} = \sqrt{\frac{k_{sp}(f_0 + s)}{m(r_0 + s)}}$$

where $s$ is displacement of the unbalanced mass from the inner to the outer stop.

From this expression for $\omega_{start}$ and $\omega_{stop}$, we determine the relation between the angular velocities of starting and stopping the vibration exciter

$$\omega_{start} = \omega_{stop} \sqrt{\frac{(r_0 + s)\lambda}{\lambda r_0 + s}}$$

where $\lambda = f_0/r_0$.

The angular velocity of the vibration exciter at which the unbalanced mass is switched over from idling to normal operation can be determined by this formula if it is known that switching from normal operation to idling must occur at speed $\omega_{stop}$. From the formula it is evident that when $\lambda > 1$, $\omega_{start} > \omega_{stop}$; when $\lambda < 1$, $\omega_{start} < \omega_{stop}$; when $\lambda = 1$, $\omega_{start} = \omega_{stop}$. For $\lambda < 1$, when the angular velocity at which the unbalanced mass moves upon starting of the vibrator is less than $\omega_{stop}$, stable positions of the unbalanced mass are possible at intermediate radii in the range from $r_0$ to $r_0 + s$. In this case the movement of the unbalanced mass will take place smoothly during some time interval until the angular velocity of the rotor rises from $\omega_{start}$ to $\omega_{stop}$ upon starting or until it is reduced from $\omega_{stop}$ to $\omega_{start}$ upon stopping.

If $\lambda > 1$, there can be no intermediate stable positions of the unbalanced mass. Indeed, the centrifugal force on the outer stop becomes less than the pressure force of the spring for the angular velocity $\omega_{stop} < \omega_{start}$ at which the unbalanced mass commences its motion during run-out. Therefore, during stopping when the radius $R = r_0 + s$ is reduced, the difference between the forces of pressure of the spring and the centrifugal force increases even if the angular velocity of the rotor remains constant but below $\omega_{stop}$. Furthermore, the unbalanced mass is displaced from the outer stop to the inner stop with increasing acceleration when the vibration exciter is brought to a halt.

Upon starting, the unbalanced mass commences its movement when the angular velocity $\omega_{stop} > \omega_{start}$ and accelerates to the outer stop when $\omega \geq \omega_{stop}$. If the angular velocity of the rotor is in the range $\omega_{stop} \cdots \omega_{start}$, the unbalanced mass can have two stable positions on the inner or outer stops. During acceleration, it is at the inner stop and when brought to a halt, it is at the outer. Therefore, when $\lambda > 1$, switching of the vibration exciter from idling to normal operation and vice versa takes place rapidly and with small impact. Impact is of course a disadvantage; however, this is compensated by the softer characteristic of the adjustable spring which, in its turn, enables generating a smaller eccentricity of the center of gravity of the unbalanced mass when the vibrator is brought to a halt.

**Mechanism with pivoted unbalanced masses** In order to elucidate in more detail the mechanism of action of the vibration exciter with pivoted unbalanced masses we shall consider its kinematic layout (Fig. 4.28).

The unbalanced masses *1* and *2* of the mechanism are mounted on a sleeve which is eccentric relative to the axis of vibration exciter shaft, and are pressed by spring *3* to stops *4* and *5* of disc *6* with a specified initial adjustable moment. The stops of the unbalanced masses are arranged on the disc in such a way that the mechanism in the static state is completely balanced relative to the axis of rotation.

The principle of operation of the mechanism of the vibrator is as follows. Axis $O_1$ of rotation of the unbalanced masses is offset relative to axis $O$ of rotation of the mechanism by $r$. The centers of the unbalanced masses are at distance $l$ from axis $O_1$. When the motor of the vibration exciter is started the unbalanced mass of the mechanism is accelerated with some angular acceleration up to the nominal angular velocity. The motion of the unbalanced masses in this case takes place in a centrifugal force field with intensity $\rho\dot{\varphi}^2$ ($\rho$ is the distance from the axis of rotation to the center of gravity of the unbalanced masses; $\dot{\varphi}$ is angular velocity of the mechanism). As a result of this, the normal $P_1^n$, $P_2^n$ and tangential $P_1^\tau$, $P_2^\tau$ inertial forces in transfer motion (rotation around axis $O$) act on the unbalanced masses until the moment they are detached from the stops.

By analyzing the forces acting on the unbalanced masses, it is easy to note that their rotation does not take place simultaneously. When the mechanism rotates counterclockwise, the unbalanced mass *1* begins to move first, since the normal $P_1^n$ and tangential $P_1^\tau$ components of the inertial force in the

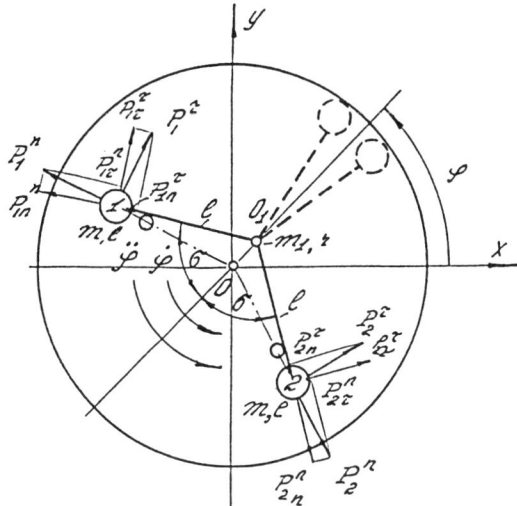

**Figure 4.28** Design schematic of a vibration exciter with pivoted unbalanced mass.

transfer motion generate a moment relative to their mounting axis $O_1$ which separates it from stop 5. Separation of the unbalanced mass occurs at an angular velocity at which the moment of the inertial forces $lP_1^n$ and $lP_1^\tau$ exceed the moment $M_{sp}$ of the pretwisted spring 3 and the moment of the frictional force $M_{fr} = (P_{1(n)}^n - P_{1(n)}^\tau) \, \mu \, d/2$ directed opposite to rotation of the unbalanced mass. Here, $P_{1(n)}^n$ and $P_{1(n)}^\tau$ are projections of the inertial forces $P_1^n$ and $P_1^\tau$ on the line joining axis $O_1$ with the center of the unbalanced mass; $\mu$ is the friction coefficient in the eccentric sleeve of the unbalanced mass; $d$ is the diameter of the eccentric sleeve.

The initial moment $M_{spi}$ of the twisted spring is so selected that the rotation of the unbalanced mass 1 would be realized at an angular velocity exceeding the frequency of natural oscillations of the vibratory machine.

Following the pivotal rotation of the unbalanced mass 1, the unbalanced mass 2 remains stationary for some time relative to the rotating disc (rotates jointly with it), since the tangential component of the inertial force generates a moment pressing it against stop 4. Rotation of the unbalanced mass commences only after the moment of the inertial force $P_2^n \, l$ relative to axis $O_1$ exceeds the sum of moments $P_2^\tau \, l$, the moment of the twisted spring, and the moment of frictional forces in the hinge of the unbalanced mass. With further acceleration of the mechanism, the rotation of both unbalanced masses takes place simultaneously until dynamic equilibrium is established.

Thus, in the mechanism under consideration, the unbalanced masses rotate in turns generating at the start a smooth load on the electric motor.

During the run-out process, the angular velocity of the unbalanced masses decreases and at a value somewhat exceeding the frequency of the natural oscillations of the vibratory machine, the spring moment overcomes the moment of the inertial forces which decreases sharply as the rotational speed of the vibration exciter shaft drops and the unbalanced masses return to their initial position. Rotation of the unbalanced masses during run-out is effected in a reversed order to starting, since the direction of the angular acceleration in this case changes to the opposite side. Further run-out of the vibrator and its transition through resonance takes place with a balanced unbalanced-mass mechanism (the kinetic moment of the unbalanced masses equal to zero).

To regulate the vibration amplitude, the vibration exciter must have an adjustment of the value of the exciting force at constant speed of rotation. The value of the exciting force is dependent on the eccentricity of the mechanism with all other parameters remaining the same. By changing the eccentricity, one can smoothly adjust the value of the exciting force and, correspondingly, the oscillation amplitude of the vibratory machine.

The adjustable unbalanced-mass mechanism has an eccentric assembly, on which the unbalanced mass is mounted, in the form of two eccentric sleeves, one placed inside the other and linked together by a threaded joint. The rotation of the sleeves relative to each other smoothly changes the value of the eccentricity of the mechanism.

The value of the total eccentricity $r$ of the mechanism depends on eccentricities of the sleeves $r_i$, $r_o$ and is determined by formula

$$r = (r_i^2 + r_o^2 + 2r_i r_o \cos \psi)^{\frac{1}{2}}$$

where $\psi$ is the angle of displacement of the sleeves relative to each other.

The design schematic of the automatic rotation of the unbalanced masses is given in Fig. 4.28. Coordinate frame $xy$, is used, whose origin coincides with the axis of rotation $O$ of the shaft of the vibration exciter.

The mechanisms are characterized by the following parameters: the unbalanced mass $m$; the eccentricity of the mechanism (the distance from the axis of rotation of the mechanism to the axis of rotation of the unbalanced mass) $r$; the distance from the rocking axis of the center of mass of the unbalanced mass $l$; the mass of the eccentric sleeve $m_1$ (the mass of the eccentric rocking sleeve of the unbalanced masses is concentrated at point $O_1$ at distance $r$ from the axis of rotation of the mechanism); the location angle of the unbalanced masses corresponding to the total balance of the mechanism $\sigma$; diameter of the eccentric sleeve $d$.

The location angle of the unbalanced masses corresponding to total balance of the mechanism is determined from the condition of static equilibrium of the static mass moments relative to the axis of rotation of the drive shaft $O$.

$$\sigma = \arccos \frac{r}{l}\left(1 + \frac{m_1}{2m}\right)$$

For the design of a vibration exciter with rotating unbalanced masses, one must know the dependence between the static moment of the mechanism and the angles of rotation of the unbalanced masses $\alpha_1$ and $\alpha_2$, which can generally be different. The resultant static moment of the mechanism is determined as the geometrical sum of the projections of the kinetic moments of each mass on the coordinate axis. When the angles of rotation of the unbalanced masses are equal $\alpha_1 = \alpha_2 = \alpha$, the resultant moment of the mechanism is determined from expression

$$M_{mech} = (m_1 + 2m)r - 2ml \cos(\sigma + \alpha)$$

The exciting forces of the vibration exciter are proportional to the kinetic moment of the mechanism and to the square of the angular rotational velocity; therefore, the resultant exciting force developed by the vibration exciter is equal to

$$P = \omega^2 \left[r(m_1 + 2m) - 2ml \cos(\sigma + \alpha)\right]$$

The moment of inertia of the unbalanced-mass mechanism relative to the axis of rotation $O$ is made up from the moment of inertia of the eccentric sleeve relative to axis $O$, the moment of inertia of the unbalanced mass $I_u$ relative to its c.g., and depends on the kinetic moment of the mechanism.

The moment of inertia of the unbalanced-mass mechanism of the vibration exciter is equal to

$$I_m = m_1 \left( r^2 + \frac{d^2}{8} \right) + 2I_u + 2m \left[ r^2 + l^2 - 2rl \cos (\sigma + \alpha) \right]$$

Dynamic analysis of the oscillation system, which is excited by an inertial vibration exciter with rotating unbalanced masses, can be conducted without significant errors by replacing the unbalanced-mass mechanism of the vibration exciter by an equivalent one-mass mechanism (Fig. 4.29).

As a criterion of equivalency of the mechanism, the equality of their kinetic moments is adopted. Let us introduce the following notations for parameters of equivalent mechanism: $m^*$ is equivalent mass, $m^* = m_1 + 2m$; $z^*$ is eccentricity,

$$r^* = z - \frac{2ml}{m_1 + 2m} \cos (\sigma + \alpha)$$

$f_o^*$ is pretension of the equivalent spring,

$$\delta_o^* = \theta \sqrt{\frac{k_{sp}}{k_{sp}^*}}$$

$k_{sp}^*$ is spring stiffness coefficient

$$k_{sp}^* = \frac{k_{sp}}{z^{*2}} \left\{ \left\{ \theta^2 + 4 \left[ \arccos \frac{m_1 + 2m}{2ml} (r + r^*) - \sigma \right] \right. \right.$$

$$\left. \left. \left[ \arccos \frac{m_1 + 2m}{2ml} (r + r^*) - \sigma + \theta \right] \right\}^{1/2} - \theta \right\}^2$$

The correlation of the eccentricity of the equivalent mechanism with the parameters of the investigated mechanism and angles of rotation of the unbalanced masses is determined from the condition of equality of the kinetic moments of both mechanisms.

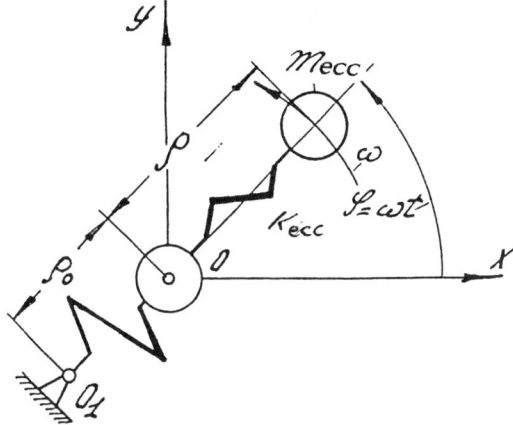

**Figure 4.29** Mathematical model of a vibration exciter with rotating unbalanced masses.

The value of the pretension $\delta_o^*$ of the spring and its stiffness coefficient $k_{sp}^*$ is determined from the condition of equality of the potential energies of the analyzed and equivalent mechanisms.

From the expressions for the spring pretention and equivalent spring stiffness, it follows that at an initial angle of twist of the spring $\theta$ equal to zero, the spring pretension of the equivalent mechanism is also equal to zero. The spring characteristic is nonlinear, since its stiffness coefficient is dependent on deformation $\delta^*$.

In the motion process of the oscillation system, the unbalanced masses of the vibration exciter mechanism are subjected to the action of an inertial force field, as a result of motion of the mechanism together with the working element of the vibratory system, and of a centrifugal force field generated upon rotation of the mechanisms. The accelerations of the inertial field force are small compared with the centrifugal accelerations acting on the unbalanced masses of the vibration exciter. This condition enables one to ignore the effect of the inertial forces when considering the system: vibration exciter-motor.

The motion of the system: drive-vibration exciter in the steady-state regime is described by the system of differential equations

$$\frac{r^*}{l}\,\omega^2 \sin(\sigma+\alpha) + \frac{\mu d}{2l}\omega^2 + \frac{\mu d r^*}{2l^2}\,\omega^2\cos(\sigma+\alpha) - \frac{k^*_{sp}}{ml^2}\,(\theta+\alpha_1+\alpha_2) = 0$$

$$\frac{r^*}{l}\,\omega^2 \sin(\sigma+\alpha) + \frac{\mu d}{2l}\omega^2 + \frac{\mu d r^*}{2l^2}\,\omega^2\cos(\sigma+\alpha) - \frac{k^*_{sp}}{ml^2}\,(\theta+\alpha_1+\alpha_2)$$

$$= 0, \quad M_m - M_r = 0$$

where $M_r$ is the dissipative resistance torque of the vibration exciter.

Analysis of the first two equations of the system reveals that the angles of rotation of the unbalanced masses in the steady-state regime are equal, i.e., $\alpha_1 = \alpha_2$. Zero difference of the moments in the third equation is a witness to the fact that the dissipative resistance torque is always balanced by the torque developed by the motor.

Since the angles of rotation of the unbalanced masses in the steady-state regime of motion are equal, the dependence of the angles of rotation of the unbalanced masses on the angular velocity and on parameters of the vibration exciter mechanism can be determined from the first equation of the system

$$(\sin\sigma + \frac{\mu d}{2l}\cos\sigma)\cos\alpha + (\cos\sigma - \frac{\mu d}{2l}\sin\sigma)\sin\alpha - \frac{2k^*_{sp}}{mrl\omega^2}\,\alpha - \frac{k^*_{sp}\theta}{mr^*l\,\omega^2}$$

$$-\frac{\mu d}{2r^*} = 0$$

This transcendental equation can be analyzed by a graphical method after it is reduced to a more convenient format (Fig. 4.30)

$$\cos\psi = c\psi + F$$

where

$$c = \frac{2k^*_{sp}}{mr^*l(\omega^2)}, \quad \psi = \alpha - \nu, \quad F = \frac{2K^*_{sp}\nu + \frac{1}{2}ml\,\mu d\omega^2 + k^*_{sp}\theta}{mr^*l\,\omega^2}$$

$\nu$ is the angle of shift of the cosinusoid with respect to the coordinate origin.

The left-hand side of the equation represents a cosinusoid, and the right-hand side is an equation of a straight line offset to the coordinate axis.

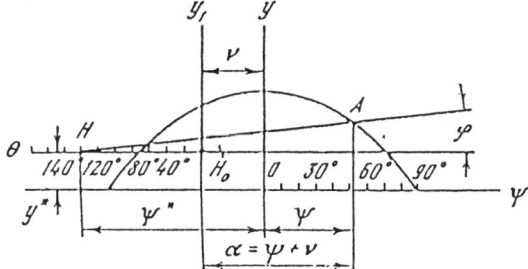

**Figure 4.30** Graphical determination of unbalanced-mass mechanism characteristics of inertial vibration exciter.

The real root of the equation is given by the abscissa of the point of intersection of the cosinusoid and the straight line. Since coefficient $c$ is the tangent of the angle of inclination of the straight line of the abscissa, we can write

$$\varphi = \arctan \frac{2k^*_{sp}}{mr^* l \omega^2}$$

Hence, it follows that angle $\varphi$ of the shaft rotation is inversely proportional to the square of angular velocity. For different angular velocities of rotation of the mechanism, the angle of inclination of the straight lines changes, but all the lines intersect at the pole (point $H$) of the graph. Let us denote the ordinate of the pole by $y^*$, and along the abscissa by $\psi^*$. Abscissa $\psi^*$ is determined by the simultaneous solution of two straight line equations corresponding to two arbitrarily selected angular velocities $\omega_1$ and $\omega_2$

$$\psi^* = \frac{F_2 - F_1}{c_1 - c_2}$$

where $c_1$, $c_2$, $F_1$, $F_2$ are coefficients for $\omega_1$ and $\omega_2$.

By analyzing the obtained expressions it can be seen that the ordinate of the pole for a specific unbalanced-mass mechanism is a constant which depends only on the mechanism parameters. The abscissa is dependent both on the mechanism parameters and on the initial angle of twist of the spring. When the angle of twist is zero, the abscissa $\psi^*$ is equal to the shift angle of the cosinusoid $\nu$.

If $\sigma = \pi/2$, a counterweight is needed for the initial balance of the unbalanced-mass mechanism. The ordinate of the pole $H$ and angle $\varphi$ of inclination of the straight line to the abscissa remain unchanged. The value of the angle of rotation of the unbalanced mass $\alpha$, in this case, is determined as the difference between angles $\psi$ and $v$ (see Fig. 4.30).

The angular velocity of the beginning of rotation of the unbalanced mass during starting is determined from the condition of dynamic equilibrium of the forces acting on the unbalanced mass relative to their axis of rotation. When the mechanism is rotated counterclockwise the first unbalanced mass separates from the stop first. The active forces acting on the unbalanced mass include the moment of the inertial forces. The rotation of the unbalanced masses is hindered by the moment of the frictional forces in the attachment hinge of the unbalanced masses and by the moment of the pretwisted spring.

From the equation of the boundary equilibrium of the forces acting on the unbalanced mass, one can determine the value of the angular velocity corresponding to the beginning of rotation of the unbalanced masses during starting. The angular velocity of the beginning of mutual rotation of the unbalanced masses during starting depends both on the parameters of the mechanism and on the angular acceleration of the vibration exciter. The higher the acceleration of starting, the lower the angular velocity at which transition of the unbalanced masses from the balanced position to the working position takes place. However, the effect of the starting acceleration on the value of the angular velocity corresponding to the rotation of the unbalanced masses is not significant.

Taking into account that the vibration exciter with rotating unbalanced masses does not require use of a drive motor with high starting moment, the starting period can be passed with small acceleration. In this case, the angular acceleration of the drive shaft can be disregarded.

The formula for the determination of the angular velocity of the beginning of mutual rotation of the unbalanced masses during starting period is

$$\omega_{start} = \sqrt{\frac{k^*_{sp}\,\theta - m\left[(l-r\cos\sigma)\,l+\frac{1}{2}\mu dr^*\sin\sigma\right]\ddot{\varphi}}{m\left[r^*l\sin\sigma-\frac{1}{2}\mu d\,)l - r^*\cos\sigma\right]}}$$

Thus, without introducing substantial simplifications, it can be ascertained that the angular velocity at the start of transition of the unbalanced mass from the idling position to the working position depends only on the parameters of the mechanism and on the initial angle of twist of the spring.

The angular velocity corresponding to the beginning of mutual rotation of the unbalanced masses during stopping process is determined from the equation of dynamic equilibrium of the unbalanced mass and has the form

$$
\omega_{stop} = \sqrt{\frac{k^*_{sp}(\theta + 2\alpha) - \left\{ l\left[ l - r^* \cos(\sigma + \alpha) \right] - \frac{1}{2} \mu d r^* \sin(\sigma + \alpha) \right\} m \ddot{\varphi}}{r^* l \sin(\sigma + \alpha) + \frac{1}{2} \mu d \left[ l - r^* \cos(\sigma + \alpha) \right]}}
$$

This expression indicates that the angular velocity depends on both the parameters of the mechanism and on the angle $\alpha$ of rotation of the unbalanced masses during starting of the vibrator.

By substituting $\alpha = 0$, we determine the angular velocity $\omega_b$ at which the unbalanced mass of the mechanism of the vibration exciter is totally balanced (switched off)

$$
\omega_b = \sqrt{\frac{k^*_{sp}\theta}{m\left[ r^* l \sin\sigma + \frac{1}{2} \mu d\,(l - r\cos\sigma) \right]}}
$$

The difference in the expressions for the angular velocity corresponding to the balancing of the unbalanced–mass mechanism during stopping and for the angular velocity determining the beginning of rotation of the unbalanced mass during starting is in the sign at the second term in the denominator. The minus sign is replaced by a plus. This means that for the same angle of twist of the spring $\theta$, the value of the angular velocity corresponding to total balancing of the mechanism during stopping is lower than the angular velocity of the beginning of mutual rotation of the unbalanced mass during starting. Therefore, detuning of the machine from resonance should be carried out for the stopping regime. It automatically provides detuning from the starting resonance.


## 4.3.3.3 Steady-State and Transient Operating Regimes of a Vibratory Machine with Rotating Unbalanced Masses

Investigations of regime parameters of the vibratory machine and the control mechanism for the kinetic moment of the unbalanced masses conducted at angular velocities of the drive shaft ranging from 0 to 2000 rpm and angles of twist of the unbalanced-mass mechanism equal to 60, 100, 140, and $200^{o}$ led to the following conclusions.

The angular velocity corresponding to total balance of the unbalanced-mass mechanism during run-out is somewhat lower than the angular velocity at which separation of the unbalanced masses from the stops takes place during starting. Accordingly, tuning of the unbalanced-mass mechanism should be conducted with respect to the run-out regime, then detuning from the starting resonance is achieved automatically.

From amplitude-frequency characteristics of the vibratory machine it follows that with increasing initial moment of the spring twist resonance, oscillation amplitudes decrease sharply. This is explained by the fact that at angles of spring twist exceeding $140^{o}$, engaging of the mechanisms in the acceleration process takes place at velocities above the resonant speed. Therefore, the vibration exciter passes the resonant frequency while balanced and without exciting the oscillation system. In the deep transresonant regime (detuning factor $z \geq 3$), the oscillation amplitude remains practically unchanged and has the same value as that of the vibratory machine without a control mechanism of the kinetic moment of the unbalanced masses.

Upon transition through resonance in conventional vibratory machines the moment of resistance to oscillations increases; therefore, for the implementation of such vibratory machines a substantial power reserve of the electric motor is needed. During the starting process of the vibratory machine with a vibration exciter equipped with an adjustable unbalanced mass, the consumed power by the electric motor upon transition through resonance also undergoes a peak change. However, with increased initial moment of spring twist, the peak power surge decreases, and under the initial spring moment corresponding to a twisting angle of $200^{o}$, it is absent altogether. In this case transition of the vibratory machine through resonance takes place in the unexcited state and without additional power consumption.

Thus, the use of vibration exciter with rotating unbalanced masses not only eliminates rsonant oscillations in transient regimes, but also reduces the starting power of the driving electric motor. This allows selection of the electric motor of the drive not with respect to the starting characteristics, but with respect to the required power in the stationary operating regime.

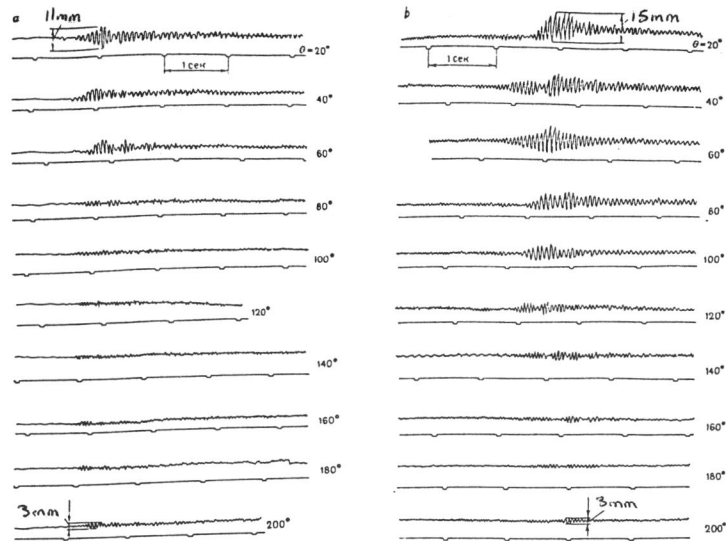

**Figure 4.31** Oscillograms of transient regimes of a vibratory machine excited by inertial vibrator with rotating unbalanced masses. a) starting; b) run-out.

Experimental study transient (starting and run-out) and steady-state regimes was performed to determine characteristic properties of the regimes of the vibratory machine driven by a vibration exciter with rotating unbalanced masses.

Figure 4.31a shows an oscillogram of displacement of the working element in a real machine. At an initial angle of spring twist of $20^o$, a maximum oscillation swing of 11 mm is obtained. With increasing angle of spring twist, the oscillation swing at the resonant frequency decreases and at angles of twist exceeding $120^o$, it remains 3 mm.

Thus, the proper tuning of the unbalanced masses of the mechanisms provides an approximately four times decrease of the resonant oscillation amplitudes upon starting of the vibratory machine.

The oscillograms of run-out are presented in Fig. 4.31b. Resonance at run-out, as expected, is more pronounced than at starting. At an angle of spring twist equal to $20^o$, the swing of resonant oscillations amounts to 15 mm as opposed to 11 mm at starting the oscillation system. As the initial moment of spring twist is increased, the amplitude of resonant oscillations decreases and the oscillation swing assumes the minimum value of 3 mm only at angles of twist that are equal to or exceeding 180°. Thus, the amplitude of resonant oscillations of the vibratory machine during run-out decreases sharply (by five times) and becomes equal to the starting value.

Analysis of oscillograms during starting and run-out indicates that detuning from resonance during run-out takes place at larger initial angles of spring twist, as was established theoretically earlier.

In order to explain the causes of this phenomenon, additional dynamic experiments of run-out of a vibratory machine were carried out. Studied were time histories of the angles of rotation of both unbalanced masses of the mechanism, effect of the initial moment of spring twist on the value of the resonant oscillation amplitude, and also angular velocities of the beginning and end of motion of the unbalanced masses into the initial balanced position.

Values of the angles of rotation of the unbalanced masses following run-out, starting, and during the steady-state regime were computed. Angular velocities of vibrator rotation corresponding to the beginning and end of motion of the unbalanced masses in the initial position were also determined by computational means.

Analysis of the oscillograms indicates that with increasing initial moment of spring tension, the motion of the unbalanced masses occurs with some time delay which is greater, the more substantial the angle of spring twist. Furthermore, the character of motion of the unbalanced masses also changes. At initial angles of spring twist equal to 60, 100, and 140°, uniform, constant-acceleration, and constant-deceleration motions in the relative displacement of the unbalanced masses were observed. However, at angles of twist of 200 and 240°, the motion of the unbalanced masses becomes strictly a constant-acceleration motion. Furthermore, with increasing angles of spring twist, the angular rotational acceleration of the unbalanced masses is increasing.

Because of significant angular accelerations of the unbalanced masses, the mechanism should have special devices to cushion impacts when the unbalanced masses return to the initial balanced position.

It also follows from the oscillograms that changing the initial angular spring twist effectively decreases resonant oscillation amplitudes of the vibratory machine at run-out. Thus, at an initial angle of twist equal to 60°, the swing of the resonant oscillations amounts to 10 mm, and at an angle of twist of $240^o$, the swing is just 2.5 mm. Thus, the resonant oscillation amplitude is reduced by a factor of four. It must be noted that at all tunings of the unbalanced-mass mechanisms, the working oscillation amplitude stays practically unchanged.

Starting regime was studied with the purpose of evaluating the character of motion of the unbalanced masses, determining their angles of rotation, angular velocities corresponding to the beginning of "take-off" of the unbalanced masses, and also determining the angular acceleration of the vibrator shaft.

The experimental results indicate that already at an initial angle of spring twist of $100^o$, the vibratory machine passes the resonant region with balanced unbalanced-mass mechanism. In this case, the swing of the resonant osciallations does not exceed 2 mm.

Motion of the left unbalanced mass occurs in the resonant region at a rotational speed of the mechanism of 910 rpm, and the right unbalanced mass at a speed of 1300 rpm. Consequently, the travel of the unbalanced masses to the working position during starting is staggered, thus generating a smooth load on the motor. The angles of rotation of both unbalanced masses in the steady-state regime in this case are equal to $30^o$.

In order to identify the effect of the oscillatory system of the vibratory machine on the parameters of the vibration exciter, i.e., the establishment of feedback parameters, additional experiments were conducted. In the first series of tests an additional viscous resistance to the motion was introduced by means of hydraulic dampers inserted between the working mass and the frame of the vibratory machine parallel to the elastic connections. The damping factor of the system amounted in this case to 1.3 1/sec. In the second series of experiments, stiffnesses of the elastic links were changed with no damping in the system.

The stiffness coefficient of the elastic links of the oscillation system was equal to 17100 kg/m, which provided resonance at a frequency of 372 1/min. In both cases, the angles of rotation of the unbalanced masses were measured as a function of the speed of rotation of the vibration exciter shaft, as well as angular velocities at which the relative motion commences of the unbalanced masses during starting and total balance is achieved of the mechanism at run-out of the oscillation system. Results of the experiments enabled one to establish that the introduction of resistances into the system, and also changing the stiffness of the elastic links do not noticeably affect the angular displacement of the unbalanced masses at the same rotational speeds of the vibration exciter.

The above stated provides the basis to conclude that the angles of rotation of the unbalanced masses are functions only of the rotational speed of the vibration exciter shaft, and the angular velocities corrsponding to switching the mechanism on during starting and to complete balance at run-out are dependent on the parameters of the unbalanced masses of the mechanisms and the initial moment of spring twist. Thus, both with introduction of resistances into the system and with the change of its natural oscillation frequency, the vibration exciter with rotating unbalanced masses ensures the elimination of resonant oscillations. Experience shows that the system provides reliable suppression of the resonant oscillations for various operating conditions of the vibratory machine.

### 4.3.4  Oscillation System with Eccentric Vibrator

### 4.3.4.1 Eccentric Vibrator with Elastic Connecting Rod

There are a number of vibratory machines in which the direction of the driving force does not coincide with one of the directions of the principal stiffness axes of the oscillation system. Such machines include, for example, vibratory conveyors with longitudinal excitation, and vibratory plants for the compaction of core sands in which the driving force acts along the axis of the working element and the elastic rubber elements are placed at an angle to the load-carrying element. The elastic supports are made as paired rubber rectangular parallelopipeds between which the bracket of the working element is mounted.

The design layout of the vibratory machine with longitudinal excitation includes an oscillating mass $m$ mounted on the frame with the aid of an elastic system working in shear along axis $x$ (stiffness in shear $k_x$, coefficient of viscous resistances $c_x$) and in compression along axis $Y$ (stiffness in compression $k_y$, coefficient of viscous resistances $c_y$). The elastic system creates a restoring force, and the damper models the hysteretic losses which are assumed proportional to the rate of strain. Drive is realized by an eccentric vibrator having an elastic connecting rod that is characterized by stiffness $k_0$ and by the coefficient of viscous resistances $c_0$ (Fig. 4.32).

The stiffness of the elastic elements along axis $X$ is a minimum (stiffness in shear), and along axis $Y$ is a maximum (stiffness in compression). Taking into account the variable force field generated by the rubber elements, the latter can be referred to as anisotropic elastic links. The force generated by the drive when the oscillator is stationary (unexcited) $F_0(t)$ is described by expression

$$F_0(t) = k_0 z_0 + c_0 \dot{z}_0 \qquad (4.40)$$

where $z_0$ is displacement of the point of the connecting rod of the eccentric mechanism, which is joined with the conveyor by an elastic link

$$z_0 = r \sin \omega t \qquad (4.41)$$

Under the action of excitation $F_0(t)$, mass $m$ is brought into oscillatory motion described by coordinates $x$ and $y$.

Let us resolve the driving force $F_0(t)$ into components along axes $X$ and $Y$

$$\begin{cases} F_{0x}(t) = (k_0 z_0 + c_0 \dot{z}_0) \cos \beta_0 \\ F_{0y}(t) = (k_0 z_0 + c_0 \dot{z}_0) \sin \beta_0 \end{cases} \qquad \begin{matrix}(4.42)\\(4.43)\end{matrix}$$

Upon excitation of the oscillation system, the driving elastic links along axis $X$ is deformed by the amount $(z_0 \cos \beta_0 - x)$, and along axis $Y$ by the amount $(z_0 \sin \beta_0 - y)$.

The components of the driving force that act from the side of the drive on the oscillatory system in operation are equal to

$$\begin{cases} F_x(t) = (k_0 z_0 + c_0 \dot{z}_0) \cos \beta_0 - k_0 x - c_0 \dot{x} \\ F_y(t) = (k_0 z_0 + c_0 \dot{z}_0) \sin \beta_0 - k_0 y - c_0 \dot{y} \end{cases} \qquad \begin{matrix}(4.44)\\(4.45)\end{matrix}$$

The differetial equations of motion of the vibratory system have the form

$$\begin{cases} m\ddot{x} + (c_0 + c_x)\dot{x} + (k_0 + k_x)x \\ \quad = r(k_0 \sin \omega t + c_0 \omega \cos \omega t)\cos \beta_0 \\ m\ddot{y} + (c_0 + c_y)\dot{y} + (k_0 + k_y)y \\ \quad = r(k_0 \sin \omega t + c_0 \omega \cos \omega t)\sin \beta_0 \end{cases} \qquad \begin{matrix}(4.46)\\ \\ (4.47)\end{matrix}$$

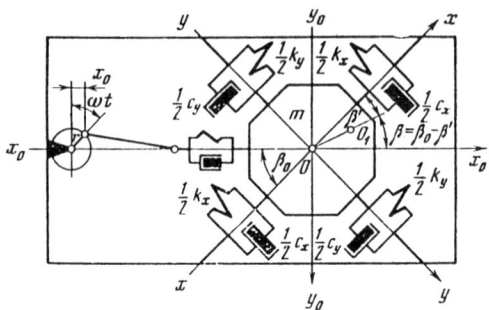

**Figure 4.32** Design schematic of a vibratory system with eccentric drive.

By dividing all the terms of the equations by the coefficient at the highest derivative $m$, we obtain

$$\begin{cases} \ddot{x} + 2n_{0x}\dot{x} + p_{0x}^2 x = P_0 \sin(\omega t + \varphi_0) \cos \beta_0 \\ \ddot{y} + 2n_{0y}\dot{y} + p_{0y}^2 y = P_0 \sin(\omega t + \varphi_0) \sin \beta_0 \end{cases} \tag{4.48}$$

Here and below the following notations are adopted (for abbreviation of the statement we denote $z = x, y$):

$P_0$ is the specific amplitude of the driving force in the unexcited (stationary) oscillation system

$$P_0 = \frac{r}{m} \sqrt{k_0 + c_0^2 \omega^2} \tag{4.49}$$

$\varphi_0$ is the angle of phase shift between the driving force $P_0$ and the displacement of the eccentric

$$\varphi_0 = \text{arctg} \frac{c_0 \omega}{k_0} \tag{4.50}$$

$n_{0z}$ is the effective coefficient of the total viscous resistances of the drive and of the elastic supports (in the direction of axes $X$ or $Y$)

$$2n_{0z} = \frac{c_0 + c_z}{m} \tag{4.51}$$

$P_{0z}$ is the natural frequency of the working element on the elastic links of the drive and the main system (in the direction of axes $X$ or $Y$)

$$p_{0z}^2 = \frac{k_0 + k_z}{m} \tag{4.52}$$

$p_z$ is the natural frequency of the working element of the vibratory machine on elastic supports (in the direction of axes $X$ or $Y$)

$$p_z^2 = k_z/m \qquad (4.53)$$

$p_0$ is the natural frequency of the working element of the vibratory machine on the elastic link of the drive

$$p_0^2 = k_0/m \qquad (4.54)$$

$n_z$ is the effective coefficient of viscous resistances of the elastic supports (in the direction of axes $X$ or $Y$)

$$2n_z = c_z/m \qquad (4.55)$$

$n_0$ is the effective coefficient of viscous resistances of the drive

$$2n_0 = c_0/m \qquad (4.56)$$

The solution of the equations is found in the form

$$z = A_z \sin(\omega t - \varphi_z) \qquad (4.57)$$

where

$$A_z = r_z \frac{k_0}{k_0 + k_z} \frac{\sqrt{1 + \dfrac{4n_0^2\omega^2}{p_0^4}}}{\sqrt{\left(1 - \dfrac{\omega^2}{p_{0z}^2}\right)^2 + \dfrac{4n_{0z}^2\omega^2}{p_{0z}^4}}} \qquad (4.58)$$

$$r_x = r\cos\beta_0 \qquad (4.59)$$

$$r_y = r\sin\beta_0 \qquad (4.60)$$

We introduce the notations:

$\lambda_0$ is the amplification factor of the elastic link of the drive

$$\lambda_0 = \sqrt{1 + \frac{4n_0^2\omega^2}{p_0^4}} \qquad (4.61)$$

$\mu_{0z}$ is the generalized dynamic coefficient of the drive and elastic supports (in the direction of axes $X$ or $Y$)

$$\mu_{0z} = \cfrac{1}{\sqrt{\left(1 - \dfrac{\omega^2}{p_{0z}^2}\right)^2 + \dfrac{4n_{0z}^2\omega^2}{p_{0z}^4}}}$$

(4.62)

Then we will have

$$A_z = r_z \frac{k_0}{k_0 + k_z} \lambda_0 \mu_{0z}$$

(4.63)

As the formulas show, the oscillation amplitudes are dependent on drive eccentricity $r$, angle of installation of the elastic elements $\beta_0$, the ratio of the stiffnesses of the drive and elastic supports, and also on the coefficients $\mu_{0z}$ and $\lambda_0$.

The phase shifts between the components of the displacement of the working element of the vibratory machine and the displacement of the eccentric are equal to

$$\lg \varphi_z = \frac{2n_{0z}\omega p_0^2 - 2n_0\omega (p_{0z}^2 - \omega^2)}{p_0^2 (p_{0z}^2 - \omega^2) + 4n_0 n_{0z}\omega^2}$$

(4.64)

The driving force as a function of time is obtained by substituting the values of $z_0$, $x$, $y$ and their derivatives into expression (4.44).

The solution is obtained in the form

$$F_z(t) = P_z \sin (\omega t - \psi_z)$$

(4.65)

where

$$P_z = r_z \frac{k_0 k_z}{k_0 + k_z} \frac{\lambda_0 \mu_{0z}}{\mu_z} = A_z k_z \frac{1}{\mu_z}$$

(4.66)

and

$$\mu_z = \cfrac{1}{\sqrt{\left(1 - \dfrac{\omega^2}{p_z^2}\right)^2 + \dfrac{4n_z^2\omega^2}{p_z^4}}}$$

(4.67)

is the dynamic coefficient of the elastic supports in the direction of the corresponding axis.

The angles of phase shift between the components of the driving force and the displacement of the eccentric is equal to

$$\lg \psi_z = \frac{A_{1z} (\lg \varphi_0 - \lg \varphi_z) - \lg \varphi_0 \sqrt{1 + \lg^2 \varphi_z}}{A_{1z} (1 + \lg \varphi_0 \lg \varphi_z) - \sqrt{1 + \lg^2 \varphi_z}}$$

(4.68)

where $A_{1z}$ denotes the dimensionless unitary amplitude of the displacement of the load-carrying element of the conveyer in the direction of the corresponding axis and is equal to

$$A_{1z} = \frac{k_0}{k_0 + k_z} \lambda_0 \mu_{0z} \tag{4.69}$$

The resultant value of the forces acting along the connecting rod is equal to

$$F(t) = P \sin(\omega t - \psi) \tag{4.70}$$

where

$$P = P_x \cos \beta_0 + P_y \sin \beta_0$$
$$= r k_0 \lambda_0 \left( \frac{k_x \cos^2 \beta_0}{k_0 + k_x} \frac{M_{0x}}{M_x} + \frac{k_y \sin^2 \beta_0}{k_0 + k_y} \frac{M_{0y}}{\mu_y} \right) \tag{4.71}$$

$$\psi = \psi_y - \psi_x \tag{4.72}$$

The dimensionless force characteristics of the system with longitudinal excitation $P/(rk_0 \lambda_0)$ as a function of the detuning factor $z$ ($z = \omega/p_{0x}$) for various values of the angles of installation of the elastic elements $\beta_0$, the ratio of the stiffnesses of the drive and main elastic links $u^2$ ($u = p_0/p_x$), and coefficients of inelastic resistance $\gamma$ ($\gamma = 2n_x/p_x = 2n_y/p_y = 2n_0/p_0$) are presented in Fig. 4.33.

The amplitude value of the driving force was computed from formula

$$\frac{P}{rk_0\lambda_0} = \cos^2 \beta_0 \frac{1}{u^2+1} \sqrt{\frac{[1 - z^2(u^2+1)]^2 + \gamma^2 z^2(u^2+1)}{(1-z^2)^2 + 4\gamma^2 z^2}}$$

$$+ \sin^2 \beta_0 \frac{6.5}{u^2+6.5} \sqrt{\frac{\left[1 - \frac{z^2(u^2+1)}{6.5}\right]^2 + \frac{\gamma^2 z^2(u^2+1)}{6.5}}{\left[1 - \frac{z^2(u^2+1)}{u^2+6.5}\right]^2 + \frac{4\gamma^2 z^2(u^2+1)}{u^2+6.5}}} \tag{4.73}$$

The relation $p_y = 6.5 p_x$, obtained during experimental investigations of elastic rubber elements, was used in the computations.

It follows from the graphs that the frequency-force characteristics have two peaks in the resonant region corresponding to the natural frequencies of the system $p_{0x}$ and $p_{0y}$. The amplitude of the driving force $P$ reaches maximum value in the resonant regime corresponding to the natural frequency $p_{0x}$, then decreases to a minimum value in the frequency regimes preceding this resonance and corresponding to the near-resonant tuning of the system at vibratory shear

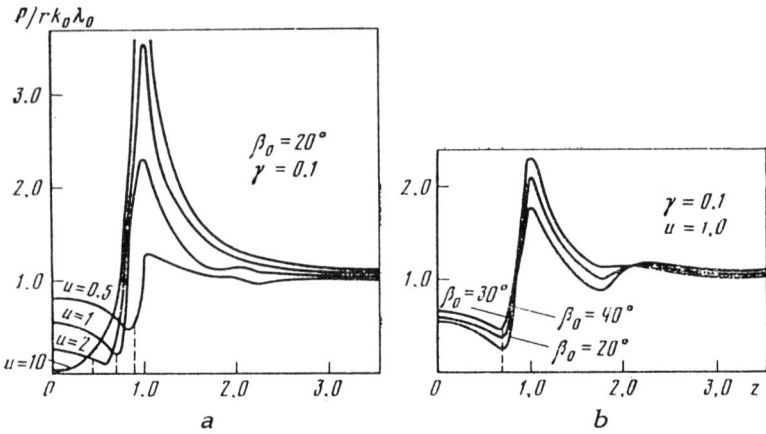

**Figure 4.33** Frequency-force characteristics of an oscillatory system with eccentric drive a) at various stiffness ratios of the drive and main elastic links; b) at various angles of installation of the elastic links.

resonance (in the graphs regime $\omega/p_\chi = 1$ is designated by dashed lines).

Not only the value of the force in the connecting rod is dependent on the ratio of the stiffnesses $k_0$ and $k_\chi$ (parameter $u$), but also the distance between the frequencies determining the regimes of tuning for a minimum driving force and for shear resonance. These regimes coincide only when $u = 1$, i.e., when the ratio of the stiffnesses $k_0/k_\chi = 1$, and such a tuning ($\omega = p_\chi$, $k_0 = k_\chi$) must be considered preferable.

The force characteristic for a rigid connecting rod, for ratio $k_0/k_\chi = 100$ ($u = 10.0$), is changing and resembles the amplitude-frequency characteristic for a system with one degree of freedom.

Increasing the value of $u$, i.e., making the drive links $k_0$ stiffer at a constant value of the total system stiffness ($k_0 + k_\chi = $ const) leads to a decrease in the amplitude values of the parameters of the system in the subresonant region.

The effect of the angle of installation $\beta_0$ of the elastic element is evident from the graphs. In the regimes encountered in practice the lower value of the force corresponds to the lower value of the angle (Fig. 4.33b).

The stiffness of the elastic supports included in the static and dynamic calculations can be determined as follows.

When the elastic system is deformed by force $P$ its application point $O$ shifts to position $O_1$.

When the joint action of both types of loadings is considered, it is assumed that the principle of independence of the action of the force (superposition) is valid. Strictly speaking, this principle is applicable only in the case of small strains of rubber elements when the influence of deformations caused by one of the loads on the resultant action of the other loads can be ignored.

If it is assumed that compression and shear are mutually independent, then the angle $\beta_{st}$ in static strain will be equal to

$$tg \ \beta'_{st} = f_y / f_x \tag{4.74}$$

where $f_y$ is the amplitude value of compression; $f_x$ is the amplitude value of shear.

Taking into account that

$$f_x = \frac{P \cos \beta_0}{k_x}, \qquad f_y = \frac{P \sin \beta_0}{k_y} \tag{4.75}$$

we have

$$tg \ \beta_{st} = \frac{k_x}{k_y} tg \ \beta_0 = \frac{G}{E} tg \ \beta_0 = \eta \ tg \ \beta_0 \tag{4.76}$$

where $G$, $E$ are the static moduli of shear and compression of the elastic rubber elements, respectively.

To determine dynamic rigidity of the elastic support in the direction of force $P$, we find the sum of the projections of the displacement components on the line of action of the force.

The oscillation amplitudes are

$$A_x = \frac{P \cos \beta_0 \mu_x}{m p_x^2} = \frac{P \cos \beta_0 \mu_x}{k_x} \tag{4.77}$$

$$A_y = \frac{P \sin \beta_0 \mu_y}{m p_y^2} = \frac{P \sin \beta_0 \mu_y}{k_y} \tag{4.78}$$

The sum of the displacment projections $A_x$ and $A_y$ on the line of action of the force is equal to

$$A_r = P \left( \frac{\cos^2 \beta_0 \mu_x}{k_x} + \frac{\sin^2 \beta_0 \mu_y}{k_y} \right) \tag{4.79}$$

Hence $P$ is equal to

$$P = \frac{A_r k_x}{\cos^2 \beta_0 \mu_x + \eta \sin^2 \beta_0 \mu_y} = A_r k_x \mu_{\beta_0} \tag{4.80}$$

where $A_\Gamma$ is the amplitude of displacement of mass in the direction of action of the force; $\mu_{\beta_0}$ is the generalized dynamic stiffness coefficient which is equal to

$$\mu_{\beta_0} = \frac{1}{\cos^2 \beta_0 \mu_x + \eta \sin^2 \beta_0 \mu_y} \tag{4.81}$$

Since stiffness of the supports in the force direction is $k_{z_0} = P/A_\Gamma$, we have

$$k_{z_0} = \mu_{\beta_0} k_x \tag{4.82}$$

For a rigid connecting rod

$$A_\Gamma = r \tag{4.83}$$

$$P = r k_x \mu_{\beta_0} \tag{4.84}$$

When $\omega = 0$, $\mu_x$ and $\mu_y$ are equal to 1, the expression $\mu_{\beta_0}$ for the static condition is

$$\mu_{\beta_{0,st}} = \frac{1}{\cos^2 \beta_0 + \eta \sin^2 \beta_0} \tag{4.85}$$

When $\omega = p_x$, deformation along axis $Y$ is absent and stiffness under shear resonance is equal to

$$k_{z_s} = \frac{k_x}{\cos^2 \beta_0 \mu_x} \tag{4.86}$$

$$P = \frac{r k_x}{\cos^2 \beta_0 \mu_x} \tag{4.87}$$

The calculations and formulas given above are obtained on the assumption of applicability of the principle of superposition to this computational model.

However, for rubber elements operating simultaneously under large compression and shear deformations, the use of the superposition principle requires experimental verification in each specific case.

Experimental investigations were conducted on a rig which enabled variation of the installation angle of the rubber elastic elements relative to the constant direction of the force acting along the axis of the connecting rod.

Dynamic tests were conducted in the regime of forced oscillations at different eccentricities ($r$ = 3 - 10 mm), oscillation frequencies ($n$ = 0 - 1200 osc/min) and installation angles of the elements ($\beta_0$ = 0 - 40$^o$) with pairs of rubber elements having the dimensions 80x50x40 mm and made of rubber mixture 1847. These elements were mounted in the vibratory conveyer KVZhG-200-P with longitudinal excitation designed by VNIIPTMASh. Figure 4.34 shows segments of experimental force-frequency characteristics of the system for different installation angles of the elements. The curves obtained from formula (4.85) are shown by the dashed lines in the same figure. The deviation between the theoretical and experimental data is no more than 15%, which is an indication of the adequate accuracy of the proposed method of calculation. The minimum value of the force in the connecting rod is obtained at shear resonance, i.e., when $\omega/p_x$ = 1.

As the plots show, by rotating the elastic supports relative to the direction of excitation, the force characteristics of the system can be varied at constant value of the oscillation frequencies.

### 4.3.4.2 Eccentric Drive Oscillatory System with Variable Stiffness of the Driving Elastic Link

Reduction of starting forces and of installed power of the motor in resonant vibratory machines is achieved by using an eccentric drive with elastic link, the stiffness of which varies in a sinusoidal law.

The fundamental design of such a drive with the said characteristic of the elastic links is depicted in Fig. 4.35a. The driving elastic link is made as a rubber-metal elastic eccentric 1 mounted on shaft 2. The elastic block of the eccentric has two through holes 3 and 4, which create zones with reduced radial stiffness in the direction I - I, at an angle $\pi/4$ with the direction of the eccentric $r$.

After the start of the machine, with rotation of the driving shaft by angle $\omega t = \pi/4$ axis I - I of the through holes coincides with the direction $x - x$ of the restoring forces $P$ of the main elastic links. Since the elastic eccentric in this position has minimum stiffness and resistance of the main system, having rather high stiffness, is large, the rubber block

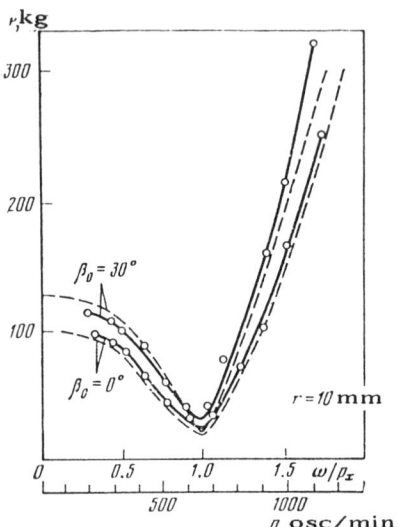

**Figure 4.34**  Experimental frequency-force characteristics.

is deformed, thus decreasing significantly eccentricity $r$. The load-carrying element of the vibratory machine receives only minor excitation. This condition enables the driving shaft to turn easily, overcoming a relatively small torque. Eventually the shaft of the motor accelerates, gradually rocking of the load-carrying element is intensified, and with the decreasing magnitude of the exciting force the effective eccentricity is increasing.

A mathematical model of a harmonic oscillator driven by an elastic eccentric is shown in Fig. 4.35$b$. The main elastic system $1$ has constant stiffness $k$, the drive system $2$ has variable stiffness which changes by the sinusoidal law $k_1 = k_0 - k_a \sin 2\omega t$ with a frequency equal to twice the frequency of rotation of the driving shaft.

The load-carrying element with mass $M$ describes motion $x$ with velocity $\dot{x}$. The elastic eccentric is strained by the value $x_1 = x_{01} - x$ at the rate $\dot{x}_1 = \dot{x}_{01} - \dot{x}$. Here, $x_{01}$ and $\dot{x}_{01}$ are projections of the displacement and velocity of the center of the outer ring of the elastic eccentric on axis $x$.

The oscillation of the vibratory system with harmonically changing stiffness of the driving elastic links is described by differential equation with periodic coefficient at displacement $x$

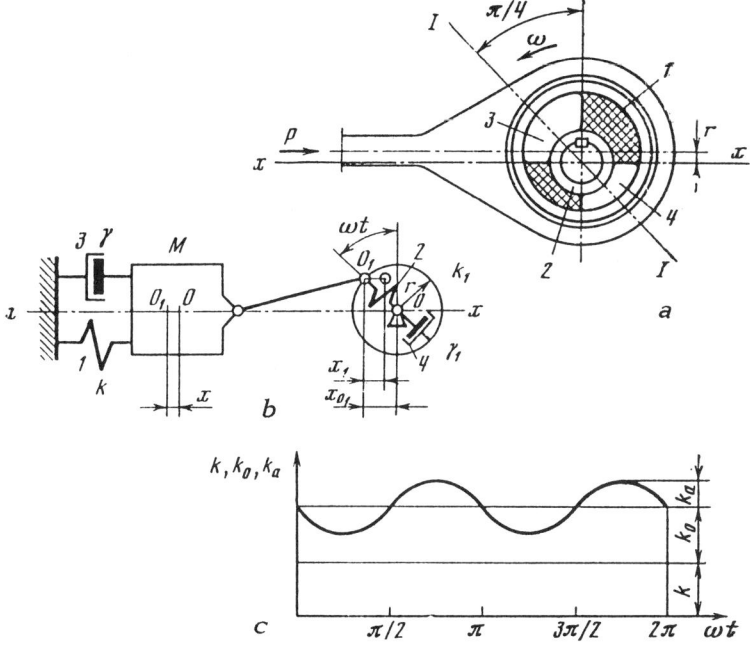

**Figure 4.35** Harmonic oscillator with elastic eccentric in the drive. *a)* design schematic; *b)* mathematical model; *c)* correlation of stiffness of the elastic links with angle of rotation of the driving shaft.

$$M\ddot{x} + \left[ \frac{\gamma}{\omega} k + \frac{\gamma_1}{\omega} k_0 \left( 1 - \frac{k_a}{k_0} \sin 2\omega t \right) \right] \dot{x}$$
$$+ (k + k_0)\left( 1 - \frac{k_a}{k + k_0} \sin 2\omega t \right) x$$
$$= k_0 r \left( 1 - \frac{k_a}{k_0} \sin 2\omega t \right)(\sin \omega t + \gamma_1 \cos \omega t) \tag{4.88}$$

Computation of energy losses to overcome resistance in the driving elastic system is based on the mean value of stiffness $k_0$ of the elastic eccentric.

By dividing all the terms of equation (4.88) by the coefficient at the second derivative, we can write it in a form convenient for analysis

$$\ddot{x} + 2n\dot{x} + p^2 (1 - \lambda \sin 2\omega t) x$$
$$= \frac{k_0}{M} r (1 - \Delta \sin 2\omega t)(\sin \omega t + \gamma_1 \cos \omega t) \tag{4.89}$$

where $n$ is the damping factor of the elastic system, $n = 1/2$ $p^2$ $(\gamma\ 1/1+\chi + \gamma_1\ \chi/1+\chi)\ 1/\omega$; $\chi$ is the ratio between the mean stiffness of the elastic eccentric $k_0$ and the stiffness of the main elastic links $k$, $\chi = k_0/k$; $p$ is the natural frequency of mass $M$ on the main and driving elastic connections, $p = \sqrt{k+k_0/M}$; $\lambda$ is the coefficient of stiffness variation of the elastic system, $\lambda = k_0/k+k_0$; $\varDelta$ is the coefficient of stiffness variation of the elastic eccentric, $\varDelta = k_a/k_0$.

The characteristic property of the motion of an elastic system with periodically changing stiffness is the fact that for a specific ratio of the natural frequency of the load-carrying element and the frequency of stiffness variation of the elastic links the restoring forces of the eccentric perform positive work over a cycle of oscillations. In this case, the elastic eccentric moves relative to the driving shaft, thus amplifying the basic oscillations of the load-carrying element, i.e., a parametric regime is created in the system. This condition can lead to the unstable operation of the vibratory machine, and for relatively small sizes of the eccentric it can also lead to intensive wear of the rubber block.

To determine conditions of generation of parametric oscillations, it is sufficient to analyze the behavior of the system whose motion is described by homogeneous equations corresponding to equation (4.90). By substituting $x = ye^{-(n/\omega)\tau}$, where $\tau = \omega t$ is dimensionless time, the equation is reduced to the Mathieu equation in the standard form

$$\ddot{y} + [v - \lambda f\ (\tau)]y = 0 \tag{4.90}$$

where for sufficiently small values of $\gamma$, $\gamma_1$ and $p/\omega < 1$, we have

$$f\ (\tau) = v \sin 2\tau; \quad v = p^2/\omega^2 \tag{4.91}$$

Substitution into equation (4.90) of two of its independent solutions $y_1\ (\tau)$ and $y_2\ (\tau)$ in the form of expansions

$$\begin{aligned} y_1\ (\tau, v, \lambda) &\approx y_{10}\ (\tau, v) + y_{11}\ (\tau, v,)\ \lambda + \ldots \\ y_2\ (\tau, v, \lambda) &\approx y_{20}\ (\tau, v) + y_{21}\ (\tau, v,)\ \lambda + \ldots \end{aligned} \tag{4.92}$$

yields, after equating the coefficients for equal powers of $\lambda$, the recurrent equation

$$\ddot{y}_{10} + vy_{10} = 0, \qquad \ddot{y}_{20} + vy_{20} = 0$$

$$\ddot{y}_{11} + vy_{11} = f\ (\tau)\ y_{10}, \qquad \ddot{y}_{21} + v\ y_{21} = f\ (\tau)\ y_{20}$$

whose solution for the initial conditions $y_1(0) = 1$, $\dot{y}_1(0) = 0$, $y_2(0) = 0$ and $\dot{y}_2(0) = 1$ enables one to write

$$y_1 (\tau, \nu, \lambda) \approx \cos \sqrt{\nu}\tau; \quad y_2 (\tau, \nu, \lambda) \approx (1/\sqrt{\nu}) \sin \sqrt{\nu}\tau$$

From the normal solutions it follows that the solutions of equation (4.90) are limited along the whole length of axis $\tau$ at $|A_y| < 1$, are unlimited when $A_y > 1$ ($A_y < -1$), and have a periodic character if $|A_y|^2 = 1$, where

$$A_y = \tfrac{1}{2} [y_1 (\pi) + \dot{y}_2 (\pi) \approx \cos \sqrt{\nu}\pi] + \tfrac{1}{2} [y_{11} (\pi) + \dot{y}_{21}\pi)] \lambda \qquad (4.93)$$

Since the regions on plane $\nu$. $\lambda$, corresponding to the limited solutions of equation (4.90) are separated from the regions of the unlimited solutions by a critical line defined by condition $|A_y|^2 = 1$, it follows (4.93) that this critical line intersects axis $\nu$ at the points $\nu = z^2$, where $z = 0, 1, 2, 3$ (natural series).

Intervals of axis $\nu$ between these points, which correspond to condition $|A_y| < 1$, belong to the regions of limited solutions.

For the approximate determination of the boundary between the regions of stable and unstable solution, which is passing through point $\nu = 1$, we use the method of undetermined coefficients and substitute the following expansions into (4.90)

$$y = y_0 (\tau) + y_1 (\tau) \lambda + y_2 (\nu) \lambda^2 + \ldots \qquad (4.94)$$
$$\nu = 1 + a\lambda + b\lambda^2 + \ldots \qquad (4.95)$$

By equating the coefficients for equal powers of $\lambda$, we obtain the following reduction equations

$$\ddot{y}_0 + y_0 = 0$$
$$\ddot{y}_1 + y_1 = - (a - \sin 2\tau) y_0 \qquad (4.96)$$
$$\ddot{y}_2 + y_2 = - (a - \sin 2\tau) y_1 - (b - a \sin 2\tau) y_0$$

Solving the system of equations (4.96) based on condition $y(0,\lambda) = 1$, when according to (4.94) $y_0(0) = 1$, $y_1(0) = 0$, $y_2(0) = 0$, we obtain $a_{1,2} = \pm 1/2$ and $b = 13/64$.

Substitution of the obtained results into (4.95) yields

$$\nu \approx 1 \pm \tfrac{1}{2}\lambda + \tfrac{13}{64}\lambda^2 \qquad (4.97)$$

We further have
$$\frac{\omega}{p} = \frac{1}{\left(1 \pm \frac{1}{2}\lambda + \frac{13}{64}\lambda^2\right)^{1/2}} \ .$$

By expanding the right-hand side of the last expression into Maclaurin series, we finally arrive at the approximate equations corresponding to two directions of the intersection of the critical line with axis $v$ when $v = z^2 = 1$

$$\frac{\omega}{p} \approx 1 \pm \frac{1}{4}\lambda - \frac{1}{64}\lambda^2 \tag{4.98}$$

The approximate equations of the critical lines passing through points $v = z^2 = 4$ and $v = z^2 = 9$ are found by an analogous method. These critical lines, based on the assumed normalization of function $y$, have only one branch each and are described respectively in the following forms

$$v \approx 4 + \frac{7}{3}\lambda^2 \tag{4.99}$$

or

$$\frac{\omega}{p} \approx \frac{1}{2} - \frac{7}{24}\lambda^2 \tag{4.100}$$

or

$$v \approx 9 - \frac{275}{64}\lambda^2 \tag{4.101}$$

$$\frac{\omega}{p} \approx \frac{1}{3} + \frac{1}{6}\lambda^2 \tag{4.102}$$

Equations (4.98), (4.100), and (4.102) enable one to construct a diagram characterizing the oscillatory motion of a vibratory machine with elastic eccentric as a function of the combinations of parameter $\lambda$ and the excitation frequency $\omega$ of the system.

In Figure 4.36a curves *1, 2, 3,* and *4* form a set of points whose coordinates in plane $\lambda$, $\omega/p$ correspond to the values of the system at which the elastic eccentric performs periodic oscillations that are limited in amplitude with respect to the axis of the driving shaft. The curves divide the diagram into a number of regions. The combinations of the values pertaining to domains *F, G, H,* and *I* correspond to the nonperiodic decaying oscillations and characterize stability of the system. Region *Q* bounded by curves *3* and *4* which intersect with axis

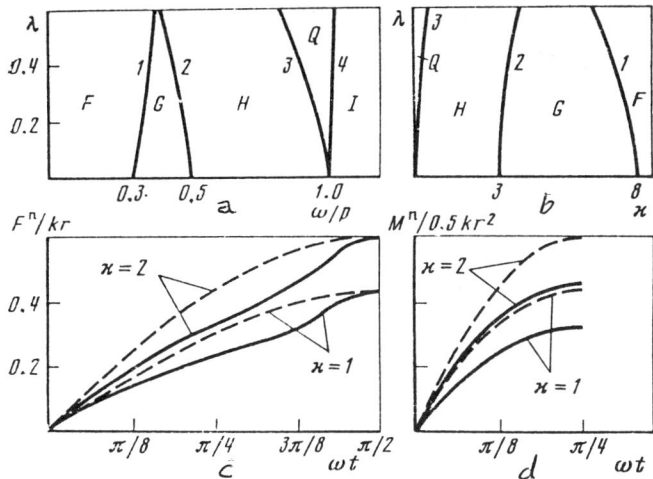

**Figure 4.36** Dependence of the character of oscillations of a vibratory system with harmonically variable stiffness of the driving elastic link (with elastic eccentric in the drive) on the ratios: the exciting $\omega$ and the natural $p$ frequencies of the system (a); stiffnesses of the driving $k_0$ and main $k$ elastic links (b); F, G, H, I are regions of stable oscillations; Q is the region of parametric resonance; 1, 2, 3, 4 are curves of parametric periodic oscillations; diagrams of correlation of the starting force $P_{start}$ (c) and moment $M_{start}$ (d) components (d) on the angle of rotation of the driving shaft with elastic eccentric (solid lines) and elastic connecting rod (dashed lines).

$\omega/p$ at point 1 corresponds to the values of the system at which parametric resonance exists. This means that the elastic eccentric in the present case, along with the working element, develop increasing unlimited oscillations.

Figure 4.36b shows a similar diagram which characterizes the oscillatory system tuned to the oscillator resonance. The diagram is plotted in the plane of parameters $\lambda$ and $\chi = k_0/k$ in accordance with equations

$$\chi = \pm \frac{1}{2} \lambda + \frac{13}{64} \lambda^2$$

$$\chi = 3 + \frac{7}{3} \lambda^2$$

$$\chi = 8 - \frac{275}{64} \lambda^2$$

obtained upon transformation of expressions (4.97), (4.98), and (4.101), taking into account the assumed ratios $p = \sqrt{(k + k_0)/M}$, $\nu = p^2/\omega^2$ and substitution $\omega = p_1 = \sqrt{k/M}$.

Figure 4.36$b$ shows that when $\lambda \leq 0.25$ and $\chi = 1 - 2$ there is no parametric resonance in the system. However, when $\chi > 2$, there is a danger of generating periodic parametric oscillations which take place in the system when its values of $\lambda$ and $\chi$ correspond to the points lying on the curves intersecting axis $\chi$ at points 3 and 8.

If the combination of parameters $\lambda$ and $\chi$ and the ratio $\omega/p$ correspond to the domains of stable solution of the homogeneous equation (4.90), then the vibratory system performs a stable motion. In this case, the approximate analytical expression of the motion of the load-carrying element, obtained with the application of the Krylov-Bogolyubov asymptotic method and the variation method of the Lagrange arbitrary constants, has the form

$$x \approx A \sin (\omega t - \varphi) \tag{4.103}$$

where the displacement amplitude of the harmonic oscillator is

$$A = r \frac{\chi}{1 + \chi} \sqrt{ \frac{ \left(1 - \gamma_1 - \frac{\Delta}{2}\right)^2 + \left(\gamma_1 - \frac{\Delta}{2}\right)^2 }{ \left(1 - \frac{\omega^2}{p^2}\right)^2 + \left(\gamma \frac{1}{1 + \chi} + \gamma_1 \frac{\chi}{1 + \chi}\right)^2 } } \tag{4.104}$$

The phase shift between the displacements of the harmonic oscillator and displacement $x_{01}$ of the elastic eccentric is

$$\varphi = \operatorname{arctg} \frac{ \left(1 - \gamma_1 \frac{\Delta}{2}\right)\left(\gamma \frac{1}{1 + \chi} + \gamma_1 \frac{\chi}{1 + \chi}\right) - \left(1 - \frac{\omega^2}{p^2}\right)\left(\gamma_1 - \frac{\Delta}{2}\right) }{ \left(1 - \frac{\omega^2}{p^2}\right)\left(1 - \gamma_1 \frac{\Delta}{2}\right) + \left(\gamma \frac{1}{1 + \chi} + \gamma_1 \frac{\chi}{1 + \chi}\right)\left(\gamma_1 - \frac{\Delta}{2}\right) } \tag{4.105}$$

It follows from formula (4.104) that for sufficiently small damping in the elastic system and $\Delta < 1$, the change of stiffness has so little influence on the amplitude that it can be neglected. This condition is an indication that during stationary operation the vibratory machine performs oscillations while realizing the mean value of stiffness $k_0$. For this reason its amplitude-frequency, force, and energy characteristics in the oscillatory resonant regime will not substantially differ from the analytical parameters of the vibratory machine with elastic connecting rod whose stiffness is constant and equal to the mean stiffness of the elastic eccentric $k_0$.

In order to represent the exciting force $F$ of the drive as a function of time, equation (4.88) is transformed, leaving in the left-hand side only the terms describing the motion of a linear oscillator

$$M\ddot{x} + \left(\gamma\,\frac{1}{1+\chi} + \gamma_1\,\frac{\chi}{1+\chi}\right)\frac{1}{\omega}\,k\dot{x} + kx = k_0 r\,(1 - \Delta\sin 2\omega t)$$
$$\times\,(\sin\omega t + \gamma_1\cos\omega t) - k_0\,(1 - \Delta\sin 2\omega t)\,x$$
$$-\left(\gamma\,\frac{1}{1+\chi} + \gamma_1\,\frac{\chi}{1+\chi}\right)\frac{1}{\omega}\,k_0\dot{x} \tag{4.106}$$

Obviously, the right-hand side of equation (4.106) is an expression of the exciting force acting from the drive to the excited oscillator.

By substituting expression (4.103) into the left-hand side of the equation and taking into account (4.104) and (4.105), we obtain

$$F = P\sin(\omega t - \psi)$$

where the amplitude of the exciting force is

$$P = kr\,\frac{\chi}{1+\chi}\sqrt{\frac{\left[\left(1 - \dfrac{\omega^2}{p_1^2}\right)^2 + \gamma^2\right]\left[\left(1 - \gamma_1\dfrac{\Delta}{2}\right)^2 + \left(\gamma_1 - \dfrac{\Delta}{2}\right)^2\right]}{\left(1 - \dfrac{\omega^2}{p^2}\right)^2 + \left(\gamma\,\dfrac{1}{1+\chi} + \gamma_1\,\dfrac{\chi}{1+\chi}\right)^2}} \tag{4.107}$$

and the phase shift between the exciting force and the displacement $x_{01}$ of the drive eccentric is

$$\psi = \operatorname{arctg}\frac{\left(1 - \dfrac{\omega^2}{p_1^2}\right)\left[\left(\gamma\,\dfrac{1}{1+\chi} + \gamma_1\,\dfrac{\chi}{1+\chi}\right)\left(1 - \gamma_1\dfrac{\Delta}{2}\right) - \left(1 - \dfrac{\omega^2}{p^2}\right)\right.}{\left(1 - \dfrac{\omega^2}{p_1^2}\right)\left[\left(1 - \dfrac{\omega^2}{p^2}\right)\left(1 - \gamma_1\dfrac{\Delta}{2}\right) + \left(\gamma\,\dfrac{1}{1+\chi} + \gamma_1\,\dfrac{\chi}{1+\chi}\right)\right.}$$

$$\frac{\left.\left(\gamma_1 - \dfrac{\Delta}{2}\right)\right] - \gamma\left[\left(1 - \dfrac{\omega^2}{p^2}\right)\left(1 - \gamma_1\dfrac{\Delta}{2}\right) + \left(\gamma\,\dfrac{1}{1+\chi} + \gamma_1\,\dfrac{\chi}{1+\chi}\right)\left(\gamma_1 - \dfrac{\Delta}{2}\right)\right]}{\left.\left(\gamma_1 - \dfrac{\Delta}{2}\right)\right] + \gamma\left[\left(\gamma\,\dfrac{1}{1+\chi} + \gamma_1\,\dfrac{\chi}{1+\chi}\right)\left(1 - \gamma_1\dfrac{\Delta}{2}\right) + \left(1 - \dfrac{\omega^2}{p^2}\right)\left(\gamma_1 - \dfrac{\Delta}{2}\right)\right]}$$

It follows from expression (4.107) that the amplitude value of the exciting force of the drive is directly proportional to the eccentricity $r$ and stiffness $k$ of the main elastic system, is dependent on the ratio $\chi$ of the mean stiffness $k_0$ of the elastic eccentric and stiffness $k$ of the main elastic links, and also on the tuning regime of the harmonic oscillator. The exciting force $P$ has a minimum value in the oscillation resonant regime when the condition $\omega = p_1 = \sqrt{k/M}$ is satisfied.

The time moment of reaching the amplitude value of the exciting force does not coincide with the time of reaching the maximum travel $x_{01}$ of the elastic eccentric and is determined by the damping characteristics $\gamma$ and $\gamma_1$ of the system and by the excitation frequency $\omega$.

The work done by the exciting force of the drive with an elastic eccentric per oscillation cycle of the system is determined from formula

$$W = \int_0^{T=2\pi/\omega} F(t)\,\dot{x}\,dt = \pi PA \sin(\varphi - \psi)$$

By substituting into this formula the amplitude values of the displacement of the load-carrying element according to (4.104) and the exciting force according to (4.107), we obtain the final expression in the following form

$$W = \pi k r^2 \frac{\chi^2}{(1+\chi)^2} \frac{\left(1 - \gamma_1 \frac{\Delta}{2}\right)^2 + \left(\gamma_1 - \frac{\Delta}{2}\right)^2}{\left(1 - \frac{\omega^2}{p^2}\right)^2 + \left(\gamma \frac{1}{1+\chi} + \gamma_1 \frac{\chi}{1+\chi}\right)^2}$$
$$\times \sqrt{\left(1 - \frac{\omega^2}{p_1^2}\right)^2 + \gamma^2} \, \sin(\varphi - \psi)$$

The power consumed by the drive with an elastic eccentric to maintain the steady-state motion of the harmonic oscillator and expended to make up for the losses in the main elastic system is equal to

$$N = \frac{W}{T} = \frac{1}{2} k r^2 \omega \frac{\chi^2}{(1+\chi^2)} \frac{\left(1 - \gamma_1 \frac{\Delta}{2}\right)^2 + \left(\gamma_1 - \frac{\Delta}{2}\right)^2}{\left(1 - \frac{\omega^2}{p^2}\right)^2 + \left(\gamma \frac{1}{1+\chi} + \gamma_1 \frac{\chi}{1+\chi}\right)^2} \qquad (4.108)$$
$$\times \sqrt{\left(1 - \frac{\omega^2}{p_1^2}\right)^2 + \gamma^2} \, \sin(\varphi - \psi)$$

Calculations show that the oscillation amplitude of the resonant vibratory machines, having elastic elements in their drives, is close but not equal to the eccentricity of the driving shaft. This is explained by the fact that the driving elastic link takes part in the work, in addition to the main elastic link whose deformation is determined by the equality

$$x_1 = |x_{0_1} - x|$$

where

$$x_{p1} = r \sin \omega t$$

Taking this into consideration, the dissipation of power in the elastic material of the eccentric on displacement $x_1$ is determined from formula

$$N_1 = \frac{\omega}{2\pi} \int\limits_0^{2\pi/\omega} F(t) \, | \dot{x}_{0_1} - \dot{x} | \, dt$$

which, accounting for the expressions derived above, becomes

$$N_1 = \frac{1}{2} kr^2 \omega \frac{\chi}{1+\chi}$$

$$\times \sqrt{\frac{\left[\left(1 - \frac{\omega^2}{p_1^2}\right)^2 + \gamma^2\right]\left[\left(1 + \gamma_1 \frac{\Delta}{2}\right)^2 + \left(\gamma_1 - \frac{\Delta}{2}\right)^2\right]}{\left(\gamma \frac{1}{1+\chi} + \gamma_1 \frac{\chi}{1+\chi}\right)^2 + \left(1 - \frac{\omega^2}{p^2}\right)^2}}$$

$$\times \left[\frac{\chi}{1+\chi} \sqrt{\frac{\left(1 - \gamma_1 \frac{\Delta}{2}\right)^2 + \left(\gamma_1 - \frac{\Delta}{2}\right)^2}{\left(\gamma \frac{1}{1+\chi} + \gamma_1 \frac{\chi}{1+\chi}\right)^2 + \left(1 - \frac{\omega^2}{p^2}\right)^2}}\right.$$

$$\left. \times \sin(\psi - \varphi) - \sin\psi \right] \tag{4.109}$$

Summation of power components according to (4.108) and (4.109) characterizes the power $N_0$, which is consumed by the drive with elastic eccentric to maintain the steady-state oscillations of the system and spent to make up the total hysteresis losses in the driving and main elastic links.

During the starting period of the vibratory machine simultaneous deformation of the rubber volume of the elastic eccentric and the main elastic links takes place. From kinematic considerations the following condition should be observed in this case

$$x^s + x_1^s = r \cos \omega t \tag{4.110}$$

where $x^s$ and $x_1^s$ are, respectively, deformations of the main elastic links and of the elastic eccentric in the initial starting period.

We assume that the rotation of the driving shaft during the first quarter of revolution takes place so slowly that the inertial forces of the load-carrying element and the internal resistance in the materials of the elastic links can be ignored.

Then, the force acting in the connecting rod from the side of the main elastic system is written as

$$F^s = kx^s \qquad\qquad (4.111)$$

or from the side of the driving shaft as

$$F^s = k_1 (2\omega t)x_1^s \qquad\qquad (4.112)$$

Equating expressions (4.111) and (4.112) and substituting into the equality the values $x^s$ from condition (4.110) and $k_1 = k_0 - k_0 \sin 2\omega t$ with consideration of the ratio $\chi = k_0/k$ yields the correlation of the exciting force of the drive with an elastic eccentric on the angle of rotation of the driving shaft

$$F^s = kr \frac{\chi}{1+\chi} \frac{1 - \Delta \sin 2\omega t}{1 - \lambda \sin 2\omega t} \sin \omega t \qquad\qquad (4.113)$$

The largest deformation of the main elastic links during the starting period of the vibratory machine is reached at a rotation of the driving shaft by angle $\omega t$ equal to $\pi/2$. At this value of angle $\omega t$, maximum force is developed on the connecting rod which is equal to

$$F^s_{max} = kr \frac{\chi}{1+\chi}$$

As a sufficiently small angle of deformation of the elastic eccentric, the arm of action of force $F^S$ can be assumed

$$r_F = r \cos \omega t \qquad\qquad (4.114)$$

The product of (4.113) and (4.114) represents an expression of the starting moment of the drive as a function of the angle of rotation $\omega t$ of the driving shaft

$$M^S = 0.5kr^2 \frac{\chi}{1+\chi} \frac{1 - \Delta \sin 2\omega t}{1 - \lambda \sin 2\omega t} \sin 2\omega t \qquad\qquad (4.115)$$

By differentiating the right-hand side of (4.115) and then equating the obtained expression to zero we determine the angle of rotation $\omega t$ corresponding to the maximum value of the starting moment $M^S_{max}$

$\omega t = \pi/4$

Consequently, the maximum value of the starting moment developed at the driving shaft of the vibratory conveyor with an elastic eccentric is equal to

$$M^S_{max} = 0.5kr^2 \frac{\chi}{1+\chi} \frac{1-\Delta}{1-\lambda} \qquad (4.116)$$

From expression (4.116) it is evident that the product of the first three multipliers determines the maximum value of the starting moment of the kinematically rigid drive. The product of the second and third partial multipliers characterizes the relative value of the reduction of the starting moment when a drive with elastic eccentric is used. Moreover, the latter, in its turn, can be regarded as the product of two factors

$$k_{e.cr} = \frac{\chi}{1+\chi}, \qquad k_{e.e} = \frac{1-\Delta}{1-\lambda}$$

which respectively determine the degree of reduction of the value of the torque on the driving shaft when an elastic connecting rod with constant stiffness characteristic $k_0$ and the elastic eccentric proper are used in the drive.

The dimensionless parameters of the starting process of the vibratory machine with elastic eccentric, namely the exciting force $F^S/kr$ and the moment on the driving shaft $M^S/0.5 \ kr^2$, are presented in Fig. 4.36c, d and compared with analogous parameters of a vibratory machine with an elastic connecting rod, which is the system most widely used presently.

The dimensionless parameters of the vibratory machine were computed by the following formulas:

for a vibratory machine with elastic eccentric:

the current value of the starting force of the drive

$$\frac{P^S}{kr} = k_{e.cr} \frac{1 - \Delta \sin 2\omega t}{1 - \lambda \sin 2\omega t} \sin \omega t$$

the current value of the starting moment of the drive

$$\frac{M^s}{0.5kr^2} = k_{e.cr} \; \frac{1 - \varDelta \sin 2\omega t}{1 - \lambda \sin 2\omega t} \; \sin 2\omega t$$

for a vibratory machine with elastic connecting rod

$$\frac{P^s}{kr} = k_{e.cr} \sin \omega t$$

$$\frac{M^s}{0.5kr^2} = k_{e.cr} \sin 2\omega t$$

It is evident from the presented graphs that for a vibratory machine with harmonically varying stiffness of the driving elastic link, a far less starting force (roughly 25 - 30%) is required than for a vibratory machine with a constant stiffness of the driving elastic link. This is achieved due to the fact that the lowest stiffness of the elastic eccentric is realized in the first quarter of shaft rotation.

As a working oscillation regime of the vibratory machine with elastic eccentric, the regime of oscillation resonance ($\omega = p_1 = \sqrt{k/M}$) can be taken at which there is no self-excitation in the system, and the exciting force of the elastic eccentric strain and the energy expenditures on maintaining the oscillations are minimum.

In the steady-state operation, the mean value of the stiffness $k_0$ of the driving link is attained and the working element performs oscillations with an amplitude which is practically equal to the design value of eccentricity $r$ of the elastic element. The best starting conditions and sufficiently large "drift" of the vibratory system from the zone of parametric oscillations occur when values of the driving $k_0$ and main $k$ stiffnesses of the elastic links are equal. The magnitude of detuning from the oscillatory resonant regime and the above mentioned ratio $\chi$ for the vibratory machine with elastic eccentric in the drive are selected on the basis of the conditions $0.9 \le \omega/p_1 \le 1.1$, $0.8 \le \chi \le 1.2$.

# DESIGN OF VIBRATORY MACHINES CONSIDERING LOADING AND DRIVE CHARACTERISTICS

## 5.1 FORMULATION OF DESIGN PROBLEMS FOR SYSTEM: VIBRATORY MACHINE-LOAD-MOTOR

The present section deals with results advancing the work of V. O. Kononenko on interaction of oscillatory systems with the source of limited power when a strong interaction of the system elements takes place. Strong interaction in heavily loaded machines occurs not only between the motor and the oscillatory system, but also with load. In machines in which the mass of the processed medium exceeds the mass of the working element the effect of load is so substantial that the machine at idling and under load are essentially two systems with different properties. In some cases, totally unexpected features are observed in these systems, for example, in vibratory crushers under large loads displacements of the crushing jaws are increasing, the crushing process is more efficient, and productivity is increasing. This example indicates that studying properties of vibratory machines under large loads is not only a question of determining the strength parameters of the machine or verifying power consumption, but also a question of increasing the effectiveness of their operation.

From the principles of rational design and a need for reduction of capital costs, the power of the drive is selected so that it would correspond to the requirements of the machine under load and not exceed it by several fold. In this case, the machine should not be designed with the assumption that the energy source is of unlimited power. Therefore, interaction between the oscillatory system and the drive will definitely take place in machines designed for maximum utilizations of resources.

Thus, the problem of design of the vibratory machine under normal operating conditions is reduced to investigation of the system: vibratory machine-load-motor.

The motion of the working element of the vibratory machine determines the regime of working of the process load; the reaction of the latter, in its turn, affects oscillations of the working element. The motor, on one hand, imparts energy to maintain the oscillations of the working element and, on the other hand, it is itself subjected to the reaction of the vibratory machine. In these conditions, operation depends on the amount and properties of the process load and also on characteristics of the energy source. Practically acceptable results can be obtained only on the basis of consideration of the entire system.

Until recently the formulation of the problems of vibratory machine design, which takes into account all the acting factors, was difficult due to the absence of dependences defining the character of the interaction of various process loads and the working element of the machine. The theoretical background developed in chapter 2 of this book with the use of phenomenological viscoelastoplastic models of the process medium provide the required dependences for the determination of the load on the working element of the vibratory machine. The use of these dependences enables one to develop the methods of design of vibratory machines for different production purposes under load.

The analysis presents some mathematical difficulties for systematic solution; therefore, dynamics of the system: vibratory machine-load-motor is best analyzed with the use of analog and digital computers. The application of the analysis techniques for vibratory machines under load is illustrated by two examples involving machines that are different in purpose and principles of operation, namely vibratory feeders and vibratory crushers.


## 5.2   HEAVY DUTY VIBRATORY CONVEYING MACHINES

### 5.2.1 Mathematical Model and Equations of Motion for a Vibratory Feeder with Inertial Drive

Let us expound the method of calculation of vibratory conveying machines under load with an energy source of limited power on the example of a heavy duty vibratory feeder with inertial elliptical drive and induction motor delivering granular load from a hopper or ore chute to a belt conveyer or to a railway car.

According to the mathematical model (Fig. 5.1), the feeder can be represented by a dynamic system with three degrees of freedom. The motion of its working element with a motor of limited power at idling is described by nonlinear autonomous system of differential equations, since the action of the non-ideal energy source on the operation of the machine is dependent on its motion regime and cannot be expressed as an explicit function of time.

$$M\ddot{x}'' + C_x\dot{x}' + K_x x' = -(m + m')\,r\dot{\varphi}^2\cos\varphi - (m + m')\,r\ddot{\varphi}\sin\varphi$$
$$M\ddot{y}' + C_y\dot{y}' + K_y y' = -m'r\dot{\varphi}^2\sin\varphi + m'r\ddot{\varphi}\cos\varphi \tag{5.1}$$

$$I\ddot{\varphi} + q_0\dot{\varphi} = L\,(\dot{\varphi}) - (m + m')r\,\ddot{x}'\sin\varphi + m'r\ddot{y}'\cos\varphi$$
$$\quad - m'gr\sin\varphi \tag{5.2}$$

where $M_0$, $m_1$ and $m_2$ are masses of the load-carrying element and the unbalanced mass, $m_1 = (m/2) + m'$, $m_2 = m/2$; $m'$ is the difference in the unbalanced masses inertia; $M = M_0 + m_1 + m_2$ is overall oscillating mass of the vibratory machines; $K_x$ and $K_y$ are the components of the stiffness of the elastic system of the vibratory machine in directions of axes $x$ and $y$; $C_x$ and $C_y$ are the components of the coefficient of viscous friction in the elastic system in directions of axes $x$ and $y$; $r$ is the distance from the axis of rotation to the center of the unbalanced mass; $I_1$ is the reduced moment of inertia of the rotor of the motor and rotating parts of the vibrator; $I = I_1 + (m + m')r^2$ is the reduced moment of inertia of the rotating parts of the vibratory feeder; $g_0$ is the coefficient of resistance to rotation of the motor shaft; $L(\varphi)$ is the torque developed by the motor.

The vibratory feeder under load is a dynamic system with five degrees of freedom. For its analysis, one must solve the system of differential equations describing the motion of the transported load in conjunction with (5.1) and (5.2). The equations of the relative motion of the load layer on the section where elastic deformation takes place are

$$m\ddot{y} + c_y\dot{y} + k_y y = -m\ddot{y}' - mg\cos\alpha \tag{5.3}$$

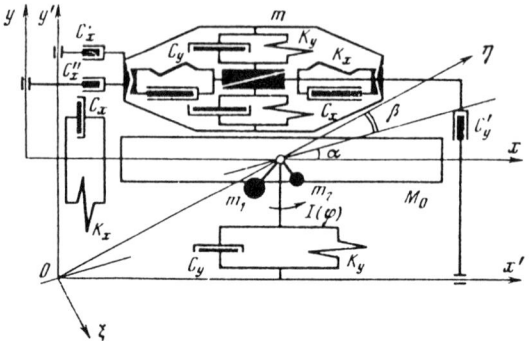

**Figure 5.1** Mathematical model of a vibratory feeder with inertial drive.

$$m\ddot{x} + (c_x + c_x)\dot{x} + k_x x = -m\ddot{x}' + mg \sin \alpha - c_x'\dot{x}' \qquad (5.4)$$

Under elastic deformation of the load layer the forces acting on the load-carrying element will be

$$F_y = k_y y + c_y \dot{y} \qquad (5.5)$$

$$F_x = k_x x + (c_x + c_x')\dot{x}' \qquad (5.6)$$

In the presence of slippage the displacement of the load layer is described by equation

$$m\ddot{x} + c_x'\dot{x} = -m\ddot{x}' + mg \sin \alpha - \text{sign}\,(\dot{x})\,\mu F_y - c_x'\dot{x}' \qquad (5.7)$$

Thus, the following force acts on the load-carrying element

$$F_x = \text{sign}\,(\dot{x})\mu F_y \qquad (5.8)$$

Displacement of the load layer on the section of free motion is described by equations

$$m\ddot{y} + c_y'\dot{y} = -m\ddot{y}' - mg \cos \alpha \qquad (5.9)$$

$$m\ddot{x} + (c_x' + c_x)\dot{x} = -m\ddot{x} + mg \sin \alpha - c_x'\dot{x}' \qquad (5.10)$$

During this period, the load carrying element is acted upon by the forces of aerodynamic resistance

$$F_y = c_y'\dot{y} \qquad (5.11)$$

$$F_x = (c_x' + c_x'')\dot{x} \qquad (5.12)$$

The equations of motion of the loaded vibratory machine will have the form

$$M\ddot{x}' + C_x\dot{x}' + K_x x' = -(m+m')\,r\dot{\varphi}^2 \cos\varphi$$
$$\qquad -(m+m')\,r\ddot{\varphi}\sin\varphi + F_x^*$$
$$M\ddot{y}' + C_y\dot{y}' + K_y y' = -m'r\dot{\varphi}^2\sin\varphi + m'r\ddot{\varphi}\cos\varphi + F_y^* \qquad (5.13)$$
$$I\ddot{\varphi} + q_0\dot{\varphi} = L(\dot{\varphi}) - (m+m')\,r\ddot{x}'\sin\varphi + m'r\ddot{\varphi}\cos\varphi - m'gr\sin\varphi$$

where $F_y^*$ and $F_x^*$ are the normal and tangential components of the load from the transported medium on the load-carrying element in the corresponding phases of motion.

Application of analog computers enables, in addition to oscillograms of all displacements, velocities and accelerations, one to directly obtain the following important characteristics of the vibratory machine:

total energy expenditures associated with the displacement of the load

$$W = \frac{1}{T}\int_0^T (F_y\dot{y}' + F_x^*\dot{x}')\,dt \qquad (5.14)$$

nonproductive expenditures due to reduction of the grain size of the load during conveying and to wear of the working element

$$W_{un} = \frac{1}{T}\int_0^T (N^*\dot{y} + F^*\dot{x})\,dt \qquad (5.15)$$

direct expenditures on load displacement are

$$W_d = W - W_{un} \qquad (5.16)$$

power consumed by the motor

$$W_{mot} = \frac{1}{T}\int_0^T L(\dot{\varphi})\,\dot{\varphi}dt \qquad (5.17)$$

power expended on the motion of the machine (on overcoming losses in the machine)

$$W_m = \frac{1}{T}\int_0^T \left(\frac{m+m}{M}\,r\ddot{x}'\dot{\varphi}^2\sin\varphi + \frac{m'}{M}\,r\ddot{y}'\dot{\varphi}^2\cos\varphi\right)dt \qquad (5.18)$$

loads on the foundation transmitted through the elastic elements

$$R_y = K_y y' + C_y\dot{y}' \qquad (5.19)$$
$$R_x = k_x x' - c_x\dot{x}' \qquad (5.20)$$

Behavior and characteristics of vibratory machines can be investigated under various operating regimes that are simulated by an arrangement of pins on the switching panel. Characteristics of the transported load, motor power and type, and parameters of the vibrator and the machine vary in each considered regime.

## 5.2.2 Study of Operation of Vibratory Feeder with Limited-Power Under Load using Analog Computers

Study of a vibratory feeder under load was carried out on the analog computer EMU-10. In order to solve equations of motion of the crusher on the analog computer they were transformed to the machine format

$$\frac{d^2\bar{x}}{d\tau^2} = -M_t \frac{C_x}{M} \frac{d\bar{x}}{d\tau} + M_t^2 \frac{K_x}{M} \bar{x} + \frac{M^2\varphi}{M_{z'}} \frac{m+m'}{M} r\left(\frac{d\bar{\varphi}}{d\tau}\right)^2$$

$$\times \sin M\varphi\bar{\varphi} - \frac{M\varphi}{M_z'} \frac{m+m'}{M} r \frac{d\bar{\varphi}^2}{d\tau^2} \cos(M\varphi\bar{\varphi})$$

$$-\frac{M_z}{M_{z'}} \frac{M_r}{M_r} \frac{M_t^2}{M_z} F^*$$

$$\frac{d^2\bar{y}}{d\tau^2} = M_t \frac{C_y}{M} \frac{dy}{d\tau} - M_t^2 \frac{K_x}{M} y' + \frac{M_\varphi^2}{M_{z'}} \frac{m'}{M_r} \left(\frac{d\bar{\varphi}}{d\tau}\right)^2 \cos(M\varphi\bar{\varphi})$$

$$+ \frac{M\varphi}{M_{z'}} \frac{m_1}{M} r \frac{d^2\varphi}{d\tau^2} \sin(M\varphi\bar{\varphi}) - \frac{M_z}{M_{z'}} \frac{M_r}{M} \frac{M_t}{M_z} N^*$$

$$\frac{d^2\varphi}{d\tau^2} = -M_t \frac{q}{I} \frac{d\varphi}{d\tau} - M\varphi \frac{m}{I} r \frac{d}{2} \mu\left(\frac{d\bar{\varphi}}{d\tau}\right)^2$$

$$-\frac{M_z}{M\varphi} \frac{m}{I} r \frac{d^2x}{d\tau^2} \cos(M\varphi\bar{\varphi}) + \frac{M_t^2}{M\varphi} \frac{L}{I} + \frac{M_z}{M\varphi} \frac{m}{I} r$$

For transformation of the coefficients into the permissible range of their variation, both parts of the last equation of the system are multiplied by 100. This is incorporated in the computational expressions for the coefficients and in the symbols on the fundamental electrical circuit of the panel.

## 5.2.3 Operation of the Vibratory Feeder under Load in the Steady-State Regime

Let us consider the results of investigation on the analog device EMU-10 of a feeder with inertial vibrator generating elliptical oscillations. The feeder is characterized by the following parameters: weight of the load-carrying element $M_0$ = 1500 kg, coefficient of viscous resistances of the elastic system $c$ = 200 kg.sec/m, coefficient of friction in the bearings of the vibratory machine $\mu'$ = 0.007, coefficient of resistance to the rotation of the rotor of the motor $q_0$ = 0.005 kg.m.sec, diameter of the vibrator shaft $d$ = 0.1 m, resonant oscillation frequency of the feeder $\omega_0$ = 17.03 rad/sec.

The total weight of all unbalanced masses of the vibrator was assumed $\Sigma m$ = 350 kg; total difference of the weights of the unbalanced masses was 100, 150, 200 kg; eccentricities of the unbalanced masses were assumed $z$ = 0.09 m. The viscoelastoplastic model of the load was characterized by the following parameters: $p$ = 200; $n_y^*$ = 220; $n_y'^*$ = 50; $n_x = n_y$ = 5; $n_y'^*$ = 2; the coefficient of friction of the load on the surface of the load-carrying element $\mu$ = 0.4. The mass of the load on the load-carrying element varied within a wide range corresponding to the value of the loading factor $k_l$ 0.5, 1.0, 1.5, and 2.0 ($k_l = M_2/M_0$).

By assuming different ratios of weights of the unbalanced masses and excitation frequencies, it was found that the amplitude-frequency characteristics and parameters of the elliptical trajectory of the load-carrying element could be varied within a wide range. Under constant total weight of the unbalanced masses of the vibrator, the amplitude-frequency characteristic of oscillations of the load-carrying element for the oscillations in the direction of axis $x$ is not changing. Amplitude-frequency characteristics of oscillations in $y$ direction are changing as a function of the difference of the weights of the unbalanced masses. Dimensions and configuration of the trajectory of the load-carrying element are also changing.

From experiments a series of amplitude-frequency and frequency-force characteristics were obtained for stationary operating regimes of the vibratory machine within a wide range of parameters, tunings, and loads.

Figure 5.2a, which pertains to the idling regime of the vibratory feeder when it is excited by a vibrator with a weight difference of the unbalanced masses 150 kg, shows the dependences on the oscillation frequencies of displacement amplitude components of the load-carrying element in the directions of $x$ and $y$ and also energy expenditures associated with overcoming resistances to the rotation of the motor and of the vibrator $W_0$ and $W_v$ and to displacement of the load-carrying element (resistances of the elastic system) $W_m$. Shown in the same figure is the dependence of the consumed power $W$ on the operating regime of the feeder. The amplitude-frequency curve has a characteristic peak in the resonant regime. Energy expenditures on overcoming the resistances to motion of the load-carrying element, associated with hysteretic losses in the elastic system, also have an extremum in the resonant regime; in the transresonant regimes energy expenditures on overcoming viscous resistances first decrease, then increase as the oscillation frequency increases. Energy expenditures on overcoming the resistances to the motor rotation increase parabolically with increasing frequency (angular velocity). It should be noted that energy expenditures on friction in the bearings units of the motor are not large due to the balanced rotor and play insignificant role in the total energy budget. Energy expenditures associated with friction in the bearings units are proportional to the square of the angular velocity (the square of oscillation frequency) and are dependent on the motion parameters of the load-carrying element. It is pertinent to note that in the transresonant regime they exceed the energy expenditures in the elastic system of the machine.

Figure 5.2b, obtained for a loading coefficient of 1.0, also shows, in addition to the characteristics considered earlier, the energy expenditures associated with the transported load on the load-carrying element (energy expenditures on transportation) $W_t$.

By analyzing amplitude-frequency characteristics of the vibratory machine under load, we can note that the peak in the resonant regime has decreased substantially, particularly during displacement in the x direction. It should also be noted that resonance has shifted to lower frequencies. Thus, if resonance was observed during idling at a frequency $\omega$ = 17.03 rad/sec, at a load of 0.5 it occurs at a frequency $\omega$ = 14.5 rad/sec. It is interesting to note that at this load the resonant oscillation amplitude in the y direction decreases insignificantly (by approximately 40%), and in the x direction by 7 times. This can be explained by the fact that, in the oscillation regimes of the load carrying element under consideration, the load travels in the non-bouncing regime, sliding along the load-carrying element over a significant part of the path. In such regimes the normal reaction of the load on the load-carrying element is small, but the reactions due to dry friction forces acting in the plane of motion are substantial. These cause a sharp decrease in the oscillation amplitudes of the load-carrying element in the direction of conveying.

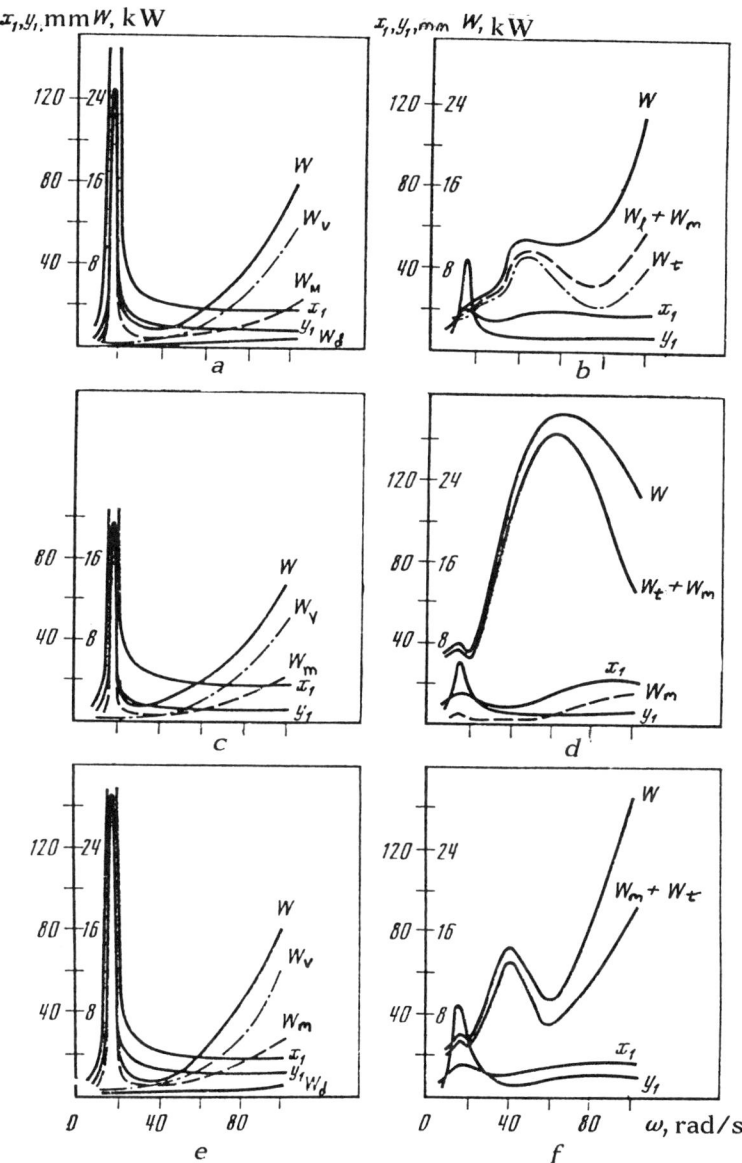

**Figure 5.2** Amplitude-frequency and frequency-force characteristics of the vibrator at various ratios of the unbalanced masses and various loads. a) $k_l = 0$, $m' = 150$ kg; b) $k_l = 1$, $m' = 150$ kg; c) $k_l = 0$, $m' = 100$ kg; d) $k_l = 1$, $m' = 100$ kg; e) $k_l = 0$, $m' = 200$ kg; f) $k_l = 1$, $m' = 200$ kg.

Such sharp and nonuniform variations of oscillation amplitudes along different axes cause considerable changes in the trajectory of motion of the load-carrying element under load at different oscillation frequencies.

It is interesting to note that at small loads, in the present case at a loading factor of 0.5, energy expenditures that are not associated with the load displacement (in elastic system, vibrator, and motor) exceed in the working regime the energy expenditures on transportation and comprises approximately 65% of the total energy expenditures by the machine.

With increasing load (loading factors 1.0 and 1.5) energy expenditure on transportation changes considerably. This is explained by the fact that when the machine is loaded the oscillation parameters of the load-carrying element are changing (effect of load on the machine operation) as a result of which another regime of transportation is established (the oscillation amplitude is decreasing, the trajectory of motion is transformed); the tossing motion commences at higher frequencies of oscillations and the flight angles of the load change. Thus, if the angle of flight $\delta = 310^\circ$ for a load factor $k_l = 0.5$, then $\delta = 294^\circ$ for $k_l = 1.0$, $\delta = 220^\circ$ for $k_l = 1.5$, and $\delta = 203^\circ$ for $k_l = 2.0$.

Under large loads, particularly with a loading factor of 1.5, new characteristic properties of the process are developing. The amplitude-frequency characteristic along axis $y$ (Fig. 5.2b) has, in addition to the main maximum, a second extremum in the region of higher frequencies at which a conveying regime is established with tossing up and falling down of the load during its downward motion. In these regimes the load causes rocking of the load-carrying element as a result of which a second extremum appears. Such a feature of the amplitude-frequency characteristics can be regarded as a manifestation of a second resonance of the two-mass oscillatory system which represents the vibratory machine-load system. Since this is a fairly complex vibration impact system, the specific features of its behavior are also significantly more complex than those for an ordinary two-mass vibratory system. Shown in Fig. 5.2c, d are the characteristics of the vibratory machine at idling and under load ($k_l = 1.0$) for a weight difference of 200 kg of the unbalanced masses. By analyzing these dependences, we can note that energy expenditures associated with hysteretic losses in the elastic system rise with decreasing weight difference of the unbalanced masses, since these losses are proportional to the square of amplitude. Power expended on overcoming friction in the bearings of the vibrator vary slightly since in the resonant regimes these expenditures are small, and in the transresonant regimes variation of amplitude only along the $y$-axis contributes little to the total sum. Change of energy expenditure on the load transportation is found to be significant. This is due to the change of its motion regime along the load-carrying element when the oscillation amplitude changes along axis $y$. Investigations show that the effect of the external load on the oscillation regime of the vibratory machine is far more significant when the exciting force of the vibrator is increased. Under substantial loads, in this case for the loading factors 1.0 and 1.5, relative energy expenditures are not associated with the transportation process are decreasing. Thus, at a load of 1.0 it amounts to 54% of the total. The fact of the matter is that with increasing load energy losses in the machine elements do not increase, and sometimes they even decrease. Therefore, from the standpoint of energy, it is better to operate vibratory machines under large loads.

From analysis of the graphs in Fig. 5.2, it can be stated that total energy expenditures of vibratory machines tend to increase with increasing frequency. Thus, in order to decrease specific energy expenditures, one must design vibratory machines with large displacements of the load-carrying element and low oscillation frequencies.

In analyzing the amplitude-frequency characteristics of a vibratory feeder with the following parameters: total weight of the unbalanced masses 187.5 kg, weight difference of the unbalanced masses 112.5 kg, the radius of the unbalanced masses $r = 0.15$ m, damping coefficient of the elastic system $C = 860$ kg.sec/m, it can be noted that the second resonance is most pronounced under large loads and predominantly in the $y$ direction. The second extremum is increasing with the increased load since it is associated with the presence of the second mass of the oscillatory system, i.e., the load. Hence, the larger is the load mass, the more significant its influence. The appearance of the second extremum only in the $y$ direction and its absence in the $x$ direction is explained by the fact that the elastic forces of the load are acting mainly perpendicular to the surface of the load-carrying element, and the interaction of the load and the load-carrying element in the plane of transportation has mainly a dry friction character.

Analysis of variation of the speed of vibratory conveyance of the load in the studied operating regimes of the vibratory feeder with different loads indicates that when the vibratory feeder operates with a loading factor of 1.0 at a frequency of 65 rad/sec, the maximum load speed of conveyance is reached (0.8 m/sec). To examine the causes of such significant increase of conveying speed, consider first of all parameters of motion of the load-carrying element. In the regime under consideration, the major axis of the ellipse is found to be rotated by $30^{\circ}$ to the plane of convergence. As test-rig investigations of elliptical regimes of vibratory conveying have shown (see Chapter 1), the greatest speed of load motion is reached at vibration angles of $20 - 30^{\circ}$.

Thus, the rotation of the elliptical trajectory of the load-carrying element, caused by the load effect, was the reason for increase of conveyance effectiveness for the same load. This is not typical; in the majority of cases the load effect leads to deterioration of the conveyance conditions. It is pertinent to note one important property of the investigated vibratory feeder. In the region of the operating regimes that are characterized by the oscillation frequency 100 rad/sec, practically constant speed is maintained for the load conveyance irrespective of the load.

To achieve optimum operating regimes of the vibratory feeder, the decrease of the specific energy expenditures per unit capacity of the conveyer is of vital interest. Since its capacity is determined by the degree of loading and by the conveyance speed of the load, the ratio of the speed to energy expenditures per unit load on the load-carrying element or per unit displacement of the load serves as an indication of the effectiveness of the operating regime. With increasing oscillation frequency the effectiveness of the regime tends to drop, although there are extremums at some frequencies. With increasing load the operating effectiveness somewhat increases. Thus, to achieve optimum operating regimes of vibratory feeders with respect to minimum energy expenditures per unit load on the load carrying element, they should be designed for low oscillation frequency and operate with large loads. If the ratio of conveyance speed to the direct energy expenditures per displacement of a unit load is taken as an indicator of regime effectiveness, then the correlations would be different. Maximum effectiveness in this case is attained at both low and high frequencies. The least effective are the operating regimes at intermediate oscillation frequencies.

Transportation effectiveness is an extremal quantity and is substantially dependent on the operating regime of the machine. Increasing the load in all cases increases operating effectiveness. The extremal character of the index of effectiveness poses the problem of optimization of operating regimes of the vibratory machine.

Vibratory machines with inertial drive are predominantly designed for transresonant operating regimes. Vibratory machines with such tuning provide stable operation in the steady-state regime with considerable load changes and do not transmit significant dynamic loads to the carrying structure. However, this regime has a substantial shortcoming, since when the machine is started or brought to a stop it passes the resonant region. During these periods large oscillation amplitudes are generated exceeding by several times the amplitudes in the steady-state regimes, and dynamic loads on the carrying structure and the stresses in the elements of the machine are increasing. Because of this, there is a need for thorough study of the transient regimes of vibratory machines with inertial drive.

The questions concerning starting and run-out of vibratory machines with inertial drive are discussed in detail below. However, some general remarks on the starting process of a vibratory machine of the considered type can be made from analysis of stationary amplitude-frequency and frequency-force characteristics.

Thus, with the aid of Fig. 5.2a, it is easy to explain why it is not possible to bring the idling vibratory machine into the operating regime using a motor of limited power (a practical case). It is evident from the graph that the machine consumes considerable power in the resonant regime during idling. However, at low rpm corresponding to resonance frequency, the motor develops low power (less than the power consumed by the machine) and, hence, cannot carry over the machine through resonance. It could turn out that for the starting process the motor power must exceed the required power in the working regime by several fold.

It is known from practice that quite often the vibratory machine which could not be started at idling is easily brought into the working regime under load. This is easily explained with the aid of the graphs of Fig. 5.2b. For a loading factor of 0.5 the consumed power in the resonant region is one half of the power consumed at idling. Hence, the motor easily overcomes the resonant regime and the increasing oscillation frequency is compensated by an increase in the power of the motor with increasing rpm.

By comparing the characteristic of the motor with the characteristics of the vibratory machine it can be noted that the characteristic of the loaded vibratory machine matches the characteristic of the motor very well. On the other hand, characteristic of the vibratory machine at idling is opposite to the characteristic of the motor. Thus, in order to facilitate easy transition through resonance, starting the vibratory machine under load is recommended. This valuable property differentiates vibratory machines radically from ordinary non-vibratory machines which are easier to start at idling but cannot be started at all in many cases under load.

However, it should be remembered that starting of vibratory machines must be carried out under moderate loads, since power consumption at low frequencies under excessive loads can also be substantial.

### 5.2.4 Transient Operating Regimes of a Vibratory Feeder with Inertial Drive

Non-stationary processes in vibratory machines at starting and run-out were studied on a vibratory machine excited by various induction electric motors of AO series at idling and under different loads with the variation of the mechanical parameters of the vibrator.

The character of the transient process at starting and run-out of the machine is determined by the character of variation in time of the angular velocity of the rotor of the motor which is formed by the difference of the total moment of the resistance forces and the moment developed by the motor. In the transient process the maximum stationary amplitudes in the resonant region are not reached; the greater the angular acceleration, the lower the amplitudes of the transient process and the farther the frequencies corresponding to them from the resonant frequencies.

Figure 5.3a shows oscillograms of the transient processes during startup of a vibratory feeder without load by a 28 kW motor. The angular velocity plot has three characteristic sections: subresonant with rapid increase in angular velocity; an almost horizontal section with respect to the maximum attainable amplitudes and, hence, to the maximum of the moment of the resistance forces; transresonant section with rapid rise of the angular velocity up to the steady-state value.

When the machine is loaded (the loading factor 0.5), the angular velocity increases almost linearly (Fig. 5.3b), and maximum values of the amplitudes decrease along axis $y$ by 30%, and along axis $x$ by almost three times. The frequencies at which maximum amplitudes are reached are higher in comparison with the corresponding frequencies in the idling regime; the total acceleration time of the machine increases by almost 30%. This is explained by the fact that the reaction forces acting on the machine from the load restrict the oscillation amplitudes of the machine during crossing of the resonant regime. In this case, the power dissipated in the elastic system decreases (the power is proportional to the square of the amplitude), and the additional power expenditures associated with the presence of load are insignificant. In the overall breakdown, energy expenditures on passing through the resonant region are decreasing; thus the speed of passage through resonance under load is increasing. Further frequency increase causes an increase in energy expenditures associated with the load presence; therefore, the angular velocity during acceleration of the loaded machine in the transresonant region rises more slowly than at idling. This leads to prolongation of the start up process.

The change of the overall consumed power and energy expenditure on transporting the load as a function of the oscillation frequency in the stationary regimes was presented above. In the transient process energy expenditures do not reach these values.

Increasing the load (loading factor 1.0) leads to doubling of acceleration time (Fig. 5.3c). In this case, after passing the resonant zone relatively rapidly, the angular acceleration decreases in the frequency range 45 – 55 rad/sec due to increase in energy expenditures on conveyance. Further increase of the load and the energy expenditures associated with it lead to a reduction of the working frequency. Thus, at a loading factor of 1.5 (Fig. 5.3d), the working frequency is equal to 62 rad/sec, at a loading factor of 2.0 it is equal to 60 rad/sec.

Let us now consider the transient processes occurring when a 40 kW motor is used. The total acceleration time of the machine at idling (Fig. 5.4a) decreases by almost 40%. Increasing the angular acceleration leads to a decrease in the maximum amplitude along axis $y$ by 1.5 times, along axis $x$ by 1.7 times; the frequency corresponding to the maximum

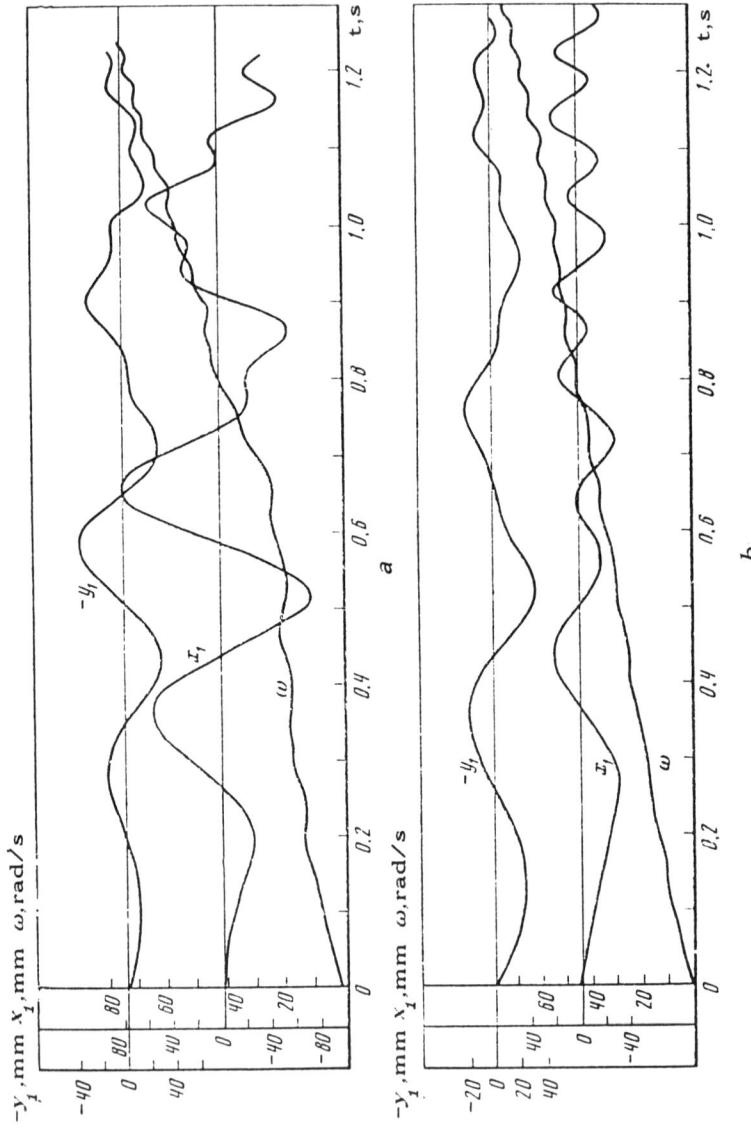

**Figure 5.3** Oscillograms of transient process during startup of vibratory feeder by a 28 kW motor under different loads *a*) $k_l = 0$; *b*) $k_l = 0.5$; *c*) $k_l = 1.0$; *d*) $k_l = 1.5$.

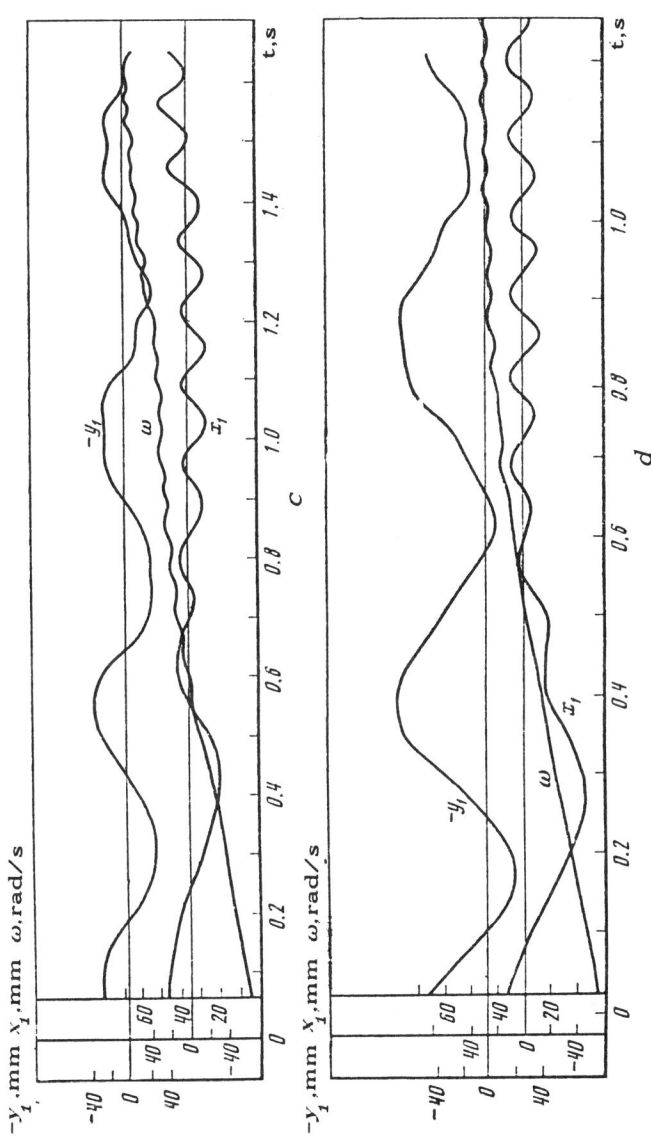

amplitude increases by 1.75 times. The characteristic horizontal section on the graph of the angular velocity, lasting nearly 0.8 sec when a 28 kW motor was used, is reduced now to 0.035 sec.

Loading of the machine (Fig. 5.4b, c) basically leads to the same effects as during starting with a 28 kW motor. However, the effect of load on the change in angular velocity is less noticeable. Thus, even under a load which is twice as large as the weight of the load-carrying element, the acceleration time of the machine increases by just 30% compared with the acceleration time at idling, and the frequency of the steady-state operating regime remains almost unchanged. Thus, increasing the power of the motor improves the conditions of machine acceleration and at the same time stabilizes the working frequency with changing loading.

When starting the idling vibratory feeder by a 20 kW motor, the power required for passing the resonant zone does not develop. The angular velocity fluctuates within 11 - 19 rad/sec. The oscillation amplitude increases over four periods to 35 mm along axis $y$ and to 92 mm along axis $x$. It is interesting to note that upon loading (loading factor 0.5) the machine is accelerated, though over a relatively long time (around 2.5 sec). As already noted, there are known practical cases whereby machines could not be started in idling but could be easily put into operating regime under load. This valuable property strongly differentiates vibratory machines from conventional non-vibratory machines, which are easier to start at idling but in some cases cannot be operated at all under load. It must be kept in mind, however, that vibratory machines should be started under moderate loads, otherwise power consumption at low frequencies can be substantial. When a loaded machine (loading factor 1.0) is started, stationary oscillations are stabilized at a frequency of 41 rad/sec, whereas the nominal frequency is 77 rad/sec.

For vibratory machines with transresonant tuning important are transient processes developing during free run-out following deenergizing. Under the action of the resistance forces, the angular velocity in this case decreases from the nominal working value to zero. Since the resonant regime is also passed in this process, it is possible for the oscillation amplitude to rise considerably, causing machine breakdown. Moreover, if during the process of machine acceleration the maximum amplitudes of the transient process are substantially dependent on the power of the motor, then

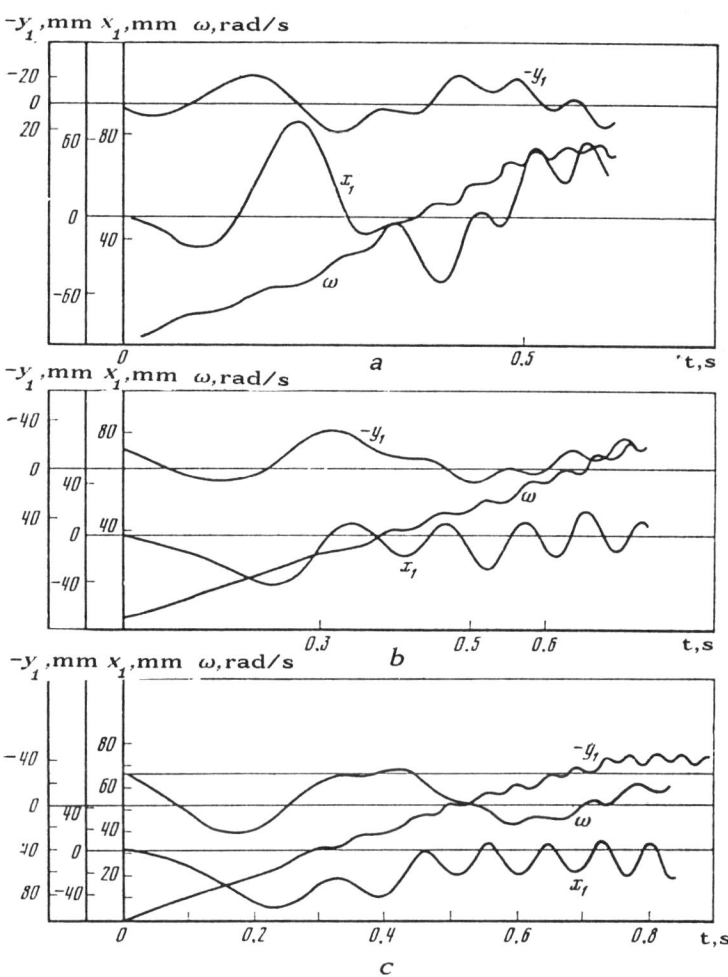

**Figure 5.4** Oscillograms of transient process during startup of vibratory feeder by a 40 kW motor under different loads. a) $k_l$ = 0, b) $k_l$ = 0.5; c) $k_l$ = 1.0.

during run-out of the machine they are determined by the machine characteristics and by magnitudes of the resistance forces. The moment of inertia of the rotor and the moment of resistance in the bearings of the motor also have some influence.

The characteristic feature of the transient process during run-out of the machine at idling is that following a frequency reduction by almost a linear law, the rate of its change decreases sharply followed by a slight increase in oscillation frequency. It is associated with the decrease in oscillation amplitude, and the slight increase of frequency takes place owing to the energy released with decreasing amplitude. During run-out of the loaded machine, part of the energy is dissipated in the load layer; therefore, the time of the transient process decreases, the rate of change of the oscillation frequency increases, and the maximum attainable amplitudes are reduced.

Thus, to facilitate the transition through resonance both in forward and backward directions, the vibratory machine must be started and stopped in the loaded state.

## 5.3  VIBRATORY CRUSHERS

### 5.3.1 Compilation of Mathematical Model and Equations of Motion of a Double-Jaw Crusher with Inertial Drive under Load

The double-jaw vibratory crusher under load is a complicated multi-mass oscillatory system with non-binding constraints. This system includes, in addition to the phenomenological model of the crushed rock, the mass of the rock in the loading hopper of the crusher and two crushing jaws driven with an oscillatory motion by inertial vibrators. The phenomenological model of the rock inside the hopper of the vibratory crusher and exerting pressure on the rock mass in the working chamber of the machine is an elastoviscoplastic inertial body.

The schematic of a jaw crusher under load is shown in Fig. 5.5. It consists of a frame resting on a carrying structure by means of vibration isolators. The crushing jaws are mounted on elastic elements in the frame. The jaws are driven by synchronized inertial self-balanced vibrators. A loading hopper is mounted on the frame to supply the rock to the working of the chamber crusher by gravity. The normal operating regime is such, in which the hopper is constantly loaded and uniform filling of the working chamber by rocks is maintained. Hence, in the normal operating regime, the rock which is in the working chamber is subjected to the action of the crushing jaws and is under the pressure of the rock mass in the hopper.

Thus, a mathematical model of the vibratory crusher under load must take into account a whole complex of the interacting components of the machine and the processed medium. For the mathematical model it is assumed that the

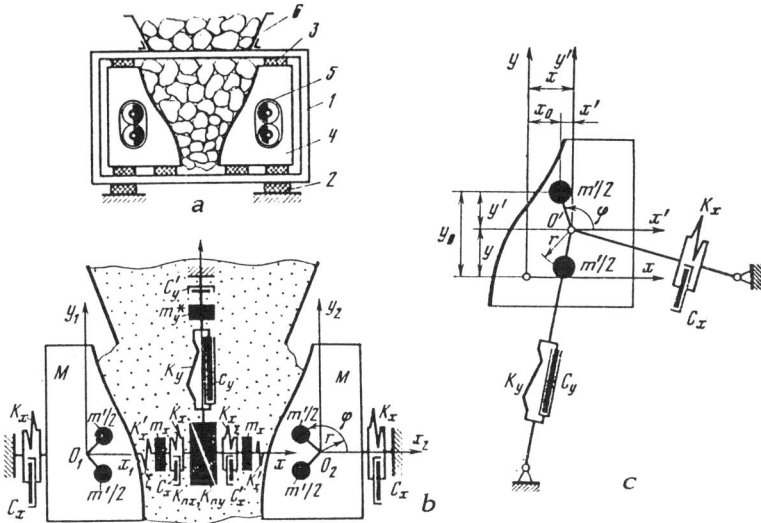

**Figure 5.5** Schematic diagram of a vibratory crusher under load
a) fundamental arrangement: *1* - frame, *2* - vibration isolator; *3*
- working elastic system, *4* - jaw, *5* - vibrator, *6* - hopper; *b*)
design layout of the crusher; *c*) design layout of the jaw.

frame    of    the    crusher    is    stationary.    This    assumption
corresponds    to    real    operating    conditions    since    due    to
equilibrium of reactions in the elastic system of the jaws and,
due to considerable weight of the frame and the hopper
mounted on it, the frame is practically stationary during
operation of the crusher.

We introduce two moving rectangular coordinate frames
$x_1 o_1 y_1$ and $x_2 o_2 y_2$, each rigidly linked with its crushing jaw and
a fixed rectangular coordinate frame *xoy* whose axes are
parallel to the moving axes and linked with the crusher frame.

Let us consider the interaction of the crushed rock with
both jaws of the machine and with the rock mass inside the
hopper. The vibratory crusher is a two-mass system. Let us
denote the mass of each crusher jaw including the vibrator
casing by *M*; the jaws are fixed to the frame of the crusher by
elastic elements characterized by stiffness *K* and by the
coefficients of viscous resistances *C*.

The jaw is brought into motion by inertial self-balanced
vibrator; the total unbalanced mass of the rotating components
of vibrator is *m'* and the eccentricity is *r*. In the general case,
the rotation velocity of the unbalanced mass of the vibrator is

$\dot{\varphi}$, in the steady-state regime it is $\omega$. The vibrator is driven by an electric motor with torque $L(\dot{\varphi})$ via Cardan shafts. From the crushed material the jaws are subjected to the reactions: $F_x$ in the direction of axis $x$ and $F_y$ in the direction of axis $y$.

Taking this into account, the differential equations of the motion of the jaws of the vibratory crusher and its drive are written as follows

$$(M + m')\, \ddot{x}_1^* + C_x \dot{x}_1^* + K_x x_1^* = m'r\varphi^2 \sin \varphi + F_{x1} \tag{5.21}$$

$$(M + m')\, \ddot{y}_1^* + C_y \dot{y}_1^* + K_y y_1^* = F_{y1} \tag{5.22}$$

$$(M + m')\, \ddot{x}_2^* + C_x \dot{x}_2^* + K_x x_2^* = m'r\varphi^2 \sin (\varphi + \gamma) + F_{x2} \tag{5.23}$$

$$(M + m')\, \ddot{y}_2^* + C_y \dot{y}_2^* + K_y y_2^* = F_{y2} \tag{5.24}$$

$$m'\,(\ddot{y}_1 r + gr - \ddot{x}_1 \mu R) \cos \omega t - m'\,(\ddot{x}_1 r + \ddot{y}_1 \mu R) \sin \omega t$$
$$+ m'\mu R\omega^2 = L_{1(\dot{\varphi})} \tag{5.25}$$

$$m'\,(\ddot{y}_2 r + gr - \ddot{x}_2 \mu R) \cos (\omega t + \gamma) - m'\,(\ddot{x}_2 r$$
$$+ \ddot{y}_2 \mu R) \sin (\omega t + \gamma) + m'\mu R\omega^2 = L_{2(\dot{\varphi})} \tag{5.26}$$

where $R$ is the radius of the bearing races of the vibrator shafts; $\mu$ is the coefficient of friction in the vibrator bearings; $\gamma$ is the phase shift between the position of the unbalanced masses of the vibrators for the right and left jaws.

Analysis of equations (5.22) and (5.24) indicates that jaw oscillations in the direction of axis $y$ are induced by the reaction of the crushed rock mass. In order to investigate the main operational features of the vibratory crusher under load, we consider its operation in the steady-state regime with a motor with sufficient power reserve and also the working stroke of the jaws in the direction of axis $x$. Parasitic oscillations of the jaws in the direction of axis $y$ induced by reactions of the crushed rock mass and methods of their abatement are expounded separately.

In this case, the equations of motion of the jaws in the direction of axis $x$, reduced to a form convenient for solution, become

$$\ddot{x}_1^* + 2n\dot{x}_1^* + p^2 x_1^* = qr\omega^2 \cos \omega t + f_{x_1} \tag{5.27}$$

$$\ddot{x}_2^* + 2n\dot{x}_2^* + p^2 x_2^{*\,\prime} = qr\omega^2 \cos (\omega t + \gamma) + f_{x_2} \tag{5.28}$$

where $n$ is the effective coefficient of viscous resistances of the elastic system of the jaw, $2n = C/(M + m')$; $p$ is the natural frequency of the jaw, $p^2 = K/(M + m')$; $C$ is the ratio of rotating and translationally moving masses of the crusher, $q = m'/(M + m')$; $f_{x1}$ and $f_{x2}$ are effective reactions of the crushed rock on the crusher jaw, $f_{x_1} = F_{x_1}/(M + m')$; $f_{x_2} = F_{x_2}/(M + m')$.

For derivation of equations (5.25) and (5.26) it was assumed that connection between the rotor of the motor and the vibrator is absolutely rigid and it can, therefore, be assumed that their angular velocities are the same.

For analysis of the steady-state regime, it should be assumed that $\ddot{\varphi} = 0$, $\dot{\varphi} = \omega$ in equations (5.25), (5.26). In this case, the equations are significantly simplified and are used to determine energy consumption by the machine in the steady-state regime. When the power of the installed motor has no significant reserve, its characteristic should be considered, for analysis.

The torque developed by the motor can be specified by its static characteristic as was done in the preceding section when considering the motion of a vibratory conveying machine. In order to consider dynamic properties of the electric motor and electromagnetic processes in it, one can use the differential equations of the motor available in the literature.

The differential equation of the induction electric motor has the form

$$\frac{1}{2M_{cr} \cdot \omega_c} \dot{M}_{mot} + \frac{s_{cr}}{2M_{cr}} = \frac{\omega_0 - \dot{\varphi}_{mot}}{\omega_0} \tag{5.29}$$

where $\omega_0$ and $\omega_c$ are angular velocity of the idling motor and supply frequency; $\omega_0 = \omega_c/2e$; $e$ is the number of pairs of poles; $\dot{\varphi}$ is the current angular velocity; $s_{cr}$ is the critical slip of rotor; $M_{cr}$ is the critical torque of the electric motor in the static regime.

The differential equation of a dc motor with independent excitation is

$$\frac{r_a}{M_{min} \, \Phi} \left( \frac{L_a}{R_a} \dot{M}_{mot} + M_{mot} \right) = \frac{\omega_0 - \dot{\varphi}_{mot}}{\omega_0} \tag{5.30}$$

where $M_{min}$ is the minimum moment of the electric motor; $r_a$ is the specific active resistance of the armature circuit; $\Phi$ is the specific magnetic flux of the excitation windings; $R_a$, $L_a$

are total active and inductive resistances of the armature
circuit of the generator and of the electric motor.

The differential equations of a dc electric motor with
series excitation are

$$M_{mot} = k_m \Phi i$$

$$\dot{U} = k_e \Phi \dot{\varphi}' + (L_a + L_{ex}) \frac{di}{dt} + (R_a + R_r)i \qquad (5.31)$$

where $k_m$ and $k_e$ are coefficients reflecting mechanical and
electrical parameters of the electric motor; $\Phi$ is the magnetic
flux; $i$ is the current in the armature circuit; $L_a$ and $L_{ex}$ are
inductance of the armature and of excitation windings; $R_a$ and
$R_r$ are active resistance of the armature circuit and of the
starter rheostat.

If the drive system includes a flexible coupling (or simply
an elastic element), then for the complete description of the
vibratory crusher the appropriate differential equations of the
motor and the coupling must be added to the system of
equations (5.21) - (5.26).

Thus, with a flexible coupling between the motor and the
vibrator, the following system of two differential equations
should be considered

$$I_{mot} \ddot{\varphi}_{mot} + (q_0 + c_c) \dot{\varphi}_{mot} + k_c \varphi_{mot} = L(\dot{\varphi}) - c_c \dot{\varphi} - k_c \varphi \qquad (5.32)$$

$$m'r^2 \ddot{\varphi} + c_c \varphi + k_c \varphi = c_c \varphi_{mot} + k_c \varphi_{mot} + m'(\bar{x}r + \ddot{y}\mu R) \sin \varphi$$
$$- m'(\ddot{y}r + gr - \ddot{x}\mu R) \cos \varphi m' \mu R \dot{\varphi}^2 \qquad (5.33)$$

where $k_c$ is the torsional stiffness of the coupling; $c_c$ is the
coefficient of viscous resistances of the coupling; $I_m$ is the
reduced moment of inertia of the motor rotor.

The reactions $f_{x1}$ and $f_{x2}$ on the crushing jaw are
determined from solution of equations of motion and
deformation of the rock mass

$$\ddot{x}_1 + \alpha R_1(x_1, \dot{x}_1) + R_1^*(x_1, \dot{x}_1, \dot{x}_1 + \dot{x}_1^*) - f = -\xi \ddot{x}_1^*$$
$$\ddot{x}_2 + \alpha R_2(x_2, \dot{x}_2) + R_2^*(x_2, \dot{x}_2, \dot{x}_2 + \dot{x}_2^*) - f = \xi \ddot{x}_2^* \qquad (5.34)$$

where $R_1(x_1, x_2)$ and $R_2(x_2, \dot{x}_2)$ are jaw pressure on the
crushed rock

$$R_1(x_1, \dot{x}_1) = -\frac{1}{\xi} f_{x1}; \qquad R_2(x_2, \dot{x}_2) = -\frac{1}{\xi} f_{x2}$$

$R_1^*$ $(x_1,\ \dot{x}_1,\ \dot{x}_1 + \dot{x}_1^*)$, $R_2^*$ $(x_2,\ \dot{x}_2,\ \dot{x}_2 + x_2^*)$ are internal stresses in the crushed rock; $\alpha$ is the factor of discontinuity taking into account the pattern of rock motion in the working chamber of the crusher (when the rock is in contact with jaw $\alpha = 1$, with loss of contact $\alpha = 0$); $\xi$ is the ratio of the masses of the crushed rock and the crushing jaw

$$\xi = (M + m')/m'$$

$f$ is the pressure from the rock in the hopper.

Let us denote the mass of the rock in the working chamber of the crusher $m_y$ and the mass of the rock in the hopper $m_y^*$. The equations of motion of the rock in the $y$ direction in the working chamber and in the hopper will have the form

$$\ddot{y} + 2n_{y2}\dot{y} + p_{y1}^2 y = -g + 2n_{y2}\dot{y} + p_{y1}^2 y' \qquad (5.35)$$
$$- \text{sign}\,(\dot{y})\mu'[\alpha R_1(x_1,\ \dot{x}_1) + \alpha R_2\,(x_2,\ \dot{x}_2)]$$

$$\ddot{y}' + (2n_{y1}^* + 2n_{y2})\,\dot{y}' + p_{y1}'^2 \dot{y}' = -g + 2n_{y1}^*\dot{y} + p_{y1}^2 y \qquad (5.36)$$

where $n_{y1}^*$, $n_{y1}$ are the reduced coefficients of viscous resistances in the crushed rock mass

$$2n_{y1} = \frac{c_{y1}}{m_y}, \qquad 2n_{y1}^* = \frac{c_{y1}}{m_y'}$$

$p_{y1}$, $p_{y1}'$ are natural frequencies of the rock mass in the working chamber and in the hopper

$$p_{y1}^2 = \frac{k_y}{m_y}, \qquad p_{y1}'^2 = \frac{k_y}{m_y'}$$

$n_{y2}$ is the reduced coefficient of the constrained motion of the rock mass in the working chamber

$$2n_{y2} = \frac{c_{y2}}{m_y}$$

$\mu'$ is the coefficient of friction of the crushed rock on the crusher jaw.

Equation (5.36) describes motion of the rock in the hopper, and (5.35) in the working chamber of the crusher. The first equation describes deformation of the rock, and the second its slippage over the crushing surfaces of the jaw.

Operation of the crusher with a loading hopper and action of rock pressure on the rock mass in the working chamber provides a more uniform flow of the crushing process owing to the continuous supply of the rock. They also create an additional outward thrust on the crushing jaws and the additional static loading of the elastic system.

## 5.3.2 Analog Computer Investigation of the Crushing Process of Rock Mass in a Vibratory Crusher

To study the laws governing the crushing process of rock mass in a vibratory crusher the differential equations describing the motion of the jaws, viscoelastic and plastic deformations of the rock mass, and also its motion in the working chamber (slippage and fall in constrained conditions) are reduced to machine format. To achieve this, the variables are replaced and the differential equations are reduced to machine format by adopting the following notations: $\mu_t$, $\mu_x^*$, $\mu_x$, and $\mu_y$ are time scales and scales for displacement of the jaws and of the rock mass in the $x$ and $y$ directions; $\overline{X}_1{}^*$ and $\overline{X}_2{}^*$ are machine (computer) displacements of the jaws; $\Omega$ is the machine oscillation frequency of the jaws; $\overline{X}$ and $\overline{Y}$ are machine displacements of the rock mass in the $x$ and $y$ directions.

Scales $\mu_t$, $\mu_x$, $\mu_x$, and $\mu_y$ are selected so that the maximum voltage of machine variables would not exceed 100 or 50 V depending on the type of the analog machine.

Let us consider the block diagram of the analog computer for analysis of interaction of one jaw with the rock mass, since the block diagram for the second jaw is analogous. In view of this, subscripts 1 and 2 referring to the first and second jaws are dropped.

Since the regime of rock-mass slippage along the crushing jaw during crushing is not realized in practice, as is observed from the analysis of photographic records, the block diagram of the analog computer envisages the analysis of viscoelastic and plastic deformations of the rock mass in addition to its motion in constrained conditions in the working chamber.

The device for modeling the crushing process of rock mass in a vibratory crusher includes the following structural elements: a generator of external actions (to obtain perturbations); a device for modeling the equation of motion of the crushing jaw; a device for modeling the system of equations describing deformation of the rock along axis $x$; a device for modeling the system of equations describing rock motion along axis $y$; logical structural control circuits in accordance with the transcendental equations of transition from one regime of deformation and motion of the rock to another.

The generator of the external excitations simulates the differential equation of the harmonic oscillator and represents the block diagram of the directional harmonic oscillations. The generator of the external excitations is assembled using a standard circuit.

The generator in which the quantity $A\Omega^2 \cos \Omega r$ is obtained at the output of the summation amplifier is used for investigation of the steady-state process of vibratory crushing. The excitation amplitude is fed into the integrator as an initial condition. With a prolonged energizing of the generator in the regime start-reset, a positive voltage feedback with diode limiter is used in the generator block diagram for elimination of the amplitude error which is caused by viscous resistances in the elements of the analog machine.

The block diagram modeling the motion of the vibratory machine jaws under the action of the exciting force and the force from the side of the crushed material is constructed according to the appropriate machine equation. The following quantities are entered into the summation amplifier: $A\Omega^2 \cos \Omega\tau$ is the acceleration of jaw motion in the absence of elastic links (a term defined by the parameters of the used inertial vibrator); $(M_t^2/M_x')F_x$ is the projection on axis $x$ of the reaction of the crushed rock on the jaw related to jaw mass; a term proportional to velocity of jaw displacement $\dot{x}'$ and taking into account viscous losses in the elastic system of the crusher; a term proportional to the jaw displacement $x'$ which takes into account the restoring forces of the elastic connections in the crusher. From the output of the amplifier, the acceleration of jaw displacement of the crusher $\ddot{x}'$ is obtained. In this case, it is acceleration of the transfer motion.

The block diagram for modeling the displacements of the crushed rock model is compiled using the corresponding machine equations and includes a logical circuit which executes all operations of transition from one stage to another according to the transcendental transition equations.

During the stage of motion of the rock in contact with the jaw the conditions of the presence of viscoelastic and plastic deformations are constantly compared. Motion of the model along axis $y$ takes place only in the stage of free motion, and the velocity of the model in the stage of joint motion is equal to zero.

The instantaneous energy consumption, both total and unproductive, are obtained in the multiplication blocks. Computations are halted when energy consumption reaches the value required for rock fracture.

Let us consider the method of calculation of the crushing process of the rock mass with the aid of digital computers. The use of computers is justified when there is a need to obtain not only the instantaneous values of the variables, their velocities and accelerations, but also the characteristics of the vibratory crusher and the crushing process, the power consumed by the electric motor

$$N_{pe} = \frac{1}{T} \int_0^T M_{pe}\dot{\varphi} dt \tag{5.37}$$

the total energy expenditures associated with the crushing process

$$N = \frac{1}{T} \int_0^T (F_x \dot{x}' + F_y \dot{y}') dt \tag{5.38}$$

the unproductive energy expenditures during the process of crushing of the material

$$N_{un\ m} = \frac{1}{T} \int_0^T (F_x \dot{x} + F_y \dot{y}) dt \tag{5.39}$$

the energy directly expended on rock crushing

$$N_{vc} = N_c - N_{un} \tag{5.40}$$

the loads transmitted to the frame of the crusher

$$R_x = K_x x' + C_x \dot{x}' \tag{5.41}$$
$$R_y = K_y y' + C_y \dot{y}' \tag{5.42}$$

the crushing forces $F_x$, $F_y$, the path travelled by a chunk in the working chamber of the crusher prior to breaking $y_c$, the time (duration) of crushing $t_c$, and so forth.

The block diagram of the calculation of the parameters of the vibratory crusher and the crushing process is formed as follows.

Block 1 effects the input of the program and the initial data including parameters of the vibratory crusher $M$, $C_x$, $K_x$, $C_y$, $K_y$, $m'$, $r$, $\omega$, $\alpha$, $\mu$, $R$, $I$, $q$ and the crushed rock $m$, $k$, $c$, $k_r$, $c'$, $c''$, and so on.

Block 2 effects the input of the initial conditions $t_j$, $x_j$, $\dot{x}_j$, $F_{xi}$, $F_{yi}$ (the computation process commences when the crushed rock lies on the jaw of the vibratory crusher).

Block 3, under the initial conditions set by block 2, finds the solution of the equations of motion of the jaw of the vibratory crusher in the direction of axis $x$ by the Runge-Kutta technique.

Block 4 finds the solution of the differential equation of the deformation of the rock mass in the phase of viscoelastic deformations.

Block 5 computes forces from the crushed rock resisting the motion of the jaw of the vibratory crusher in the direction of axis $x$.

Block 6 executes the logical operation of comparing the force of resistance to jaw motion with zero. If the condition is satisfied, which means loss of contact of the jaw with the crushed rock, control is transferred to block 14, otherwise control is transferred to block 7.

Block 7 sends parameters corresponding to the motion in the ith cycle based on the results of computations at the end of the $(i-1)$-th cycle to the register for initial conditions.

Block 8 executes the logical operation of comparing the force of the viscoelastic resistance to the motion of the jaw with the plastic strain limit in the ith cycle. If the condition is satisfied, which corresponds to the transition from viscoelastic deformations to plastic deformations, control is transferred to block 9, if the condition is not satisfied then control branches to block 3.

Block 9 finds the solution of the differential equation of the deformation of the crushed rock in the phase of plastic deformations. The block computes the resistance force to the motion of the vibratory crusher jaw from the crushed rock in the stage of plastic strain.

Block 11 executes the logical operation of comparing the force of resistance to the motion of the vibratory crusher jaw in the plastic deformation phase with zero. If the condition is satisfied, which means a loss of contact of the jaw with the crushed material, control is transferred to block 12, if the condition is not satisfied, control is branched to block 15.

Block 12 finds the solution of the differential equation of the crushed rock chunk in the stage of free fall in the crusher working chamber.

Block 13 puts a zero into the register for the force of resistance to the motion of the jaw.

Block 14 executes the logical operation of comparing displacement of the jaw and of the crushed rock in the direction of axis $x$. If the condition is satisfied, which corresponds to the jaw coming into contact with the crushed rock, control is transferred to block 3, if the condition is not satisfied, control branches to block 12.

Block 15 executes the logical operation of comparing the crushing forces acting on the rock in the plastic strain stage in the $(i-1)$-th and $i$-th cycles. If the condition is satisfied, which means the beginning of discharge of the crushed material, control is transferred to block 3, otherwise to block 13.

The conclusion of the crushing process is determined either on the basis of the conditions of reaching fracture forces, or (when using the energy theory of crushing) when the crushed rock absorbs energy that is comparable with the crushing energy. In the first case, block 16 executes a logical operation comparing the crushing force in the plastic deformation phase with the breaking force. If the condition is satisfied, which corresponds to the fracture of the rock inside the working chamber of the crusher, control is transferred to block 17, otherwise to block 3. In the second case, block 16 executes a logical operation comparing useful energy expenditures on the rock crushing process with the crushing energy.

Block 17 computes parameters of the vibratory crusher and the crushing process.

Block 18 prints the input data and the results of the computations.

### 5.3.3 Study of Dynamics of Double-jaw Vibratory Crusher with Inertial Drive under Load on Test Rigs, Industrial Prototypes, and Computers

This section mainly deals with interactions in the load-vibratory crusher system and with principal correlations of vibratory crusher operation under load. Figure 5.6a shows oscillograms of the processes in vibratory crusher under load obtained at the Skochinskii IGD (Mining Institute) on the vibratory crusher VD-2M [18]. The oscillograms depict the displacements, velocities, and accelerations of the crushing jaws, and also the current of the motor. The recordings pertain to operation of the machine within the frequency range 1000 - 1800 osc/min. Analysis of high speed movie recordings and oscillograms enables identification of the following characteristic features of a vibratory crusher under load. At idling the crusher operates with a detuning from resonance of $z = 1.5 - 2.5$. The displacements, velocities, and accelerations of the crushing jaws have sinusoidal character. When the rock mass is fed into the working chamber, the character of jaw motion changes significantly: the displacements, velocities, and accelerations are no longer of harmonic character. Displacements of the jaws increase gradually and the neutral line shifts outside of the working chamber. The maximum jaw opening in the crushing regime of hard rocks at high degree of filling of the chamber can exceed the idling amplitude by 2 - 2.5 times. It is observed that the jaw opening in the final cycles of breaking the rock chunk stabilizes at maximum values, which indicates an increase in material resistance. The velocities, and particularly accelerations of jaw oscillations become "spiky", on the acceleration time histories sharply defined peaks are observed, the current is rising (energy consumption increases). The increase in jaw displacements is usually followed by substantial decrease to magnitudes lower than at idling. This decrease is accompanied by further displacement of the neutral line outside of the working chamber, and energy consumption drops in this case slightly. It has been established that crushing of the rock is most effectively carried out with large amplitudes of the jaws. Depending on the properties of the rock, the degree of filling of the crushing chamber, and on other factors the increase in energy consumption by 5 - 100%, and in some cases by 200 - 300%. As experiments show, some rock chunks are crushed in the vibratory crusher in 10 - 20 or more working cycles. When rock mass is being crushed with a high degree of filling of the chamber, the process is more stable than when crushing large individual rock chunks. When crushing a flow of rock mass the magnitude of jaw deviations from the neutral line at idling and the energy consumption are relatively stabilized and tend to average values. The time the jaw is acted upon by the reaction of the crushed rock amounts to 0.2 - 0.4 of the oscillation period in various regimes. The magnitude of rock reactions exceeds the amplitude values of the inertial forces acting on the jaws at idling by 2 - 5 times and, sometimes, even more. The reactions of the rock act on the jaw usually when it moves from the neutral line into the crushing chamber. With nonuniform supply of the rock mass to the chamber, a certain operating regime is feasible whereby the rise in amplitude of jaw oscillations is followed by their decrease to values corresponding to idling. In such a case, the neutral line returns to its initial position and the current diminishes. The velocities and accelerations of jaw oscillations become harmonic. Similar variations in the operating regime of the crusher take place due to the fact that one batch of the rock inside the chamber is crushed prior to delivery of the next batch. In this case, the crusher constantly shifts from operation under load to idling. This operating regime cannot be regarded as acceptable either from the standpoint of ensuring effective crushing of the rock, or with respect to normal operation of the machine. The crusher must be equipped with special devices that provide uniform feed of the rock. Only then can one expect effective operation of the plant. It should be noted that transition from elevated amplitudes to nominal amplitudes at idling frequently has an alternating character with large and small displacement amplitudes following each other (see Fig. 5.6$b$) - this is an indication of oscillation breakdowns.

It is noted that increase of jaw oscillations is significantly dependent on the amount of the rock mass in the crushing chamber and on the magnitude of the exciting force of the vibrator. In the crusher of the investigated type, increasing the degree of loading of the crushing chamber causes an increase in the swings of jaws. Increasing the exiting force of the vibrator at nominal loading of the chamber leads to a decrease in the swings of jaws within the investigated range of regimes. When the degree of crusher loading is increased, intensification of the jaw oscillations is followed by their sharp decline. The displacement of the jaws under load from the neutral position at idling increases with increasing degree of chamber filling. Crushing of hard rocks causes more intensive increase of oscillations; while crushing soft rocks the process is more stable and the amplitude and opening of the jaws are less. Experiments indicate that when excessively large and very hard rock chunks enter the crusher, large jaw displacements occur in some cases. In this case, the crushing process might stall and torsional oscillations of the jaws take place around the jammed chunk. Most often, such disruptions take place if the crusher is not correctly tuned and might result, for example,

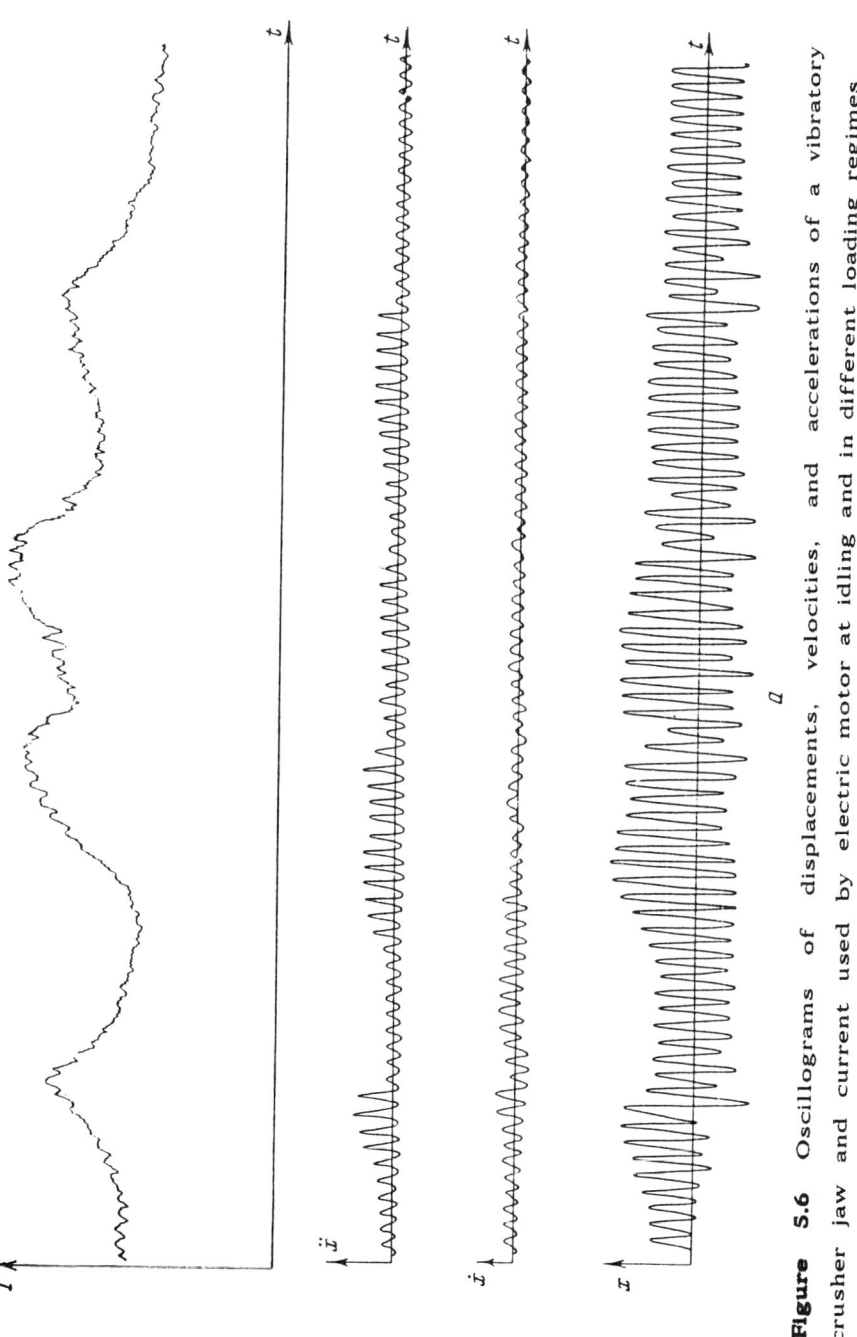

**Figure 5.6** Oscillograms of displacements, velocities, and accelerations of a vibratory crusher jaw and current used by electric motor at idling and in different loading regimes

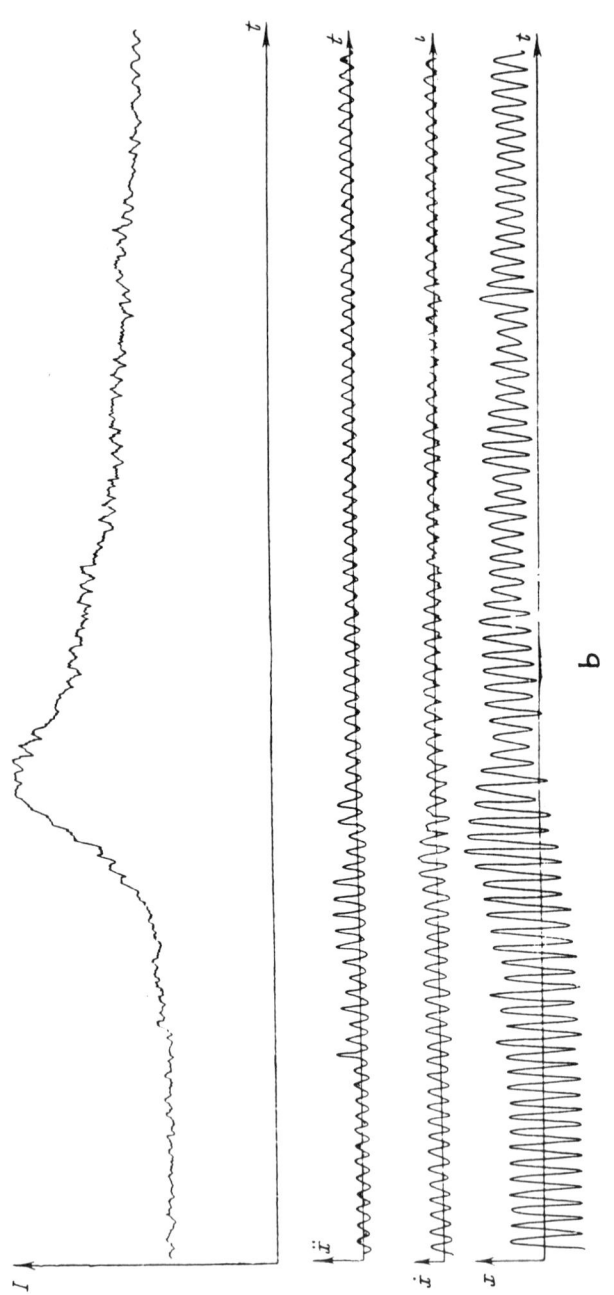

in the transition to transresonant regimes or mismatch of the frequency and oscillation amplitudes of the jaws with the angle of inclination of the crushing plate. When the jaws are thrust outward by an uncrushable rock mass, the dc component of the restoring forces of the elastic system becomes so large that the exciting force of the vibrator cannot overcome it and the jaws perform torsional oscillations instead of translational displacements. When a continuous flow of rock mass is crushed, similar phenomena are not observed, the working process is of a stable character, and jaw displacements are stabilized and are approaching average values.

The hypothesis, assumed when developing the phenomenology of the crushing process, stating that the crushed material during operation under load performs the role of an additional nonlinear elastodamping system, enables one to explain effects observed in real installations.

When the rock mass is fed into the crushing chamber, the effective stiffness of the machine elastic system increases in accordance with the hardness and quantity of the rock in the chamber. System tuning approaches resonance and can also, under specific conditions, become resonant or even subresonant. With a sufficient power reserve of the drive, which was always available in the tested crushers, the drive energy is transferred to the oscillatory system: shock absorber-crusher jaw-crushed rock, as a consequence of which the oscillation swing increases. This takes place both in the near-resonant and subresonant characteristics of the crusher under load. In the latter case, this is associated with the fact that in order to achieve the nominal (in the given case subresonant) operating regime the crusher must pass through the resonant region. The larger the power reserve of the motor, the more rapidly transition through resonance takes place and the less the resonant amplitude will be. It should be noted that owing to the fact that the principal stiffness of the crusher-load system is concentrated in the crushed rock, the greatest part of the input energy is spent on straining and fracturing the rock mass. As a rule, satisfactory crushing of the rock mass is achieved in the regimes accompanied by an increase in the oscillation swing of the jaw. If the rock is hard, with low internal resistances, the oscillation swing in the resonant region increases the most. When a weak rock with high internal resistances is crushed, the increase in oscillation swing is not as large. When the crushing chamber is nonuniformly loaded, the effective stiffness of the crusher increases with the fracture of sharp protuberances on the crushed rock pieces. The system, meanwhile, can change over to subresonant tuning with time. In the subresonant operating regime the oscillation amplitudes of the jaws are decreasing and, additionally, their opening is increasing, thus energy consumption drops. Thus, the subresonant operating regime under load is less effective, since the crushed rock receives less of the energy required for its deformation and fracture.

In such regimes, also with a large magnitude of the exciting forces and with the necessary reserve of motor power, the crusher crosses the subresonant operating regime more rapidly and with less jaw swings if the stiffness of the rock mass is substantial. This is associated with reduction in the effectiveness of the crushing process. If the motor power is limited, then the crusher does not receive the required energy to overcome resonance and operates beyond the resonant regime in a region close to the resonant despite the large stiffness of the elastic system corresponding to subresonant tuning. Thus, the tendency towards increasing the vibrator exciting force and using a motor with large power reserve is not always useful. The drive power must be adopted in close correlation with the other parameters and operating conditions.

The presence of jumps in the amplitude of oscillations upon transition of the system from the resonant regime to the transresonant regime, which takes place with the conclusion of the crushing process and commencement of idling, indicates that the crushed rock inside the working chamber has progressively increasing stiffness vs. deformation characteristic (with increasing strain rate the area of the rock pieces engulfed by deformations increases and the effective stiffness coefficient of the system rises proportionately). As a result, the amplitude-frequency characteristic of the crusher under load is deformed towards large excitation frequencies. Therefore, upon transition from one tuning regime to another, part of the amplitude-frequency characteristic is not realized and abrupt amplitude change of jaw oscillations takes place (which is recorded in the presented oscillograms). If the effective stiffness varies little with time during crushing of the rock mass which is uniformly and totally filling the crushing chamber, then the stiffness during crushing of a single chunk rises with the progress of the crushing process reaching maximum prior to its fracture.

The above stated enables one to regard the vibratory crusher under load as an oscillatory system with a nonlinear elastic connection. The variable stiffness is formed both as a result of the presence of clearances between the crushed rock and the jaw and owing to the increase of the surface area of the deformed material during the crushing process (chipping of the pieces).

The effective stiffness of the rock mass for the given standard-size crusher is proportional to the degree of filling of the chamber by the rocks. Maximum stiffness is attained when it is completely full. Therefore, the volume of the chamber and its parameters - width, depth, and the dimension of the feed opening - must be selected so that the natural frequency of the jaw under nominal loading would not exceed frequency of the exciting force and be close to it. This enables the realization of the near-resonant operation of the crusher during deformation of the rock mass. This regime is the most effective from the standpoint of transferring energy which is required for deformation and crushing. When selecting parameters of the chamber, it must be born in mind that increasing its width and depth will cause an increase in the effective stiffness of the load, and increasing the delivery and discharge openings causes a decrease in this stiffness. All this is valid for the same degree of filling crushing chambers of different sizes.

The jaw mass, parameters of the vibrator and the motor must be selected in such a way that for given reduced elastodamping properties of the rock mass, steady oscillations would be ensured in the near-resonant region with the necessary amplitude.

Analysis of the recorded oscillograms of the crushing process of rock mass in a vibratory crusher obtained as a result of simulation on analog computers and numerical analysis on digital computers enables, on the one hand, ascertaining good qualitative and quantitative correlation of the recorded parameters with the data obtained in real objects and to identify, on the other hand, many important characteristic features of the vibratory crushing process which are not as yet accessible for investigation with physical experiments.

When analyzing oscillograms of displacements, velocities and accelerations of the vibratory crusher, and also the relative (with respect to the jaw) deformations, velocities and accelerations of the crushed rock, it is apparent that the harmonic character of displacements, velocities, and accelerations of the jaw is changing over several oscillation cycles when rock mass is fed into the crushing chamber of the vibratory crusher operating at idling. The amplitude of displacements begins to increase and the neutral line of the jaw oscillation under load shifts gradually relative to the neutral line of idling outward from the crushing chamber. Peaks appear on the plot of oscillation velocities. Clearly defined impulses, corresponding to the moments of impact between the jaw and the crushed rock, appear on the acceleration plot. Deformations of the rock mass delivered to the crushing chamber begin to gradually rise from zero to a maximum over several cycles. Deformations occur relative to an axis which is considerably offset from the neutral line. This indicates that during the process of vibratory crushing the rock is under constant load on to which periodic dynamic loads are superimposed. The character of deformation of the rock in the stages of interaction with the crushing jaw is "spiky". During motion without contact with the jaw it approaches the harmonic character. It is interesting to note that vibratory accelerations of the rock are less than vibratory accelerations of the jaw, which is explained by the fact that the stiffness of the elastic system of the crusher is less than the effective stiffness of the rock.

Let us consider specifics of a vibratory crusher under load when the crushing chamber is filled with rocks even prior to starting. Study of the crusher behavior in such conditions is of practical interest. To obtain more complete information we shall first consider the starting and operation of the vibratory crusher in the steady-state regime at idling. Figure 5.7a shows oscillograms of the displacements, velocities and accelerations of the jaw of the vibratory crusher at idling in transient and steady-state regimes. The natural frequency of jaw is $p = 60$ 1/sec, the coefficient of viscous resistances of the elastic system $n - 10$ kgf/m, the vibrator frequency $\omega = 120$ 1/sec, i.e., the detuning factor of the crusher is $z = 2$. It follows from the presented oscillograms (Fig. 5.7a) that displacements, velocities and accelerations of the jaws at idling are of sinusoidal character. Since a motor of sufficient power reserve has been used, the crusher very quickly changes over to the steady-state operating regime. With start-up under load (Fig. 5.7b), characterized by the expected value of the natural frequency $p = 250$ 1/sec, the damping factor $n = 300$ kgf/m, and effective ultimate plastic strength $f = 30$/sec, the character of oscillations of the crushing jaws undergoes some change. Since the effective natural frequency of the system: load-jaw-elastic system is higher than the natural frequency of

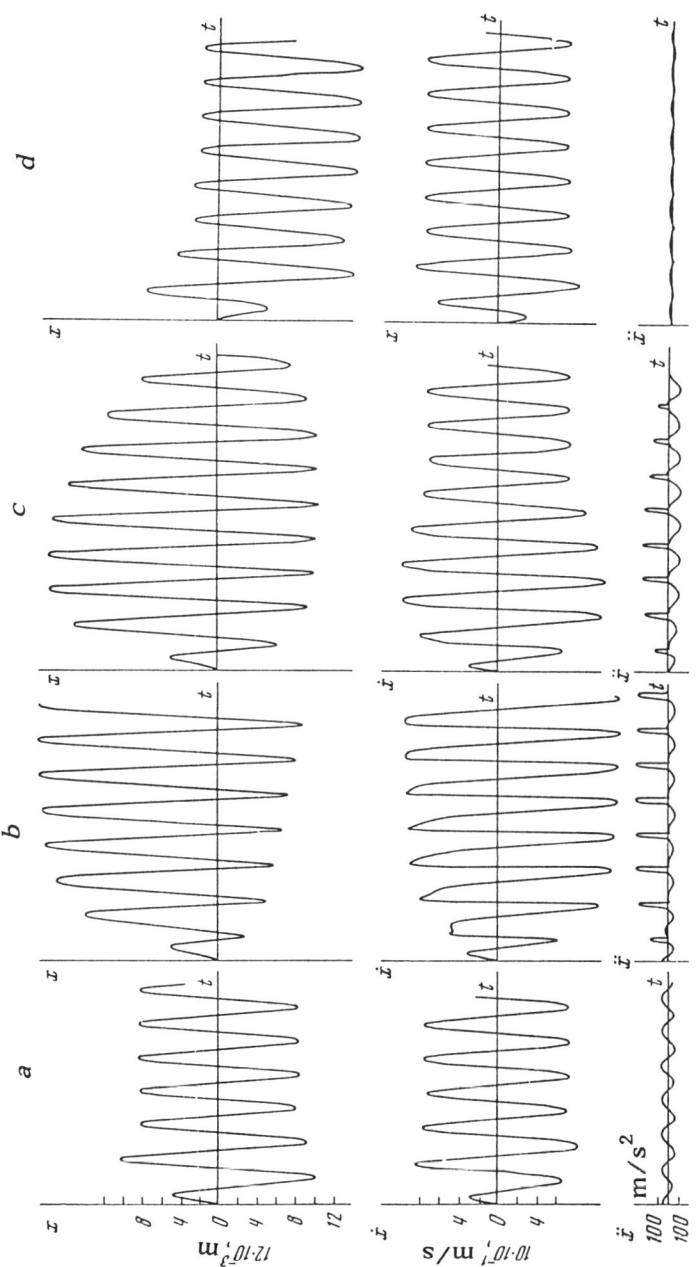

**Figure 5.7** Oscillograms of jaw oscillations and deformations of the rock mass ($z = 2.0$) a) displacement, velocity, and acceleration at idling; b) displacement, velocity, and acceleration under load; c) deformations, velocities, and accelerations of deformations of the rock during the crushing process; d) same for the rock interaction with a jaw of unlimited mass.

the jaw, the detuning factor of the crusher under load is lower than at idling, i.e., under load the crusher is tuned at the subresonant regime close to the near-resonant regime. Hence, during start-up the system passes through resonance which is seen in the oscillograms as an increase of vibratory displacements and velocities of jaws in the transient regime. Acceleration shows peaks that appear at interactions of the jaw with the rock.

For comparison, Fig. 5.7c shows oscillograms of deformations, their velocities, and accelerations of the rock. Essentially, the crusher under load (jaw-rock system) is a system with two oscillating masses, namely the jaw and the effective part of the rock mass, joined together by a system of nonlinear elastic connections with clearances. Such a system has two natural oscillation frequencies.

Effect of the properties of the crushed rock on the character of oscillations of the machine is of interest. Let us consider the crusher under similar regime parameters and tuning with two different types of rocks that are characterized by different limits of inelastic deformations ($f = 10$ m/sec$^2$ and $f = 30$ m/sec$^2$). The limit of irreversible strains for the first material is 3 times lower than for the second. Let us start with the consideration of the idling regime of the crusher. Parameters of crusher oscillations $x$, $\dot{x}$, $\ddot{x}$ at idling are presented in Fig. 5.8a. The characteristics of the crusher and the crushing process of the elastic material are given in Fig. 5.8c, of the material with large irreversible strains in Fig. 5.8e. By comparing these characteristics it can be noted that at the beginning of the deformation process, velocity and acceleration of deformation of both rock masses are approximately the same. However, with time and with the development of the deformation level, significant differences begin to emerge. Thus, the displacements of the jaw and deformation of the elastic rock, after reaching a certain maximum level, stabilize and are maintained within the same bounds during the steady-state process. The deformation of the rock mass with large internal resistances first increases, then starts to decrease, and stabilizes only at a level of 50 - 60% of the maximum. This is explained by the action of high resistances in the rock and by the characteristics of the oscillatory system: rock-jaw-shock absorber which decreases the oscillation swing with increasing resistances. We can thus explain the known fact that rock mass of low hardness with large internal resistances are crushed worse in vibratory crushers than hard and elastic rock masses.

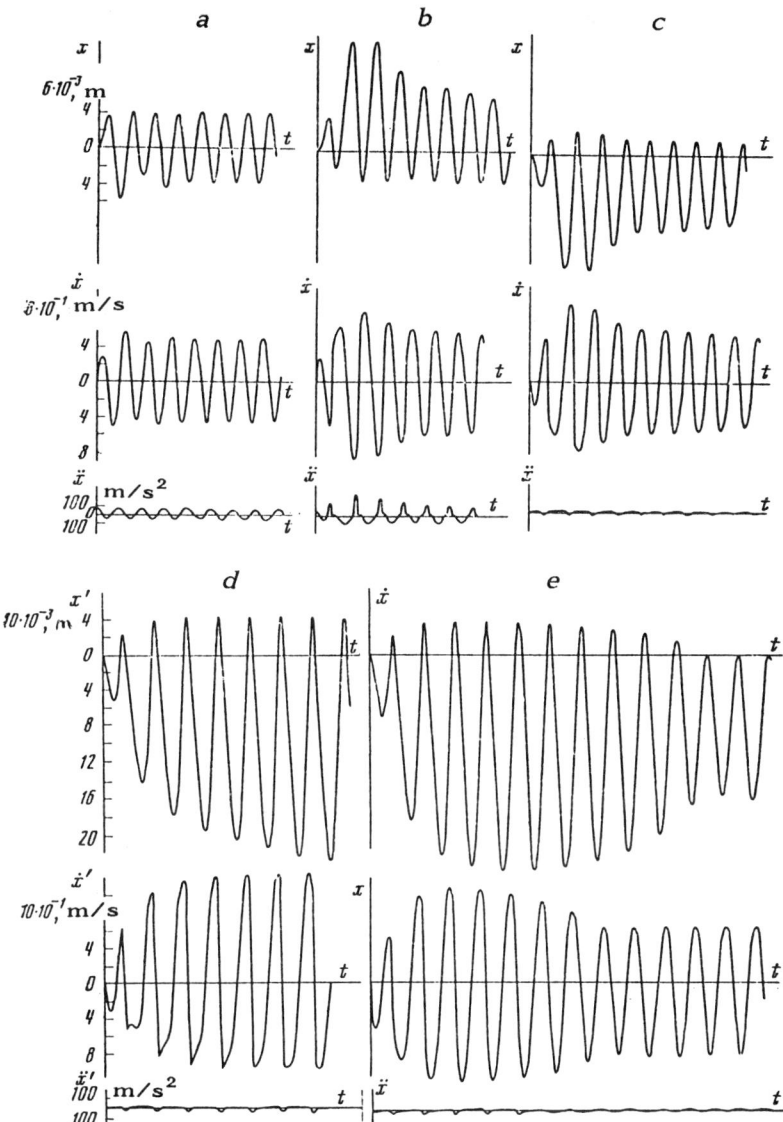

**Figure 5.8** Oscillograms of jaw oscillations and strains of crushed rock mass ($z = 1.5$). a) displacements, velocities and accelerations of the jaw at idling; b) displacements, velocities, and accelerations of the jaw when crushing elastic rock mass; c) deformations, their velocities, speeds, and accelerations of the elastic rock mass; d) displacements, velocities, and accelerations of the jaw when crushing rock mass with high internal resistances; e) deformations, their velocities, and accelerations of rock mass with high internal losses.

In the same way, the oscillation amplitudes of the jaws are adjusted at the maximum level when crushing elastic rock mass, while for crushing rocks with large internal resistances the amplitude does not exceed 60% of the first case and, more importantly, the jaw pressure on the crushed rock in the steady-state regime is significantly higher in the first case.

Thus, when designing crushers to crush rock masses with small elasticity and large internal resistances, it is recommended to increase the mass of the jaw (or decrease the degree of filling of the crushing chamber by the rock mass) and decrease the degree of detuning of the crusher from resonance (to establish an elastic system of high stiffness) in order to improve the process effectiveness.

Let us consider the effect of the detuning factor of the crusher on the crushing process. We analyze the operating regime of the machine while crushing elastic rock mass. We had examined earlier the operation under load of the crusher whose idling detuning factors are $z = 2$ (see Fig. 5.7) and $z = 1.5$ (see Fig. 5.8). Let us consider its operation under load at the detuning factors $z = 2.5$ (Fig. 5.9) and $z = 1.0$ (Fig. 5.10). When the crusher is tuned for the transresonant regime of operation (detuning factor $z = 2.5$), at idling the jaws perform harmonic oscillations (Fig. 5.9). With the delivery of the rock to the crushing chamber the character of oscillations changes sharply: their amplitude rises (approximately by 2.5 times) and the neutral line is shifted. The displacements of the jaw become asymmetric and the impact of the jaw against the crushed rock is realized once every two cycles of oscillations. Hence, the period of jaw oscillations increases twofold (see Fig. 5.9). The period of rock deformation also increases twofold.

It is interesting to note that in the stable crushing regimes, jaw impact against the rock usually takes place with maximum or near-maximum relative speed. Furthermore, the maximum relative speeds of collision of the jaw and the rock are close to the maximum absolute velocities of jaw oscillations.

The impact conditions of the jaw with the rock deteriorate sharply in the unstable regimes when the period of jaw oscillations increases twofold. This occurs in the transresonant region near resonance. Thus, during one half of the cycle, when a larger amplitude value of the relative velocity is observed, impact occurs at the collision speed which is equal to 0.2 - 0.4 of the amplitude value. In the second half of the

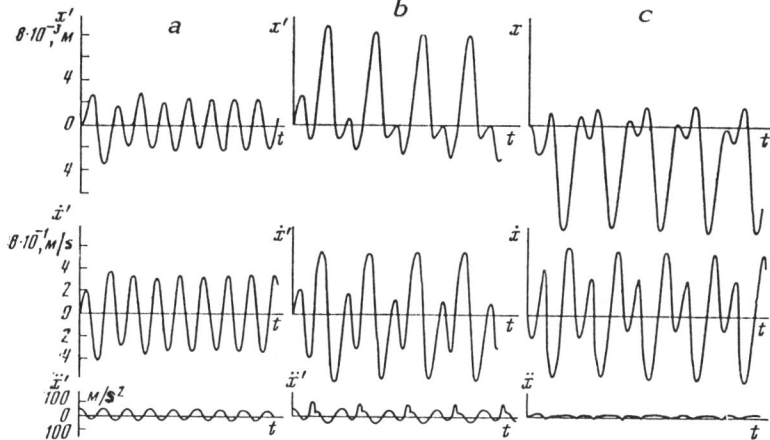

**Figure 5.9** Oscillograms of jaw oscillations and deformations of the rock ($z$ = 2.5). a) displacements, velocities and accelerations of the jaw at idling: b) displacements, velocities and accelerations of the jaw under load; c) deformation, its velocity, and acceleration of the crushed rock.

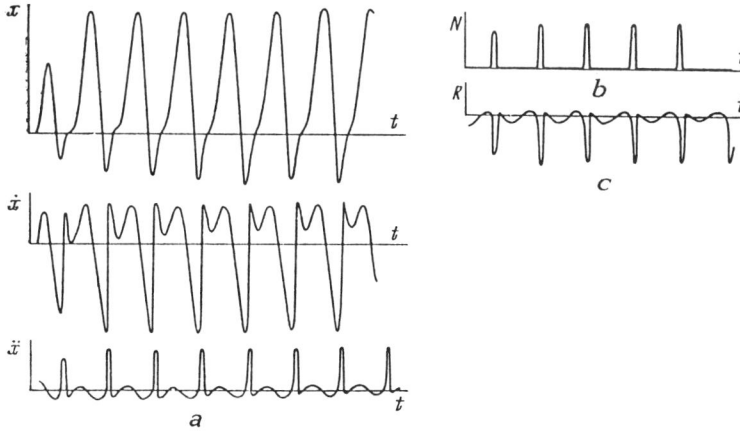

**Figure 5.10** Oscillograms of jaw oscillations and deformations of the rock ($z$ = 1.0). a) displacements, velocities, and accelerations of jaw oscillations under load; b) loads on the jaw from the crushed rock; c) internal stresses in the crushed rock.

cycle impact takes place at the maximum value of the relative speed but the amplitude of speed in the second cycle comprises only 0.5 - 0.55 of the amplitude of speed in the first cycle. Thus, both collisions occur at low relative speeds as a result of which the collision forces are also small (see Fig. 5.9). At resonance tuning of the crusher under load, jaw oscillations also have a sharply defined non-linear character (Fig. 5.10).

Let us analyze the way the character of interaction of the crushed rock and the jaw affects the effectiveness of the crushing process. We shall compare the characteristics of the crushing process in the absence of influence of the crushed material on the character of jaw oscillations (see Fig. 5.7c, d). This regime can be established when a jaw having a substantial mass crushes a small amount of rock mass. This regime is easily realized on an analog computer by excluding the load (reaction of the crushed rock) from the equation describing motion of the crusher jaw.

The deformation parameters of the rock mass in the absence of interaction between the crushed rock and the jaw is shown in Fig. 5.7d, and in the presence of interaction the parameters are shown in Fig. 5.7b, c. The comparison shows that in the presence of interaction between the rock and the crusher the oscillation regime of the jaws intensifies and the deformation of the rock, as well as velocities and accelerations of collision of the jaw with the rock, increase. In short, with the interaction of load with the jaw, the crushing process is intensified. Since the interaction of the rock mass and the crusher positively influences the flow of the process, one must take into consideration this circumstance when designing vibratory crushers and avoid excessive increases of the mass of the jaw. One must seek a rational combination of the masses of the rock and the jaw that provides optimal operation of the system. It must be noted that for the same parameters of the crusher and the rock mass and under constant exiting force the stroke of the jaw increases with decreasing oscillation frequency, deformations of the rock during the crushing process increase, and the forces with which the jaw acts on the rock mass rises sharply. From comparison of the operational parameters of the crusher at frequencies $\omega$ = 90, 120, and 150 rad/sec and the characteristics of the crushing process it follows that with decreasing oscillation frequency the amplitude of jaw oscillations at idling increases reaching 2.25 mm at $\omega$ = 150 rad/sec, 4 mm at $\omega$ = 120 rad/sec, and 8.2

mm at $\omega$ = 90 rad/sec, i.e., the amplitude increases by 1.8 and 3.6 times. Furthermore, the maximum deformations of the rock are 0.95, 2.0, and 10.0 mm, respectively, i.e., 2 to 10 times increases. The maximum collision velocities of the jaw with the rock amount in this case to 4.4, 6.0, and 8.0 m/sec, which shows an increase by 1.35 and 1.8 times. The increase in the force by which the jaw presses against the crushed rock mass is of particular significance. Thus, at the oscillation frequency $\omega$ = 150 rad/sec, the maximum pressure on the rock is equal to 80 m/sec$^2$ per unit jaw mass, at $\omega$ = 120 rad/sec the pressure is 160 m/sec$^2$ and at $\omega$ = 90 rad/sec, it reaches 300 m/sec$^2$, i.e., it rises by 2 and 3.7 times. These specific features of the operation of the vibratory crusher must be considered for selecting the operating regimes.

In order to assess the crushing process in a vibratory crusher, one can consider such indicators as reaction of the rock on the jaw, internal stresses in the rock, the character of deformation and reduction of the dimensions of the chunks in the course of crushing, and energy consumption on straining and breaking the rock. Figure 5.11 shows reactions of the rock on the crushing jaw $F$, internal stresses in the rock which are increasing during the crushing process $R_x$, energy consumption on straining and breaking the rock $W_R$, and also the absolute strains (considering irreversible strains) and decrease of the dimensions of the rock chunks $x^{**}$.

From the plots $F = f(t)$, it is evident that the rock resists the motion of the jaw for approximately 0.2 of the oscillation period. The action and reaction of the rock on the jaw is of pulsating character with spikes. Under the action of periodic impacts of the jaw, internal stresses are generated in the rock, causing it to oscillate. It is interesting to note that as a result of oscillations of the rock, both compressive and tensile stresses are generated in the jaw. As the plot $R(t)$ shows, the tensile stresses are roughly one order of magnitude less than the compressive stresses but act for significantly longer periods of time. The tensile stresses in the rock during the crushing process arise only in vibratory crushers as a result of the high-frequency application of the crushing forces. Generation of tensile stresses in the rock is a very important principal feature of vibratory crushing. They enhance effectiveness of the process, since rock mass has low resistance to tensile stresses. Additionally, alternate-sign loading (tension-compression) of the rock generates fatigue stresses and facilitates the crushing process. It is pertinent to

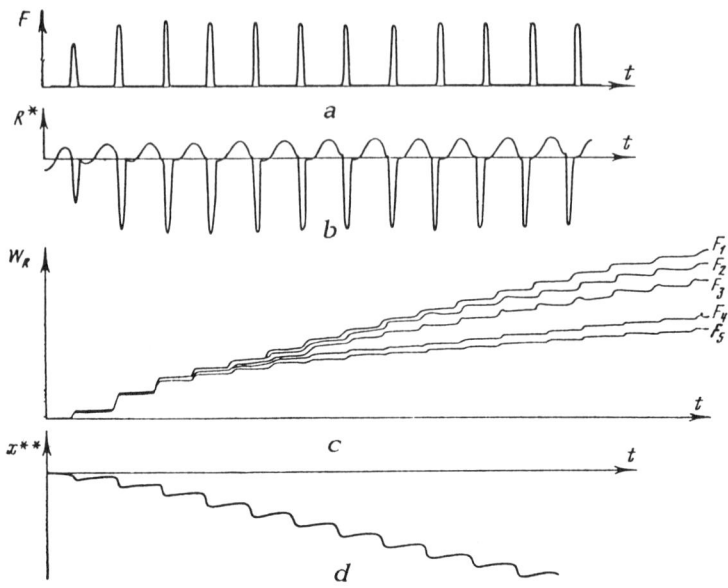

**Figure 5.11** Characteristics of the vibratory crushing process of rock mass. a) reaction of the rock on the crushing jaw; b) internal stresses in the rock; c) energy expenditures on deformation and fracture of the rock; d) absolute deformation of the crushed rock.

mention that the level of internal stresses in the rock increases as the frequency of excitation approaches the natural frequency of the rock. The plot shows time history $W_R(t)$ of energy expenditures on straining and breaking the rock. It is evident from the graph that the effectiveness of energy absorption by the rock in the vibratory crusher increases with increasing of its elasticity ($F_1 > F_5$) and decreasing of internal resistances. This fact explains why the process of fracturing hard rocks is more effectively achieved in vibratory crushers.

One of the main advantages of double-jaw inertial crushers is that the main forces, generated during the crushing process, are acting between the jaw and the rock; they are locked on the rock mass and are not transmitted to the frame. The crushing forces considerably exceed the reactions of the elastic system of the jaws on the crusher frame.

The patterns of motion of the rock in the crushing chamber from the charging opening to the discharge door, considering the pressure of the material in the hopper, are illustrated by the plots of the displacement, velocity, and acceleration of the rock mass presented in Fig. 5.12. It is evident from the graphs that the rock is periodically displaced towards the discharge opening when it is not in contact with the jaws. When it comes into contact with the jaws it slows down and significant accelerations are generated. The most heavily loaded and most often subjected to breakdowns are the bearings supports of the unbalanced-mass shafts of the vibrator. The low durability of the bearings units of vibratory crushers is caused by the considerable dynamic loads that are generated by both the centrifugal forces of the unbalanced masses and by the high accelerations of the jaws (hence, of the vibrator housing) resulting from the crushing of rock masses having large chunks of high hardness.

The loads on the bearing units of the inertial vibrator can be found from the following expression (4.31). Here, the first term in parentheses determines loads from the centrifugal forces generated by the unbalanced masses, and the second terms account for the loads generated by accelerations of the crusher jaws. Figure 5.13$a$ shows values of the loads on the bearings units of a crusher performing oscillations in the $x$ direction at idling, which are obtained at various detuning factors $z$ and ratios of the rotating and total oscillating masses of the machine $q$. It is evident from the graphs that the loads on the bearings units at transresonant tuning of the machine ($z > 1$) decrease and at subresonant tuning ($z < 1$), they increase.

The loading diagrams of the bearings units when the crusher is operating under load (in the process of crushing of rock mass) are presented in Fig. 5.13$b$. Analysis of the diagrams shows that, in this case, the loads on the bearing units increase sharply, considerably exceeding the centrifugal forces generated by the unbalanced masses. It is evident from the diagrams that as a result of the continuous change of the character of the crushing process, the loads on the bearings units also change. The large loads arising in the process of crushing are causing frequent breakdowns of bearing unit in vibratory crushers.

Thus, the most serious problem associated with the design of large vibratory crushers is to provide adequate durability of the bearing units. The loads on them can be reduced by joining the bearings with the frame of the vibrator via an elastic ring, for example, made from wire mesh.

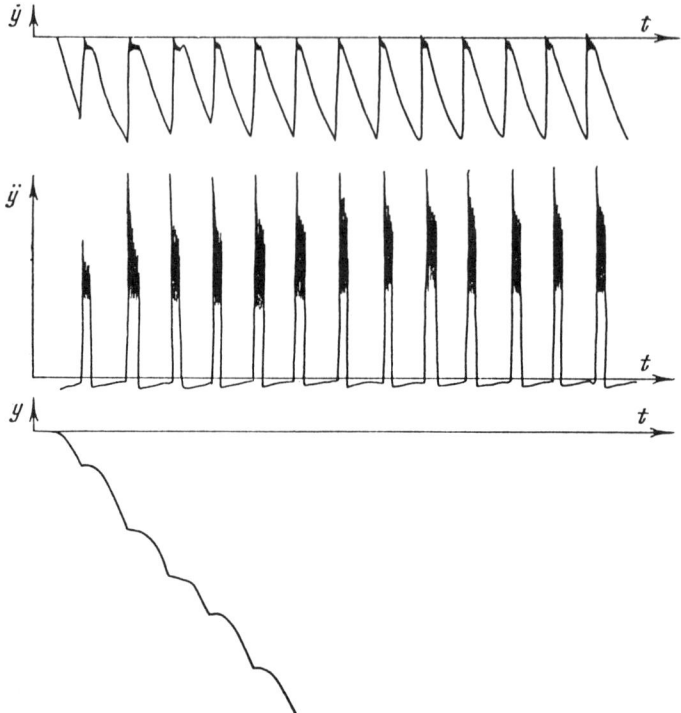

**Figure 5.12** Parameters of motion of the rock in the crushing chamber.

The mathematical model of a vibratory crusher with an inertial vibrator of the considered design is shown in Fig. 5.14. The working element has mass $M$ and is mounted on a stationary frame by means of elastic elements having stiffness coefficients $K_x$, $K_y$ and coefficients of hysteretic losses $C_x$, $C_y$. Excitation of the machine is effected by an inertial exciter whose total mass and eccentricity of the unbalanced rotating parts are equal to $m$ and $r$, respectively, and the angular velocity of the shaft of the unbalanced mass is $\omega$. During operation of the crusher the unbalanced mass generates a rotating exciting force $mr\,\omega^2$. The mass of the bearing unit of the vibrator is $m_1$. This unit is mounted on elastic elements which are characterized by coefficients of stiffness $k_x$, $k_y$ and viscous resistance, $c_x$, $c_y$.

Two coordinate frames $xoy$ and $x_1o_1y_1$ are used. The origin of the stationary coordinate system is placed at the center of gravity of mass $M$ while it is in static equilibrium on elastic

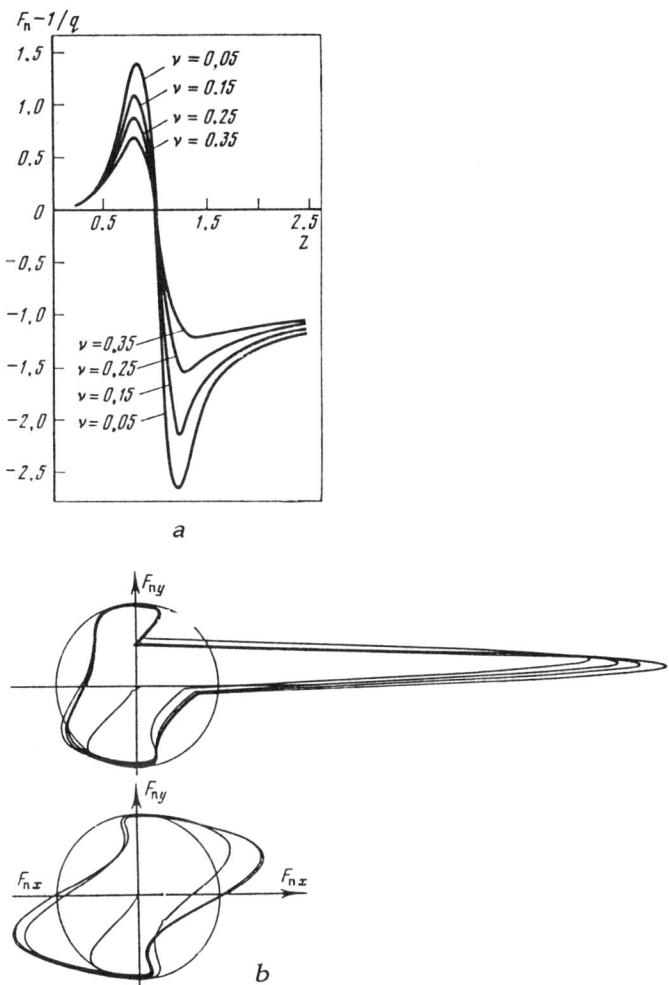

**Figure 5.13** Loads on the bearing units of a vibratory crusher. *a)* variation of the load as a function of crusher tuning at idling; *b)* under load.

elements and axes $x$ and $y$ are in the directions of the extremal stiffness values. The origin of the moving coordinate frame coincides with the center of gravity of mass $m_1$. Thus, coordinates of the center of gravity of mass $M$ in the stationary coordinate system will be $x$, $y$; coordinates of mass $m_1$ in the same system will be $x_1$, $y_1$; coordinates of the center of gravity of the unbalanced mass in the moving coordinate

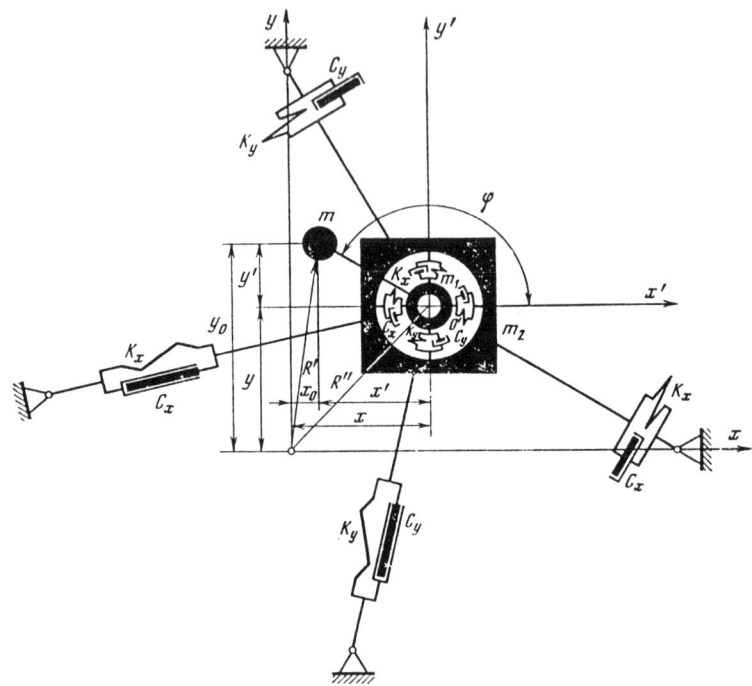

**Figure 5.14** Mathematical model of inertial vibrator with shock-absorbing bearing units.

system will be $x'$, $y'$; the angle of rotation of the unbalanced mass is $\omega\, t_1$.

The differential equations of motion of the vibratory crusher with the considered inertial vibrator are

$$\ddot{x} + 2\xi\,(n_x + n_{1x})\,\dot{x} + \xi\,(p_x^2 + p_{1x}^2)\,x - 2\xi\,n_{1x}\dot{x}_1$$
$$- \xi\,p_{1x}^2 x_1 = 0$$
$$\ddot{y} + 2\xi\,(n_y - n_{1y})\,\dot{y} + \xi\,(p_y^2 + p_{1y}^2)\,y - 2\xi\,n_{1y}\dot{y}_1$$
$$- \xi\,p_{1y}^2 y_1 = 0$$
$$\ddot{x}_1 + 2\,n_{1x}\dot{x}_1 + p_{1x}^2 x_1 + 2n_{1x}\dot{x} - p_{1x}^2 x = qr\,\omega^2\sin\omega t$$
$$\ddot{y}_1 + 2n_{1y}\dot{y}_1 + p_{1y}^2 y_1 - 2n_{1y}\dot{y} - p_{1y}^2 y = qr\omega^2\cos\omega t$$

$$(5.43)$$

Here, the following notations are assumed:

$p_x$, $p_y$ and $p_{1x}$, $p_{1y}$ are the natural frequencies of the bearing unit of the vibrator on its own elastic connections,

$$p_x^2 = \frac{k_x}{m_1 + m}, \qquad p_y^2 = \frac{k_y}{m_1 + m}$$

and on the elastic connections of the crushing jaw

$$p_{1x}^2 = \frac{K_x}{m + m_1}, \qquad p_{1y}^2 = \frac{K_y}{m + m_1}$$

$n_x$, $n_y$ and $n_{1x}$, $n_{1y}$ are damping factors of the elastic connections of the bearing unit of the vibrator,

$$2n_x = \frac{c_x}{m + m_1}, \qquad 2n_y = \frac{c_y}{m + m_1}$$

and of the crushing jaw

$$2n_{1x} = \frac{C_x}{m + m_1}, \qquad 2n_{1y} = \frac{C_y}{m + m_1}$$

$\xi$ is the ratio of masses of the vibratory crusher bearing unit and the crushing jaw, $\xi = (m + m_1)/M$; $q$ is the ratio of the unbalanced mass and the mass of the bearings unit $q = m/(m + m_1)$; $p^* = \sqrt{\xi}\, p_1$ is the natural frequency of the jaw of the vibratory crusher (mass $M$) on its own elastic links, $p_x^* = \sqrt{\xi}\, p_{1x}$ in the $x$ direction, $p_y^* = \sqrt{\xi}\, p_{1y}$ in the $y$ direction; $p^{**}$ is the natural frequency of the crushing jaw on the elastic links of the bearing unit and of the crushing jaw, $p_x^{**} = (K_x + k_x)/M$, $p_y^{**} = (K_y + k_y)/M$.

Solving equations of jaw motion, we find the displacements in the steady-state regime of the crushing jaw and of the bearing unit of the vibrator $x = A \sin(\omega t + \varphi)$, $x_1 = A_1 \sin(\omega t + \varphi_1)$. Since industrial crushers have vibrators of the self-balanced type, oscillations of the jaw are induced in the $x$ direction.

The ratio of the oscillation amplitudes of the bearing unit of the vibrator and of the crushing jaw can be determined from expression

$$\frac{A_1}{A} = \frac{\sqrt{[\xi(1 + x^2) - z^2]^2 + 4\xi(1 + \delta)\,v^2 z^2}}{\xi\sqrt{1 + 4v^2 z^2}} \qquad (5.44)$$

Here, $z$ is the detuning factor of the bearing unit of the vibrator on its own elastic elements, $z = \omega/p$; $v$ is the damping factor of the elastic elements of the bearing unit of the vibrator, $v = n/p$; $\chi$ is the ratio of the natural frequencies of the bearing unit of the vibrator, $\chi = p/p_1$; $\delta$ is the ratio of the coefficients of resistance of elastic systems of bearing unit of the vibrator, and the suspension of the crushing jaw, $\delta = n/n_1$.

By determining the velocity and displacements of the working element of the vibratory crusher and the bearing unit of the vibrator, the loads on the bearings that are flexibly mounted by means of elastic rings can be found as

$$F = c_x (\dot{x} - \dot{x}_1) + k_x (x - x_1) \tag{5.45}$$

Figure 5.13b shows the loading diagrams of flexibly mounted bearing units when the crusher is operating under load. It is evident from the diagrams that the bearing units in this case operate under more favorable conditions.

## 5.4  OPTIMIZATION OF STRUCTURAL AND PROCESS PARAMETERS OF VIBRATORY EQUIPMENT

### 5.4.1  Fundamental Principles of the Method of Multi-Criterial Optimal Design of Structural and Process Parameters of Vibratory Machines

Modern vibratory machines represent complex systems whose operational quality is determined by the character of interaction of drive, working element, processed product, and human operator. The effective interaction with the processed medium, which can be either orderly or stochastic action, and the coordinated operation with the human operator are the basics for productive work of the machine.

The performance of the modern vibratory machine and its correlation with the state-of-the-art is determined by the degree it satisfies the many requirements made of the machine and which are often contradictory. These requirements include high power to weight ratio with low electric consumption per unit capacity, low specific metal content with simultaneous high reliability and performance, small dimensions combined with high durability, operating stability, high efficiency, protection of the environment from vibrations, noise, harmful emissions, and other unfavorable effects.

The design of modern vibratory machines or auxiliary equipment is a very demanding and many-sided process. At the present time, attention is paid not only to the object being designed, but increasingly to those changes that industry will undergo during the process of acquisition and utilization of the new machines and plants. Modern mass production and high cost of machines result in significant losses due to the manufacture and use of non-optimal designs in industry. Hence, the design of a new machine must utilize a significant element of scientific prediction of those events and variations that will occur during its introduction into the production system. Only correct estimate of the development tendencies and allowance for the growth requirements of the industry can ensure that the introduction of the new machine would yield the desired result: the planned progress in the production system, substantial increase in productivity and improvement of the working conditions, the realization of new production processes, protection of the environment, and so on.

The traditional design methods also seem to be less effective when solving the problem of construction of vibratory machines of a qualitatively new level of development, based on fundamentally new and far more progressive principles which differ significantly from the existing machines. This makes the use and interpolation of the available data more difficult.

The construction of well-designed machines requires the designer to approach the task from different viewpoints; first of all, the design must be linked with the tendencies of development of the industry and the consequences of the introduction of the new machine must be predicted and estimated. The design itself must satisfy many requirements that are in the majority of cases contradictory.

The solution of such tasks by traditional design techniques is usually found by studying several alternative versions of the design. This process is laborious and requires considerable expenditures of resources and time. In practice, not more than two or three alternative models of the design can be prepared. Thus, the traditional design methods cannot, in principle, provide the designer with a complete picture of the machine capabilities, since they require the consideration of a very large number of versions and, therefore, unacceptably large expenditures of effort and time in their development.

For development of new machines, various optimization techniques are being increasingly used in the design practice. Systems of automated design are being introduced. In the design practice, optimum design is still conducted in many cases with respect to one quality criterion which, in the designer's opinion, is the most critical. Constraints, on the other hand, are imposed on the other quality criteria. For example, when designing a large vibratory grizzly-feeder for loading a main conveyor, the set task was to ensure maximum hourly output of the system. Constraints, such as weight, cost, installed power of the drives, and so on, were imposed on the other quality criteria.

This approach suffers from the unsubstantiated selection of the most important quality criterion and from great difficulties in designating criterial constraints. The essence of the matter here is not only that the subjective decisions of the designer are essentially flawed. Subjective evaluations in making decisions are present in other methods of optimal design and it is unlikely that they will be superceded in the foreseeable future. The important thing is that traditional techniques of optimal design with respect to one criterion do not provide the designer with sufficient objective information on which basis he could competently make his decision, which is still a subjective decision.

It has been established by experience that the decisions taken on the basis of inadequate information are characterized by very similar errors.

Thus, four of the most characteristic errors that are admissible when using traditional methods of optimal design were noted in [44].

First of all, the term "optimal" often does not mean the best solution, rather a solution which is preferred by the designer or the customer or the consumer. Meanwhile, the optimal design solution must be understood to mean an admissible design whose realization ensures the construction of a machine satisfying the technical objective as much as possible.

The second error is a very serious one, whereby the design is technically optimal but its effectiveness function (quality criterion) is incorrectly selected. For example, in the foregoing example when designing a vibratory grizzly-feeder for a main conveyor, the maximum output was taken as the effectiveness function, while for the quarry operator its maximum reliability is vital.

The other characteristic error in the optimal design which is often made is that the designer does not fully assess, or ignores, some essential constraints such as the conditions of comfort for the operating personnel or the effect of plant operation on the environment.

There is often also a tendency to achieve a global optimum at any cost, i.e., by means of unacceptable reduction of constraints. Furthermore, it is frequently overlooked that cheaper and simpler solutions sometimes differ little from the ideal optimum.

The indicated errors of the traditional method of optimal design follow from the fact that the designer prior to the analysis and solution of the problem must determine the nomenclature or, as mathematicians say, the domain of acceptable models of the machine, i.e., those satisfying the technical requirements or imposed constraints. The acceptable models are sometimes referred to as the "satisfactory" models. If the imposed constraints are too rigid, then there are few acceptable models, and if the constraints and not too rigid, many different solutions are obtained.

Without performing special analysis, as is practiced traditionally, the designer cannot competently determine the domain of acceptable models. And if errors are made at the initial stages of the optimal design, a really good solution cannot be found later even with application of the most perfect techniques. Ignoring this factor often leads to the fact that extensive and costly optimization efforts yield unsatisfactory solutions. The unsubstantiated use of optimization methods can lead not only to the best optimal solution, but also to the worst solution.

Presently, the success of the optimal designing is determined not so much by application of one method or another in search of the optimal solution, but rather by the correct formulation of the problem and, most importantly, by the substantiated determination of the admissible set of solutions (machine models). The technique for solving multi-criteria optimal problems, which is adopted in this book as the main one, is thoroughly developed [38].

This method enables one to take into account as many quality criteria as required for a really comprehensive and sufficiently complete estimate of the quality of the designed machine. The method provides an effective inspection of the multi-dimensional parameter space which facilitates the establishment of the domain of admissible models of the machine. The method stipulates that problem formulation and machine design be carried out in the "designer-computer" dialogue regime.

At the onset of the design with the use of the optimal parameter selection technique, the designer assumes a kinematic layout of the vibratory machine and by varying its parameters he arrives at a new model (modification) of the machine. Each machine model is assessed by the local quality criteria corresponding to it. For this purpose, the designer must have at his disposal all functional dependencies (formulas, equations, regression correlations) linking parameters of the machine and its quality. By using these dependencies for each machine model characterized by a fully defined set of parameters, the corresponding set of quality criteria is determined. The numbered machine models, with the corresponding set of quality criteria, are usually entered into tables known as the test tables.

If there are only two quality criteria, the data can be represented graphically. Figure 5.15 shows a graph on which the machine models, denoted by circles, are plotted in the coordinates of quality criteria $K_1$ and $K_2$. The number inside the circle denotes the number of the machine model. The limit values of the quality criteria which are acceptable in the opinion of the designer are designated $K_{1max}$ and $K_{2max}$ The machine models in the region of the graph which is bounded by straight lines plotted perpendicular to the coordinate axes through points $K_{1max}$ and $K_{2max}$ are, in the opinion of the designer, acceptable. These are usually referred to as the admissible models. The aggregate of these models is known as the admissible set of models.

The admissible set includes machine models of two categories: those that can be improved simultaneously with respect to both quality criteria, and those that cannot be improved simultaneously with respect to all quality criteria.

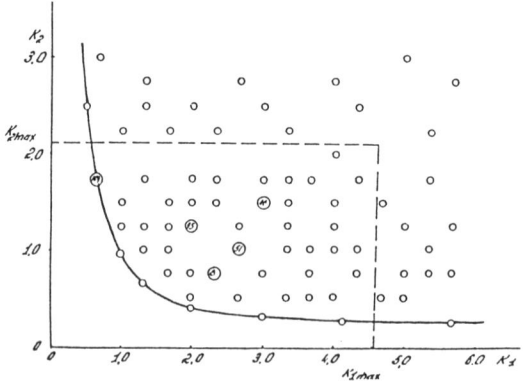

**Figure 5.15** Graphical representation of the tables of test models.

Assume that there really exists a machine which is constructed according to the model design 46. The question is whether its quality can be improved simultaneously with respect to criteria $K_1$ and $K_2$. The answer is affirmative since models 51 and 95 will be characterized by higher characteristics due to the fact that $K_1^{95} < K_1^{55} < K_1^{46}$ and $K_2^{51} < K_2^{95} < K_2^{46}$.

Let us now consider a machine prepared according to design 104 and attempt to improve it with respect to both quality criteria. From the graph it is evident that this is not feasible. Let us take, for example, model 8 whose quality index $K_2^8$ is much better than $K_2^{104}$, but on the other hand its quality index $K_1^8$ deteriorates in comparison with $K_1^{104}$. The same applies to the other remaining models. A conclusion can be drawn that models 8, 12, 61, 104 cannot be improved with respect to both quality criteria simultaneously. Improvement of one criterion always leads to the deterioration of the other. The aggregate of these unimprovable models form a Pareto set. This set is sometimes referred to as the compromise set.

Thus, we arrive at the very important conclusion that the models which can be improved are of no practical interest. Untapped reserves are implemented in the machines constructed according to the designs based on these models.

Only the Pareto machine models must be put in production, since all internal resources are used in their designs.

A machine satisfying requirements of the acceleration of the technological progress can be designed only on the basis of analysis of the Pareto set of models. One of the most vital tasks of optimal designing of machines is the construction of admissible models, an undertaking which is associated with the competent establishment of parametric constraints, identification of Pareto models, and selection of their compromise solutions.

However, in such a formulation, i.e., when only combinations of machine parameters are subjected to analysis, one can make a machine in which all its internal resources have been "squeezed out". In view of this, it was proposed in work [19] to conduct not only parametric analysis but also structural analysis of the machine. This approach presents an opportunity to change over from building very good traditional machines to building very good machines on a new level of development.

On the face of it, it may appear that the application of modern computer technology and the use of the corresponding methods of optimal design diminishes the role of the designer. However, the situation is quite the contrary. More precisely, with the introduction of the techniques of automated optimal design, inadequately qualified designers become obsolete. On the other hand, with the application of the techniques of optimal parametric multi-criteria designing, the role of very skilled designers increases immensely. It can be stated without exaggeration that the success of optimal designing is largely determined by the presence of a highly-qualified designer.

Generally the role of the designer in parametric multi-criteria optimum designing becomes a determining factor. The skills level of the designer totally determines the success of the project.

When using the multi-criterial optimal design method, the question arises as to how many machine models are to be adopted for further analysis and how to select parameters of each machine so that there would be confidence that the selected models are sufficiently representative of all the possible variations. Elimination of vital groups of machines from consideration can lead to the fact that the selected versions might turn out to be far from useful specimens.

Such a method of selecting parameters of the representative machine models does exist. It assumes that the properties of the machines change fairly smoothly with the variation of their parameters. Based on this assumption it can be concluded that the selected parameters of the machine must be distributed more uniformly in the general space of parameters whose boundaries are defined by parametric constraints. The model parameters are selected with the aid of the points of the so-called $LP_\tau$ sequence, thus assuring a uniform search of the parameter space. It is shown in work [38] that the $LP_\tau$ sequence exhibits the best uniformity characteristics among the presently known uniformly distributed sequences.

The method of multi-criterial optimal design in the regime of designer dialogue with the computer using $LP_\tau$ search enables limitation of the search to a specific number of test machine models providing at the same time a uniform review of the main variations of the design, and guarantees obtaining exhaustive information (obviously, upon implementation of the condition of monotonous dependence of the qualitative characteristics of the machine on their parameters). This accelerates the design process and ensures selection of the best arrangement.

Increasing the effectiveness of the process and reducing the time for analysis of alternative design versions provides the opportunity for the search of the most successful solutions, and enables not only expansion of the review region, but also application of a second original technique, namely the investigation of alternative versions of the object structure. If versions of the machine satisfying the requirements are not revealed in the allowable region of the parameters of the given machine design layout, transition to fundamentally new layout takes place. The possibility of the search for structural versions of the machine opens the way to develop products of a qualitatively new level.

The wide application of computers enables not only examination of the alternative versions of the design, but also effectation of the formulation of the problem of selection of the optimum machine parameters. When traditional techniques are used to design structures which satisfy multiple contradictory requirements, the designer frequently faces indeterminate difficulties in substantiating the formulation of the problem - what to regard as quality criteria, how to specify functional constraints, to what extent one criterion can be sacrificed for the sake of another, how to determine the allowable set of machine design versions, and so on. When the method of selection of the optimum parameters is used, lifetime possibilities, effectiveness, degree of fulfillment of the requirement of the customer are identified in the process of dialogue of the designer with the computer. Determined in particular are the parametric, structural, functional, and criterial constraints, and also the allowable set of versions of the machine design.

As mentioned earlier, many and, quite often, contradictory requirements are targeted for modern vibratory machines and installations that are intended for operation in various production conditions. However, traditional design methods, in the majority of cases, allow the improvement of one criterion selected as the most vital, for example, the power to weight ratio is increased, whereas constraints are imposed on other indexes such as dynamic stability, vibratory loads and so on, so that their values would not exceed the acceptable layers. The existing design methods do not enable the creation of machines that are simultaneously superior with respect to many indices. The capabilities of vibratory machines and auxiliary equipment designed on their basis are not totally utilized with respect to the majority of criteria and, frequently, the existence and possibilities of improvement of the product are not even evident. It is not always evident which criterion to adopt as the main one. When the multi-criterial optimal design technique is used, the number of criteria is not limited.

By applying the multi-criterial optimal design technique, the designer can create vibratory machines that simultaneously satisfy many scores of contradictory requirements and select the best design among numerous probable design versions. The practical application of the multi-criterial technique for designing vibratory equipment is achieved by the designer-computer interactive methods presented below and developed specifically for these purposes.

The design process begins as the computer "designs" the variations of the vibratory machine in accordance with the developed program and evaluates them with respect to each quality criterion. The design versions are presented as test tables headed by the best model for the given criterion, i.e., the quality criteria are arranged in a descending order. Since a specific "set" of parameters, encoded in the number of the tests, corresponds to each model, the designer can decide what to do by inspecting the table in order to obtain the best machine with respect to one quality criterion or another.

As a rule, there are no vibratory machines that are simultaneously optimal with respect to all criteria. Therefore, the results of analysis of the tables are used for the substantiated selection of a compromise solution. Given the data on the design possibilities with respect to each quality criterion, the designer together with the customer can specify with confidence the constraints for each criterion which are, on the one hand, practically attainable and, on the other hand, satisfy the customer requirements. Further, the computer verifies whether there are designs simultaneously satisfying all constraints. Such designs make up the allowable set of models. From the domain of the allowable set of designs the designer together with the customer can select the optimal model. If as a result of computations it becomes evident that none of the vibratory machines "designed" by the computer satisfies all the requirements, the designer can, with the consent of the customer, somewhat ease the constraints. If the requirements are rigid and cannot be modified, the designer assists the computer in finding an acceptable solution on account of widening the domain of machine parameters. When even this measure does not achieve the result, the structure of the designed machine is reconsidered. As a rule, this technique not only enables the generation of acceptable models, but it also cardinally improves their indices. The investigation and transformation of the structure of the object carried out with the use of the given method enable the designing of vibratory machines and auxiliary equipment at a new qualitative level.

## 5.4.2   Background of Multi-criterial Optimal Design

### 5.4.2.1   Formulation of the Problem

In order to proceed with the design of an optimal machine or production process, sufficiently good functional dependences between the parameters of the designed object and its service properties are needed in the first place. These dependences can be used either as functional constraints or as quality criteria.

Since all subsequent operations of the multi-criterial optimal design technique are based on these functional dependences, it is vividly clear that the success of the matter will be decisively determined by the quality of these relations. If the relations are insufficiently accurate approximations, then the machine designed on their basis will only be conditionally optimal.

In brief, the successful outcome is contingent on the availability of a well-developed theory of the machine or functional dependences of the production process or, as a last resort, of adequately representative experimental material.

In simple cases the required information can be presented as formulas, graphs, recommendations, regression correlations. For complex dynamic objects these are usually differential equations.

Given the functional dependences linking the parameters and properties of the designed object, one can proceed to the optimal design process proper.

When selecting the fundamental approach to the optimal design, it must be taken into consideration that the current stage of accelerated scientific and technical development requires devices and production processes of a qualitatively new level based on the latest achievements of the basic sciences instead of the ordinary technology.

Consequently, it is pointless to conduct simple parametric optimization of the designed object with the use of ordinary traditional arrangements. The design should commence with consideration of new structures of the object. Moreover, the wider the representation of the structures and the more diverse the new physical effects that they incorporate, the higher the probability of designing an object of a qualitatively new level.

Two approaches are at least possible to specify the structure of the designed object. Firstly, the simpler approach stipulates the development of the structural layout and computational algorithm for each principal model.

Secondly, the more general approach involves the development of a generalized structure and computational algorithm of the designed object including the review of special cases.

In such an approach some of the parameters of the designed object will assumed zero values. In this case the computation algorithm will operate in the "yes-no" regime. In short, when using the generalized structural schemes some parameters of specific models can assume zero values.

### 5.4.2.2   Design Specification

The specification must include a list and values of the main parameters of the designed object which enter into the initial mathematical dependences (differential or finite equations in the case of designing a vibratory machine) describing its motion. These are inertial, rigidity, dissipative, force, and other characteristics of the designed object. A specific set of parameters corresponds to each designed model.

For each parameter $\alpha_i$ the compiler of the spec sheet must, on the basis of his experience, assume acceptable minimum $\alpha_i^*$ and maximum $\alpha_i^{**}$ values. Thus, the specification must include parametric constraints for each of $r$ parameters. Point $\alpha = f(\alpha_i, \ldots, \alpha_r)$ in the $r$-dimensional space of parameters corresponds to each set of parameters. A specific model of the designed object corresponds to each such point.

In the majority of cases, in order to provide the required design quality in addition to the parametric constraints, one must also assume the constraints for some generalized characteristics $f(\alpha)$ whose values are dependent on all parameters of the designed object. For example, this might be the level of the dynamic loads arising in the links of the machine being designed or the velocity of motion of the rollers in the bearings and so on. These are functional constraints and the minimum $f_x^*$ and maximum $f_x^{**}$ values of the functional are assumed.

Thus, the parametric constraints $\alpha_i^* \leq \alpha_i \leq \alpha_i^{**}$ where $i = 1, ..., r$ and the functional constraints $f_k^* \leq f_k \leq f_k^{**}$ where $k = 1, ..., t$ must be assumed.

Specifications must include the quality indices of the designed object. These indices known as the local quality criteria $\varphi_1 (\alpha), ..., \varphi_k (\alpha)$ are specified by numerical characteristics which can be functionals of the integral curves when designing dynamic objects. It is important to note that to use the existing optimal design technique the local quality criterion must be a numerical characteristic linked with the quality of the designed object by a monotonic dependence. It is usually assumed that the lower or the larger the quality indices, the better the designed object.

The number of machine models that can be "designed" within the framework of the imposed constraints will be determined by the selection of the numerical values of the parameters we assume in the range of the allowed parameters. The number of the models will be the larger, the smaller the step with which we select each parameter. The minimum number of models is obtained if for each parameter we assume just one value. By increasing the number of the values of each parameter we can obtain an increasing number of models. The models will differ to the greatest extent from each other for a minimum set of parameters. With increasing sets of parameters the models will become less distinguishable between each other and more alike. Therefore, there is no point in "designing" too many models. Meanwhile by limiting oneself to the minimum number, one can lose good or, perhaps, the best models. The nomenclature of the models must be assumed on compromise basis.

Thus, under the assumed selection step of the parameter, the number of the models of the designed object is determined by the parametric constraints. However, if the designer additionally imposes functional constraints, then not all the models that are "designed" within the bounds of the parametric constraints are of interest for further analysis. The models whose generalized characteristics do not correspond with the functional constraints ought to be immediately excluded. The remaining models form the allowable set, i.e., these are the design objects which do not leave the bounds of the parametric and functional constraints.

### 5.4.2.3    Selection of the Test Models

For the selection of the test models the points of the $LP_\tau$ sequence are used. The model selection process (the points $\alpha$ in the $r$-dimensional space) is carried out as follows. From the coordinates of each successive point of the $LP_\tau$ sequence the parameters of the $i$-th model are determined (the coordinates of point $\alpha_i$). From these parameters the object is designed and tested whether it corresponds with the functional constraints. If the model does not go beyond the functional constraints, it is selected as a test version and all the quality criteria are calculated for it. Otherwise the model is discarded. When selecting the numbers of the test models the length of the calculation of the parameters of each system is also taken into consideration in addition to the above stated. If a single or small-scale production object for moderately important purposes is being designed, only a small number of models should be considered.

When designing specimens of large-scale and mass production it is expedient to increase the number of test models. For the optimal design of vibratory machines, one can limit oneself to the consideration of 32 - 128 models of the designed object.

### 5.4.2.4    Analysis Method for the Test Models

The investigations of the test models selected as mentioned is normally carried out in three stages.

First of all, a tests table is computed with the aid of a computer. For each test model of the designed object all the local quality criteria $\varphi_\nu (\alpha_i)$ are computed, and disordered and ordered test tables are compiled for each criterion. In the test table the number of the model to which the given criterion belongs is indicated next to the criterion. The value of each criterion is arranged in the ordered test tables in a descending order of quality, i.e., the first is the best model with respect to the given quality criterion.

The test tables can be constructed not only with respect to quality criteria. In the case when the functional constraints are not rigid and their variation is allowed, the functional constraints can be regarded as quality pseudo-criteria. In this case, additional test tables featuring functional constraints are added to the ordinary test tables of quality criteria.

In the second stage, the designer, by analyzing the obtained test tables, specifies the criterial constraints $\varphi_\nu^{**}$. The tendency is always to adopt the most rigid constraints in order to increase the quality of the product. However, in this case, it might turn out that none of the assumed models satisfies the set requirements.

The analysis of the tables enables the designer to assess roles of individual quality criteria and carry out a substantiated arrangement which facilitates the selection of the critical quality criteria.

In the last stage, executed by a computer, the models corresponding to the criterial constraints are selected. The selected models make up the allowable set of models. It might turn out that none of the models satisfies the required quality criteria.

A similar situation is encountered owing to the fact that unjustifiably high requirements are made of the designed object. In this case, the criterial constraints ought to be relaxed and the procedure of selection of the allowable set of models repeated.

The best specimens might be missed as a result of selection of insufficient number of test models. To avoid such a situation the number of test models must be increased and the entire calculation procedure repeated.

### 5.4.2.5  Identification of the Pareto Models

The set is called Pareto set if it comprises all the models that are optimal according to Pareto. Since not all models are included in the selected allowable set but only that part equal to $N$ models, then it is natural that one cannot identify the Pareto set from this set. Therefore, from the allowable set only the so-called quasi-Pareto set can be identified. The algorithm for the determination of quasi-Pareto models is carried out by the program PARET. The principle of operation of the program is to the effect that the numbers $i = 1$ and $j = 2$ are ascribed to two of the models with number $i$ and number $j$ from a total number of $\nu$ models. The model $\alpha^{(i)}$ is then compared with model $\alpha^{(j)}$. Further, we determine the set to which the model belongs. For this purpose, condition (I) is tested: if for all quality criteria $\varphi_\nu$ $(\alpha)$ from 1 to $K$ inequalities $\varphi_\nu(\alpha^{(i)}) \le \varphi_\nu (\alpha^{(j)})$ are fulfilled and condition (II) – $\varphi_\nu (\alpha^{(i)}) < \varphi_\nu(\alpha^{(j)})$ is fulfilled for at least one quality criterion $\varphi_\nu(\alpha)$, then set $D\alpha\backslash\alpha^{(j)}$ is generated and the next model index $j$ is formed.

If case (I) is not fulfilled, condition (II) is tested: if for all quality criteria $v$ from 1 to $K$ the following inequality is fulfilled

$$\varphi_v(\alpha^{(i)}) \geq \varphi_v(\alpha^{(j)})$$

and condition

$$\varphi_v(\alpha^{(i)}) > \varphi_v(\alpha^{(j)})$$

is fulfilled for at least one quality criterion $\varphi_v(\alpha)$, then set $D\alpha\backslash\alpha^{(i)}$ is generated and a new index $i$ is formed.

When both conditions (I) and (II) are simultaneously fulfilled, $j$ is formed. A set in this case is not generated.

The rules of formation of indices $j$ and $i$ are as follows.

For the formation of index $j$ the latter is assigned the value $j = j + 1$. If $j \leq N$, transfer occurs to the start of the cycle; if $j > N$, formation of index $i$ is commenced.

For the formation of index $i$ the latter is assigned the value $i = i + 1$. If $i \leq N$, $j$ is assigned the value $j = i$ and index $j$ is formed. If $i \geq N$, the execution of the algorithm is halted.

## 5.4.2.6 Improvement of Quality Criteria by Changing Parametric Constraints

If the quality criteria for the selected model do not satisfy the designer, an attempt can be made to improve them by relaxing the constraints.

It is natural that the quality criteria of the designed object are dependent on its parameters; therefore, by widening the range of these parameters and relaxing the parametric constraints, one can improve the quality criteria. Relaxation of the parametric constraints can lead to the violation of specifications, conditions, and standards. In a number of cases, this might even be inadmissible. In this case, a radical approach is of help, leading to restructuring of the designed object. Furthermore, the desired parametric changes are attained in the new structural scheme.

Expansion of the parameters of the investigated object and relaxation of the parametric constraints at a constant step of parameter selection leads to an increase in the number of test models. The allowable set of models and the Pareto set of models $P^*(\alpha)$ increases. Furthermore, the new subset of the Pareto models is considered to be improved if it includes models with parameters $\alpha_{(i)}$ which, when compared with respect to the quality criteria with models $\alpha_{(j)}$ of the old Pareto subset $P(\alpha)$, satisfy inequality

$$\varphi_\nu(\alpha_{(i)}) \leq \varphi_\nu(\alpha_{(j)})$$

for any quality criteria $\nu$ starting from 1 to $K$ and inequality $\varphi_\nu(\alpha_{(i)}) < \varphi_\nu(\alpha_{(j)})$ for at least one quality criterion.

On the basis of the selected vital (decisive) rule, one can determine the optimal model $\alpha^P$ which is adopted as the initial model during the execution of the procedure of improving the quality criteria of the optimal model by relaxing the parametric constraints. It is assumed that $\alpha^P \equiv \alpha^0_{(1)}$, where the subscript in parentheses denotes the number of the step in the calculations.

Further, search is carried out for additional test models in the region of parameters close to the first selected model. At each subsequent step ($i$-th step) of the calculations additional $N_{(i)}$ test models $\alpha^j_{(j)}$ are selected, i.e., $j = 1, \ldots, N_{(i)}$. Usually a small number of additional test models are assumed. For each model all $\nu$ local quality criteria $\varphi_\nu(\alpha^j_{(i)})$ are computed. Also all Pareto models are identified. Depending on the results of the calculations at the first step, further analysis is conducted as follows.

If a model which is an improvement on the initial model is found, it serves as the reference model for the next step of the calculations.

When an entire set of improved models is found, then the following procedure is followed.

The number of test models $\alpha^j_{(i)}$ is increased. If the computation time of values of the quality criteria is short, the number of test models can be increased significantly, for example doubled.

The number of test models around the initial model is increased slightly.

If either approach does not enable one to find a better model, any of the better known decision-making methods can be used for the selection of the initial model for the next step of the calculations. From the set of Pareto models one model must be selected and adopted as an initial model for conducting the next calculation step. When deciding whether to adopt the newly found model as the optimal one the designer must, on the basis of analysis of all local quality criteria, decide whether the attained improvements of the values of the quality criteria justify the parametric violation of the specifications. If the new model is found to be of better quality, further search is resumed. Otherwise, the Pareto models of the preceding calculation step are taken as the solution of the problem of the search for a better model on the basis of reconsideration of the parametric constraints.

The practical recommendation to the designer on the application of this search method for the best solutions by relaxing the parametric constraints is as follows. When the value of any parameter of the Pareto model coincides or approaches one of the corresponding parametric constraints, it is expedient to investigate the question of improving the quality criteria by searching in the vicinity of these constraints.

### 5.4.2.7   Determination of Essential Quality Criteria

One of the most important operations of the optimal multi-criterial design technique is the correct specification of the quality criteria. Quality criteria greatly affect the transformation of the allowed and Pareto model sets of the designed object.

If the quality criteria are few and insufficiently rigid, then the allowable and Pareto model sets are highly representative. The number of models entering the sets differs little from the overall number of models which can be generated within the bounds of the established parametric constraints.

When the quality criteria are rigid and there are many of them, the number of models in the allowable and Pareto sets is insignificant in comparison with the total number of models within the bounds of the assumed parametric constraints.

The ratio of the allowable model sets and the model sets within the bounds of the existing parametric constraints is characterized by the selection effectiveness factor $\gamma$.

Thus, for diverse and rigid criterial constraints the selection effectiveness factor is small, but in the case of nonrigid constraints with respect to a low number of criteria it increases considerably. The low values of the selection effectiveness factor is an indication of the unproductive waste of time on computations and inadequate preparation of the problem.

Meanwhile, the unjustifiable reduction of the number of local quality criteria leads to the fact that the allowable model set becomes insufficiently informative. Furthermore, there are superfluous models that do not conform to the requirements of the remaining not considered quality criteria. For an inadmissibly low number of quality criteria, the computed allowable set ceases to correspond to the meaning implied in this concept.

Thus, the fact that the optimal models are selected from the allowable set of models points to the importance of carrying out this operation correctly. In order for the allowable set of models to be correctly determined, a whole set of quality criteria must be used for which the use of any additional quality criterion would change neither the allowable nor the Pareto set of models of the object being designed.

The apprehension of the designer of leaving out any significant criterion can lead to the fact that they might include superfluous criteria that exert influence on the formation of the allowable and Pareto model sets. The test for criteria redundancy for the formation of the allowable model sets is carried out automatically with the aid of the program DOPUS.

The test for criteria redundancy for the formation of the Pareto model set is carried out by a number of methods. In particular, the test can be effected on the basis of the analysis of the strength of the correlation between the quality criteria with the aid of program KORRE.

### 5.4.3  Resonant Vibratory Table for Forming Reinforced Concrete Products

### 5.4.3.1  Dynamic Model of the Table

In addition to the many designs of vibration compactors, resonant table vibrators are used in the construction material industry.

Resonant table vibrators are designed on the basis of nonlinear elastic systems which increase the operational stability of the machine, enabling the control of oscillations of the working elements as a means for their optimization. The design problems of resonant table vibrators are reduced to the selection of the parameters of the nonlinear drive system for which the periodic motion of the working element best satisfies the production requirements and design constraints thereby ensuring stability of the working regimes with the change of the production load.

The vibratory table consists of a two-mass oscillatory system in which the working element and the balancing frame are the oscillating masses. The main elastic links include the supporting elements and the bumpers, which are mounted between the masses, arranged with a clearance, and which collide during opposite motion of the masses. As a result, different accelerations of the working element are generated during upward and downward motion, inducing asymmetric cycle of oscillations. In order to reduce the dynamic loads on the foundation the working element and the balancing frame are mounted on rigid supports by means of vibration isolators. Vibration isolators both of working element and balancing frame are mounted on each support bracket along the same vertical axis. Oscillations are induced by eccentric drive fixed on the frame and linked with the working element by connecting rods via elastic links. The drive shaft is driven by an electric motor via V-belt. The molds with concrete are attached to the working element by electromagnets.

The dynamic problem of the resonant vibratory table in general formulation is reduced to analysis of the system: drive-machine-worked medium (see section 5.1). However, in the present case, for the parameter synthesis of the nonlinear system and drive of the resonant vibratory table, a simplified computational model has been used based on the propositions and assumptions usually made when solving applied problems of vibration theory (Fig. 5.16a). The main moving masses are assumed to be absolutely rigid and must satisfy conditions guaranteeing their uni-directional motion. The masses of the elastic links are not taken into account in the calculations, the characteristic of the restoring force of the main elastic links is assumed nonlinear (Fig. 5.16b). Energy dissipation in the elastic links is accounted for on the basis of the viscous friction hypothesis (Fig. 5.16c). The effect of the production load on the system dynamics is accounted for by adding an associated mass and some additional damping. It is assumed that the eccentric shaft rotates with a constant angular velocity (this is equivalent to the assumption of ideal motor with infinitely large power). In the design stage of the machine these assumptions are fully admissible.

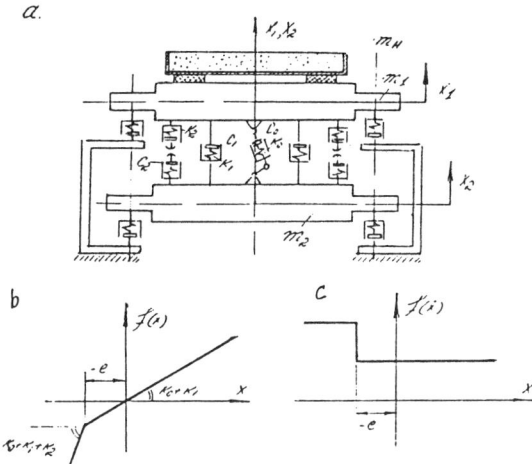

**Figure 5.16** Mathematical model of a vibratory table for compaction of concrete. a) basic arrangement; b) stiffness of elastic links; c) hysteretic losses of elastic links.

Taking into account these assumptions, the oscillation of the resonant asymmetric vibratory table is described by the following nonlinear differential equations

$$m_1^* \ddot{x}_1 + f(x)\dot{x} + \mu k_3 \dot{x}_1 + P(x) + k_3 x_1 = k_0 r \sin \omega t + \mu k_0 r \omega \cos \omega t$$

$$m_2 \ddot{x}_2 - f(x)\dot{x} + \mu k_2 \dot{x}_2 - P(x) + k_2 x_2 = - k_0 r \sin \omega t - \mu k_0 r \omega \cos \omega t$$

Here

$$x = x_1 - x_2, \qquad P(x) = (k_1 + k_0)x + \sigma(x)k^*(x + l)$$

$$f(x) = \mu \left[ k_1 + k_0 + \sigma(x)k^* \right]$$

$$\sigma(x) \begin{cases} 0 \text{ when } x \geq - e \\ 1 \text{ when } x < - e \end{cases}$$

$m_1^* = m_1 + k^* m$ is the reduced mass of the working element;

$m_l$ is the mass of the useful load;

$k^*$ is the association factor of the mass of the pay load;

$\mu$ is the coefficient of internal resistances in the rubber elastic connections;

$\mu_1 = \mu + \mu_l$ is the effective coefficient of resistances.

The dynamics and method of calculation of such systems are considered in sufficient detail in [3, 19] where the Krylov-Bogolyubov method is used to construct approximate periodic solutions.

### 5.4.3.2   Formulation of the Design Problem

The computational analysis and design of resonant asymmetric vibratory tables is carried out in stages.

Specifications include data on the range of variation of the maximum dimensions of the formed products and masses of the pay loads $m_{l min} \le m_l \le m_{l max}$. Additionally, the main characteristics of the working regime are stipulated in accordance with the process requirements with respect to effective compaction of the concrete mixes: the "upper" acceleration limit of the working element in the upward motion (for general purpose vibratory tables $0.9g \le W_{up} \le 2.2g$); the limits of variation of the acceleration of the working element in the downward motion (the range $4g \le W_d \le 10g$ is usually used); the range of working frequencies (investigations show that the most rational range is $50 \; \text{sec}^{-1} \le \omega < 100 \; \text{sec}^{-1}$).

The working regimes are realized on the ascending branch of the amplitude-frequency characteristic (in the subresonant region).

The design commences with the preliminary design of the main components of the vibratory table. Data on the dimensions of the products and the maximum values of the accelerations of the working element enable the determination of the least possible magnitudes of the main moving masses $m_1$ and $m_2$ from the conditions of their dynamic strength. Then, selection of the parameters of the nonlinear elastic suspension and drive is conducted using standard techniques. Measures are taken to reduce forces in the drive and dynamic loads transmitted to the foundation. At this stage of the design the requirements of production stability of the machine with the changing mass of the useful load is not taken into account.

Efforts are then directed towards the working design of the machine during which the system parameters are taken into consideration. Subsequently, dynamic computation of the vibratory table are carried out and the amplitude-frequency and frequency-force characteristics are plotted at given input parameters. The results are analyzed from the viewpoint of process stability of the machine which is ensured by automatic tuning, which is based on changing characteristics of the elastic links with increasing mass of the working element by additional compression of the supporting links and reduction of clearances in the bumpers.

In this stage the clearance in the bumpers is selected and the oscillation frequency at which the parameters of the working regimes satisfy the specifications are verified in the widest range of variation of the pay load mass. Thus, an attempt is made at each stage to select the parameters of the machine from the conditions of improvement of the individual local criteria.

As already noted, the optimal design of the vibratory table envisages the construction of the allowable and then the Pareto sets of models and the identification of the optimal model in the latter. The allowable set is determined by finding the models which simultaneously satisfy the parametric constraints.

Model $\alpha^P$ is called Pareto model (optimal according to Pareto) if model $a \in D$ does not exist that $\varphi_\nu(\alpha) < \varphi_\nu(\alpha)^P$ for $\nu = \overline{1, k}$ and, at least, for only one $\varphi_\nu(\alpha) < \varphi_\nu(\alpha)^P$.

Here, $\varphi_\nu(\alpha)$ is the local quality criterion; $\alpha = \alpha_{1p}, ..., \alpha_r$ is the $r$-dimensional parameter vector.

Thus, the determination of the Pareto set is reduced to finding all optimal models according to Pareto. The allowable set of solutions are constructed from a set of local criteria using the optimal parameter selection technique. Then the optimal (according to Pareto) solutions are determined in different situations.

For sufficiently large number of tests $N$, the complexity and labor intensity of the problem is characterized to a certain extent by the generalized and functional factors of the selection effectiveness $\gamma = N''/N' \cong V''/V'$ and $\bar{\gamma} = N'/N \cong V'/V$, where $V''$ and $V'$ are the $r$-dimensional volumes of the sets in the space of parameters, from which the first satisfies the functional and criterial constraints simultaneously, and the second satisfies only the functional constraints; $N''$ and $N'$ are the number of models encountered in $V''$ and $V'$, respectively;

$V$ is the $r$-dimensional volume of the parameter space being searched.

Case $\gamma \ll 0.1$ corresponds to very strong constraints, for example, when the customer imposes rigid conditions or sets a goal to improve an already good existing machine design. If information is lacking in this case with respect to the connectivity of the search domains, the determination of the allowable set becomes a complex task. In the following example this most interesting case is analyzed.

### 5.4.3.3  Parametric and Functional Constraints, Local Quality Criteria

Table 5.1 shows the values of initial (input) parameters of an existing design of vibratory table VRA-8. Their potential ranges of variation are considered below. Thus, the parameter vector is equal to $\alpha = \alpha_1, \ldots, \alpha_{10}$.

Further, the functional constraints $F_j(\alpha)$ $j = 1.7$ are formulated. These constraints ensure the conformity of the time history of oscillation of the working element to the specifications: $f_1(\alpha)$ and $f_2(\alpha)$ limit magnitude of the exciting force; $f_3(\alpha)$ determines the load on the rubber isolators; $f_4(\alpha)$ and $f_5(\alpha)$ assure stiff bilinear characteristic of the elastic suspension under maximum process load; $f_6(\alpha)$ and $f_7(\alpha)$ guarantee the realization of the working regimes on the ascending branch of the frequency characteristic (in the subresonant region). The constraints $f_1(\alpha)$ and $f_2(\alpha)$ are functionals that are dependent on the integral curves of the differential equations, the remaining constraints are directly expressed in terms of the initial parameters of the system

$$0.9g \le f_1(\alpha) = w_{1up} \le 2.2g$$

$$4g \le f_2(\alpha) = w_{1d} \le 10g$$

$$f_3(\alpha) = k_0 r \le 120$$

$$f_4(\alpha) = m_1 + m_{lmax} \le \frac{Q}{gk^1}\left(k_1 + k_3\right)$$

$$f_5(\alpha) = m_1 + m_2 + m_{lmax} \le \frac{Q}{gk^1}\left(k_1 + k_2\right)$$

$$f_6(\alpha) = l_H \ge k_r m_{lmax} \, g - k_0 r (k_2 + k_3)$$

$$f_7(\alpha) = \omega \le p(\alpha)$$

Here:

$w_{1up}$ and $w_{1d}$ are maximum accelerations of the working element during upward and downward motion;

$m_{lmax} = 8t$ is the maximum payload mass;

$k'$ is the stiffness of one rubber isolator, kN/cm;

$Q$ is the limit force on one isolator, kN;

$p(\alpha)$ is the natural frequency of oscillations of the linearized system, $sec^{-1}$.

The presented limits of the variation of the "upper" and "lower" accelerations ensure the conformity of the oscillation time history of the working element with the specifications for effective compaction of concrete mixes and also ensures the dynamic strength of the machine. These constraints are specified on the basis of process studies and strength tests and of course are not rigid. Depending on conditions of the specific industry, the ease of concrete handling, specifications to reinforced concrete products, the admissible values of the acceleration can somewhat vary. The maximum acceleration $10g$ determines the dynamic strength of the machine and can be varied depending on the used materials, the manufacturing process of the structural components, and on other factors.

The constraint $f_3(\alpha)$ is also not rigid, and depends on the starting characteristics of the selected electric motor and on other factors.

It is proposed to assess the quality of the designed vibratory table with respect to six local criteria. These are the functionals characterizing machine weight, dynamic forces in the drive, dynamic forces transmitted to the foundation, asymmetry of the oscillation time histories, and stability of the working element with changing process load. If the mass of the vibratory table is expressed directly in terms of the initial parameters of the system, then all the remaining local quality criteria are functionals of the integral curves. These criteria depend on the weight of the payload, which can assume any value in the range 2 - 8 t. Practically, several levels of payloads are considered. In the present example the mass of the payload was varied in four levels in steps of $\Delta m_l = 2t$; $m_{li} = i\,\Delta m_l$; $i = 1, 2, 3, 4$.

**Table 5.1**

| Parameter | Symbol | Values |
|---|---|---|
| Stiffness of driving elastic elements, kN/cm | $\alpha_1 = k_0$ | 48 |
| Stiffness of linear principal elastic elements, kN/cm | $\alpha_2 = k_1$ | 96 |
| Stiffness of bumpers, kN/cm | $\alpha_3 = k^*$ | 1600 |
| Stiffness of supporting isolators under the frame, kN/cm | $\alpha_4 = k_2$ | 25 |
| Stiffness of supporting isolators under the working element, kN/cm | $\alpha_5 = k_3$ | 25 |
| Drive eccentricity, cm | $\alpha_6 = r$ | 0.7 |
| Mass of working element, t | $\alpha_7 = m_1$ | 3 |
| Mass of balancing frame, t | $\alpha_8 = m_2$ | 5 |
| Clearance in the bumpers without process load, cm | $\alpha_9 = l_H$ | 0.2 |
| Working frequency, $\text{sec}^{-1}$ | $\alpha_{10} = \omega$ | 97 |

It as assumed that the association coefficient of the mass of the useful load $k^* = 1$.

The local quality criteria were presented in the following form:

$\varphi_2(\alpha) = m_1 + m_2$      – characterizes mass of the machine;

$\varphi_2(\alpha) = \max \dfrac{w_{1upi}}{w_{1li}}$      – characterizes asymmetry of the oscillation time history of the working element;

$\varphi_3(\alpha) = \min Q_1$      – characterizes dynamic loading of the drive;

$\varphi_4(\alpha) = \min Q_2$      – characterizes dynamic loading of the carrying structures;

$\psi_5(\alpha) = \dfrac{w_{1upmaxi}}{w_{1upmini}}$      – characterizes stability of the "upper" and "lower" accelerations of the working element, i.e., the stability of the effectiveness of compaction with variation of the process load.

$\varphi_6(\alpha) = \dfrac{w_{1lmaxi}}{w_{1lmini}} = 1$

Here, the notations are: $Q_1 = k_C A_0$ are the dynamic forces in the drive; $A_0$ is the amplitude of the driving elastic links, $Q_2 = (k_2 A_2) - (k_3 A_1)$ is the dynamic load transmitted to the foundation; $A_1$ and $A_2$ are oscillation amplitudes of the working element and balancing frame.

For analysis of the space of the initial parameters special programs are used, including a subprogram for calculation of the amplitude-frequency and frequency-force characteristics of the considered nonlinear system.

## 5.4.3.4 Determination of the Optimal Model during Vibratory Table Modernization

During the modernization of a vibratory table, the objective is to find a better model by means of relatively small changes in the operating model. The designer assumed small variations of the parameters of the model

$44 \leq \alpha_1 \leq 52$                           $0.5 \leq \alpha_6 \leq 0.9$

$92 \leq \alpha_2 \leq 100$                           $\alpha_7 = 3.0$

$1400 \leq \alpha_3 \leq 1800$                    $3.0 \leq \alpha_8 \leq 7.0$

$20 \leq \alpha_4 \leq 30$                         $0.0 \leq \alpha_9 \leq 0.4$

$20 \leq \alpha_5 \leq 30$                         $94 \leq \alpha_{10} \leq 100$

These parameters specify a parallelepiped whose center is comprised of the parameters of the design VRA-8 to be improved which will, henceforth, be denoted $\alpha^1$.

A priority sequence of the local criteria is specified representing an ordered set of criteria indexes $v = 1.6$.

$$I = \left\{1, \; [3,4], \; 2, \; [5,6]\right\}$$

The priority sequence $I$ reflects purely qualitative relationships of the criteria dominance: the weight of the machine $\varphi_1$ is more important than the dynamic forces in the drive $\varphi_3$ and the dynamic loads transmitted to the foundation $\varphi_4$; $\varphi_3$ and $\varphi_4$ are assumed equal in importance; then follows the characteristic of asymmetry of the oscillation time history of the working element of the machine $\varphi_2$ and, finally, the characteristics of stability of the oscillation character of the working element with the variation of the process load $\varphi_5$ and $\varphi_6$ which are equal in importance.

The quantitative side of the dominance was not specified. However, it is known that there is a priori information and it is quite possible that the number of criteria and the qualitative correlations of their dominance might change depending on the obtained analytical results and the requirements for the specific machines by the conditions of their exploitation.

In the region of machine parameters $N = 256$ tests were conducted, taking into account the assumed functional constraints. The number of tests satisfying the constraints amounted to $N' = 24$; with the functional selection effectiveness factor $\gamma = 24/256 \approx 0.093$.

The selection effectiveness factor is a criterion of the complexity and labor intensity of the problem. The lower this factor, the higher the labor intensity of the problem solution.

Taking into consideration the low value of the selection effectiveness factor, the designer opted for relaxation of the functional constraints with the purpose of increasing the selection factor. Since constraint $f(\alpha)_{4-7}$ cannot be relaxed, changing the limits of the constraints $f(\alpha)_{1-3}$ had been considered.

A decision was made to opt for small changes of the functional constraints $f(\alpha)_{1-3}$ providing that this will be compensated by improvements of the main local quality criteria. To this end, the functional constraints $f(\alpha)_{1-3}$ were converted to quality pseudo-criteria $\varphi(\alpha)_8$, $\varphi(\alpha)_9$, $\varphi(\alpha)_7$, respectively.

In view of the low value of the selection effectiveness factor, the problem is regarded as a nine-criteria task $\varphi(\alpha)_\nu$, $\nu = \overline{1,9}$ and the allowable set is determined with the account of all nine criteria. Criteria $\varphi(\alpha)_{1-9}$ are not featured in the priority series and do not take part in the construction of the decisive rule. This is exactly why they are regarded as pseudo-criteria. This enabled the relaxation of the constraints on $f(\alpha)_{1-3}$ and the formulation of the new functional constraints

$$0.85g \leq f(\alpha)_1 \leq 2.25g \qquad\qquad f(\alpha)_3 \leq 123$$

$$3.70g \leq f(\alpha)_2 \leq 10.3g \qquad\qquad f(\alpha)_{4-7} \leq 0.00$$

In accordance with the dynamic properties of the vibratory table for each model belonging to the allowable set of solution, function $\varphi_8(\alpha)_8$ in the given range $[0.85; 2.25]$ assumes the values $0.85g \leq \varphi'(\alpha)_8 \leq \varphi^2(\alpha)_8 \leq \varphi^4(\alpha)_8 \leq 2.25g$. The above is true also for criterion $\varphi(\alpha)_9$: $3.70g \leq \varphi'(\alpha)_9 \leq \varphi^2(\alpha)_9 \leq \varphi^3(\alpha)_9 \leq \varphi^4(\alpha)_9 \leq 10.3g$. Hence, if for model $\alpha^i$ at least one of the values $\varphi(\alpha)^i_7$ is in the ranges

$$120g \leq \varphi(\alpha)_7 \leq 123g, \quad 0.85g \leq \varphi(\alpha)_8 \leq 0.90g, \quad 3.7g \leq \varphi(\alpha)_9 \leq 4.0g$$

$$2.20g \leq \varphi(\alpha)_8 \leq 2.25g, \quad 10.0g \leq \varphi(\alpha)_9 \leq 10.3g$$

and the remaining functional constraints are within the specified bounds, then the designer estimates the attained values in this case $\varphi(\alpha)\nu$, $\nu = \overline{1, 6}$. He can either agree with them or reject them. In the former case, $\alpha^i$ enters into the allowable set of solutions.

We introduce the notation $\Delta_1 = \varphi(\alpha)_8 - 0.9g$, $\Delta_2 = 2.2g - \varphi^4(\alpha)$, $\Delta_3 = \varphi^1(\alpha)_9 - 4.00g$, and $\Delta_4 = 10.0g - \varphi^4(\alpha)$, and specify as a resource with respect to $\varphi(\alpha)_8$ or $\varphi(\alpha)_9$ the quantities $\min\{\Delta_1, \Delta_2\}$ or $\min\{\Delta_3, \Delta_4\}$, respectively. In table 4 which is presented below we write $\varphi^1(\alpha)$ if the resource is equal to $\Delta_1$ and $\varphi^4(\alpha)$ if the resource is $\Delta_2$. By analogy we have $\varphi^1(\alpha)_9$ for resource $\Delta_3$ and $\varphi^4(\alpha)_9$ for resource $\Delta_4$. When $\Delta_i < 0$, $i = \overline{1, 4}$, we shall assume that the corresponding resource is equal to zero.

Since the resource with respect to $\varphi(\alpha)_7$, $\Delta_5 = 120 - \varphi(\alpha^j)_7$ is sufficiently high for models $\alpha^j$ presented in table 4, the value of $\Delta_5$ is not listed.

Within the bounds of the parametric constraints under the assumed new functional constraints, $N = 1024$ tests (1024 models constructed) were conducted. From this number, 108 models are featured (passed the constraint test) in the test table (table 5.2). The functional selection effectiveness factor amounts in this case to $\gamma = 108/1024 = 0.105$ (this was earlier 0.093). Presented in the test table are only the first six local criteria.

Verification of the criterial constraints is carried out with the aid of the test tables. This is one of the vital stages of the process of multicriterial optimal design. In selecting the criterial constraint the designer is guided by the desire to make minimum concessions for best results with respect to all local quality criteria. However, there is the danger in this case that none of the models will simultaneously satisfy all quality criteria.

Big concessions in regard to lowering criterial requirements can cause great difficulties in the analysis of the obtained results since low-quality models find their way into the allowable set.

The resolution of these difficulties is facilitated by the dialogue regime with the computer during specification of the criterial constraints. During the dialogue procedure, by changing the constraints the designer can efficiently estimate the cost effectiveness of each criterion. The qualification of the designer at this stage is extremely critical in excluding the loss of models that can be of interest.

The first designer-computer dialogue was intended to determine the Pareto models from 108 models finding their way to the test table. This procedure reduces the criterial constraints to the lower limits.

From the selected 108 models 59 models, including $\alpha^1$ which is the existing machine design, found their way to the Pareto set. It hence follows that the existing vibratory table cannot be improved with respect to all six local criteria. This conclusion can, to a certain extent, be regarded as natural, since a fairly good machine design was being investigated. This conclusion enables the verification of the statement of the problem. Since the improvement of all criteria simultaneously

Table 5.2

| $\alpha_i$ | $\varphi_1$ | $\alpha_i$ | $\varphi_2$ | $\alpha_i$ | $\varphi_3$ | $\alpha_i$ | $\varphi_4$ | $\alpha_i$ | $\varphi_5$ | $\alpha_i$ | $\varphi_6$ |
|---|---|---|---|---|---|---|---|---|---|---|---|
| 1 | 2 | 3 | 4 | 5 | 6 | 7 | 8 | 9 | 10 | 11 | 12 |
| 248 | 7.02 | 445 | 0.233 | 319 | 28.25 | 475 | 3.70 | 588 | 0.047 | 214 | 0.051 |
| 353 | 7.04 | 475 | 0.241 | 248 | 28.34 | 1022 | 3.74 | 418 | 0.106 | 295 | 0.120 |
| 132 | 7.14 | 141 | 0.241 | 794 | 28.46 | 957 | 4.10 | 910 | 0.133 | 853 | 0.120 |
| 673 | 7.30 | 40 | 0.241 | 47 | 28.94 | 593 | 4.15 | 863 | 0.144 | 910 | 0.122 |
| 47 | 7.31 | 78 | 0.242 | 935 | 29.38 | 253 | 4.37 | 214 | 0.155 | 231 | 0.130 |
| 528 | 7.32 | 831 | 0.242 | 176 | 29.52 | 768 | 4.52 | 134 | 0.170 | 807 | 0.163 |
| 905 | 7.37 | 183 | 0.243 | 498 | 29.59 | 143 | 4.55 | 853 | 0.176 | 819 | 0.177 |
| 733 | 7.39 | 667 | 0.243 | 13 | 30.15 | 643 | 4.62 | 295 | 0.180 | 445 | 0.183 |
| 379 | 7.41 | 925 | 0.243 | 40 | 30.34 | 116 | 4.66 | 637 | 0.197 | 40 | 0.186 |
| 836 | 7.42 | 438 | 0.243 | 528 | 30.62 | 204 | 4.77 | 947 | 0.214 | 116 | 0.188 |
| 141 | 7.42 | 655 | 0.244 | 922 | 30.82 | 588 | 4.91 | 347 | 0.233 | 638 | 0.214 |
| 690 | 7.43 | 396 | 0.245 | 475 | 30.94 | 552 | 4.93 | 396 | 0.236 | 588 | 0.217 |
| 620 | 7.47 | 319 | 0.246 | 905 | 316.62 | 176 | 5.16 | 203 | 0.243 | 406 | 0.221 |
| 421 | 7.49 | 913 | 0.246 | 682 | 31.68 | 733 | 5.20 | 231 | 0.256 | 475 | 0.227 |
| 922 | 7.50 | 719 | 0.248 | 141 | 31.69 | 667 | 5.27 | 925 | 0.256 | 790 | 0.243 |
| 498 | 7.52 | 682 | 0.248 | 819 | 31.70 | 661 | 5.42 | 406 | 0.256 | 655 | 0.245 |
| 787 | 7.57 | 794 | 0.248 | 1 | 31.72 | 379 | 5.43 | 116 | 0.271 | 362 | 0.260 |

**Table 5.2** (cont'd)

| 1 | 2 | 3 | 4 | 5 | 6 | 7 | 8 | 9 | 10 | 11 | 12 |
|---|---|---|---|---|---|---|---|---|---|---|---|
| 218 | 7.58 | 488 | 0.249 | 1022 | 31.83 | 224 | 5.54 | 819 | 0.281 | 203 | 0.266 |
| 78 | 7.59 | 176 | 0.250 | 831 | 32.00 | 368 | 5.63 | 619 | 0.283 | 36. | 0.266 |
| 181 | 7.61 | 935 | 0.250 | 433 | 32.41 | 619 | 5.68 | 153 | 0.294 | 619 | 0.274 |
| 552 | 7.63 | 619 | 0.250 | 421 | 32.54 | 406 | 5.69 | 913 | 0.298 | 78 | 0.279 |
| 990 | 7.65 | 134 | 0.252 | 379 | 32.80 | 433 | 5.70 | 648 | 0.298 | 224 | 0.281 |
| 319 | 7.70 | 922 | 0.252 | 445 | 33.62 | 794 | 5.77 | 807 | 0.303 | 149 | 0.282 |
| 768 | 7.70 | 615 | 0.252 | 181 | 33.95 | 682 | 5.78 | 224 | 0.303 | 667 | 0.285 |
| 336 | 7.73 | 879 | 0.252 | 368 | 34.77 | 533 | 5.81 | 40 | 0.305 | 48 | 0.286 |
| 879 | 7.73 | 115 | 0.255 | 353 | 34.79 | 13 | 5.81 | 445 | 0.305 | 628 | 0.286 |
| 13 | 7.75 | 1 | 0.256 | 218 | 35.01 | 408 | 5.87 | 628 | 0.310 | 925 | 0.287 |
| 831 | 7.76 | 243 | 0.256 | 787 | 35.06 | 874 | 5.89 | 183 | 0.316 | 416 | 0.194 |
| 605 | 7.79 | 600 | 0.258 | 874 | 35.18 | 498 | 5.92 | 520 | 0.322 | 706 | 0.304 |
| 153 | 7.80 | 996 | 0.260 | 643 | 35.34 | 859 | 5.96 | 996 | 0.335 | 615 | 0.306 |
| 40 | 7.81 | 143 | 0.261 | 768 | 35.64 | 637 | 5.96 | 362 | 0.336 | 847 | 0.313 |
| 293 | 7.82 | 103 | 0.261 | 688 | 35.70 | 913 | 5.97 | 78 | 0.342 | 1022 | 0.313 |
| 794 | 7.82 | 643 | 0.262 | 836 | 35.75 | 922 | 6.03 | 115 | 0.343 | 347 | 0.315 |
| 643 | 7.86 | 368 | 0.262 | 341 | 36.43 | 293 | 6.13 | 48 | 0.346 | 513 | 0.324 |
| 433 | 7.87 | 874 | 0.262 | 847 | 36.43 | 787 | 6.18 | 1022 | 0.353 | 336 | 0.332 |

Table 5.2 (cont'd)

| 1 | 2 | 3 | 4 | 5 | 6 | 7 | 8 | 9 | 10 | 11 | 12 |
|---|---|---|---|---|---|---|---|---|----|----|----|
| 910 | 7.87 | 13  | 0.262 | 253 | 36.62 | 879 | 6.25 | 667 | 0.354 | 880 | 0.339 |
| 952 | 7.93 | 163 | 0.263 | 132 | 36.67 | 248 | 6.36 | 475 | 0.357 | 1   | 0.359 |
| 192 | 7.95 | 661 | 0.263 | 879 | 36.69 | 935 | 6.37 | 36  | 0.366 | 115 | 0.360 |
| 767 | 7.95 | 990 | 0.263 | 224 | 37.06 | 615 | 6.48 | 790 | 0.374 | 408 | 0.380 |
| 619 | 7.97 | 421 | 0.263 | 192 | 37.18 | 295 | 6.62 | 149 | 0.375 | 688 | 0.383 |
| 418 | 7.99 | 953 | 0.264 | 103 | 38.11 | 243 | 6.67 | 513 | 0.385 | 668 | 0.390 |
| 925 | 8.00 | 347 | 0.264 | 655 | 38.20 | 1   | 6.70 | 615 | 0.386 | 35  | 0.393 |
| 1   | 8.00 | 293 | 0.265 | 143 | 38.44 | 690 | 6.82 | 706 | 0.386 | 600 | 0.402 |
| 819 | 8.01 | 204 | 0.265 | 78  | 38.85 | 319 | 6.82 | 655 | 0.395 | 957 | 0.431 |
| 593 | 8.04 | 48  | 0.265 | 243 | 39.08 | 905 | 6.90 | 336 | 0.399 | 103 | 0.432 |
| 149 | 8.05 | 224 | 0.266 | 185 | 39.15 | 78  | 7.01 | 847 | 0.400 | 368 | 0.451 |
| 682 | 8.05 | 149 | 0.266 | 733 | 39.27 | 218 | 7.03 | 600 | 0.418 | 293 | 0.463 |
| 408 | 8.05 | 203 | 0.266 | 406 | 39.57 | 47  | 7.08 | 35  | 0.420 | 948 | 0.474 |
| 935 | 8.06 | 787 | 0.267 | 990 | 39.63 | 790 | 7.16 | 1   | 0.421 | 183 | 0.481 |
| 36  | 8.06 | 520 | 0.267 | 513 | 39.82 | 853 | 7.17 | 668 | 0.439 | 863 | 0.485 |
| 790 | 8.07 | 253 | 0.268 | 593 | 40.05 | 48  | 7.17 | 880 | 0.441 | 253 | 0.486 |
| 628 | 8.10 | 153 | 0.268 | 204 | 40.06 | 952 | 7.22 | 948 | 0.447 | 433 | 0.492 |
| 176 | 8.11 | 668 | 0.268 | 925 | 40.76 | 341 | 7.28 | 103 | 0.470 | 170 | 0.501 |

**Table 5.2** (cont'd)

| 1 | 2 | 3 | 4 | 5 | 6 | 7 | 8 | 9 | 10 | 11 | 12 |
|---|---|---|---|---|---|---|---|---|----|----|----|
| 655 | 8.11 | 406 | 0.269 | 163 | 40.96 | 831 | 7.32 | 688 | 0.474 | 648 | 0.502 |
| 445 | 8.12 | 648 | 0.269 | 690 | 41.56 | 996 | 7.34 | 408 | 0.481 | 163 | 0.506 |
| 368 | 8.16 | 353 | 0.269 | 661 | 41.75 | 513 | 7.35 | 368 | 0.488 | 153 | 0.506 |
| 847 | 8.17 | 362 | 0.270 | 628 | 42.13 | 648 | 7.39 | 253 | 0.507 | 396 | 0.508 |
| 134 | 8.17 | 1022 | 0.270 | 755 | 42.14 | 181 | 7.54 | 957 | 0.509 | 520 | 0.511 |
| 948 | 8.18 | 807 | 0.271 | 952 | 42.14 | 103 | 7.57 | 673 | 0.524 | 755 | 0.512 |
| 520 | 8.19 | 341 | 0.271 | 880 | 42.25 | 836 | 7.61 | 163 | 0.533 | 952 | 0.517 |
| 204 | 8.20 | 418 | 0.271 | 619 | 42.98 | 847 | 7.61 | 293 | 0.540 | 673 | 0.518 |
| 755 | 8.21 | 513 | 0.272 | 615 | 43.12 | 141 | 7.63 | 143 | 0.540 | 143 | 0.522 |
| 1022 | 8.21 | 628 | 0.272 | 767 | 43.32 | 36 | 7.63 | 905 | 0.543 | 905 | 0.525 |
| 615 | 8.22 | 588 | 0.272 | 36 | 43.38 | 605 | 7.70 | 170 | 0.546 | 690 | 0.529 |
| 341 | 8.23 | 637 | 0.272 | 673 | 43.38 | 163 | 7.78 | 185 | 0.563 | 204 | 0.541 |
| 874 | 8.23 | 116 | 0.272 | 520 | 43.48 | 418 | 7.78 | 952 | 0.569 | 533 | 0.543 |
| 163 | 8.23 | 790 | 0.273 | 668 | 43.70 | 807 | 7.83 | 204 | 0.578 | 996 | 0.559 |
| 668 | 8.24 | 533 | 0.273 | 404 | 43.87 | 620 | 7.93 | 533 | 0.582 | 661 | 0.568 |
| 913 | 8.25 | 947 | 0.273 | 620 | 44.34 | 185 | 7.95 | 755 | 0.584 | 831 | 0.577 |
| 243 | 8.27 | 847 | 0.275 | 552 | 44.44 | 767 | 8.02 | 690 | 0.599 | 879 | 0.607 |
| 103 | 8.28 | 36 | 0.275 | 48 | 44.61 | 688 | 8.05 | 661 | 0.614 | 767 | 0.614 |

Table 5.2 (cont'd)

| 1 | 2 | 3 | 4 | 5 | 6 | 7 | 8 | 9 | 10 | 11 | 12 |
|---|---|---|---|---|---|---|---|---|---|---|---|
| 600 | 8.29 | 218 | 0.275 | 996 | 44.78 | 203 | 8.12 | 243 | 0.617 | 185 | 0.615 |
| 362 | 8.29 | 863 | 0.275 | 396 | 43.12 | 132 | 8.30 | 433 | 0.637 | 859 | 0.618 |
| 853 | 8.29 | 408 | 0.275 | 648 | 45.78 | 755 | 8.43 | 879 | 0.641 | 134 | 0.621 |
| 214 | 8.33 | 673 | 0.276 | 605 | 45.84 | 421 | 8.51 | 831 | 0.649 | 243 | 0.626 |
| 475 | 8.34 | 231 | 0.276 | 948 | 45.91 | 353 | 8.55 | 767 | 0.687 | 947 | 0.660 |
| 996 | 8.34 | 690 | 0.276 | 790 | 46.01 | 528 | 8.63 | 859 | 0.687 | 176 | 0.665 |
| 637 | 8.35 | 768 | 0.277 | 35 | 46.30 | 990 | 8.71 | 421 | 0.696 | 682 | 0.719 |
| 880 | 8.36 | 185 | 0.277 | 293 | 46.31 | 863 | 8.81 | 176 | 0.713 | 379 | 0.720 |
| 185 | 8.36 | 755 | 0.277 | 600 | 46.46 | 192 | 8.83 | 922 | 0.715 | 643 | 0.728 |
| 224 | 8.39 | 910 | 0.277 | 533 | 46.51 | 40 | 8.87 | 836 | 0.739 | 922 | 0.742 |
| 116 | 8.41 | 905 | 0.278 | 295 | 47.47 | 673 | 8.93 | 379 | 0.740 | 552 | 0.758 |
| 143 | 8.42 | 295 | 0.279 | 362 | 47.70 | 819 | 8.96 | 132 | 0.743 | 421 | 0.766 |
| 688 | 8.43 | 528 | 0.279 | 667 | 48.11 | 183 | 9.07 | 319 | 0.756 | 218 | 0.768 |
| 957 | 8.43 | 192 | 0.280 | 706 | 48.73 | 706 | 9.20 | 552 | 0.767 | 319 | 0.773 |
| 513 | 8.44 | 379 | 0.280 | 149 | 48.83 | 655 | 9.21 | 248 | 0.768 | 836 | 0.782 |
| 170 | 8.48 | 859 | 0.280 | 203 | 49.09 | 115 | 9.25 | 643 | 0.772 | 935 | 0.805 |
| 661 | 8.49 | 170 | 0.280 | 336 | 49.55 | 336 | 9.32 | 935 | 0.774 | 593 | 0.815 |
| 253 | 8.52 | 688 | 0.280 | 170 | 50.26 | 347 | 9.33 | 682 | 0.777 | 132 | 0.844 |

Table 5.2 (cont'd)

| 1 | 2 | 3 | 4 | 5 | 6 | 7 | 8 | 9 | 10 | 11 | 12 |
|---|---|---|---|---|---|---|---|---|----|----|----|
| 706 | 8.52 | 880 | 0.281 | 183 | 50.48 | 520 | 9.36 | 593 | 0.780 | 353 | 0.871 |
| 859 | 8.54 | 35 | 0.281 | 134 | 50.81 | 445 | 9.36 | 528 | 0.823 | 248 | 0.871 |
| 35 | 8.56 | 767 | 0.282 | 153 | 50.85 | 362 | 9.44 | 353 | 0.844 | 528 | 0.891 |
| 183 | 8.61 | 248 | 0.283 | 859 | 52.01 | 153 | 9.59 | 218 | 0.849 | 768 | 0.894 |
| 648 | 8.61 | 593 | 0.283 | 418 | 54.35 | 170 | 9.59 | 733 | 0.881 | 733 | 0.897 |
| 807 | 8.64 | 952 | 0.284 | 957 | 54.73 | 880 | 9.64 | 787 | 0.886 | 913 | 0.903 |
| 396 | 8.68 | 948 | 0.285 | 637 | 55.93 | 149 | 9.65 | 990 | 0.901 | 47 | 0.927 |
| 947 | 8.68 | 706 | 0.285 | 910 | 56.25 | 947 | 9.87 | 181 | 0.908 | 787 | 0.933 |
| 48 | 8.69 | 47 | 0.287 | 115 | 57.10 | 134 | 9.93 | 768 | 0.913 | 874 | 0.936 |
| 203 | 8.70 | 214 | 0.287 | 214 | 57.89 | 668 | 9.96 | 605 | 0.919 | 990 | 0.940 |
| 667 | 8.74 | 836 | 0.291 | 116 | 58.56 | 628 | 9.98 | 874 | 0.928 | 620 | 0.943 |
| 406 | 8.80 | 181 | 0.292 | 347 | 58.82 | 948 | 10.33 | 47 | 0.929 | 341 | 0.961 |
| 533 | 8.82 | 132 | 0.292 | 913 | 58.87 | 214 | 10.42 | 192 | 0.934 | 192 | 0.989 |
| 295 | 8.82 | 957 | 0.295 | 863 | 61.19 | 925 | 10.79 | 620 | 0.944 | 181 | 0.993 |
| 231 | 8.89 | 552 | 0.297 | 853 | 61.79 | 231 | 11.15 | 341 | 0.959 | 605 | 1.000 |
| 115 | 8.91 | 733 | 0.297 | 807 | 62.56 | 396 | 11.20 | 794 | 0.971 | 794 | 1.000 |
| 588 | 8.91 | 336 | 0.297 | 588 | 63.84 | 910 | 11.50 | 141 | 1.066 | 141 | 1.109 |
| 347 | 8.98 | 605 | 0.305 | 231 | 64.18 | 35 | 11.82 | 13 | 1.118 | 498 | 1.164 |
| 863 | 9.04 | 620 | 0.313 | 947 | 70.62 | 600 | 12.51 | 498 | 1.143 | 13 | 1.222 |

is impractical, screening of the most important and of the secondary non-important quality criteria is performed. Thus, optimization was henceforth conducted taking into account only the most important criteria

$$I = \left\{ 1, \ [3,4], \ 2 \right\}$$

as to $\varphi(\alpha)_5$ and $\varphi(\alpha)_6$, they were transferred to the rank of functional constraints.

    Thus, the allowable set of solutions is constructed in the following calculations with respect to nine criteria, whereas the search for optimal models is effected only with respect to four criteria.

    The purpose of further investigations was to identify the vibratory table models $\alpha^j$ which are superior by the main four quality criteria to the original model $\alpha^1$. This problem is solved in the process of designer-computer dialogue. Furthermore, the actual values of the real table vibrator $\varphi^{**}(\alpha)_{1,4} = \varphi(\alpha^1)_{1,4}$ are assumed to be the worst values of the quality criteria. Constraints $\varphi_{7,9}$ are taken into account later when analyzing the allowable set of models constructed with respect to six quality criteria.

    Two new models $\alpha^{922}$ and $\alpha^{794}$ and the existing vibratory table $\alpha^*$ ended up in this set. For these models the designer considered the obtained values of the quality criteria $\varphi(\alpha)_{7,9}$ to be acceptable.

    Taking the assumed priority sequences into account, models $\alpha^{922}$ and $\alpha^{794}$ are found to be Pareto models. The existing vibratory table did not find its way to the Pareto set. This points to the fact that it can be improved with respect to all four quality criteria.

    Both models are summarily close in quality. The mass of model $\alpha^{922}(\varphi(\alpha)_1 = 7.50)$ is less than the mass of model $\alpha^{794}$ $(\varphi(\alpha)_1 = 7.82)$. However, this feature is counterbalanced by lower dynamic forces in the drive of model $\alpha^{794}$ $(\varphi(\alpha)_3 = 28.46)$ compared with model $\alpha^{922}$ $(\varphi(\alpha)_3 = 30.82)$ and lower dynamic load on the foundation $\alpha^{794}(\varphi(\alpha)_4 = 5.77)$, instead of $(\varphi(\alpha)_4 = 6.03)$ for model $\alpha^{922}$. Furthermore, both models have practically identical vibration characteristics of the working table $\alpha^{922}$ $(\varphi(\alpha)_2 = 0.252)$, $\alpha^{794} = (\varphi(\alpha)_2 = 0.248)$.

    The designer gave preference to model $\alpha^{922}$, since it has a better stability and acceleration characteristic in the course of motion of the working table upward and downward – $\alpha^{922}$ $(\varphi(\alpha)_5 = 0.715, \ \varphi(\alpha)_6 = 0.742, \ \alpha^{794} = \varphi(\alpha)_5 = 0.971, \ \varphi(\alpha)_6 = 1.000)$.

The purpose of further investigations was to find good models in which deterioration of one of the four criteria in comparison with the actual vibratory table is compensated by relative improvement of the remaining criteria.

Accordingly, the following criterial constraints were assumed

$$\varphi^{**}(\alpha)_1 = \varphi(\alpha^{996})_1 = 8.34$$

$$\varphi^{**}(\alpha)_2 = \varphi(\alpha^{836})_2 = 0.291$$

$$\varphi^{**}(\alpha)_3 = \varphi(\alpha^{768})_3 = 35.64$$

$$\varphi^{**}(\alpha)_4 = \varphi(\alpha^{847})_4 = 7.61$$

$$\varphi^{**}(\alpha)_{5,6} = 1.00$$

Under the established constraints 21 models found their way into the allowable set among which 16 models were Pareto models with respect to six quality criteria (table 5.3).

The most interesting and unexpected result of the analysis of the Pareto models is the presence of model $\alpha^{475}$. The value of the dynamic force on the foundation $\varphi(\alpha)_4$ for this model is one half of that for model $\alpha^1$. Model $\alpha^{475}$ is somewhat worse than $\alpha^1$ with respect to weight (by 0.34 t), and better with respect to the remaining five criteria (table 5.4 and table 5.5).

Newly designed installations and, particularly, future installations can be built as multistory buildings in which vibratory tables can be mounted not only on the first, but also on higher floors. In these conditions, the most important factor is the minimization criterion $\varphi(\alpha)_4$; therefore, the designer is guided by the priority sequence

$$I = \left\{4, \ 1, \ 3, \ 2, \ [5, \ 6]\right\}$$

Following the inspection of the Pareto models from the standpoint of selecting the best for existing installations, the preference was given to model $\alpha^{258}$ (table 5.4 and table 5.5) having a weight, which is the most significant criterion, of 1 t less than for the active machine design $\alpha^1$.

It is pertinent to mention that model $\alpha^{248}$ under the previous functional constraints would not have found its way to the allowable set of solutions, since the designer would not have realized the existence of this model. With transfer of

Table 5.3

| $\alpha^i$ | $\varphi_1$ | $\alpha^i$ | $\varphi_2$ | $\alpha^i$ | $\varphi_3$ | $\alpha^i$ | $\varphi_4$ | $\alpha^i$ | $\varphi_5$ | $\alpha^i$ | $\varphi_6$ |
|---|---|---|---|---|---|---|---|---|---|---|---|
| 248 | 7.02 | 248 | 0.283 | 248 | 28.34 | 248 | 6.36 | 248 | 0.768 | 248 | 0.871 |
| 905 | 7.37 | 905 | 0.278 | 905 | 31.62 | 905 | 6.90 | 905 | 0.543 | 905 | 0.525 |
| 379 | 7.41 | 379 | 0.280 | 379 | 32.80 | 379 | 5.43 | 379 | 0.740 | 379 | 0.720 |
| 922 | 7.50 | 922 | 0.252 | 922 | 30.82 | 922 | 6.03 | 922 | 0.715 | 922 | 0.742 |
| 319 | 7.70 | 319 | 0.246 | 319 | 28.25 | 319 | 6.82 | 319 | 0.756 | 319 | 0.773 |
| 831 | 7.76 | 831 | 0.242 | 831 | 32.00 | 831 | 7.32 | 831 | 0.649 | 831 | 0.577 |
| 794 | 7.82 | 794 | 0.248 | 794 | 28.46 | 794 | 5.77 | 794 | 0.971 | 794 | 1.000 |
| 433 | 7.87 | 433 | 0.243 | 733 | 32.41 | 433 | 5.70 | 433 | 0.637 | 433 | 0.492 |
| 682 | 8.05 | 682 | 0.248 | 682 | 31.68 | 682 | 5.78 | 682 | 0.777 | 682 | 0.719 |
| 176 | 8.11 | 176 | 0.250 | 176 | 29.52 | 176 | 5.16 | 176 | 0.713 | 176 | 0.665 |
| 1022 | 8.21 | 1022 | 0.270 | 1022 | 31.83 | 1022 | 3.74 | 1022 | 0.352 | 1022 | 0.313 |
| 475 | 8.34 | 475 | 0.241 | 475 | 30.94 | 475 | 3.70 | 475 | 0.357 | 475 | 0.227 |

$f(\alpha)_{1,3}$ into pseudo-criterion $\varphi(\alpha)_{7,9}$ and new formulation of the problem, $\alpha^{248}$ not only found its way to the allowable set, but was also identified as the optimal model.

Investigation of the optimum data enabled the determination of the role of functional constraints blocking 916 models which could not find their way to the test table. The functional constraints $f(\alpha)_1$ were not satisfied by 659 models or 72%, $f(\alpha)_2$ by 63 models or 6.9%, $f(\alpha)_6$ by 30 models or 3.2%, and $f(\alpha)_7$ by 164 models or 17.9%.

Thus, constraints $f(\alpha)_1$ and $f(\alpha)_2$ are responsible for 80% of the models that fail to make it to the test table in the new formulation for the problem (in the initial formulation their share was nearly 86%). This once again emphasizes the expendience of transferring the functional constraints $f(\alpha)_1$ and $f(\alpha)_2$ into the quality pseudo-criteria $\varphi(\alpha)_8$ and $\varphi(\alpha)_9$, respectively. Analysis of the correlation between the local criteria indicated that there is a correlation between criteria $\varphi(\alpha)_5$ and $\varphi(\alpha)_6$ which is close to linear.

It follows from the test table that the weight criterion actually varies within the bounds $7.02 \leq \varphi(\alpha)_1 \leq 9.04$ instead of the anticipated $6 \leq \varphi(\alpha) \leq 10$. Section [6,10] consists of the nonworking zones [6; 7.02] and [9.04; 10] and the working zone [7.02; 9.04],. This takes place owing to the fact that due to the action of the functional constraints the actual bounds of variation of the mass of the balancing frame comprise $4.02 \leq \alpha_8 \leq 6.04$ (in the formulation of the problem the range $3 \leq \alpha_8 \leq 7$ was envisaged).

Analysis of the Pareto models enabled one to conclude that the resources have been exhausted with respect to the functional or parametric constraints taken either each separately or simultaneously. Further development of these models was restricted by the appropriate constraints.

Under the established parametric and functional constraints the values of all the main local criteria could not be significantly improved. Further relaxation of functional constraints is inadmissible. Consequently, the quality of the vibratory table VRA-8 cannot significantly be improved without fundamental design changes. The resources for its improvement must be sought in changing the dynamic layout or by expanding the boundaries of variation of the initial parameters.

Resonant machines are made, as a rule, on the basis of two-mass arrangements which are distinguished by simplicity of design and which provide effective vibration isolation. In a number of cases, they are even distinguished by total dynamic

Table 5.4

| Number of experiment | Model $\alpha_i$ | $\varphi_1$ | $\varphi_2$ | $\varphi_3$ | $\varphi_4$ | $\varphi_5$ | $\varphi_6$ | $\varphi_7$ | $\varphi_8$ | $1/\underline{g}\ \varphi_9$ | Resource÷Resource pc no. $1/\underline{g}\ \varphi_8$ | no. $1/\underline{g}\ \varphi_9$ |
|---|---|---|---|---|---|---|---|---|---|---|---|---|
| I | $\alpha^1$ | 8.00 | 0.256 | 31.72 | 6.70 | 0.421 | 0.359 | 33.60 | 1.051 | 4.236 | 0.151 | 0.236 |
|  | $\alpha^{248}$ | 7.02 | 0.283 | 28.34 | 6.36 | 0.768 | 0.871 | 38.15 | 1.063 | 3.976 | 0.163 | 0.0 |
|  | $\alpha^{475}$ | 8.34 | 0.241 | 30.94 | 3.70 | 0.357 | 0.227 | 42.23 | 0.962 | 7.404 | 0.062 | 0.404 |
| II | $\alpha^{1452}$ | 9.64 | 0.232 | 47.56 | 9.72 | 0.412 | 0.371 | 35.09 | 1.238 | 5.40 | 0.338 | 1.40 |
|  | $\alpha^{116}$ | 6.00 | 0.234 | 26.44 | 5.68 | 0.895 | 0.953 | 30.87 | 2.226 | 3.85 | 0.0 | 0.0 |
|  | $\alpha^{2406}$ | 9.13 | 0.260 | 44.36 | 10.94 | 0.545 | 0.605 | 56.08 | 2.089 | 5.336 | 0.111 | 1.336 |
|  | $\alpha^{12}$ | 6.00 | 0.240 | 28.21 | 5.40 | 0.674 | 0.749 | 52.05 | 0.943 | 4.232 | 0.043 | 0.232 |

Table 5.5

| Number of experiment | Model | $\alpha_1$ | $\alpha_2$ | $\alpha_3$ | $\alpha_4$ | $\alpha_5$ | $\alpha_6$ | $\alpha_7$ | $\alpha_8$ | $\alpha_9$ | $\alpha_{10}$ |
|---|---|---|---|---|---|---|---|---|---|---|---|
| I | $\alpha^1$ | 48.00 | 96.00 | 1600.00 | 25.00 | 25.00 | 0.700 | 3.00 | 5.00 | 0.200 | 27.00 |
| | $\alpha^{248}$ | 44.96 | 99.03 | 1757.81 | 22.07 | 29.49 | 0.848 | 3.00 | 4.01 | 0.398 | 95.10 |
| | $\alpha^{475}$ | 50.85 | 99.42 | 1489.84 | 20.09 | 26.54 | 0.830 | 3.00 | 5.33 | 0.110 | 94.20 |
| II | $\alpha^{1452}$ | 36.75 | 61.46 | 1429.58 | 22.24 | 46.38 | 0.554 | 4.52 | 5.12 | 0.425 | 80.59 |
| | $\alpha^{116}$ | 31.79 | 59.04 | 1448.82 | 20.42 | 48.64 | 0.971 | 3.00 | 3.00 | 0.438 | 91.48 |
| | $\alpha^{2406}$ | 52.05 | 85.51 | 2402.68 | 21.27 | 59.44 | 1.007 | 5.03 | 4.10 | 0.789 | 84.48 |
| | $\alpha^{12}$ | 48.87 | 83.13 | 2395.62 | 20.00 | 46.87 | 1.065 | 3.00 | 3.00 | 0.637 | 90.63 |

balance of the machine. The characteristic structural layouts are shown in Fig. 5.17. The task of determining the rational structure is reduced to the selection of the best among the known dynamic structures in accordance with the requirements for the machine. The traditional algorithm of the solution of the problem of optimal structural synthesis includes the investigation of the space of parameters of each of the possible layouts, combination of all test tables into a single summary table, identification of the allowable set of solutions, and determination of the optimal model in the latter. However, this approach is fairly laborious and requires substantial computer time. In view of this, an heuristic technique for the solution of the structural synthesis problem has been proposed based on the investigation of the parameter spaces and the hypothesis on the structural preference. For comparable vectors of parametric constraints and sufficiently large number of tests, out of the two compared structures preference is given to the one which has large values of the functional generalized selection effectiveness factor and, simultaneously, less suboptimal values of all or of the main local criteria. The investigated goal functions are assumed piecemeal continuous.

### 5.4.3.5 Determination of the Rational Structure of Vibratory Tables and its Optimal Parameters at the Initial Design Stage

The first task of investigations is to select the best dynamic structure of the resonant vibratory table among the existing standard layouts (see Fig. 5.17). The layouts shown in Fig. 5.17a and 5.17b are used with horizontal and inclined oscillations of the working elements. Resonant vibratory tables with vertical oscillations of the working element are constructed on the basis of the last three layouts (Fig. 5.17c, d, e). The best arrangement must be chosen from these three structural layouts. For this purpose a wide variation range for the initial parameters is assumed

$20 \le \alpha_1 \le 100;$    $30 \le \alpha_2 \le 150;$    $400 \le \alpha_3 \le 3000$

$0.5 \, \alpha_6 \le 2;$      $3 \le \alpha_7 \le 7;$      $0.3 \le \alpha_9 \le 1.5$

$50 \le \alpha_{10} \le 100$

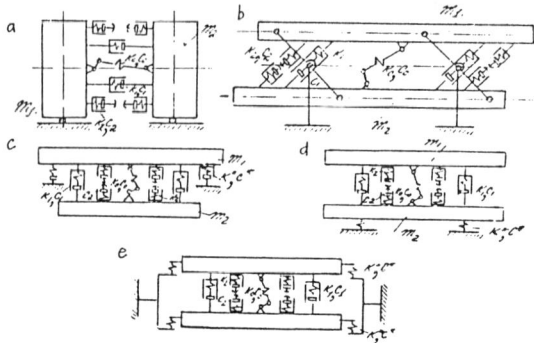

**Figure 5.17** Fundamental layout of vibratory tables for concrete compaction.

for layout "c"  $\alpha_4 = 0$;   $35 \leq \alpha_5 \leq 70$;   $3 \leq \alpha_8 \leq 7$

for layout "d"  $35 \leq \alpha_4 \leq 70$;   $\alpha_5 = 0$;   $5 \leq \alpha_8 \leq 12$

for layout "e"  $20 \leq \alpha_4 \leq 60$;   $20 \leq \alpha_5 \leq 60$;   $3 \leq \alpha_8 \leq 7$

Within the ranges of these parametric and functional constraints, 4096 tests were conducted. This required considerable computational time. The functional constraints were satisfied by two of the arrangements of Fig. 5.17 in accordance with 12, 23, and 100 models. Table 5.6 shows the values of the functional selection effectiveness factor $\gamma$ and suboptimal values of the local criteria for each of the vibratory table layouts under consideration.

**Table 5.6**

| Layout index | $\bar{\gamma}$ | $\varphi^*(\alpha)_1$ | $\varphi^*(\alpha)_2$ | $\varphi^*(\alpha)_3$ | $\varphi^*(\alpha)_4$ | $\varphi^*(\alpha)_5$ | $\varphi^*(\alpha)_6$ |
|---|---|---|---|---|---|---|---|
| c | 0.0026 | 7.35 | 0.132 | 22.30 | 15.60 | 0.223 | 0.235 |
| d | 0.0055 | 9.05 | 0.129 | 28.27 | 21.60 | 0.037 | 0.157 |
| e | 0.0245 | 6.95 | 0.147 | 19.54 | 4.63 | 0.050 | 0.054 |

The analysis of the obtained data on the basis of the structural preference hypothesis enables one to come to the conclusion that the dynamic layout of the existing vibratory table (Fig. 5.15a) is the best. The resources of improving the quality of vibratory table VRA-8 can be revealed, if they exist, only by thorough investigation of the space of initial parameters of the structural layout within wide ranges.

These investigations can suggest fundamental changes to the design of the existing vibratory table. These must be applied at the initial design stage of the machine.

The purpose of this experiment is realization of the attempts to substantially improve the values of the main quality criteria and to find the models satisfying conditions $\varphi(\alpha) \leq 6.5$; $\varphi(\alpha)_2 \leq 0.25$; $\varphi(\alpha)_3 \leq 30$; $\varphi(\alpha)_4 \leq 6$.

During the process of the experiment and global investigation of the parameter space of the structural layout under consideration $N = 4096$ tests were conducted, from which 100 models found their way to the test table. Six versions of the criterial constraints were formulated. Furthermore, the allowable set was varied from 0 - 200 models which corresponds to the change of the selection effectiveness factor $\gamma$ from 0 - 0.005.

The low selection effectiveness factor of the model is associated with the fact that although four times more tests than earlier were conducted in the experiment, the volume of the new parallelepiped is $10^7$ times larger than the volume of the previous parallelepiped of parameters. Therefore, the investigation of the space of parameters can be continued. Further tactics of the experiment concerned the improvement of the obtained results; moreover, the models finding their way to the test table were considered to be the ones to be improved.

It is known that the success of application of the local search is largely determined by the appropriate selection of the initial models. The designer turned his attention to two factors: firstly, constraints $f(\alpha)_{1,2}$ play a decisive role in the formation of the "unacceptable" set of solutions; secondly, analysis of the optimal models with respect to Pareto in the preceding experiment indicated that they lie, in the majority of cases, at the boundaries of the set of models satisfying constraints $f(\alpha)_{1,2}$. In accordance with the previously assumed notations, this is equivalent to the case when the Pareto optimal models have no resources with respect to $\varphi(\alpha)_8$ and $\varphi(\alpha)_9$. In view of this, the designer suggested the selection of those models having a resource of improvement with respect to $\varphi(\alpha)_8$ and $\varphi(\alpha)_9$ as the initial models.

From a thousand models satisfying all functional constraints the designer selected four, namely $\alpha^{1452}$, $\alpha^{2049}$, $\alpha^{2406}$, and $\alpha^{3686}$. Firstly, these models have a resource of improvement with respect to $\varphi(\alpha)_8$ and $\varphi(\alpha)_9$. Secondly, the already attained values of $\varphi(\alpha)_{1,6}$ for these models were considered as acceptable.

The neighborhoods of the improved models were determined approximately; the improved model was taken as the center of each neighborhood with the exception of parameters $\alpha_7$ and $\alpha_8$. In the vicinity of these models 128 tests were carried out.

The models selected by the designer were substantially improved: models were found with minimal possible weight $\varphi(\alpha)_1$ = 6t and sufficiently good values of the other local criteria. Tables 4 and 5 show two of the improved models: $\alpha^{1452}$ and $\alpha^{2406}$, and also two best models found in the corresponding vicinity of the models, namely $\alpha^{116}$ and $\alpha^{12}$. These models are taken as equivalent.

The solution of complex problems similar to the one considered in this experiment entails the increase of expenditure on the determination of the Pareto set over the entire investigated space. However, one can avoid the determination of the Pareto set. In this case, the search for optimal solutions is best conducted in two stages: global (rough) survey of the space of parameters, and local investigation of the vicinities of the models finding their way to the test table. In similar situations the determination of the improved models constitutes a very important task. In the conditions of lack of information on the connectivity of the allowable set of solutions, which is characteristic for the problem of optimal design of machines, the fundamental factor is the uniform survey of the space of parameters and construction of the test tables. The models belonging to the test tables can, without exception, be used as improved models for local investigation.

Analysis of the Pareto models indicates that, in the majority of cases, they lie on the boundaries of the set of the models satisfying the functional as well as some of the parametric constraints. This is the reason behind the acceptance of the hypothesis of LPR concerning the selection of the initial models depending on the availability of resources prior to the onset of the constraint.

## 5.4.4  Vibratory Conveyors*

The successful use of vibratory transport devices in a number of branches of the national industry, particularly in cases of simultaneous processing of the transported product, has caused great interest in the construction of improved vibratory conveying installations. In view of the fact that traditional types of vibratory conveyors, which are characterized by substantial weight and large dimensions, do not always satisfy the requirements, the need arose for development of new vibratory conveyors that are fundamentally different from the existing ones for horizontal and vertical conveying and are compact.

A similar new type of vibratory conveyors was developed by the VNIIPTMASh and Skochinskii IGD Institutes, and a number of the developed models are presently being manufactured.

The characteristic features which differentiate these types of vibratory conveyors from the conventional machines are the longitudinal application of the exciting forces, the use of a rubber elastic system operating in shear, and compression and generating of two-component oscillations of the load-carrying element, and equilibrium of the structure. This design enables the significant reduction of parasitic oscillations of the load-carrying elements and the frame and eliminate or substantially decrease transmission of dynamic loads to the carrying structure. The specific material consumption of the plants has been significantly reduced and the opportunities have been provided to design compact installations which are needed in many applications in which there are space constraints for conveyors.

The load carrying element in such a conveyor is also a force transmitting element through which the exciting force is transmitted from the drive. The direction of this force coincides with the axis of the load-carrying element, but does not coincide with any of the principal axes of the other supports (with the extremal values of their stiffnesses). In the general case, elliptical oscillations of the load-carrying element are induced which, with proper tuning of the oscillation system, can be transformed into rectilinear oscillations directed at the desired angle to the longitudinal axis of the

---

*M.L. Izrailevich participated in writing section 5.4.4.

load-carrying element. Thus, the mechanism of operation of the conveyor consists of the following. Under longitudinal excitation of the load-carrying element, which is the vibration exciter for all elastic supports, elliptical oscillations are generated in the supports and with special tuning the oscillations can become rectilinear. For the coordinated operation of all supports along the length of the conveyor, the stiffness parameters at any section of the flight must be the same. In this case, all points of the load-carrying element travel along identical trajectories.

Eccentric vibrators with elastic connecting rod are used to drive conveyors of this type. In special cases, the vibrator has an elastic eccentric made in the shape of rubber-metal block of variable stiffness in different directions.

The characteristic property of the operation of the drive with elastic eccentric is in the following. At the moment of start the drive has minimum stiffness; therefore, under the action of the forces of the main elastic system, which has higher stiffness, the rubber block is deformed considerably, reducing the working eccentricity. Accordingly, at the initial moment the load-carrying element receives only a slight disturbance. This enables the driving shaft to rotate easily overcoming relatively small torque. Then the drive shaft increases its rotational speed, rocking of the load carrying element is gradually increasing, the forces required to maintain oscillations are decreased, and with the decrease of the magnitude of the exciting force, the eccentricity of the drive gradually increases.

The design of horizontal and vertical vibratory conveyors on the basis of the new design schematics required of the designer to solve multicriterial problems, in the majority of cases, with contradictory goal functions. The optimal parameters of vibratory conveyors of this type are characterized by many criteria pertaining to the design of the machine and to the performed production process. Some of these criteria are contradictory, for example, the requirement, on the one hand, to intensify oscillations of the working member of the machine in order to increase the effectiveness of the production process and, on the other hand, to reduce dynamic loading of the structural elements.

The problem of the optimal design of horizontal and vertical vibratory conveyors is reduced to the selection of the most perfect dynamic layout and parameters of the structural elements of the conveyor for which the motion of the working elements best satisfies the process requirements and the loading conditions of the elements satisfy the structural constraints. For the parameter optimization of horizontal and vertical vibratory conveyors with respect to several local criteria the method of selection of optimal machine and structural parameters is used. The problem of construction of the domains of the allowable values of the vector of the initial parameters and formation of local constraints are discussed on the example of synthesis of horizontal compact balanced conveyors with the discussion of the obtained study results.

Thus, the first stage of solving the optimization problem is associated with the selection of the best structural layout of the machine. The task is set as follows: on the basis of analysis of all existing and practically utilized layouts and also new and prospective designs in this field, a generalized structural layout of the vibratory conveyor is constructed including the special structural layouts conceivable at the current level of development in this field. Dynamic equations are written for the generalized structural layout. The obtained system of differential equations describes the motion of the structural elements of the generalized layout of the vibratory conveyor. The equations describing the special layouts are obtained by equating to zero some terms of the generalized system of equations. For this, in the appropriate terms of the system of equations coefficients are introduced, which are regarded as system parameters (they enter into the vector of the parameters of the investigated set of models) and can assume zero or nonzero values.

To prepare for the investigation of the first stage - the selection of the best structural layout of a conveyor - the following alternative layouts, their main merits, and shortcomings are examined.

The dynamic system which is generalized on the basis of synthesis of all the developed structural layouts and which represents a vibratory conveyor under load is a four-mass system where $m_1$ and $m_2$ are the masses of the load carrying elements; $m_3$ and $m_4$ are the masses of the frame. The load-carrying elements (masses $m_1$ and $m_2$) are joined with the frame (masses $m_3$ and $m_4$) by elastic links whose rheological properties are functions of their deformation and rate of deformation

$$f(x_1 - x_3, \dot{x}_1 - x_3), \quad f(x_2 - x_4, \dot{x}_2 - \dot{x}_4)$$
$$f(y_1 - y_3, \dot{y}_1 - y_3), \quad f(y_2 - y_4, \dot{y}_2 - \dot{y}_4)$$

Oscillations to the load-carrying elements with phase shift $\pi$ are imparted by an eccentric drive; components of the exciting force acting on the right and left load-carrying elements in the direction of axes $x$ and $y$ are, respectively

$$f_x(\omega t),\ f_y(\omega t)\quad \text{and}\quad f_x(\omega t+\pi),\ f_y(\omega t+\pi)$$

In addition to the exciting forces, the load-carrying elements are acted upon by the reactions of the conveyed load $F'_x,\ F'_y,$ and $F''_x,\ F''_y.$

The left-hand and right-hand sides of the frame of the conveyor are acted upon by reactions of the adjacent section $f_x,\ f_y,$ and $f_x'',\ f_y'',$ and by the reactions of the isolating elastic links $f(x_3,\dot{x}_3),\ f(y_3,\dot{y}_3),\ f(x_4,\dot{x}_4),\ f(y_4,\dot{y}_4).$

In accordance with the adopted notations, the motion of the vibratory conveyor in coordinates $x,\ y$ (the coordinate axes coincide with directions of the extremal stiffnesses of the elastic system) is described by the system of nonlinear differential equations

$$m_1\ddot{x}_1 + f(x_1-x_3,\ \dot{x}_1-\dot{x}_3) = f_x(\omega t) + F'_x$$
$$m_1\ddot{y}_1 + f(y_1-y_3,\ \dot{y}_1-\dot{y}_3) = f_y(\omega t) + F'_y$$
$$\cdot\ \cdot\ \cdot\ \cdot\ \cdot\ \cdot\ \cdot\ \cdot\ \cdot\ \cdot\ \cdot\ \cdot\ \cdot\ \cdot\ \cdot\ \cdot\ \cdot\ \cdot\ \cdot$$
$$\cdot\ \cdot\ \cdot\ \cdot\ \cdot\ \cdot\ \cdot\ \cdot\ \cdot\ \cdot\ \cdot\ \cdot\ \cdot\ \cdot\ \cdot\ \cdot\ \cdot\ \cdot\ \cdot$$
$$m_y\ddot{x}_y + f(x_4-x_2,\ \dot{x}_4-\dot{x}_2) = \ddot{f}_x + f(x_4,\ \dot{x}_4)$$
$$m_4\ddot{y}_4 + f(y_4-y_2,\ \dot{y}_4-\dot{y}_2) = \ddot{f}_y + f(y_4,\ \dot{y}_4)$$

The deformations and the motion of the transported load in the system $x^*,\ y^*$, represented by an inertial viscoelasto-plastic model, are described by the system of nonlinear differential equations

$$m_y\ddot{y}^* + \begin{vmatrix} c_y^* \\ 0 \\ c_y^{*'}+c_y^{*''} \end{vmatrix}\dot{y}^* + \begin{vmatrix} k_y^* \\ k_y^{**} \\ 0 \end{vmatrix}y^* = -\,m\ddot{y}_{1,2} - \begin{vmatrix} 0 \\ 0 \\ c_y' \end{vmatrix}\dot{y}_{1,2} - mg$$

$$(m+m_0)\,\ddot{x}^* + \begin{vmatrix} c_x^* \\ c_x^* + c_x^{*''} \end{vmatrix}\dot{x}^* + \begin{vmatrix} k_x^* \\ 0 \end{vmatrix}x^* - \begin{vmatrix} c_x^* \\ 0 \end{vmatrix}\dot{x}_0 - \begin{vmatrix} k_x^* \\ 0 \end{vmatrix}x_0$$
$$= -\,(m+m_0)\,\ddot{x}_{1,2}$$

$$m_0\ddot{x}_0 + c_x^*\dot{x}_0 - k_x x_0 - c_x^*\dot{x}^* - k_x x^*$$
$$= -\,m_0\ddot{x}_{1,2} - \begin{vmatrix} 1 \\ 0 \end{vmatrix}F_x^*\mu\,\mathrm{sign}\,(\dot{x}_0)$$

where $y^*$, $x^*$, and $x_0^*$ are displacements of the model of the load relative to the load carrying element (right or left) in the coordinates $x^*$, $y^*$; $x_{1,2}$, $y_{1,2}$ are displacements of the halves of the load-carrying element; $c_y^*$, $c_y^{*'}$, $c_y^{*''}$, $c_y'$, $c_x^*$, $c_x^{*'}$, $c_x^{*''}$ are coefficients determining the hysteretic characteristics of the rheological model of the load; $\mu$ is the coefficient of friction of the load on the load-carrying element; $m$ and $m_0$ are masses of the inertial elements of the rheological model of the load. The solution algorithm of the system of equations describing the motion of the conveyed load is automatically formed by the logical correlations which are represented by the system of transcendental equations determining the transition moments during load travel from one characteristic stage to another.

The design of the vibratory conveyor begins with the selection of the most perfect structural layout which fully corresponds to the structural requirements and production needs. The selection of the layout can be carried out with conventional design methods or with the method of selection of optimal parameters. In the latter case, investigations are conducted on the basis of global study of the generalized structural layout within a wide range of variation of the initial parameters. Among the layouts finding their way to the allowable set of models, the models with potential for further improvement are selected.

Optimal design stipulates the construction of the allowable and Pareto (optimal with respect to Pareto) sets of solutions and the identification of the optimal solutions in the latter. Models that simultaneously satisfy the Pareto and criterial constraints belong to the allowable set of solutions. Following the selection of the structural layout of the vibratory conveyor and its initial design, the designer can, from the design concepts, determine and specify the limits of variation of each parameter on which conveyor characteristics can depend. The latter are referred to as quality criteria and must be functionals of the parameters of the conveyor. Further, the equation of motion of the conveyor are solved and the quality criteria in the range of its standard parameters are determined. The selection of specific values of the parameters in the assumed range is carried out on the basis of $LP_\tau$ sequence.

The presented system of equations generally reproduces the interaction of the load-carrying elements of the conveyor with the load and support structures.

When designing the machine, the following parameters were assumed as variables: oscillation frequency of the load-carrying element $n(\omega)$, eccentricity of the vibrator $\eta$, installation angle of the working elastic elements $\beta$, masses of the load-carrying elements $m_1$, $m_2$, masses of the conveyor frame $m_3$, $m_4$, stiffness of the working elastic elements $k$, stiffness of the driving elastic elements $k_0$, the ratio of the minimum and maximum stiffnesses of the driving elastic links $k_{0max}/k_{0min}$ which is determined by the type of the vibrator (with constant stiffness of the elastic link, with stiffness that varies by periodic law), the design of the load-carrying element (continuous, sectioned).

The following parametric constraints were assumed

$$0.75\ n^* \leqslant n = \alpha_1 \leqslant 1.25\ n^*$$
$$0.50\ r^* \leqslant r = \alpha_2 \leqslant 1.5\ r^*$$
$$0.8\ \beta^* \leqslant \beta = \alpha_3 \leqslant 1.2\ \beta^*$$
$$0.75\ m^*_{3,4} \leqslant m_{3,4} = \alpha_5 \leqslant 1.25\ m^*_{3,4}$$
$$0.60\ k^* \leqslant k = \alpha_6 \leqslant 1.4\ k^*$$
$$0.8\ m^*_{1,2} \leqslant m_{1,2} = \alpha_4 \leqslant 1.2\ m^*_{1,2}$$
$$0.70\ k^*_0 \leqslant k_0 = \alpha_7 \leqslant 1.5\ k^*_0$$
$$0.7\ \frac{k^*_{0\ max}}{k^*_{0\ min}} \leqslant \frac{k_{0\ max}}{k_{0\ min}} = \alpha_8 \leqslant 1.3\ \frac{k^*_{0\ max}}{k^*_{0\ min}}$$

Here, parameters $n^*$, $\beta^*$, $r^*$, $m^*_{1,2}$, $m^*_{3,4}$, $k^*$, $k^*_0$, $k^*_{0max}$, $k^*_{0min}$ pertain to the nominal model which was assumed as a reference design model.

Thus, the vector of the parameters of the investigated set of models $\alpha = \{\alpha_1,\ \alpha_2,\ \alpha_3,\ \alpha_4,\ \alpha_5,\ \alpha_6,\ \alpha_7,\ \alpha_8\}$ is determined by the values of the nonzero parameters in it. In addition to the parametric constraints, the following functional constraints were assumed

$$1.2\ g \leqslant F = f_1\ (\alpha) \leqslant 1.8\ g$$
$$3\ mm \leqslant A = f_2\ (\alpha) \leqslant 8\ mm$$
$$18° \leqslant \beta' = f_3\ (\alpha) \leqslant 30°$$
$$k_{0r} = f_4\ (\alpha) \leqslant P^*_{max}$$

Here, $P^*_{max}$ is the maximum force in the drive of the original model during the starting process.

The first three functional constraints determine the correlation of the time history of oscillations of the load-carrying element with the requirements of the regime of vibratory conveying, the fourth restricts the magnitude of the exciting force according to the starting conditions. The quality of the designed vibratory conveyor is assessed by six local criteria. These are functionals that assess weight of the plant, dynamic forces in the drive, dynamic loads on the carrying structure, ratio of the starting and consumed in steady regime powers, speed of transportation of the load, and the minimum possible length of the load carrying element per drive. All installations are characterized in terms of the initial parameters of the system; the remaining criteria are functionals of the integral curves. These criteria are dependent on the degree of loading of the load-carrying element by the transported material and by its transportability. Calculations were conducted for the nominal load by a standard load and for the maximum possible loading.

As local quality criteria, the majority of which must be minimized, the following were assumed:

weight of the installation

$$\varphi_1(\alpha) = \frac{\min}{\underline{\alpha}} \; \Sigma m$$

where $m$ are the masses of the structural elements of the conveyor (load-carrying elements, frame, elastic system, and the drive);

dynamic forces in the drive

$$\varphi_2(\underline{\alpha}) = \frac{\min}{\underline{\alpha}} \; | Ae^{j\psi}(k_0 + j\omega c_0) |$$

dynamic loads on the load-carrying element

$$\varphi_3(\underline{\alpha}) = \frac{\min}{\underline{\alpha}} \; \Sigma f(x, \dot{x}) + f(y, \dot{y})$$

ratio of the starting and consumed power

$$\varphi_4(\underline{\alpha}) = \frac{\min}{\underline{\alpha}} \; N_n/N$$

mean conveying speed

$$\varphi_5(\underline{\alpha}) = \frac{\max}{\underline{\alpha}} \; v$$

length of installation per one drive

$$\varphi_6(\underline{a}) = \frac{\max}{\underline{a}} L$$

The main outcome of the selection of the optimal parameters was the design of vibratory horizontal and vertical conveyors of fundamentally new structural arrangement with longitudinal application of the exiting force. These machines are characterized by compactness of design, increased length per drive, reduced material and energy consumption, and by high degree of balance.

The length of the plant per drive was increased, depending on its modification, by 40 - 50%, its weight reduced by 25 - 30%; the dynamic loads on the load-carrying structures were reduced: the horizontal components by 80 - 90%, the vertical components by 40 - 50%; the dynamic forces in the drive reduced by 15 - 20%; the ratio of the starting and consumed powers decreased by 15 - 20%; the conveying speed rose by 10 - 12%.

The economic effectiveness from the application of the optimal designs of vibratory horizontal and vertical conveyors is attained by both improving the technical and operating characteristics in comparison with the reference model and as a result of expansion of the possible effective range of application by replacing less progressive conveying devices (scraper, apron, and screw conveyors). Compared with the replaced conveyor systems, vibratory horizontal and vertical conveyors of optimal models are characterized by lower material and power consumption, less maintenance, high durability, feasibility of combining in one installation both conveying and processing operations, and by the creation of comfortable working conditions in industries with harmful environments.

## 5.4.5  Vibratory and Grizzly Feeders

When vibratory feeders are designed, the assumed variable parameters are: oscillation frequency of the working element $\omega$ = $2\pi n$; eccentricity $r$ for an eccentric vibratory drive, or kinetic moment of the unbalanced masses $mr$ for an inertial vibratory drive; angle of installation of the working elastic elements $\beta$; the masses of the load-carrying element $m_1$, the reactive part $m_2$ (in balanced type machines), and the frame $m_3$; stiffness of the working elastic elements $k_1$; stiffness of the driving elastic elements of the eccentric vibratory drive $k_2$; stiffness of the isolating elastic elements $k_3$; the installed power of the motor (for high-capacity machines) $N_i$; the ratio of the maximum and minimum stiffnesses of the driving elastic links (for an eccentric vibratory drive of parametric excitation) $k_{1max}/k_{1min}$.

The following parametric constraints are assumed:

$$0.75n^* \leqslant n = \alpha_1 \leqslant 1.25n^*, \quad 0.50r^* \leqslant r = \alpha_2 \leqslant 1.50r^*$$
$$0.7\,(mr)^* \leqslant (mr) \leqslant \alpha_2' \leqslant 1.3\,(mr)^*, \quad 0.8\beta^* \leqslant \beta = \alpha_3 \leqslant 1.2\beta^*$$
$$0.8m_1^* \leqslant m_1 = \alpha_4 \leqslant 1.2m_1^*, \quad 0.8m_2^* \leqslant m_2 = \alpha_5 \leqslant 1.25m_2^*$$
$$0.75m_3^* \leqslant m_3 = \alpha_6 \leqslant 1.25m_3^*, \quad 0.7k_1^* \leqslant k_1 = \alpha_6 \leqslant 1.5k_1^*$$
$$0.75k_2^* \leqslant k_2 = \alpha_7 \leqslant 1.25k_2^*, \quad 0.5k_3^* \leqslant k_3 = \alpha_8 \leqslant 1.5k_3^*$$
$$0.5N_i^* \leqslant N_i = \alpha_9 \leqslant 1.5N_i^*, \quad 0.7(k_{1\,max}/k_{1\,min})^* \leqslant (k_{1\,max}/k_{1\,min})$$
$$= \alpha_{10} \leqslant 1.3\,(k_{1\,max}/k_{1\,min})^*$$

The parameters $n^*$, $r^*$, $(mr)^*$, $\beta^*$, $m_1^*$, $m_2^*$, $m_3^*$, $k_1^*$, $k_2^*$, $k_3^*$, $N_i^*$, $(k_{1max}/k_{2min})^*$ pertain to the leading model which is taken as the reference model during design process.

Along with the parametric constraints the following functional constraints are established:

for machines operating under small loads $1.2g \leq \Gamma = f_1(\alpha)$ $\leq 1.8g$, $3.0$ mm $\leq A = f_2(\alpha) \leq 8.0$ mm

for machines operating under rock burst $1.5g \leq 1' = f_1'(\alpha)$ $\leq 2.5g$, $2.0$ mm $\leq A = f_2'(\alpha) \leq 5.0$ mm; $18^O \leq \beta_1 = f_3(\alpha) \leq 30^O$, $F = f_4(\alpha) \leq F^*_{max}$, $F_b = f_5(\alpha) \leq F^*_{max}$.

$F^*_{max}$ and $F^*_{bmax}$ are the maximum force in the connecting rod of the eccentric drive and the maximum pressure on the bearings of the inertial drive of the reference model.

The first three functional constraints determine the correlation of the time history of oscillations of the load-carrying element of the feeder with the requirements of the regime of vibratory conveying, the fourth and fifth limit the magnitude of the forces in the eccentric and inertial drives.

The following are taken as local quality criteria:

the mean conveying speed $\varphi_1(\alpha) = \max v$;

intensity of oscillations of the load (for machines operating under rock burst $\varphi_2(\alpha) = \max \alpha_1 \omega^2$, where $A_1$ is the mean oscillation amplitude of the load;

weight of the machine $\varphi_3(\alpha) = \min (m_1 + m_2 + m_3)$;

dynamic forces in the drive $\varphi_4(\alpha) = \min F$;

dynamic loads on the frame $\varphi_5(\alpha) = \min F_f$;

ratio of the starting and consumed power $\varphi_6(\alpha) = \min N_{st}/N_i$.

The main outcome of the selection of the optimal parameters was the construction of universal vibratory feeders of VP type from components that are unified with the components of vibratory conveyors of the KVZhG type with longitudinal application of the exciting force and having a compact design (the height of the unit 40 - 50% less than for feeders of other types). The weight is reduced by 25 - 30%, energy consumption is reduced and the capacity is simultaneously increased. Capital expenses are reduced by 30 - 40%, operating costs by 50% as a result of ensuring high reliability and ease of maintenance. It is the first time that large vibratory feeders and grizzly feeders of high output (up to 2000 t/h) of KVG and SVG types have been designed in the Soviet Union for movement of large chunks of rock (up to 1.2 m in cross section) with large carrying capacity (up to 25 t).

### 5.4.6  Vibratory Installations for Charging and Discharging Mineral Ores

Designing vibratory installations for charging and discharging mineral ores is faced with the complex problem of selecting the effective regimes of action on the rock mass with the purpose of enhancing its mobility, preventing jamming in discharge devices, and increasing the discharge velocity. Therefore, the regimes of action of vibration on broken rock mass was investigated in a wide range.

The following are assumed as the main variable parameters: the main frequency of the working element $\omega_i = 2\pi n$; the order of high harmonics $\xi = n_i/n$; configuration of the trajectory of the working element of the installation (phase shift between the longitudinal and transverse components of the exciting force) $\gamma$; the kinetic moment of the unbalanced masses $(mr)_j$; the angle of installation of the elastic elements $\beta$; the mass of the load-carrying element $m_1$ and frame (foundation) $m_2$; stiffnesses of the working elastic elements $k_x$, $k_y$; the distance between the elastic elements $l$ (the moment of the elastic system); the depth of the rock burst $l_p$; the installed power of the motor $N_j$.

The following parametric constraints are introduced;

$$800 \, 1/min \leq n_1 = \alpha_1 \leq 1400 \, 1/min, \; 1 \leq \xi = \alpha_2 \leq 3, \; 0 \leq \gamma = \alpha_3 \leq 90°, \; 15° \leq \beta = \alpha_4 \leq 30°, \; 0 \leq lp \leq \alpha_5 \leq l.$$

The remaining parametric constraints are the same as for vibratory feeders.

The functional constraints are assumed within the following ranges

$$1.5g \leq \Gamma = f_1(\alpha) \leq 3.0g, \quad 2.0mm \leq A = f_2(\alpha) \leq 5mm$$
$$10° \leq \beta_1 = f_3(\alpha) \leq 90°, \quad F_{ymax} \leq 10mg$$

The presented functional constraints determine the correlation of the time history of oscillations of the working element of the machine with the regime of vibration discharge and the magnitude of the maximum load-carrying capacity.

As local quality criteria the following is assumed: mean speed of vibration discharge $\varphi_1(\alpha) = $ max $v$.

Oscillation intensity of the rock mass on the vibratory machine is $\varphi_2(\alpha) = $ max $A_1 \omega_1^2$, where $A_1$ is the mean oscillation amplitude of the rock mass, the weight of the installation $\varphi_3(\alpha) = $ min $(m_1 + m_2)$, the dynamic forces in the drive $\varphi_4(\alpha) = $ min $F$ max, the power requirement $\varphi_5(\alpha) = $ min $N$.

As a result of the conducted investigations, vibratory installations were designed for discharging minerals, capable of operating under rock burst, intensively exciting the rock mass and preventing it from sticking in the discharge openings and providing uniform high output.

### 5.4.7 Vibratory Machines for 3-D Product Processing

Due to large volumes of production of products requiring vibratory hardening, requirements are set forth to establish the effective regimes for lowering the working time and increasing

the final level of hardening. At the same time, requirements
are set forth to increase the output and improve the technical
and operating characteristics of the designed equipment. This
entails, in the first place, increasing the operational lifetime,
lowering operational costs, and decreasing the level of the
dynamic loads on the carrying structure.

Development of vibratory machines for hardening of
various products satisfying these requirements necessitates the
solution of complex multi-criterial problems of optimal design.

When designing installations for hardening products from
structural steels and hard alloys, one has to seek the effective
hardening regimes and the parameters of the machine that
ensure the maintenance of these regimes and attaining high
reliability at the same time.

The main variable parameters are the following: main
(carrying) oscillation frequency of the container $\omega_1 = 2\pi n_1$;
kinetic moment of the unbalanced masses of the first $m_1 r_1$ and
second $m_2 r_2$ stages; the transmission ratio between the first
and the second stages $i$; angle of shift of the unbalanced
masses of the second stage relative to the first stage $\gamma$; mass
of the container $m_1$; mass of the frame $m_2$; stiffness of the
elastic links $k_x$, $k_y$; mass of the working medium and the
products $m_3$.

The parametric constraints were as follows:

$$800 1/\text{min} \le n_1 = \alpha_1 \le 1400 1/\text{min}, \ 1 \le i = \alpha_2 \le 3, \ 0 \le r$$
$$= \alpha_3 \le 90°, \ 0.8 m_1^* \le m_1 = \alpha_4 \le 1.2 m_1^*, \ 0.8 m_2^* \le m_2^* = \alpha_5$$

The parameters superscripted with a star pertain to the
leading model which was assumed as the reference model
during design process.

In addition to the parametric constraints, the following
functional constraints are taken into account:

the maximum vibratory acceleration of the container $\ddot{x}_{max}$
$\ddot{y}_{max} \le 10g$,

the maximum strain acceleration of hard alloys $\ddot{x}_{max}$
$\ddot{y}_{max} \le 25g$,

the maximum strain acceleration of steels $\ddot{x}_{max}$ $\ddot{y}_{max} \le$
$50g$.

The first functional constraint determines the correlation
of the operating regime of the machine with the conditions of

reliability, the second and third determine the correlation of the working regime with the condition of maintaining the worked products not damaged.

As local quality criteria the following are adopted:

processing time (determined from the value of the energy of irreversible strains) $\varphi_1(\alpha) = \min t_w$;

uniformity of product working (determined by the value of the energy of irreversible strains in the direction of axes $x$, $y$) $\varphi_2(\alpha) = w_{x_u}/w_{y_u} = 1$, where $w_{x_u}$, $w_{y_u}$ are the useful expenditure of energy on product working in the direction of axes $x$, $y$;

energy intensity of the process $\varphi_3(\alpha) = \max W/t$;

energy efficiency $\varphi_4(\alpha) = \max W_u/W$, where $W_u$ is the useful energy expenditures, $W$ is the total energy expenditure;

dynamic loads in the drive $\varphi_5(\alpha) = \min F_e$;

dynamic loads on the foundation $\varphi_6(\alpha) = \min F_f$;

mass of the machine $\varphi_7(\alpha) = \min (m_1 + m_2)$.

As a result of the optimal multi-criterial design, the time of product working was reduced by 25%, the degree of hardening (specific energy absorption) increased by 18 - 20%, the weight of the machine reduced by 20%.

The economic effectiveness of introducing hardening by vibratory machines of optimal design is attained as a result of increasing the durability of the vibration-hardened, rock-cutting hard-alloy tool of cutter-loaders by not less than 50% and the strength of jack hammer tips by not less than 60%, which provides an economy of 14 - 18% of the hard alloy.

## 5.4.8  Optimal Multi-criterial Design of Vibratory Crushers

The construction of highly-effective designs of vibratory crushers accounting for the dynamic parameters of the oscillatory system, the characteristics of the motor, and the properties of the crushed material require from the designers the solution of multi-criterial problems with contradictory goal functions. This problem can be solved most effectively with the use of the method of selection of the optimal machine and design parameters [19].

The optimal design of vibratory crushers is in the selection of the best dynamic layout and the parameters of the structural elements and crusher drive at which the movements of the crushing jaws best satisfy the crushing process of rocks of specified properties and the loading conditions of the crusher elements provide high design reliability.

The design of a vibratory crusher begins with the selection of the latest structural layout which fully satisfies the structural and process requirements. The selection is made on the basis of a global study of the generalized structural layout of the vibratory crusher within wide ranges of variation of the initial parameters. From the layouts finding their way to the allowable set of models, the models having the potential for further improvement are selected. Local search is then used for the selection of the best design. Following the selection of the structural layout of the vibratory crusher and its draft study, the designer, on the basis of design concepts, determines and specifies the limits of variation of each parameter of the crusher. These characteristics are the quality characteristics of the crusher. The main quality criteria of the crusher which must be minimized or maximized are:

weight of the crusher

$$\leq 1.2 \ m_2^*, \quad 0.7 \, (m_1 r_1)^* \leq (m_1 r_1) = \alpha_6 \leq 1.3 \, (m_1 r_1)^*, \quad 0.7 \, (m_2 r_2)^*$$
$$\leq (m_2 r_2) = \alpha_7 \leq 1.3 \, (m_2 r_2)^*, \quad 0.7 \ k_{x,y}^* \leq k_{x,y} = \alpha_8 \leq 1.5 \ k_{x,y}^*$$
$$0.5 \ m_1 \leq m_3 = \alpha_9 \leq 1.5 \ m_1$$

where $m$ denotes the masses of the structural elements of the vibratory crusher;

dynamic forces in the bearings of the vibrator

$$\varphi_2 \, (\underline{\alpha}) = \min \, F$$

dynamic loads on the crusher frame

$$\varphi_3 \, (\underline{\alpha}) = \min \, F_p$$

stability of operation with the change of characteristics of the crushed material

$$\varphi_4 \, (\alpha) = \min \, \frac{A_{max}}{A_{min}}$$

where $A_{max}$ and $A_{min}$ are the maximum and minimum oscillation amplitudes of the jaws when the crusher is operating under load;

specific energy consumption on crushing

$$\varphi_5(\underline{a}) = \min W_{sp}$$

technical output with respect to crushing

$$\varphi_6(\underline{a}) = \max Q$$

The determination of crusher characteristics which are used as quality criteria is carried out as a result of solving the equations of motion of the crusher jaws under load in the range of established parameters of the crusher and the crushed material. It is important to bear in mind that the reliability of the design parameters is largely dependent on the accuracy by which the rheological model of the crushed rock identifies the actual crushed rock. Two contradictory requirements are made of the rheological model: maximum simplicity (the inclusion of the least number of fundamental rheological bodies) and maximum accuracy of reproduction of the main properties of the crushed rock. In view of this, the structure and parameters of the rheological model of the crushed rock are also established by the multi-criterial optimal design technique.

The following are assumed as quality criteria:

scatter of experimental and theoretical values of the reaction of the rock on the crushing jaw

$$\varphi_1(\underline{\beta}) = \min \frac{F_{exp}}{F_r}$$

energy expenditure on the crushing process

$$\varphi_2(\underline{\beta}) = \min \frac{W_{en}}{W_r}$$

number of fundamental rheological bodies

$$\varphi_3(\underline{\beta}) = m_x = 0, \qquad \varphi_7(\underline{\beta}) = c'_y = 0$$
$$\varphi_4(\underline{\beta}) = m_y^* = 0, \qquad \varphi_8(\underline{\beta}) = k_y = 0$$
$$\varphi_5(\underline{\beta}) = k'_x = 0, \qquad \varphi_9(\underline{\beta}) = c_y = 0$$
$$\varphi_6 = (\underline{\beta}) = k_{pxy} = 0$$

The parameters of the model $m$, $k_x$, $c_x$, $m_x$, $m_y^*$, $k'_x$, $k_{pxy}$, $c'_x$, $k_y$, $c_y$ are varied during calculations in the range

from zero values to maximum. Moreover, the boundary values
are corrected by the designer in the course of calculations.
When designing a crusher, the assumed variable parameters are:
oscillation frequency of the crushing jaws $n$, kinetic moment of
the vibrator $mr$, masses of the jaw and the frame of the
crusher $m_j$, $m_f$, stiffness of the working and isolating elastic
elements $k$, $k_a$. For the varied parameters the following
allowable range of variations is assumed:

$$0.75n* \leqslant n = a_1 \leqslant 1.25n*$$

$$0.50\,(mr)* \leqslant (mr) = a_2 \leqslant 1.50\,(mr)*$$

$$0.75m_j^* \leqslant m_j = a_3 \leqslant 1.25m_j^*$$

$$0.75m_f^* \leqslant m_f = a_4 \leqslant 1.25m_f^*$$

$$0.50k* \leqslant k = a_5 \leqslant 1.50k*$$

$$0.50k_a^* \leqslant k_a =: a_6 \leqslant 1.50k_a^*$$

Here, parameters $n$, $(mr*)$, $m_j^*$, $m_f^*$, $k$, and $k_a$ pertain to the
nominal model which is taken in the design as the initial
model.

The values of the parameters generate a vector of the
parameters of the investigated set of model crushers

$$\underline{a} = \{a_1, a_2, a_3, a_4, a_5, a_6\}$$

Along with the parametric constraints, the following
functional constraints are also assumed;

amplitude of linear oscillations of the jaw

$$5_{mm} \leqslant A = f_1(\underline{a}) \leqslant 20_{mm}$$

angle of torsional oscillations of the jaw

$$\theta = f_2(\underline{a}) \leqslant 3°$$

The selection of the test points by the used method
provides uniform view of the entire multi-dimensional space of
parameters for relatively small number of tests and, therefore,
guarantees reliability of the calculations. From the determined
values test tables are compiled in which each quality criterion
is arranged in a descending order, i.e., the best crusher model

with respect to the given criterion comes first. Since a specific set of parameters encoded in the number of tests corresponds to each crusher model, the designer decides by looking at the test table what course of action should be taken in order to obtain the best crusher with respect to one quality criterion or another. By implementing the regime of designer–computer dialogue, the designer can confidently specify the constraints with respect to each criterion that are attainable and satisfy the customer. Further, a computational test is made to verify whether there is a design simultaneously satisfying all the constraints. Similar designs comprise the allowable set of solutions. The designer selects the optimal model from this set.

# REFERENCES

1. Goncharevich, I. F. Vibratory rheology in mining. Moscow, Nauka, 1977, 144 pp.
2. Severdenko, V. P., Klubovich, V. V., and Stepanenko, A. V. Ultrasound and plasticity. Minsk, Tekhnika, 1976, 316 pp.
3. Kryukov, B. I. Dynamics of vibratory machines of resonant type. Kiev, Nauk. dumka, 1967, 212 pp.
4. Blekman, I. I. Vibratory conveyance. Moscow, Nauka, 1964, 488 pp.
5. Goncharevich, I. F. and Dokukin, A. V. Dynamics of mining machines with elastic links. Moscow, Nauka, 1975, 212 pp.
6. Ragul'skis, K. M. and Gribauskas, I. K. Excitation of transverse motions in a moving flat body. Mezhvusovskii tematichskii sbornik nauchnykh trudov "Vibrotekhnika", Vol. 2, No. 23, Vilnius, Kaunass polytechnical institute, 1978, pp. 153-160.
7. Goncharevich, I. F. On the theory of vibratory conveyance of bulk loads. In: Problems of improving processes and hardware of underground mining transport. Moscow, Nauka, 1967, pp. 173-182.
8. Goncharevich, I. F. Dynamics of vibratory conveyance. Moscow, Nauka, 1972, 212 pp.
9. Frolov, K. V., Tselikov, A. I., Sergeev, V. I., and Goncharevich, I. F. Determination of optimal parameters of vibratory molds in continuous casting machines: Abstracts of the conference "Problems of nonlinear oscillations of mechanical systems". Kiev, In-t problem mekhaniki AN UkSSR, 1978, p.18.
10. Goncharevich, I. F., Ur'ev, N. B., and Toleisnik, M. A. Vibration engineering in the food industry. Moscow, Food

industry, 1977, 280 pp.

11. Abramov, O. V. Molding of metals in ultrasound field. Moscow, Nauka, 1972, 282 pp.

12. Frolov, K. V., Goncharevich, I. F., Sergeev, V. I., Chernyavskii, I. T., Gusev, V. P., and Viznyuk, A. V. Investigation on analog computer of the ingot motion in the mold of continuous casting machine. In: Methods of solution of the problems in machine science on computers. Moscow, Nauka, 1979, pp. 104-114.

13. Sofinskii, P. I., Tselikov, A. A., Charnyi, A. Kh., and Shustorovich, V. M. The effect of inner shell thickness variation of a hollow ingot on conditions of its formation in continuous casting. In: Machines of continuous casting of metals and rolling casting units. Moscow, Nauka, 1975, pp. 59-64.

14. Frolov, K. V., Goncharevich, I. F., and Koloskov, V. F. Investigations of the laws of formation of a hollow ingot on a vibrating mandrel in continuous casting. Methods of solution of the problems in machine science on computers. Moscow, Nauka, 1979, pp. 114-119.

15. Frolov, K. V. Vibrations of machines with limited power of the energy source and variable parameters. In: Nonlinear oscillations and transient processes in machines. Moscow, Nauka, 1972, pp. 5-17.

16. Lavendel, E. E. Synthesis of optimal vibratory machines. Riga, Zinatne, 1970, 252 pp.

17. Kononenko, V. O. Vibratory systems with limited excitation. Moscow, Nauka, 19645, 248 pp.

18. Goncharevich, I. I. Computer-aided and experimental investigation of the laws of operation of double-jaw vibratory crushers. Mezhvuzovskii tematicheskii sbornik nauchnykh trudov "Vibrotekhnika", 1979, Vol. 2, No. 36.

19. Materials for the selection of optimal parameters of machines and structures. Moscow, IMASh named after Academician A. A. Blagonravov, 1980, 202 pp.

20. Stoev, S. Vibroacoustic technology for processing mineral ores, Sophia, Technik, 1979 (in Bulgarian).

21. Poturaev, V. N., Franchuk, V. P., and Chervonenko, A. G. Vibratory conveying machines. Moscow, Mashinostroyeniye, 1964.

22. Murtskhvaladze, R. M. and Goncharevich, I. F. Some experimental correlations of vibration conveyance under elliptical oscillations. In: Transportation in mining works. Moscow, Nedra, 1964, pp. 44-53.

23. Tondl, A. Self-excited oscillations of mechanical systems. Moscow, Mir, 1979.

24. Vukolov, E. A. and Goncharevich, I. F. On dynamics of vibratory machines with a drive for elliptical oscillation excitations. In: Nonlinear oscillations and transient processes in machines. Moscow, Nauka, 1972, pp. 56-66.

25. Goncharevich, I. F. Investigation of vibratory conveying machines with limited excitation. In: Nonlinear oscillations and transient processes in machines. Moscow, Nauka, 1972, pp. 25-38.

26. Frolov, K. V. Some problems of parametric oscillations of machine elements. In: Oscillations and stability of machines. Moscow, Nauka, 1968, pp. 5-20.

27. Zemskov, V. D. and Panin, V. F. Vibratory grizzly driven by motor vibrators with liquid lubrication. In: Design and development of vibratory machines. Moscow, 1979, pp. 46-54.

28. Spivakovskii, A. O. and Goncharevich, I. F. Vibratory conveyors, feeders, and auxiliary devices. Moscow, Mashinostroyeniye. 327 pp.

29. Gani'yev, R. F. and Kononenko, V. O. Vibrations of solid bodies. Moscow, Nauka, 1976, 432 pp.

30. Ragul'skene, V. L. Vibratory impact systems. Vilnius, Mantis, 1974.

31. Babichev, A. P. Vibration processing parts. Moscow, Mashinostroyeniye, 1974, 134 pp.

32. Patent, Great Britain, No. 1168287.

33. Patent, USA, No. 3465976.

34. Patent, Canada, No. 1534669.

35. Frolov, K. V. Nonlinear resonant effects in mechanical systems considering properties of the energy source. Vestnik AN SSSR, No. 10, Moscow, 1987.

36. Frolov, K. L. Methods of improving machines and current problems of machine science. Moscow, Mashinostroyeniye, 1984, 224 pp.

37. Artobolevskii, I. I., Genkin, M. D., Sergeev, V. I., and Statnikov, P. B. On use of computers in the formulation of problems of optimal machine design. Dokl. AN SSSR. 1977, Vol. 233, No. 4, pp. 567-570.

38. Sobolev, I. M. and Statnikov, R. B. Selection of optimal parameters in problems with multiple criteria. Moscow, Nauka, 1981, 110 pp.

39. Kryukov, B. I., Litvin, L. M., Sobolev, I. M., and Statnikov, R. B. Optimal design of resonant vibratory machines. Machine science, 1980, No. 5, pp. 31-39.

40. Statnikov, R. B., Russman, I. B., and Fridman, S. I. On methods of decision making in machine science problems. In: Dynamic characteristics and oscillations of the elements of energy devices. Moscow, Nauka, 1980, pp. 5-13.
41. Frolov, K. V. Vibration: friend or foe? New York, Hemisphere Publishing Corp., in press.
42. Goncharevich, I. F. Vibration - nonstandard path. Moscow, Nauka, 1986, 208 pp.
43. Spivakovskii, A. O. and Goncharevich, I. F. Vibratory and wave conveying machines. Moscow, Nauka, 1983, 288 pp.
44. Jones, J. K. Engineering and artistic design. Moscow, Mir, 1976, 374 pp.
45. Zhgulev, A. S. Field of trajectories of a vibratory machine driven by synchronously rotating unbalanced rotors. Vibrotekhnika, Vol. 4, No. 28, Vilnius, 1979.

Abrasion, 217
Abrasive granular materials, 101
Acceleration, 8, 37, 55, 58, 252, 299
Acceleration amplitude, 8, 120
Accelerometers, 26, 40
Acoustic:
    oscillations, 318
    reflectors, 313
Acoustic waves, 318
Adhesion between particles, 56
Aerodynamic:
    forces, 127
    pressure, 127
    resistances, 11, 99, 112, 124
Air:
    bed, 11
    boundary of granular body, 30
    gap(s), 117, 307
    jet, 113
    permeability, 11
Albumen-water layer, 41, 44
Amplification factor, 75
Amplitude, 10, 408
    frequency, 423
    ratio, 13, 18
Angle(s):
    of displacement, 380
    of fall, 100

of inclination, 48
of vibration, 48
Angular:
    frequency, 103
    oscillations, 286
    velocity, 209, 275, 289, 298, 340, 375, 376, 386
Anisotropic:
    elastic connections, 357
    properties, 40
Annular crushing chamber, 234
Antiphase, 235, 238, 249
Arch roofing, 35
Atomic coagulation, 122
Average hardness, 54

Barium powder, 23
Batching, 241
Bearings, 354, 366, 461
Belt conveyer, 102
Biharmonic:
    law, 275, 302
    oscillations, 17, 18, 19, 160, 243, 278
    vibrator, 280
Blasted ore, 214
Blown-through air, 62

Boiling:
    layer, 56
    state, 57
Bounded soil, 213
Brittle:
    fracture, 155, 158
    rock mass, 150, 152
Bulk load(s), 89, 95, 120


Cardan:
    joints, 233
    shaft, 234
Carrying element, 141, 142
Casting, 163
    of metals, 163, 166
    moulds, 62
    -rolling aggregates, 163
    speed, 164
Catalytic reaction, 1
Calm regimes (low intensity), 6
Caved:
    ore, 138
    rock mass, 138
Cavitation, 166
Centrifugal force, 234, 284, 291, 339
Centrifuging, 2
Chains, 262
Chalk, 57
    layer, 58
Chamber trajectory, 217
Chipping, 146
Chunk motion, 154
Circular oscillations, 270
Circulatory motion of particles, 49
Cleaning, 255
Closed-loop magnetic circuit, 315
Coagulation, 124
Coal, 58
Coarse sandstone, 50
Collision(s), 27, 128, 153
Column-shaped crystals, 165
Combustion, 1
Compaction, 58
Compression, 128
    ratio, 59
Concrete, 2
Conical nozzle, 316
Connecting rods, 249
Constant-magnitude accelerations, 25
Continuous-casting machines (CCM), 163
Compaction, 57
Compressed air, 255
Contaminants, 165

Constraints, 502, 504, 508–510
Convective heat transfer, 242
Conveyance, 1, 2, 112, 117, 213, 241
    speed, 2, 3, 97
Conveyer:
    belts, 232
    surface, 107
Cooled core, 177
Cord-reinforced rubber cylinders, 255
Core, 181
    sands, 62, 253
    vibratory crusher, 233
    penetration, 161
Coriolis force, 291
Coupling(s), 262
    conditions, 28
Cracking, 145
Crushed ore, 9
Crusher chamber, 145, 148
Crushing, 1, 2, 113, 151, 157, 203
Crystal:
    growth speed, 165
    lattice, 188, 194
    -liquid boundary, 165
Crystallization, 2, 166, 178
    of metals under vibration, 163
Crystallized structures, 124, 182
Curves, 12
Curvilinear surfaces, 31
Cycles of oscillations, 52
Cyclic pulsations, 148
Cylindrical chamber, 269
Cylindroids, 29


Dampers, 90, 167, 298, 374
Damping factors, 94, 122, 134, 368, 404
Deformation(s), 79, 128, 146, 195, 452
    amplification, 80
    of an elastovertical body, 74
    processes, 69
    vector, 323
    waves, 124
Deforming force, 80
Dehumidifiers, 214
Density, 145, 259
Design problem formulating, 486
Detuning factor, 75, 162
Dewatering, 2, 235
Diaphragms, 213
Dimensionless displacement, 162
Discharge channels, 246
Discharge manifold, 304
Dispersed media, 120, 123, 133

Displacement(s), 25, 27, 32, 42, 51, 69, 71, 136, 153, 160, 167, 200, 210, 265, 291, 298, 343, 411, 445
    diagrams, 24
    magnitude, 29
Dissolution, 1
Double-jaw vibratory crusher, 146, 434, 444
Drive characteristics, 415
Driving elastic link, 401
Dropping phase, 98
Dry friction, 90, 93, 128, 147, 154, 156, 167, 323
Drying, 2
Dry resistances, 122, 133

Eccentric:
    bushings, 225
    drive, 394
    vibrators, 294, 392
Edge-wetting angle, 165
Elastic:
    block, 401
    bodies, 70
    connecting rod, 266, 392
    damper, 91
    forces, 75, 296
    link(s), 91, 241, 395
    properties, 78, 125
    pulse, 27, 30, 40
    resonators, 102
    rheological bodies, 106
    rubber elements, 393
    strains, 81
    stresses, 74
    suspension, 223
    waves, 187
Elastodissipative properties, 131
Elastoinertial body, 74, 76, 77
Elastoplastic media, 83
Electric vibrations, 265
Electroacoustic converters, 310
Electromagnetic:
    field, 118
    grizzlies, 265
    vibrators, 102, 236, 309
Electrostatic field, 306
Elevated surfaces, 1
Elevators, 214
Ellipse, 14
Elliptical:
    oscillations, 12, 51
    trajectories, 131, 346
    vibrations, 15
Elongated oval, 41

Embedded sensors, 26
Embryo formation, 165
End-face surfaces, 190
Energy:
    consumption, 42, 73
    density, 314
    dissipation, 129, 484
    expenditures, 151
    losses, 80
    parameters of vibratory machines, 383
Equiaxial crystals, 165
Equilibrium, 37, 92
Ethyl cellulose, 40
Excitation frequencies, 132
External friction, 100
External harmonic excitation, 174
Extraction, 1

Feed:
    chamber, 157
    chutes, 232
    hopper, 155
Ferromagnetic rod, 314
Ferrostatic pressure, 167, 169, 180, 183
Film coating, 11
Filming, 23
    zone, 24
Filtering, 57
Fine grinding, 216
Fine sandstone, 50
Fish, 41
    length, 47
    processing machines, 41
Fishing industry, 42
Five-chamber vibratory mill, 219
Flexible carrying element, 110
Flight, 106
    angles, 96
    stage, 20, 22, 100, 101
Flow disorder phenomenon, 46
Flywheel, 297
Force impulses, 124
Forced vibrations, 21, 79
Formation processing, 209
Fourier series, 77
Four-shaft vibrators, 278
Fracture, 153, 195
    process of rock masses, 146
    stress(es), 53, 149
Free displacements, 94
Free fall acceleration, 60, 120, 121
Free motion, 94, 106, 116, 167
Frequency-force characteristics, 398

Friction, 119, 123, 128, 188, 215, 252, 259
  pairs, 93
Frozen load, 215


Gas, 61
  dynamic emitters, 319
  -jet emmiter, 316
  media, 316
  permeability, 62, 66, 127
  pressure, 20
Gelatin-glycerin, 40
Grain-size reduction, 213
Granular body, 33, 40
Granulation, 2
Granulometric composition, 21, 60, 217
Gravitational forces, 130, 167
Gravity, 14
Greifers (clam shells), 214
Grinding, 255
Grizzlies, 261
  feeders, 518


Half-wave vibrations, 16
Hard-alloy balls, 193
Hardening, 2, 81, 174, 255
  process, 256
  stress, 84
Hardness, 54, 145
Harmonic functions, 355
Harmonic:
  laws, 41
  oscillations, 18, 107, 182, 253, 402
  periodic force, 76
  rectilinear vibrations, 15
  vibrations, 2
Heaping arms (claws), 2
Heat:
  flux, 165
  generation, 306
  transfer, 182, 242
Helical spring, 294
Helium flash lamp, 24
High-frequency:
  harmonics, 111
  oscillations, 162
  vibrators, 310
High hardness, 54
High-intensity oscillations, 169
High noise level, 257
High speed films, 28, 41

Hollow:
  billet, 179
  ingot, 178
  workpieces, 177
Hoppers, 47
Humidity, 17, 50, 60, 128, 145
Hydraulic:
  cylinder, 301, 304
  vibrators, 300
Hydrodynamic emitters, 318, 319
Hyrovibrators, 102
Hysteresis, 80, 128, 131, 149
  losses, 190
  properties, 122


Idling, 55, 417, 450
Impact buffers, 228
Inclination, 6
  angles, 173
Inertial:
  eccentric crusher, 162
  eccentric drive, 159
  forces, 80
  elastic body, 75
  properties, 131
  vibrators, 24, 362, 382
Infrared rays, 242
Infrasound, 2
Ingot, 163, 166
  layer, 182
  motion, 167
Internal:
  cracks, 53
  dampers, 97
  friction, 81
  shear strains, 35
  stresses, 348
Interphase boundary, 123
Isochores, 40


Jaw(s):
  acceleration, 210
  displacement, 441
  oscillations, 53, 154, 158, 453, 457
  stroke, 160
  wear, 160
Jet pulsations, 317
Joint motion, 92, 94, 101
Jolting, 31
  regime, 4, 13

Kinematic vibratory parameters, 28
Kirpichev-Kick energy, 159
Knuckle pie, 223

Lag, 179
Lateral pressure, 31
Laws of deformation, 79
Layer:
    height, 21
    porosity, 61, 100
    thickness, 6, 18, 125, 126
Leaching, 1
Lens, 24
Linear oscillations, 263
Lipschitz category/function, 198, 200, 202
Liquidation, 189
Liquid metal, 178
Load displacement, 100
Loading characteristics, 415
Lobes, 258
Longitudinal:
    axis, 41
    oscillations, 110
    strain, 178
    traveling waves, 102
Louvers, 245
Low:
    frequency oscillations, 162
    -pressure vapor, 243
    -speed regimes, 10, 17
    speed zone, 43
Lump loads, 235
Lump parameter, 146

Machine design specification, 476
    Pareto models, 479
    test models, 478
Maclaurin series, 406
Manganese ore, 15, 17
Magnetic:
    attractive forces, 118
    core, 315
    fields, 309
    traction, 117
Magnetostriction, 314
Magnitude, 37
Mass transfer, 1
    processes, 12, 124
Measuring beaker, 62
Mechanorheological phenomenology, 70, 89
Melt, 165, 168
Membrane, 319
Metal cutting, 213
Metallurgical production cycle, 163

Metallurgy, 214
Microcracks, 149, 152
Microheterogeneous media, 121
Migration of sand and sodium silicate, 365
Mineral ores, 520
Mixing, 2
Mobility, 56, 188
Moisture, 242, 245
Mold, 181
    oscillations, 171, 177
    walls, 170, 176, 184
Molding box, 253
Molding cylinder, 63
Monoharmonic vibrations, 17
Monolayer boundaries, 30
Motor-vibrator drive, 248
Multi-criterial optimal machine design, 475
Multi-cycle loading, 83
Multi-lobe locus, 286
Multi-phase media, 213
Multiple regression, 21

Natural frequencies, 88
Nonlinear elastodamping system, 449
Nonlinear mechanics, 23
Nonmetallic inclusions, 166
Nozzle(s), 245, 319

Oblong ellipse, 12
Ore chute, 417
Oscillating:
    masses, 241, 298
    plane, 42, 44
    surfaces, 41
Oscillations, 2, 9, 35, 41, 89, 166, 371
    acceleration, 7
    amplitude, 3, 4, 13, 101, 131, 351
    excitation, 253
    frequency, 219
Oscillogram(s), 23, 32, 97
Oval, elongated, 41
Overcooling, 165

Parabolic law, 13
Paraffin, 25
    impregnation, 28
Pareto models, 479
Particle trajectory, 96
Part strengthening, 255
Pendulum, 275
Pendulum vibrator, 248
Petroleum jelly, 40
Phenomenological rheology, 123

Physicomechancial properties of rock, 149
Piezoelectric converters, 310
Piezoexciters, 311
Piston dampers, 26
Pivoted unbalanced masses, 378
Planar stress distribution, 33, 84
Plastic:
    bodies, 70
    deformations, 26, 81, 156
    flow(s), 35, 77
Plasticity:
    strains, 81, 114, 147, 148
    stresses, 74
Plastoinertial body, 74, 77
Plotting screen, 24
Pneumatic cylindrical rollers, 218
Polyfrequency vibratory field, 182
Polyharmonic oscillations, 243
Polymer powders, 61
Pores, 120
Porosity, 61
Poured mixtures, 2
Pressure:
    diagrams, 38, 39
    fluctuations, 20
    jams, 20
    processing, 2
Probability density, 85
Process-conveying machines, 235
Pseudoliquidity, 56, 120
Pseudoboiling, 125
Pulsating force, 307
Pulse:
    energy, 49
    load application, 190
Punch, 192

Quality criteria, 480
Quartz ballast, 11, 12
Quartzite, 14
Quasi-liquefaction, 120, 125, 136
Quasi-Pareto models, 479

Radial oscillations, 311
Radial viscoelastic elements, 148
Radius vector, 293
Random natural frequency, 85
Random reduced plasticity, 85
Random reduced viscosity, 85
Rarefaction, 21, 128
Rational structure of vibratory tables, 507
Rayleigh method, 147
Rectangular cells, 206

Rectilinear:
    displacements, 234
    harmonic oscillations, 3, 7, 48
    oscillations, 13, 89, 270, 280
    vibrations, 2
Relative equilibrium, 92
Residual stresses, 83
Resistance forces, 322, 350, 428
Resonant vibratory machines, 330
Restoring force, 334
Rheological bodies, 72, 182
Rock(s), 120
    fracture, 213
    mass, 145, 146, 438
        slippage, 440
    stiffness, 162
    strip, 193
Rotary crusher, 233
Rotating chamber, 217
Rotating unbalanced masses, 387
Round particles, 6
Rubber:
    brushings, 275
    hinges, 268
    -metal bushing, 276
    pad mounting, 210, 216
    vibration isolators, 255
Rubbing, 215
Run-out process, 380
Rupture of casting crust, 164

Sand, 3, 4, 11, 253
    layers, 28
Sandstone, coarse and fine, 50
Scale factor, 34
Scaly skin cover, 44
Schwab criterion, 20
Screen feeders, 214
Screening surface, 268
Self-synchronizing vibrators, 241
Shear resistance, 34
Semi-harmonic vibrators, 2
Semi-viscous lubrication, 43
Separation phase, 98
Shaft rotation, 294
Shear resistance, 30
Shearing force, 156
Shells, 178
Shock:
    absorber(s), 222, 227, 232, 241, 298
    -absorbing bearings, 226
    -absorbing elastic links, 248
Sieve(s), 69, 213
Single-drum vibratory mill, 216
Skin rupture, 167

Sliding friction, 6
Slip, 93
Slip regimen, 95
Slippage, 105, 114, 158, 192, 418
Slotted screen, 250
Small-chunk media, 120
Small-lump load, 3, 4
Sodium silicate core sand, 62
Sodium silicate sand mixture, 63
Soil mechanics, 33
Solid:
    crust, 182
    -phase reactions, 1
    workpieces, 177
Spatial:
    displacements, 29, 125
    polyfrequency oscillations, 182
Spill of liquid metal, 167
Spiral load-carrying element, 1
Springs, 255
Squeezing, 217
Stainless steel surface, 42
Static stress distribution, 37
Steady-state regime, 420
Stiffness, 139, 147, 175, 229, 335, 376, 383
    coefficient, 36
Strain, 76
    -calibrated beam, 43
Straining force, 77
Straining load, 76
Static:
    friction, 93
    pressure, 113
    strain, 94
Sticking, 167
Sticky loads, 128
Stratification, 66
Stratified shear strains, 30
Strength, 145
Stress:
    circle, 33
    distribution, 37
    interaction, 84
    magnitude, 80
    oval, 34
Stroboscope, 23
Strouhal criterion, 20
Surface roughness, 151
Suspended (boiling) layer, 245
Swinging jaw, 230

Tachometer, 23
Tensile strains, 169
Thickness, 15, 18, 97, 121
Thick sand layer, 7

Three-dimensional vibrations, 2
Tie-rod(s), 233, 240, 267
Torque, 278, 299, 365, 437
Torsional oscillations, 249, 311, 359, 526
Torsional vibrations, 2
Trajectory, 201
    of motion, 2, 97, 331, 337
    traversing, 14
Transfer motion, 340
Transient process, 433
Transient regimes, 373
Transitional oscillations, 221
Transmission ratio, 287
Transresonant regimes, 371
Transresonant vibratory oscillations, 373
Transverse oscillations, 110
Transverse travelling waves, 102
Travel velocities, 196
Two-beam oscillograph, 26
Two-chamber crusher, 231
Two-component vibrations, 2
Two-entry helical trough, 250
Two-joint pendulum vibration, 282
Two-piston pulsator, 302
Two-screen grizzly, 250
Two-shaft vibrator, 362
Two stationary jaws, 230

Ultrasound, 165, 187
    frequencies, 166
Unbalanced-mass kinetic moment, 375
Unbalanced-mass mechanism, 385
Unbalanced-mass type, 325

Vaporized moisture, 243
V-belt drive, 297
V-belt transmission, 255
Velocity, 160, 210
    amplitude, 8
    dependent resistances, 355
    transfer, 10, 16
Vertical:
    acceleration, 39
    displacement, 39
    oscillations, 34
    plane, 222
Vibrating:
    crystallizer, 166
    vertical tube, 89
Vibration:
    amplitude, 21
    angle(s), 4, 7, 10
    biharmonic, 17
    conveyance, 4

Vibration (*Cont.*):
  crushing of rock masses, 145, 148
  exciter(s), 121, 125, 130, 273, 320, 376
  frequency, 9, 79
  harmonic, 2
  intensity, 28
  liquefaction, 60
  monoharmonic, 17
  motion, 6
  propagation, 129
  regimes, 1
  technology, 1
Vibrator shaft, 282, 391
Vibratory:
  acceleration, 7, 8, 10, 127, 180
  boiling layer, 56
  bunkering (hoppers), 47, 130
  compaction, 56
  compression process, 60
  conveyance, 2, 8, 22
    of bulk loads, 89
    regimes, 3
  conveying machines, 416, 437
  conveyors, 511
  crushers, 434, 440, 523
    jaw, 51
  cutting, 213
  discharge, 139
  displacements, 56, 99
  drawing, 186, 188, 191
  dryers, 242
  elevator, 249
  environment, 163
  feeder(s), 418, 428, 518
  field, 187
  fracture, 213
  grizzly-feeder, 468
  hardening, 186
  jaw crushers, 146
  loading, 1
  machine(s), 69, 320, 368
    schematics and design of, 213
  mills, 129
  mixers, 259
    for granular media, 269
  oscillations, 177
  rolling, 186, 191
  separators and mixers, 259
  table modernization, 491
  technology, 213
  velocity, 8, 49, 59

Vibrohydraulic machines, 153
Vibroimpact, 190
Vibromagnetic machines, 113
Vibropneumatic machines, 113
Vibrotraction, 104
  machines, 113
  plants, 113
Viscoelastic:
  deformation, 152
  resistance, 443
  strains, 184
Viscoelastoplastic strains, 106
Viscoplastoinertial system, 122
Viscosity, 88, 145
Viscous:
  bodies, 70
  damping, 122
  fluids, 78
  friction, 185
  properties, 39, 78
  resistance(s), 44, 79, 97, 122, 143, 395, 422
  stresses, 74

Wall friction, 33, 35
Wall-thickness variation, 177
Washing, 2
Wattled screens, 260
Wave:
  conveyer, 102, 104
  energy, 312
  -length, 186
  -like motion, 268
  oscillations, 101
Wear assessment, 101
Wedge element, 84, 191
Wedging, 214
  of jaws, 55
  system, 160
  stresses, 214

X-rays, 23, 24

Yield point, 81

Zone:
  of low speeds, 43
  of plastic deformation, 81
  of vibration propagation, 120